GELS Handbook

Fundamentals, Properties and Applications

GELS Handbook

Fundamentals, Properties and Applications

Volume 2: Applications of Hydrogels in Regenerative Medicine

Mohammad Reza Abidian, Umut Atakan Gurkan
and **Faramarz Edalat** *Volume Editors*

Editors

Utkan Demirci
Stanford

Ali Khademhosseini
Harvard

World Scientific

NEW JERSEY • LONDON • SINGAPORE • BEIJING • SHANGHAI • HONG KONG • TAIPEI • CHENNAI • TOKYO

Published by

World Scientific Publishing Co. Pte. Ltd.
5 Toh Tuck Link, Singapore 596224
USA office: 27 Warren Street, Suite 401-402, Hackensack, NJ 07601
UK office: 57 Shelton Street, Covent Garden, London WC2H 9HE

British Library Cataloguing-in-Publication Data
A catalogue record for this book is available from the British Library.

GELS HANDBOOK
Fundamentals, Properties and Applications
(In 3 Volumes)

Volume 1: Fundamentals of Hydrogels
Volume 2: Applications of Hydrogels in Regenerative Medicine
Volume 3: Applications of Hydrogels in Drug Delivery and Biosensing

ISBN 978-981-4656-10-8 (Set)
ISBN 978-981-4656-13-9 (Vol 1)
ISBN 978-981-4656-14-6 (Vol 2)
ISBN 978-981-4656-15-3 (Vol 3)

In-house Editors: Rhaimie Wahap/Dipasri Sardar

Typeset by Stallion Press
Email: enquire@stallionpress.com

Printed in Singapore

Contents

Chapter 1 Hydrogels in Regenerative Medicine 1

Nesrin Hasirci, Cemile Kilic, Aylin Kömez, Gökhan Bahcecioglu, and Vasif Hasirci

Chapter 2 Determining Stem Cell Fate with Hydrogels 53

Aylin Acun, Andreana Panzo, and Pinar Zorlutuna

Chapter 3 Applications of Hydrogels in 3D Functional Tissue Models 87

Tamer Çırak, Tayfun Vural, Doğa Kavaz, and Emir Baki Denkbaş

Chapter 4 Engineering Regenerative Dextran Hydrogels for Acute Skin Wound Healing 111

Jie Cheng, Ying Jin, David M. Owens, Jeremy J. Mao, and Guoming Sun

Chapter 5 Application of Hydrogels in Ocular Tissue Engineering 137

Vipuil Kishore, Yunus Alapan, Ranjani Iyer, Ryan Mclay, and Umut A. Gurkan

Chapter 6 Hydrogels in Bone Tissue Engineering: A Multi-Parametric Approach 165

Silvia M. Mihaila, Rui L. Reis, Alexandra P. Marques, and Manuela E. Gomes

Chapter 7 Hydrogels in Intervertebral Disk (IVD) Repair 199

Cem Bayram, Murat Demirbilek, and Emir Baki Denkbaş

Chapter 8 Hydrogels in Cartilage Tissue Engineering 215
Antonella Motta, Mariangela Fedel, and Claudio Migliaresi

Chapter 9 Application of Hydrogels for Tendon and Ligament Repair and Tissue Engineering 271
Matteo Stoppato, Friedrich von Flotow, and Catherine K. Kuo

Chapter 10 Hydrogels in Bone Tissue Engineering 295
Minh Khanh Nguyen, Julia E. Samorezov, and Eben Alsberg

Chapter 11 Hydrogels in Cardiac Tissue Engineering 323
Shauna M. Dorsey and Jason A. Burdick

Chapter 12 Application of Hydrogels in Heart Valve Tissue Engineering 363
Jesper Hjortnaes and Frederick J. Schoen

Chapter 13 Hydrogels in Vascular Tissue Engineering 385
Y. J. Blinder, D. J. Mooney, and S. Levenberg

Chapter 14 Use of Hydrogels in the Engineering of Lung Tissue 397
Joaquin Cortiella, Jean A. Niles, Stephanie P. Vega, Lissenya Argueta, Adriene Eastaway, and Joan E. Nichols

Chapter 15 Hydrogels for Hepatic Tissue Engineering 427
Kerim B. Kaylan and Gregory H. Underhill

Chapter 16 Agarose Hydrogel Beads for Treating Diabetes 463
Nguyen Minh Luan, Yuji Teramura, and Hiroo Iwata

Chapter 17 Hydrogels in Urogenital Applications 479
James Turner and Jiro Nagatomi

Index 501

Chapter 1

Hydrogels in Regenerative Medicine

Nesrin Hasirci,[*,†,§] **Cemile Kilic,**[†,‡,§] **Aylin Kömez,**[†,§]
Gökhan Bahcecioglu[†,‡,§] **and Vasif Hasirci**[†,‡,§,¶]

[*] *Department of Chemistry, Middle East Technical University, Inonu Blvd. No: 1, 06800 Ankara, Turkey*

[†] *Department of Biotechnology, Middle East Technical University, Inonu Blvd. No: 1, 06800 Ankara, Turkey*

[‡] *Department of Biological Sciences, Middle East Technical University, Inonu Blvd. No: 1, 06800 Ankara, Turkey*

[§] *BIOMATEN, Center of Excellence in Biomaterials and Tissue Engineering, Inonu Blvd. No: 1, 06800 Ankara, Turkey*

[¶] *vhasirci@metu.edu.tr*

Hydrogels are soft, jelly-like, polymeric networks with very high water retention capacity. They are made from natural materials or synthetic polymers. They are very important for the biomedical field because of the similarity between their mechanical and chemical properties and that of the extracellular medium, the microenvironment of a cell. They can be used in cell loaded form. High water content of the hydrogels is very suitable even without the cells in mimicking the hydrated tissues or viscous bodily fluids where they serve as a lubricant or a shape forming material. Actually cells do not like to adhere to very highly hydrophilic structures so they are very useful in applications where non-adherence is sought, like prevention of adhesion of tissues after a surgery. On the other hand, when we need the cells to attach to the hydrogels we make some chemical modifications on the material such as attaching cell adhesive arginine–glycine–aspartic acid (RGD) amino acid sequences. In this chapter the types, the sources of their starting materials, the methods of insolubilization, the network formation, the methods to use to control their properties, and their biomedical applications have been presented. Among the applications discussed are drug delivery systems, tissue engineering scaffolds, wound dressings, anti-adhesive membranes, and meshes.

1. Introduction to Hydrogels

Hydrogels are hydrophilic polymeric networks that absorb and retain significant amount of water. Hydrophilic side or pendant groups attached to the polymeric backbone give hydrogels the ability of water absorption and the various types of contacts between the chains of the network provide resistance to dissolution in water.[1,2] The first synthetic hydrogel for medical purpose was produced by Witchterle and Lim in 1954[3] and since then this technology has been employed in a wide range of applications as biomedical materials, pharmaceuticals, biosensors, drug carriers, and food additives.[4] Hydrogels can absorb water up to thousands of times of their dry weight. The ones that contain more than 95% water by weight are called as super absorbent which are also highly biocompatible because of their high water content. Water inside the hydrogel is needed for the diffusion of water soluble molecules and the polymeric network is there in order to hold the water inside the structure. The form of "gel" mentioned in this definition is not completely solid nor liquid but between these two states that provides excellent relaxation and mechanical properties to the scaffold.[5] This property is the reason for their compositional and mechanical similarity to the native soft tissues and the extracellular matrix (ECM).[6] The interfacial tension is also low and this decreases protein adsorption. Low protein adsorption decreases the inflammatory responses, an essential prerequisite for biomedical applications.[7] Hydrogels offer significant advantages for tissue engineering purposes, such as high biocompatibility, ease of modification, and injectability into voids at injury sites. It is also possible to fabricate smart or stimuli responsive hydrogels that respond to environmental stimuli like pH, temperature, exposure to light, and concentration of the metabolites. Responsibility makes them suitable for controlled release of the bioactive agents including drugs and growth factors. Hydrogels, however, have several disadvantages including high cost, difficulty of sterilization, non-adherent nature, and low mechanical properties. Crosslinking is also challenging if attempted after loading the gels with cells and drugs.[2,4]

Hydrogels can be classified into several categories depending on their various properties. Based on their charges they can be categorized into four groups: neutral, anionic, cationic, or amphoteric hydrogels. Based on their structure, the network can be completely amorphous or semicrystalline. Finally, depending on their network porosity, they may be macroporous, microporous, and non-porous.[8,9]

1.1. Formation of hydrogels

Formation of a gel, namely gelation, is the creation of a three-dimensional (3D) network structure via linking of macromolecular chains together. Gelation can occur by physical linking (physical gels) or by chemical linking (initiated by chemical agents or by radiation). Depending on the functional groups attached to the main chains, hydrogels may have responsive property.

1.1.1. *Physical hydrogels*

In the physical hydrogels the network is held together by forces like ionic interactions, hydrophobic forces, H-bonding, and chain entanglements, and therefore, network formation is reversible.[2] In the recent years, physical hydrogels are preferred more than chemical ones because crosslinking agents are not needed to form the structure. This is important because crosslinking with chemical agents or radiation can alter the activities and properties of the proteins, enzymes, and the cells if crosslinking follows entrapment. Moreover, chemical agents are usually toxic due to their high reactivity and the excess agents have to be removed with additional washing steps.[10] Physical crosslinking, on the other hand, has disadvantages such as inhomogeneity due to clusters of molecular entanglements, and ionic or hydrophobic domains. Additionally, transient network defects can be observed because of chain loops or free chain ends. Another limitation of reversible hydrogels is the risk of disruption of the molecular interactions due to changes in environmental conditions like pH, temperature, ionic strength, and application of stress.[11]

1.1.2. *Chemical hydrogels*

Chemical hydrogels with covalently crosslinked networks are called "permanent" gels. They are usually produced by polymerization or parallel crosslinking through covalent bond formation. The crosslinks may either form during the polymerization reaction or afterwards.[12] Condensation and addition polymerizations are two main polymerization categories and vulcanization is another covalent interaction via sulfur bonds which generally occurs after the formation of the polymer.[6,10,13] Covalent bonds are much stronger than the physical ones and therefore gels formed with covalent bonds have higher stability.

Four molecular interaction types, namely ionic bonding, entanglement, interpenetrating networks (IPN) and covalent bonding, are involved in hydrogel formation (Fig. 1).

1.1.3. *Stimuli responsive hydrogels*

Stimuli responsive (or intelligent) hydrogels can experience significant property changes (e.g., volume change) in response to small changes in physiological or biological environment.[14] For example, a reversible hydrogel which was formed using ionic groups will dissolve if the ionic groups lose their charges with a change in the

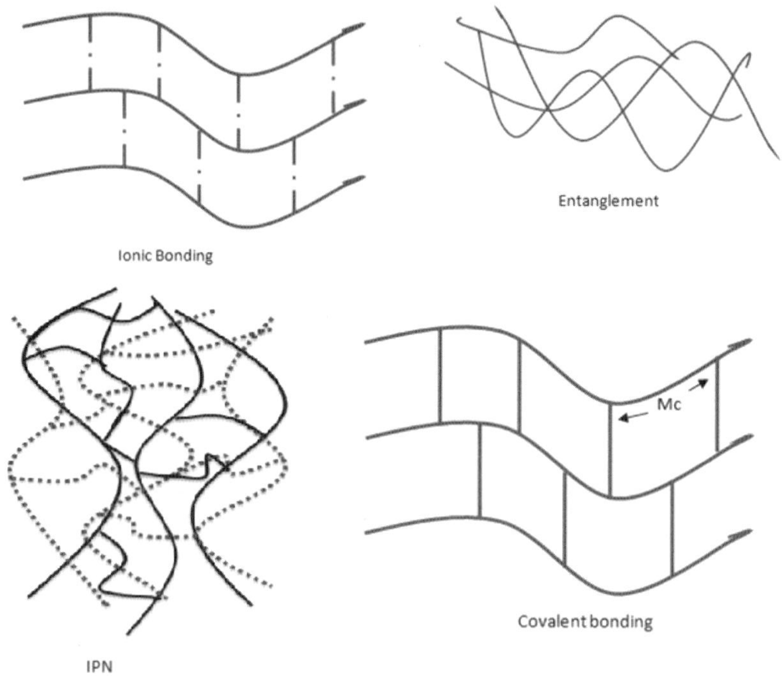

Fig. 1. Molecular interactions involved in the formation of hydrogel networks.

pH of the medium. pH-dependent swelling behavior of poly(acrylic acid) hydrogels obtained from crosslinking with poly(L-glutamic acid)-g-(HEMA is 2-hydroxyethyl methacrylate) (PGA-g-HEMA) was reported in the literature.[15] The hydrogels exhibited pH-dependent swelling behaviors where in acidic pH they had lower swelling ratio and in basic pH the swelling ratios reached 5 to 10 times more than those in acidic medium. Lower swelling ratio at acidic pH is due to the protonation of both acrylic acid and glutamic acid residues and higher swelling ratio is because of ionization of these groups above their pKa ($pH > 4$). In another study pH-sensitive hydrogels were developed for drug release by Vaghani and Patel.[16] They utilized chitosan and poly(N-vinyl pyrrolidone) (PVP) blends crosslinked with glutaraldehyde in order to obtain a semi-IPN structure. The hydrogels showed high swelling at acidic pH due to protonation of primary amino groups of chitosan and low swelling behavior was observed under alkaline conditions.[16] Light sensitive hydrogels may be another example for stimuli responsive gels which change conformation upon exposure to light. In their study,[17] they produced photoresponsive hydrogels in the design of microfluidic devices. They obtained hydrogels from poly(N-isopropylacrylamide) (PNIPAM) functionalized with chromophore spirobenzopyran. The hydrogels were fabricated into microchannels by *in situ* polymerization, and upon exposure to blue light shrinkage of the gels was observed, a response that may serve as a useful tool in the control and fabrication of integrated microfluidic devices.[17] Photoresponsive hydrogels are widely used in drug delivery applications. Wells *et al.*[18] reported a hyaluronan-based hydrogel for controlled delivery of ophthalmic drugs upon ultraviolet (UV) exposure. They used two types of drugs with low and high molecular weights in order to see the pattern of delivery and concluded that the drug release could be manipulated with size of the drug and the exposure to UV radiation. One advantage of this system is that release rate can be decreased or totally turned off in case any unwanted reactions occur.[18]

1.2. Hydrogel types

ECM is the best model in the construction of the tissue engineering scaffold designs because the goal is to create an environment where the cells are able to attach and stretch as in ECM. Any scaffold should fulfill the basic requirements like biocompatibility, allowing cell ingrowth, proliferation, and differentiation.[19] Thus, the biomaterials to be used in hydrogel construction should be chosen by taking these requirements into consideration. The following are the materials used in scaffold formation grouped according to their sources.

1.2.1. *Natural materials*

Natural polymers are obtained from various sources including mammals, bacteria, plants, and crustaceans, and they are polysaccharides or polypeptides in nature.[20] Hydrogel networks based on natural materials offer significant benefits over synthetic materials due to their high biocompatibility and ECM-like chemistry, and as a result, they are in widespread use. Biodegradability is an inherent property of natural polymers and this is important for tissue engineering applications. Natural polymers are metabolized in the biological systems via hydrolysis or via enzymatic degradation and the typical products are amino acids or saccharide units if the macromolecules are polypeptides or polysaccharides, respectively.[20,21] The main concerns about the natural polymers and the resultant hydrogels are the risk of induction of immune responses, batch to batch variability, differences in the source, and limited availability. However, new techniques and processing methods for purification and production offer better materials for applications.[22] In the following sections, some information about the most commonly used natural polymers is provided.

1.2.1.1. Collagen

Collagen is the major constituent of the mammalian tissues like skin, cartilage, cornea, bone, and blood vessels and it is very popular in biomedical applications due to its role in cell attachment, proliferation, anchorage, and differentiation (Fig. 2).[23] There are more than 27 different types of collagen identified so far[24] and all share the same basic structure of three left-handed polypeptide chains assembled into a right-handed triple helix.[25] Collagen can form fibers through self-aggregation and crosslinking and these fibers provide the mechanical strength and stability to the scaffolds. Collagen can be used as it is or often is stabilized by using physical and/or chemical methods in order to decrease the degradation rate which is important for many applications like drug delivery. Glutaraldehyde, carbodiimides, formaldehyde, chromium ions, and acyl azide are some of the chemical agents used to stabilize the collagen fibers. UV and gamma ray irradiation, and dehydrothermal treatments are examples for physical methods to crosslink the collagen matrix.[19,26] Collagen hydrogels have been used in many applications like corneal tissue repair,[27–29] cartilage defects,[30,31] cardiovascular tissue regeneration,[32,33] and meniscus repair.[34] They are semisolid and viscoelastic at rest but can be liquefied under stress. Injectable suspensions of collagen fibers or non-fibrillar collagen solutions carrying drugs or proteins can be applied directly to the defected site of the body.[35]

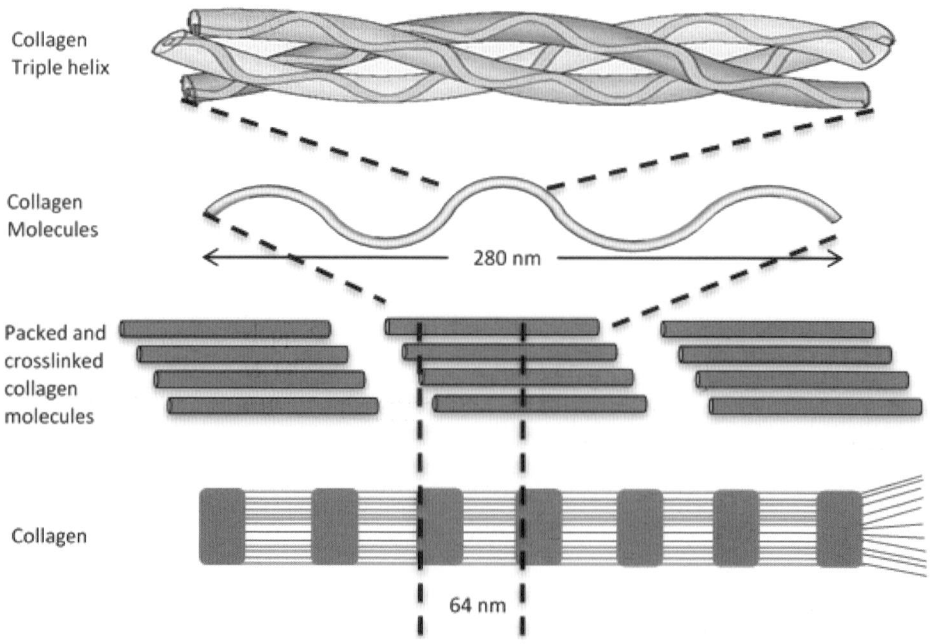

Fig. 2. Hierarchical organization of collagen fibers.

1.2.1.2. Hyaluronic acid (HA)

HA or hyaluronan is a linear, high molecular weight, and non-sulfated polysaccharide found in the ECM of many connective tissues like skin, vitreous humor, and umbilical cord.[19,36] It consists of the repeating disaccharide units of N-acetyl-D-glucosamine and D-glucuronic acid, that are linked with alternating β-1,3 and β-1,4 glycosidic bonds (Fig. 3(A)). HA is a glycosaminoglycan (GAG) of the body that is essential for matrix organization, cell signaling, mechanical integrity, and in wound repair. The hydrolysis of HA in the body is achieved rapidly by hyaluronidase and reactive oxygen species, within hours or days, the rate varying according to the tissue type.[37,38] Chemical modification of the HA may be done at three functional groups of the disaccharide: hydroxyl groups, the carboxylic acid of glucuronic acid, and the N-acetyl group of N-acetylglucuronic acid.[37] In recent years, HA has become a very popular biomaterial for tissue engineering and regenerative medicine since it can be modified to obtain the desired properties such as proper biological activity, hydrophobicity, and even crosslinkable groups. Injectable HA hydrogels loaded with drugs or bone morphogenic protein-2 (BMP-2) were used at the defect site to stimulate bone and cartilage formation.[39,40] Besides these applications HA is also widely used for cosmetic purposes like tissue augmentation in order to fill the wrinkles resulting from aging, sun exposure, and gravity, and it is one of the most popular non-surgical dermal filler approved by the Food and Drug Administration (FDA) (USA). One disadvantage of HA-based fillers is its short half-life and this could be solved by using different chemical crosslinkers.[41,42] HA can be bound to other polymers in order to obtain stimuli-responsive behaviors. In one study, researchers used PNIPAM to obtain a thermoresponsive injectable hydrogel that would both have biodegradability and also phase transition behavior with temperature change. This approach was shown to be promising for adipose tissue engineering as demonstrated in mouse *in vivo* studies where it was concluded that this hydrogel can be used for a variety of soft tissue engineering applications.[43]

1.2.1.3. Fibrin

Fibrin is another natural material used in the preparation of hydrogels. It is a biopolymer formed from the monomer fibrinogen and these two compounds are very critical for the homeostasis of the body. Fibrin results from the covalent crosslinks among the fibrinogen monomers and the fibrin network is quite resistant to protease degradation. Additional crosslinks can be introduced by using molecules like genipin to introduce more resistance and stability.[44] Fibrin can be purified from the blood of the patient; therefore, it does not pose any risk of infection or foreign body reaction. It is also widely used as a tissue adhesive in the medical field. Fibrin-based tissue adhesives mimic the last stages of the blood clotting scheme by utilizing fibrinogen and thrombin that form a clot when mixed in the presence of calcium ions.[45] Fibrin sealants are used commonly following procedures like cardiovascular, vascular, neurological, abdominal, and thoracic surgeries.[46] Fibrin gels are also used together with other gels like HA-based ones in order to improve their mechanical properties to be used in different applications involving cartilage tissue repair.[47]

1.2.1.4. Chitosan

Chitosan is another naturally occurring polysaccharide obtained from shrimp or crab shells and consists of a series of linear copolymers of N-acetyl-D-glucosamine and β-1, 4-linked-D-glucosamine (Fig. 3(B)). Chitosan is derived from chitin which is composed mainly of only N-acetyl-D-glucosamine units and some of these units are deacetylated.[19] It is biocompatible, biodegradable, and has a pH-dependent cationic property. Chitin is stable and degrades upon heating instead of melting,[48] and this limits its processability. However, the main limitation of chitin is its poor solubility in many solvents. For chitosan, an acidic environment of around pH 6.2 is suitable for dissolution and upon neutralization (at pH over 6.2) it forms a gel.[49] Chitosan is one of the most widely used polysaccharides in biomedical applications such as, dermal tissue repair,[50] cell encapsulation,[51] bone reconstruction,[52–55] cartilage tissue engineering,[56] and drug delivery.[57,58]

1.2.1.5. Agarose

Agarose is a polysaccharide derived from red algae and seaweed, and has a structure of alternating 1,3-linked D-galactose and 1,4-linked 3,6-anhydro-L-galactose residues (Fig. 3(C)). Agarose is highly soluble in water at temperatures around 65°C and gels upon cooling in the temperature range 37–40°C. At lower temperatures it is stable and does not liquefy again until heated to 65°C. The gelation temperature is dependent on the degree of substitution of hydroxymethyl groups on the side chains.[59,60] Agarose hydrogels have been used in different applications including cartilage repair and bone regeneration. In their study Khanarian *et al.*[61] produced hydrogel-ceramic composite scaffold made up of agarose and hydroxyapatite (HAp) in order to achieve calcified cartilage formation. Their studies showed that biomimetic ceramic-hydrogel composite is a promising scaffold for osteochondral interface regeneration. For bone regeneration, Puértolas *et al.*[62] used biphasic calcium phosphate (BCP) and agarose in order to enhance the mechanical properties of the agarose-based hydrogels. They concluded that these ceramic-agarose systems offer advantages for bone tissue applications due to their biocompatibility, biodegradability, high mechanical properties, and affinity for proteins.

1.2.1.6. Alginate

Alginates are the second most abundant polysaccharides on earth after cellulose and they are produced by bacteria and seaweeds (Fig. 3(D)).[20] They are linear block copolymers of 1,4-linked β-D-mannuronic acid (M) and α-L-guluronic acid (G) residues. These residues in the blocks are either similar or alternating (MMMM, GGGG, or MGMG). L-guluronic acid blocks or the alternating blocks are stronger than D-mannuronic acid blocks due to diaxial linking among the residues. Thus, the physical and chemical properties vary with the M/G ratio and structure of the alternating regions. Alginate solution becomes a gel in the presence of divalent cations like calcium, lead, barium, and copper due to chelation of the carboxylic acid groups in the sugar. The interactions with these divalent cations are significantly affected by the M/G ratio.[63] Therefore, the viscosity of the alginate solution depends on the polymer concentration, molecular weight distribution and M/G ratio. Temperature is another factor that affects the crosslinking level of the structure; at lower temperatures more ordered gels are obtained because of slow diffusion of calcium ions. Covalent crosslinking of the alginate can also be achieved in the presence of diamines and dihydrazines; however, these methods are not so widely used as the chelation with the multivalent cations.[19,63] Alginate is widely used in tissue engineering applications due to its non-immunogenic and hydrophilic nature. It can be used as an injectable solution that forms a gel in the presence of multivalent cations at the defect site. Injectable alginate hydrogels are also used for controlled drug delivery

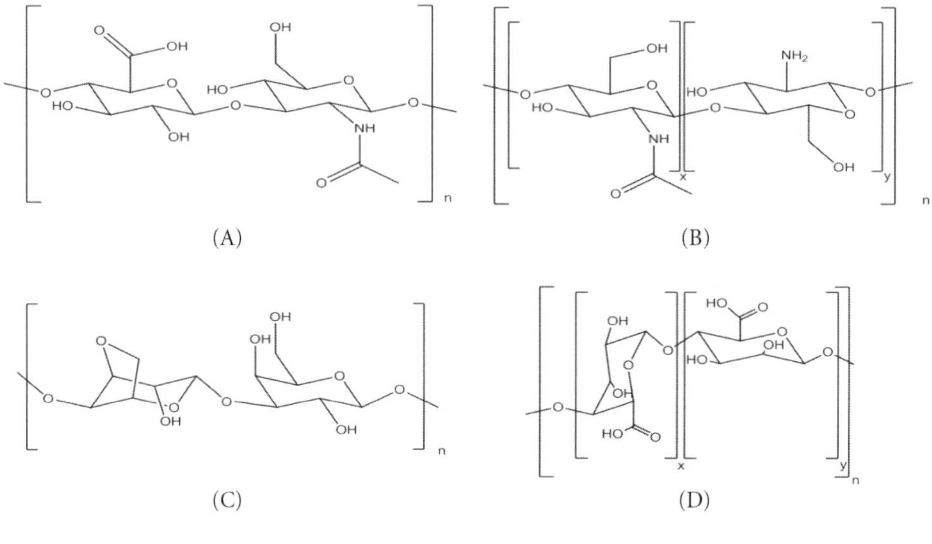

Fig. 3. Molecular structure of typical natural macromer units used in hydrogels for tissue engineering. (A) HA, (B) Chitosan, (C) Agarose, and (D) Alginate.

purposes.[64] For instance, these hydrogels were used in myocardial repair studies as a dual delivery system for hepatocyte growth factor (HGF) and insulin-like growth factor.[65] Alginate hydrogels are also widely used in the formation of interpenetrating polymer networks.[63] One of the limitations of the alginate hydrogels is their unregulated degradation because they are not degraded enzymatically. Moreover, chemical or biological modifications are needed to make alginate hydrogels suitable for cell attachment; otherwise cells do not adhere on them.[19,63]

1.3. Synthetic materials

Synthetic materials are attractive alternatives of the naturally originating compounds in building hydrogels because of their following advantages: (i) a large number of biodegradable synthetic polymers are biocompatible and do not induce any immunological responses; (ii) their mechanical properties and degradation rates can be altered by changing the production and process conditions and components; and (iii) their sources are not limited as the natural ones are.[66,67] However, sometimes harsh chemical methods are needed for their synthesis. The various additives and ingredients needed during the synthesis and the unreacted remnants could be detrimental to the cells and tissues.[1,19] Thus, it is very important to make sure no residues remain behind after polymerization before use in clinical applications.[19] A wide variety of synthetic polymers are used in the production of hydrogels. PHEMA, poly(vinyl alcohol) (PVA), poly(ethylene glycol) (PEG), and poly(N-vinylpyrrolidone) (PVP) are among the most frequently employed synthetic polymers.

1.3.1. *PHEMA*

PHEMA networks are obtained by free radical polymerization of monomer of 2-hydroxyethyl methacrylate.[19] PHEMA is used in various biological and medical applications due to its high swelling capacity, non-degradability, and biocompatibility.[68] It is one of the most widely known synthetic polymers used in hydrogel production (Fig. 4(A)). PHEMA is the earliest hydrogel material that is used in biomedical applications such as soft contact lenses, artificial cornea, and keratoprostheses.[69] PHEMA hydrogels are also extensively employed in drug delivery applications,[70] breast reconstruction,[71] cartilage repair,[72] nerve guide,[73] and wound dressings.[74]

1.3.2. *PVA*

PVA which is obtained from hydrolysis of poly(vinyl acetate) is another major synthetic material widely used in hydrogel preparations[19] (Fig. 4(B)). These hydrogels are similar to PHEMA; they are highly stable and elastic, and can be formed either through chemical or physical crosslinks. Physical methods include lyophilization and the hydrogels produced by this way are erodible. PVA can be crosslinked by various chemicals like acetaldehyde, glutaraldehyde, and formaldehyde[75]; however, due to their toxic effects on the cells, researchers try to avoid using chemical crosslinking methods. The alcohol groups on the structure of PVA are attractive sites for attachment of some biological molecules and its elasticity is very suitable for cell orientation. One limitation of PVA, like other synthetic, neutral hydrophilic materials, is its being non-adhesive towards proteins and cells, and modification with functional groups is needed to create surfaces attractive for biological molecules.[75,76] In the study of Schmedlen *et al.*,[77] photocrosslinkable PVA hydrogels that could have cells entrapped in the structure was reported. To enhance cell attachment and proliferation they further functionalized the hydrogels with adhesion peptide (RGD) and achieved significant improvement in cell proliferation, migration, and differentiation. PVA was also employed with positive results in vascular tissue regeneration,[78] meniscus repair,[79] and wound dressing studies.[75]

1.3.3. *PEG*

PEG is one of the most commonly used and most successful synthetic materials employed in hydrogel formation that is highly biocompatible, hydrophilic, and non-toxic (Fig. 4(C)).[19] Michael-type addition, a conjugate addition reaction between acrylated and thioylated macromers, can be used in the production of PEG hydrogels. This reaction is advantageous over other chemical methods like free radical polymerization for gel preparation since it does not include any initiators or exposure to radiation. Moreover, this method does not need harsh conditions to form hydrogels; the hydrogels can be obtained at physiological conditions with no side products. This gel formation reaction occurs very rapidly and it makes the injection of PEG solution directly into the defect site possible.[80] However, this method can be risky for the cell injection applications because the side reactions involving the cells can occur due to competing nucleophiles.[19] Other methods like photopolymerization with short exposure to long wavelength UV is another alternative approach to formation of hydrogels loaded with cells. PEG does not carry any functional groups suitable for cell adhesion or any biological activity; however, it can be modified with proteins and bioactive peptide sequences like RGD to enhance cell attachment and proliferation.[81] Francisco *et al.*[82] developed PEG hydrogels functionalized with laminin molecules for use in the regeneration of nucleus pulposus (NP) of intervertebral disc. They reported that this system enhanced GAG production and cell clustering when compared to PEG-only hydrogels.[82] PEG was also used in a variety of areas including bone repair,[83] drug delivery,[84] and tissue adhesive studies.[85]

1.3.4. *PVP*

PVP is a linear, water soluble, and non-ionic polymer applied in a wide range of areas including cosmetics, foods, pharmaceuticals, medicine, textiles, and inks because of its non-toxicity, biodegradability, compatibility, and good solubility (Fig. 4(D)).[86] Hydrogels of PVP can be obtained by crosslinking under ionizing radiation which results in a highly biocompatible transparent hydrogels.[12] Besides these advantages they have several disadvantages such as highly fragile nature which limit their use in biomedical applications especially when lightly crosslinked. However, PVP can be successfully employed in wound dressings and temporary skin covers after blending with other polymers, like polysaccharides. Wang *et al.*[87] employed a blend of PVP with carboxymethyl cellulose (CMC) and crosslinked by ^{60}Co gamma irradiation. They reported that CMC/PVP offered enhanced mechanical strength, transparency, and flexibility when compared to pure forms of both CMC and PVP. They concluded that this blend can be a good material to be used in wound dressing applications.[87] In another study, PVP was used together with another material called sodium alginate-g-poly(sodium acrylate) (NaAlg-g-PNaA). The researchers obtained a semi-IPN by free radical polymerization

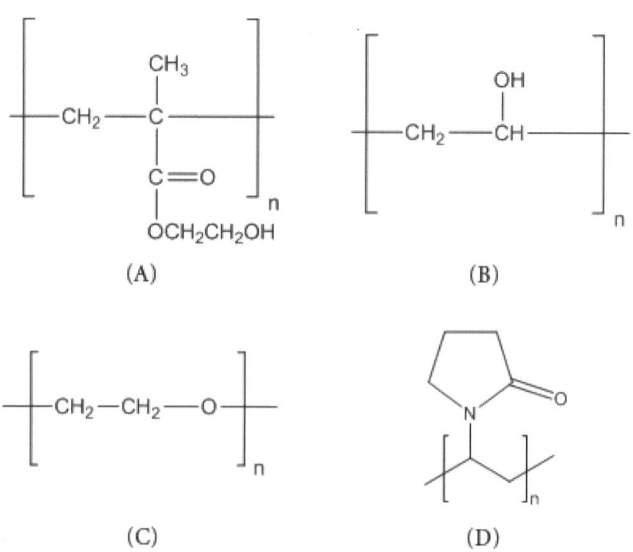

Fig. 4. Representative chemical structures of synthetic polymers used in hydrogel designs. (A) PHEMA, (B) PVA, (C) PEG, (D)PVP.

using N, N-methylene bisacrylamide (MBA) and ammonium persulfate (APS). They reported that, these semi-IPN hydrogels presented a pH-dependent sensitivity to the external stimuli and showed an on/off swelling switch due to stimulation. They concluded that this system can be employed in drug delivery systems and devices controlled by swelling with water.[86]

2. Hydrogel Processing Techniques

In tissue engineering applications, design of hydrogel constructs is very important in terms of regulating cell behavior and resultant tissue formation. Many researchers have tried to develop processing techniques for the engineering of hydrogel constructs with desired architecture, and certain aspects regarding the hydrogels, such as their low mechanical strength, biocompatibility of the chemicals and fabrication process used, had to be taken into account. Here, we present the hydrogel processing techniques that can be used to introduce the desired topographical properties and appropriate distribution of cells and bioactive molecules in the hydrogel structures fabricated.

2.1. Solvent casting and particulate leaching

Solvent casting and particulate leaching are methods widely used in the fabrication of 3D, porous scaffolds. As the particulates, mainly salt particles are used because of their availability and ease of handling.[88] In addition to salt, sucrose and sacrificial beads of similar materials can be dispersed in the hydrogel preparation solution and are entrapped in the hydrogel network during polymerization and/or crosslinking. After immersion of the hydrogels into a solvent of the particles, generally water, pores are left behind upon the dissolution of particles. Generally, the hydrogel pore sizes are in the range of 40–400 μm and these can be adjusted by changing the size and the concentration of the porogens.[89] A modification of the traditional salt leaching method was applied by Chiu *et al.*[90] to form 3D, porous PEG hydrogels. Since aqueous solvents are used in the preparation of PEG hydrogel, salt leaching method could not be used before some modifications. In this procedure, the hydrogel precursor was prepared in a saturated salt solution (instead of a solution) containing porogens of different sizes and then the polymerization was initiated with UV exposure and the porous PEG hydrogels were formed. Upon introduction of cell adhesion peptide sequence RGD onto the porous scaffolds, cell adhesion could be observed. This technique can now be reproducibly used to fabricate scaffolds with different pore sizes. The main advantages of the solvent casting process are the control over porosity, pore size, and pore shape using different particle sizes, concentration, and shape. On the other hand, this method is limited to construction of thin hydrogels generally less than 500 μm thick because of the need for complete removal of porogen particles from the hydrogel structures. However, assembly of these thin hydrogel sheets in multilayer stacks could be used to create thicker 3D scaffolds.[91]

2.2. Freeze drying

Freeze drying (or lyophilization) is a method extensively used in the fabrication of porous hydrogel structures. The approach mainly involves rapid freezing of the solution at a very low temperature that causes thermodynamic instabilities, and as a result, phase separation occurs between the solvent and the polymer. During the drying stage, the solvent is evaporated by vacuum sublimation. Hydrogels of both natural and synthetic origin can be constructed by using freeze drying. This process creates physical crosslinks, it is advantageous over other techniques such as chemical crosslinking, photocrosslinking, and irradiation due to absence of crosslinking process remnants. The main drawback of this method is the difficulty in controlling the pore size (varies in the range 10–120 μm), since it is the freezing kinetics that determine the pore size distribution.[89] Freeze dried oligo(polyethylene glycol fumarate) (OPF) hydrogel scaffolds were fabricated by Lim *et al.*[92] for use in the repair of osteochondral defects. The scaffolds were rehydrated with phosphate buffered saline (PBS) carrying bone marrow mesenchymal stem cells (BMSCs) before implanting into the chondral defects created in eight-month old Prestige World Genetics micropigs. Defect filling (99%) and hyaline-like cartilage formation (84%) were observed after four months of implantation. Another drawback of freeze drying is the incomplete phase separation in the first freezing step since the polymer solution is frozen rapidly. This causes the generation of some regions low in polymer content. To remedy this situation, repeated freeze-thaw cycles was advised. With repeated freezing, polymer-poor phases form within pores and the polymer-rich phases are aggregated and generate pores.[93] Liu *et al.*[94] fabricated PVA hydrogels with freeze-thaw technique. In this study PVA was blended with chitosan, gelatin, and starch, and among them PVA/gelatin hydrogels showed the highest bovine vascular endothelial cell proliferation. Mechanical properties of PVA hydrogels matched those of the vascular tissue and were found to be promising as scaffolds for vascular tissue engineering. Isikli *et al.*[53] constructed 3D, porous, freeze dried scaffolds from chitosan and gelatin. They added either nonsintered hydroxyapatite (nsHAP) or sintered hydroxyapatite (sHAP) particles with sizes of 16 μm and 6 μm, respectively, into the scaffolds to create composite structures. Young's modulus and compressive strength of the constructs were very similar to those of human trabecular bone in the presence of both nsHAP and sHAP. Chitosan/gelatin (C–G) scaffolds consisting of sHAP showed the highest cell adhesion since the sHAP particles carried less electrical charge and had lower particle sizes in comparison to nsHAP. They concluded that, freeze dried C-G/sHAP scaffolds could be used in the trabecular bone engineering. In another study, Kinikoglu *et al.*[95] prepared lyophilized collagen–glycosaminoglycan–chitosan (CGC) foams to mimic a non-keratinized oral mucosal equivalent. Firstly, lamina propria equivalent (bottom layer of oral mucosa that contains dense connective tissue) was created by seeding human oral fibroblasts on the porous scaffolds for 3 weeks. After that, human oral epithelial cells were seeded on these cultured scaffolds to create the oral mucosal equivalent.

2.3. Gas foaming

Gas foaming method in hydrogel preparation involves generation of gas bubbles during the chemical reaction of an acid and sodium carbonate and these bubbles form interconnected pores within the spongy structure. In this method, gas foaming and hydrogel polymerization take place simultaneously, and therefore, entrap the gas in the scaffolds. It produces 3D scaffolds without the use of toxic, organic solvents (dichloromethane or chloroform) or salts.[96] Carbon dioxide (CO_2) is widely used as the gas in this procedure because of its attractive features such as low toxicity, non-flammability, and low cost. Additional properties such as gas like diffusivity and liquid-like density of CO_2 enables one to control the properties of the product easily with pressure.[97,98] 3D hydrogel constructs were fabricated by Fathi *et al.*[99] by using poly(lactide-co-ethylene oxide fumarate) (PLEOF) and gelatin with this technique to create pores with an average diameter of 250 μm. IPN approach was used to achieve homogenous mixing of these two polymers. Foaming was achieved by the reaction of ascorbic acid and sodium bicarbonate. They demonstrated that these hydrogel constructs support human osteoblast cell adhesion, proliferation, and growth. In addition, the constructs were found to be suitable for bone tissue engineering applications especially after increasing the compression modulus two-fold with the addition of gelatin.

Some disadvantages of solvent casting/salt leaching, freeze drying and gas foaming include the presence of toxic solvents, labor intense processing, solvent removal with evaporation, unequal pore formation, low interconnectivity, and the difficulty of culturing cells within the hydrogel constructs.[98]

These processing techniques generate isotropic porous forms that do not resemble the fibrillar structure of the natural ECM. Actually, native ECM consists of fibrillar collagen microstructures and such fibrillar structures (such as the electrospun ones stated in the following section) could be better constructs to investigate in terms of the morphology, differentiation and polarity of cells seeded on them.

2.4. Electrospinning

Electrospinning is an attractive method in terms of the ability to create interconnected highly porous scaffolds. Electrospinning is based on electrically charging of a polymer solution by applying a high potential and ejecting the solution as a thin jet towards an oppositely charged collector plate through the electrostatic force. The fiber diameter, porosity, and morphology can be changed by modifying the electrospinning parameters like potential, concentration, temperature, length of trajectory, and conductivity of the polymer solution. PVA, poly(acrylic acid), collagen, silk fibroin, and fibrinogen have all been used in electrospinning to form non-woven mats. However, there are some drawbacks in the fabrication of hydrogels with electrospinning; these are low mechanical properties, limited control on porosity and inability to construct 3D complex shapes with hydrogels.[91] Hydrogel microfibers were fabricated by Zhang *et al.*[100] by using combinations of alginate, fibrin, gelatin, and HA by using both electrospinning and mechanical stretching techniques. They projected a polymer jet onto a rotating bath containing a crosslinking agent to align the polymer chains by means of mechanical shear. In addition, cells were mixed with the polymer solution before electrospinning. As a result, they created microfiber mats with high water content, porosity, and good mechanical properties originating from effective alignment. Fibers served as an ideal scaffold with their biodegradability, specific geometry, and alignment that influenced cellular orientation and differentiation. These versatile scaffolds could be suitable for engineering microvascular tissue with desired diameters.

2.5. Lithography

Use of lithographic techniques in the fabrication or modification of materials gained great attention especially in the last decades. Fabrication of biomedical materials in the desired shape and preferred pattern became possible with application of lithography. Some types of lithographic techniques are summarized in this part.

2.5.1. *Photolithography*

Photolithography is also called photopatterning when used in the production of microfabricated hydrogels. Originally, this method emerged from the development of micro-electromechanical system (MEMS) applications. In photolithography, firstly, a mask that has a desired pattern is produced. Some areas of the mask are opaque preventing the light to pass and other areas are transparent allowing the light through. With irradiation, regions of substrate corresponding to the transparent region of the mask are either crosslinked or degraded. During washing, the uncrosslinked or degraded parts are removed. Depending on the type of the photoresist, either the positive or the negative image of the mask is obtained. In general, radiation in the UV range is used to provide the energy required for the initiation of a photosensitive chemical reaction.[101] To make the polymers photo-crosslinkable, researchers used acrylamide or acrylate-based groups conjugated to a prepolymer chain; some of these were PEG diacrylate (PEGDA), PEG dimethacrylate (PEGDM), gelatin methacrylamide (GelMA), methacrylated HA (MeHA), and methacrylated tropoelastin (MeTro). A photoinitiator is always added into the gel precursor solutions to create the free radicals and initiate polymerization. In photolithography, the crosslinking depth is not regulated. Diffraction feature is mainly related with the wavelength of the light source. In microelectronic industries, resolution may be decreased to a few tens of nanometers, but in cell culture studies, it could be in the range of a few hundreds of nanometers to micrometers. The cells can be cultured on photolithographically engineered hydrogels, but in general, these substrates are used in soft lithographic processes as master templates. In these applications, the aspect ratio (the ratio between their height and width) of the hydrogel should not be too high, because hydrogels are not mechanically strong and could collapse if the patterns on them are too tall.[102]

Hydrogel colloids generated with photolithographic techniques may be used in the encapsulation of molecular or colloidal objects. To achieve this, the desired material is added into the hydrogel precursor fluid and after crosslinking with UV the hydrogel construct gains a new feature. For example, addition of magnetic particles into the hydrogel precursor creates hydrogels responsive to magnetic field. Bioactive molecules can also be entrapped in hydrogel colloids. Lithographic techniques have critical advantages for entrapment of bioactive molecules in the hydrogel since these molecules only interact with the hydrogel precursors and not with the stabilizers, carrier fluids, or microfluidic materials.[103] Cells can also be entrapped in photolithographically engineered hydrogel. For this, a cell suspension is added in the prepolymer solution before irradiation to achieve crosslinking. Nickol *et al.*[104] created cell-laden, micropatterned hydrogels by using a gelatin methacrylate (GelMA). They both seeded the cells on micropatterned GelMA constructs and entrapped them in GelMA hydrogels. Cell proliferation, elongation, and migration were observed in both substrates. They demonstrated that properties of hydrogel can be changed with the degree of methacrylation and concentration of gel precursor and thus could be tailored for individual uses.

HA hydrogels were constructed by Khetan and Burdick[105] by sequentially applying multiple modes of crosslinking. Primary addition reaction introduced thiol groups of peptide crosslinks that are degradable by matrix metalloproteinases (MMP) to acrylated HA in the presence of human mesenchymal stem cells (hMSCs). Then, non-degradable crosslinks were formed in selected areas with UV exposure. In this way, some areas on the hydrogel allowed remodeling by cells and the remaining UV crosslinked areas did not. The stiffness of the photopatterned hydrogels was increased from 6 kPa to 15 kPa after the second crosslinking. Entrapped cells in the first region showed high spreading because they degraded the matrix locally. On the other hand, cells entrapped in the second region appeared round.

2.5.2. *Soft lithography*

Soft lithography is the name of a group of microfabrication methods that include printing and molding using patterned elastomeric stamps. It carries certain advantages over photolithography, especially, the capability to control the molecular structure of surfaces, to create channels for studying microfluidics, to generate complex molecular patterns related with biology, and to control cellular behavior.[106] In addition, expensive materials or special equipment are not needed in soft lithographic techniques. For example, soft polymeric stamps are used to imprint molecular solutions or biospecies on a substrate and create patterns that have feature sizes between 30 nm and 100 μm.

In the past two decades, various patterning methods involving imprinting, molding, and embossing with an elastomeric stamp emerged. Among these methods are microcontact printing (μCP), micromolding in capillary (MIMIC), microtransfer molding (μTM), replica molding (REM), solvent assisted micro-molding (SAMIM), phase shifting edge lithography, decal transfer lithography, nanotransfer printing, dip pen lithography (DPL), and nanoskiving. Figure 5 describes the main steps of soft lithographic techniques.[107]

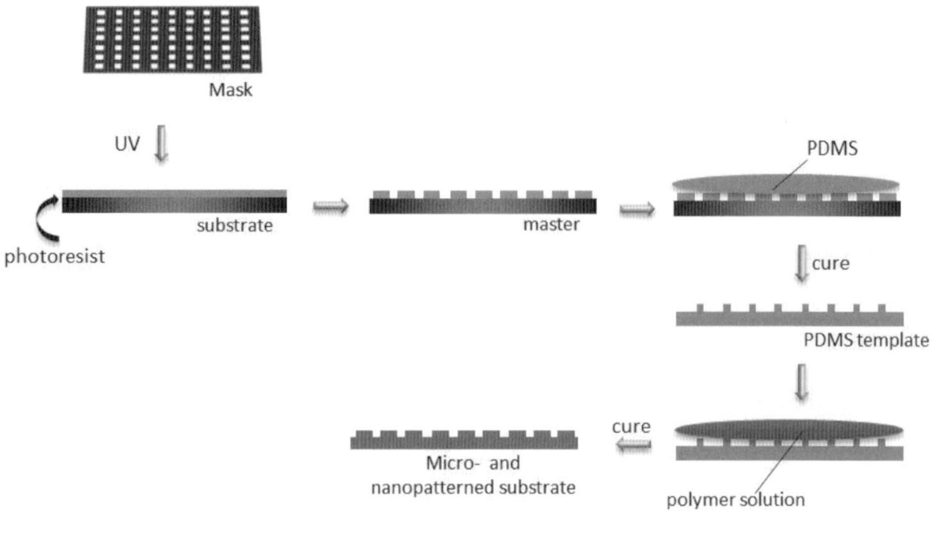

Fig. 5. Schematic illustration of soft lithography.

There are four main steps in soft lithography: (i) pattern design, (ii) creation of mask and master, (iii) generation of the polydimethylsiloxane (PDMS) stamp, and (iv) construction of micro- and nanostructures with the fabricated stamp by printing, molding, or embossing. For the first step, the design of the pattern, there are several computer aided design (CAD) software programs to use. For example, Adobe Freehand® and Illustrator® are among these programs that are easy to use and practical for creating photomasks on transparency films with feature sizes down to 1 μm. Second step comes after designing the pattern and photomasks are fabricated by using a high resolution printer. Photolithography and e-beam lithography (EBL) are mainly used to construct the master carrying the design. Photolithography is a classical method used to generate patterns with feature sizes larger than 1 μm. EBL, on the other hand, can form patterns with feature sizes of 20–30 nm. In the third step, a stamp is fabricated by casting a precursor liquid on the patterned master and then crosslinking is performed. At this stage, the mechanical properties of the stamp are crucial in the accurate generation of the pattern. Any elastomer can be used to generate a relief patterned stamp but generally crosslinked PDMS or similar silicone-based rubbers are used.[108] Another popular material is poly(methyl methacrylate) (PMMA), which has attractive features such as biocompatibility, ease of microfabrication and hydrophobicity that is useful in the removal of crosslinked hydrogel from the mold.[102] In the last, the molding step, a prepolymer for the hydrogel is poured on the crosslinked PDMS master and excess liquid scraped off. This method generates constructs with resolution from a few μm to 150 μm.[101]

Soft lithography and patterning techniques were applied by You *et al.*[109] in the fabrication of four hydrogel microstructures that have distinct micropatterned surfaces to culture primary hepatocytes. These hydrogels were PEG microwells patterned on glass, heparin hydrogel microwells patterned on glass, PEG microwells patterned on heparin hydrogel coated substrates, and heparin hydrogel microwells patterned on heparin hydrogel coated substrates. The last two hydrogels were created in two steps. First, a bottom layer of heparin gel was fabricated with photopolymerization, and then micromolding and microcontact printing were used to fabricate the second upper layer carrying microwells. By microcontact printing collagen type I was printed under the hydrogel to provide cell adhesion. The first three micropatterned surfaces: PEG microwells on glass substrate, heparin gel microwells on glass substrate, and PEG microwells on heparin gel layer were used as controls. It was reported that cells in all heparin hydrogel microwells showed the highest hepatic phenotype and production of albumin.

Zorlutuna *et al.*[110] prepared nanopatterned collagen films by using PDMS templates. X-ray interference lithography was used to create photoresists with nanochannels (groove widths 332.5, 500, and 650 nm) and PDMS replicas were constructed by using these photoresists. Then, films were prepared by solvent casting on PDMS molds. They seeded human microvascular endothelial cells (HUVECs) on both the nanopatterned and smooth (control) collagen films. Nanopatterned films with parallel channels did not affect cell proliferation and alignment; on the other hand, under fluid flow, they had improved cell retention. Cell seeded, nanopatterned films showed high O_2 and 4-acetaminophenol diffusivity. These demonstrated that cell viability could be maintained when 3D scaffolds are built from these films since they allowed O_2 and nutrients transfer. These properties make these nanopatterned films promising scaffold substrates for vascular tissue engineering.

Zheng *et al.*[111] fabricated a microfluidic vascular network (μVN) by seeding HUVECs into microfluidic channels that were generated by soft lithography in collagen type I gel. The lithographic technique allowed the creation of an endothelium through the channels by the interaction of the cells with the collagen substrate. The critical step of this processing technique is the design and engineering of the micropatterned silicone stamp to mold the collagen with microfluidic channels. They studied angiogenic remodeling, interactions between endothelial cells and perivascular cells, and interactions between endothelium and blood constituents under flow conditions in various co-culture systems. In the co-cultures, human umbilical arterial smooth muscle cells (HUASMCs) and human brain vascular pericytes (HBVPCs) entrapped in collagen hydrogels and HUVECs seeded into channels were used. HUVECs were shown to form a cylindrical lumen that shows the appropriateness of the approach.

2.6. Rapid prototyping (RP)

Soft lithography and micromolding are simple hydrogel processing techniques. With layer-by-layer approach it is possible to fabricate complex structures. Molding is a fast, scalable, and biocompatible process, but, formation of 3D biomimetic constructs such as artificial vessels cannot be easily achieved by using these techniques. They generally are used to engineer planar structures.[102] Engineering of sophisticated 3D structures is, however, very important as the fabrication of complex 3D tissues becomes a promising approach to treat tissue defects and to study pathologic cell behavior.[112] Moreover, there is a 3D interaction between the ECM and the cells that creates a dynamic regulatory system in tissue morphogenesis. As 2D constructs cannot mimic the natural microenvironment of cells, static and flat substrates cannot represent *in vivo* heterogeneity. The engineering of patterned 2D substrates and native ECM mimicking 3D scaffolds are possible, but transferring the patterned 2D substrates to complex 3D models is challenging.[113]

CAD-based RP is becoming a versatile technique to fabricate spatially controllable scaffolds. Its main principle is the bottom up construction of the complex structure deposition of layers on layers, which are produced sequentially by using computer controlled devices.[114]

RP is an additive fabrication technique that creates 3D scaffolds with computer controlled layer-by-layer processing. This is an automated technology and the external design of the construct can be made by CAD software or from reconstruction of imaging data (Magnetic resonance imaging (MRI), Computed tomography (CT), X-ray). The internal design is generated by means of processing CAD data into a *Taci Pixia palette* (STT) file. Then, the STT data is sliced to obtain machine parameters. This shows the suitability of RP to fabrication of scaffolds that carry patient specific external geometry with controllable internal design. In addition, RP leads to high interconnectivity, has the ability to use various materials, is rapid due to automatization, has high cost efficiency and few process steps.

In the laser-based RP systems, radiation energy is deposited repetitively in a desired pattern unlike the deposition of material like in the nozzle and printer-based systems. Photocrosslinkable polymers can be used in laser-based systems to generate crosslinked 3D hydrogel structures. Stereolithography (SLA) and 2-photon polymerization are examples of laser-based systems. There are several examples of nozzle-based RP systems but only some of them are suitable for hydrogel fabrication. These are pressure assisted microsyringe (PAM) technique, low temperature deposition modelling (LDM), 3D Bioplotter, and Direct ink writing (DIW). Finally, 3D printing is the major RP technique suitable for hydrogel fabrication.[115]

2.6.1. *SLA*

SLA is the oldest laser-based RP technology. By using this technique, hydrogel constructs are generated layer after layer. SLA uses UV crosslinking of the polymer by following the sliced model. In this approach, a polymer layer on the focal plane is UV crosslinked according to a predefined pattern and then the focal plane is lowered and the next layer is created.[102] There are two types of SLA, maskless SLA and mask-based SLA. In maskless SLA, the desired shape designed by using CAD software is fabricated in a layer-by-layer fashion by means of polymerization and crosslinking of a photocrosslinkable solution. With this technique, various cell types can be entrapped in different or the same polymer layer to study cellular interactions using these 3D hydrogels. This technique allows accurate patterning and low feature sizes. In mask-based SLA, a photomask is used to create the patterns. Hydrogel precursor solution with photoinitiator is crosslinked in the presence of the photomask and the transparent regions of the mask allow solution right underneath to be crosslinked, while the remaining parts are rinsed. This technique does not need printers, and therefore, cytocompatibility and clogging of the dispenser needles is not a problem. However, there are several disadvantages of this method such as low resolution of the mask, the requirement of as many masks as there are different layers and the use of a non-automated system.[113]

Zorlutuna *et al.*[112] fabricated a well-defined 3D hydrogel structure containing poly(ethylene glycol) methyl ether methacrylate and oxidized methacrylic alginate (OMA) modified with optional cell adhesion oligopeptides containing RGD sequence by using automated maskless SLA. This hydrogel allowed spatial organization of primary hippocampus neurons (HNs) and skeletal muscle myoblast cells (MCs). HNs showed increasing cholinergic activity when they were co-cultured with MCs by means of the factors released by MCs. In this way, they studied the spatial interaction of two cell types and controlled their phenotypic properties by creating a novel polymer blend that supported the viability and spreading of cells. Both the factors released by MCs and the 3D microenvironment of cells determined the interaction between the two cell types. In addition, this study demonstrated that the spatial cellular interaction is not special to native skeletal muscle-motor neuron couple, but, different cell types like hippocampal neurons could show the same results.[112]

Although there are several studies reporting the fabrication of cell entrapment in 3D hydrogel structures using stereolithographic technique, long term cell viability is an issue due to the slow processing with SLA. A Digital Light Projection (DLP) technology has emerged to perform high speed fabrication and resolution. In this technique, the projection of the pattern is done by using a digital micro mirror device or a liquid crystal display (LCD). The application is rapid since two-dimensional (2D) pixel pattern is projected onto a transparent plate and a complete layer of photocrosslinkable polymer on the focal plane is crosslinked at once. Construction time of a layer is only related with its thickness, and this directly depends on the exposure time. On the other hand, size of the construction in the *x–y* plane and the number of layers being fabricated do not affect build times.[101,116] In a study, Lin *et al.*[117] fabricated a hydrogel scaffold by using PEGDA and visible light to avoid UV exposure that can damage cells. They entrapped human adipose derived stem cells (hADSCs) into the hydrogel and demonstrated that the hydrogels had designed geometries and internal structures, and the cells showed high viability in the hydrogels for up to 7 days.

2.6.2. *The two photon polymerization (2PP)*

The 2PP is an alternative to SLA. With the SLA, the pore sizes of hydrogel scaffolds are relatively higher than the cell dimensions. This can cause isolation of cells, but clustering of cells may be required for proper function. The 2PP systems crosslink photopolymers by activating a photoinitiator by femtosecond laser pulses of 800 nm wavelength. This creates structures with feature sizes in the range 65 nm–1 cm. The 2PP systems use a focal spot, whereas SLA systems expose a focal plane, and therefore, the precision of 2PP is much higher than SLA. Moreover, there are some drawbacks of SLA system with regard to surface development when compared with 2PP method such as the use of high viscosity solutions that cause surface tension, the requirement of supportive material, the need of recoating and oxygen prevention. Furthermore, 2PP systems are done under relatively mild conditions by using light in the near-infrared (NIR) range in contrast to SLA that uses UV exposure.[119] In a study, Ovsianikov *et al.*[118] used the 2PP system in the engineering of 3D hydrogel constructs from photosensitive gelatin as the hydrogel precursor material that included living cells. They found that the human osteosarcoma cells (MG63) within the laser exposed regions were damaged, but the cells located in the immediate vicinity and within the 2PP-fabricated structures survived and proliferated. The researchers concluded that cell damage was a result of reactive species formed during 2PP instead of laser itself.

2.7. Bioprinting

Regenerative medicine aims to repair or substitute non-functional or damaged tissues with fabricated scaffolds. Recent studies report the construction of tissues like trachea,[120] bladders,[121] and skin[122] by culturing cells on scaffolds. However, these artificial tissues do not have the critical features found in ordinary biological tissues like spatial interaction of various cell types, presence of the complex ECM, and microvasculature to provide nutrients. Regenerative therapies, therefore, need to fabricate 3D constructs that carry the required combination of multiple cell types, the ECM, and the integrated microvasculature. Bioprinting is a method that might help to create the multidimensional biohybrid systems.[123]

Bioprinting, also known as biofabrication, is the generation of constructs that can mimic 3D tissues by using cells together with biomaterials in a single processing technique. The combination of hydrogels and cells used in biofabrication are called "bioink".

The physicochemical features of the hydrogels such as crosslinking method and rheological parameters such as viscosity, shear thinning, and yield stress determine the printability of hydrogel. Some process parameters like fabrication time, nozzle gauge, polymeric network properties, and gelation time also affect the success of bioprinting. The hydrogel properties like concentration, molecular weight, and chemical composition affect viscosity, gelation procedure, and rate of gelation. All these materials and bioprinting process parameters affect the cell viability. Cell viability and proliferation in natural hydrogels such as alginate, gelatin, collagen, chitosan, fibrin, and HA are very satisfactory in contrast to synthetic hydrogels such as PEG and pluronics due to absence of cell binding sites. Synthetic hydrogels require bioactive compounds like growth factors and peptide sequences to attract cell binding.

There are three types of hydrogel bioprinting systems: (a) inkjet bioprinting, (b) laser induced forward transfer, and (c) direct write bioprinting. In inkjet bioprinting (Fig. 6(A)), very small volumes (1–100 picoliters) of bioink consisting of PBS, cell culture media and hydrogel is dispensed on a substrate. Thermal inkjet printing creates pulses, via microheater to vaporize the printing liquid that helps the ejection of droplets from the printer head. The drawback of this method is the creation of stress on cells due to evaporation and it results in the generation of temporary pores on cell membrane. In piezoelectric inkjet printing, a piezoelectric actuator creates a mechanical force on the liquid and the liquid is expelled from the nozzle.[124] In laser induced forward transfer (Fig. 6(B)), there is a laser transparent print ribbon on which the cells are coated and a receiving film on which the biopolymer or the culture medium is coated to provide adhesion and growth of cells. By applying pulsed laser beam, cells attached on the ribbon are transferred to the receiving substrate.[125] Direct write bioprinting (Fig. 6(C)) uses computer-controlled actuators that use mechanical or pneumatic power to generate dots or lines of materials on a substrate. This type of systems use needles and syringes to extrude bioinks. Layer-by-layer construction is done by the movement of the stage in *x*, *y*, and *z* directions.[123]

Duan *et al.*[126] fabricated tissue engineered heart valve conduits with living leaflets via 3D bioprinting. They used MeHA and methacrylated gelatin (Me-Gel) carrying human aortic valvular interstitial cells (HAVIC). Fabrication of trileaflet valves was achieved and they observed that HAVIC entrapped within the bioprinted heart valve had high cell viability. In another study, Wüst *et al.*[127] generated a hydrogel composite with alginate and gelatin hydrogel precursors and

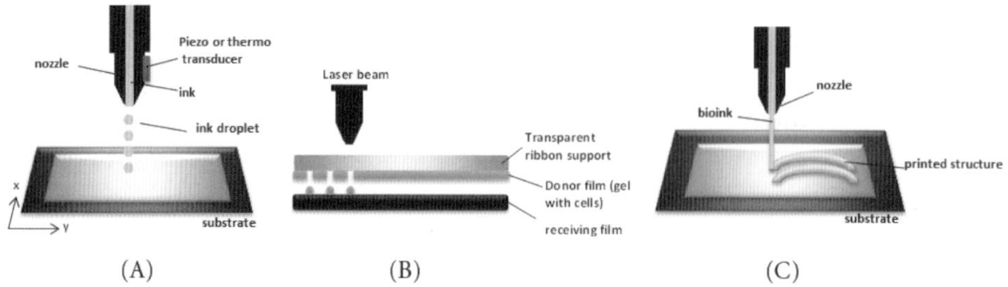

Fig. 6. Hydrogel printing systems. (A) Inkjet bioprinting, (B) Laser induced forward transfer, (C) Direct write bioprinting.

hydroxyapatite (HAp). Long term structural stability of constructs was achieved by a two-step process based on reversible thermal gelation of gelatin on a cold substrate and chemical gelation of alginate by $CaCl_2$. HAp was used to increase the visibility of the hydrogel in μCT. hMSCs were mixed into hydrogel precursor before bioprinting and high cell viability was observed. It was found to be promising for the construction of a 3D hydrogel-based bone tissue.[127]

Bioprinting is a versatile technique since it can be combined with conventional laboratory techniques to provide effective, scalable, and high throughput fabrication of scaffolds. It offers an effective method for many applications, not only for tissue engineering and regenerative medicine, but also for cancer studies. In addition, it plays an important role in printing and differentiation of stem cell by spatially controlling the microenvironment of cells. Besides *in vitro* studies, bioprinting is an emerging hydrogel processing method for *in vivo* applications.[128]

3. Applications of hydrogels in regenerative medicine

Hydrogels have been used in various fields such as electronics, diagnostics, and in customer products. The first hydrogel with potential biomedical uses was the synthetic poly(2-hydroxyethyl methacrylate), PHEMA, and it was used in soft contact lens production.[2,3] The main advantage of that revolutionary biomaterial was its stability under varying pH, temperature, and tonicity conditions.[129] Besides, due to low protein adsorption which is generally associated with low cell attachment, it avoided inflammation and foreign body response.[3,20] These properties were exploited in various biomedical applications such as contact lenses, absorbable sutures, and treatment of some diseases such as diabetes mellitus, osteoporosis, asthma, heart diseases, and neoplasms.[129]

The second breakthrough came with the use of biological macromolecules such as calcium alginate for encapsulation of islet cells,[130] and collagen and GAGs as burn dressings.[131] These hydrogels created optimal conditions for wound healing due to their high oxygen permeability and water content, in addition to serving as a barrier to protect the wounds from external agents.[20] The extensive use of hydrogels in controlled release and later in cell entrapment studies led the scientists to exploit hydrogels in directing cell attachment, migration, and differentiation, and eventually as scaffolds in tissue engineering and regenerative medicine. Tissue engineering is relatively a new approach in the biomedical field which aims to restore, maintain, or enhance tissue and organ function by substituting damaged tissues and augmenting tissue regeneration through use of biodegradable polymeric scaffolds, cells and bioactive agents such as growth factors.[132] Use of cells alone (cell therapy) or in the context of tissue engineering is collectively called regenerative medicine.[133] In many regenerative medicine applications stem cells are used.

3.1. Hydrogels for regenerative medicine applications

The function of a damaged tissue may be recovered using various methods such as tissue repair (using sutures), regeneration (using cells and growth factors), and replacement (using transplants or prostheses). Tissue repair may lead to fibrous scar which impairs the function of the tissue. Transplants have the disadvantage of limited availability and the risks of tissue rejection and disease transmission. Tissue engineering, thus, appears to be more promising in meeting the demand for healthy tissues and organs.

Hydrogels are applied as cell free systems, as space filling agents and delivery vehicles for bioactive molecules, and in cellular systems as cell carriers and 3D scaffolds that guide cell organization and direct the formation of a desired tissue.[134]

3.1.1. *Acellular systems*

Hydrogels have been used in many applications without incorporation of cells as space filling agents and bioactive molecule delivery vehicles.

3.1.1.1. Hydrogels as space filling agents

Hydrogels constructed of alginate, chitosan, and collagen, have been used as bulking agents (volume increasing substances), anti-adhesion agents, and bioadhesives.[134] Bulking agents may be applied as scaffolds after sol-gel transition, or as *in situ* gelling agents injected into the body before or after gelation. For example, crosslinked collagen structures have been used as bulking agents for the treatment of urinary incontinence,[26,135,136] in plastic and reconstructive surgery,[137,138] treatment of corneal damages[139] and vesico-ureteral reflux.[140,141]

A widely used space filling application of the hydrogels is wound dressings. Wound dressings create a moist microenvironment to the dry wound which enhances healing by promoting granulation tissue and epithelial tissue formation, and by stimulating monocytes to release cytokines and proinflammatory factors.[142–144] A hydrogel composed of alginate-chitin/chitosan-fucoidan blends was used on the wounds on rats.[138] The dressing was applied on the wound in its dry state to absorb the wound fluid and keep the wound site dry which protected it against bacterial infection.[145] After gelling, it supplied water to the wound site which promoted healing. Chitosan hydrogel was applied on full thickness burn sites and was shown to be well tolerated.[50] Skin regeneration, new tissue formation, and vascularization were observed and the newly synthesized tissue resembled the native skin esthetically and in flexibility. Another dressing made of gelatin hydrogel and loaded with growth factors at different concentrations were studied on full skin defects on rabbits and faster healing compared to empty hydrogel applied wounds were reported.[146]

In situ gelling hydrogels are injected in fluid state and they then transform into gels afterwards. They have the advantage of minimal invasion and the ability to fill irregular shaped defect sites. For instance, methacrylated HA hydrogels were gelled with exposure to a laser beam, after application of a hydrogel precursor solution on corneal wounds of rabbits.[139] The gels sealed the corneal perforations and also provided an opportunity of sutureless wound healing and tissue reconstruction.

3.1.1.2. Hydrogels as adhesives and antiadhesives

Bioadhesives are widely used as tissue sealants and hemostatic agents to reduce leakage of air and loss of blood and body fluid, to shorten surgery time, and to help in post operative healing of the wounds.[147] Examples of conventional adhesives are fibrin glue and cyanoacrylates, but hydrogels such as chitosan and other chitin derivatives are also being used. For example, photocrosslinked chitosan gels have successfully been applied in sealing wounds in various tissues in mice.[148]

Most recently bioadhesives-based on PEG and 3,4-dihydroxy-L-phenylalanine (DOPA) which is originally used by mussels to adhere on substrates under water are being recommended. These PEG-DOPA-based bioadhesives were shown to be superior to fibrin glue in moist environments in terms of adhesiveness and durability.[149] A study, in which PEG-DOPA hydrogels were used to seal pancreatic islets on the epididymal fat pad in diabetic mice, showed that these bioadhesives were stable for 4 months with minimal tissue response and good vascularization.[150] The main drawback of these hydrogels is that they are not biodegradable. Derivatives of these hydrogels carrying elastase-cleavable Ala–Ala dipeptide sequences,[151] or citrate[152] were shown to be biodegradable.

Hydrogels are also being widely used as anti-adhesives. Their low protein adsorption and the lack of integrin binding sequences are the reasons for the poor cell attachment.[20] A PEG hydrogel was used to prevent post operative adhesions,[153] a PVA hydrogel was used to prevent adhesion of tendon,[154] and a hydrogel of PLL-PEG graft copolymer was made to serve as a tissue adhesive on the one (tissue) side, and as anti-adhesive on the other side, to prevent post surgical abdominopelvic adhesions due to the inherent hydrophilicity.[155] Recently, a novel thermosensitive hydrogel, poly(ε-caprolactone)-poly(ethylene glycol)-poly(ε-caprolactone) (PCEC) was produced and tested in a Wistar rat model.[156] This biodegradable PCEC copolymer had low toxicity *in vitro* and *in vivo*. Most of the hydrogel treated rats did not develop adhesions after a sidewall defect (cecum abrasion), while the untreated rats developed strong adhesions.

3.1.1.3. Hydrogels as delivery vehicles for bioactive molecules

Bioactive molecules such as growth factors are an important component of tissue engineering and regeneration. Growth factors may influence cell proliferation, spreading, migration, and differentiation. They may guide adult cells to secrete ECM or stem cells to differentiate into the desired cells.

Among the various systems developed for controlled release of bioactive agents, implantable or injectable hydrogels that are biodegradable or bioresorbable are used in the field of tissue engineering and regenerative medicine.[20] The release of the agents may be modulated upon external stimuli such as temperature, pH, enzymes, or ultrasound, light, or electrical field. The material may also be decorated with specific domains which preferentially bind to some bioactive molecules but not others. This may enable localized and/or sequential release of bioactive molecules. For example, coupling of heparin to hydrogels resulted in better interaction of heparin-binding growth factors.[157]

Hydrogels are widely used in the release of bioactive molecules. For example, the delivery of vascular endothelial growth factor (VEGF) from alginate displayed a diffusion controlled[158,159] and mechanical stimuli controlled[160,161] release, and from collagen displayed a degradation controlled release.[162] All the systems proved useful in promoting angiogenesis both *in vitro* and at the implant site *in vivo*. VEGF and platelet-derived growth factor-BB (PDGF-BB) were sequentially delivered from alginate gels injected to rats with myocardial infarction[163] and mice with ischemic limb hind.[164] Hao and colleagues[163] exploited the higher affinity of PDGF-BB to alginate gels, to achieve a release of PDGF-BB slower than that of VEGF. Sun and colleagues,[164] however, entrapped PDGF-BB in poly(lactide-co-glycolide) (PLGA) microspheres which were then incorporated to the VEGF-containing gels. This enabled slower release of PDGF-BB because PLGA decreased its release rate. In another study, sequential release of IGF-1 followed by HGF from alginate gels was shown to reduce infarct expansion and scar fibrosis, and to promote angiogenesis at the infarct site.[65]

The high affinity of alginate gels to dibutyryl cyclicadenosine monophosphate (dbcAMP), a regulator of human keratinocyte proliferation, was exploited to achieve sustained release of dbcAMP to the wound site in a rat model.[165] The hydrogel system carrying dbcAMP promoted reepithelialization and thus enhanced healing of wound. Gelatin and polyurethane (PU)-based hydrogels were loaded with free and entrapped epidermal growth factor (EGF) to full thickness skin defects in rabbits.[166] The hydrogels showed no foreign body reaction, and those containing EGF-loaded gelatin microspheres led to higher wound healing with new tissue very much resembling the native skin. The same hydrogels were used in diabetic and healthy rats and led to better healing than gels without EGF.[167] The scaffolds with EGF-loaded gelatin microspheres were effective in healing the wound in both diabetic and non-diabetic rats.

Hydrogel scaffolds loaded with bioactive molecules have been used in bone regeneration. BMP-2 was incorporated in porous HA scaffolds[168] and in photo-crosslinked poly(D,L-lactic acid)-polyethylene glycol (PLA-PEG) and poly(D,L-lactic acid)-p-dioxanone-polyethylene glycol (PLA-DX-PEG) systems.[169] Ectopic bone formation on PLA-DX-PEG scaffolds was observed in the muscle of the mouse model. In another study, a small interference RNA (siRNA) against Noggin (the antagonist of BMPs) gene was incorporated in the PLA-DX-PEG scaffolds to induce ectopic bone formation in the mouse model.[170] Higher bone formation was observed in the hydrogel containing both siRNA and BMP-2 than that containing only BMP-2. Alginate gels grafted with RGD motifs were also used to deliver BMP-2 to critical size femoral defects in rodents.[171] Bone regeneration with consistent bony bridging of the critical size defects was recorded. When perforated poly(ε-caprolactone) (PCL) scaffold was used together with the alginate/BMP-2 delivery system, the newly formed bone like tissue showed mechanical properties equivalent to the intact bone.

3.1.2. *Cellular systems*

Incorporation of cells in hydrogels dates back to the work of Lim and colleagues,[130] in which pancreatic islets were entrapped in alginate gels. Since then, many studies were performed in which hydrogels were used either to entrap cells or to engineer tissues.

3.1.2.1. Hydrogels as cell carriers

Alginate is one of the most commonly used hydrogels in the entrapment of cells, because it can be crosslinked with different cations under very mild conditions, and due to its high hydrophilicity it does not evoke reactions in the immune system.[20] Polyethylene glycol (PEG) hydrogels are also widely used to entrap cells. PEG conjugated with negatively charged compounds allows for coating with polycations, including poly(L-lysine) and polyornithine, which help regulate the release kinetics and introduce the ability to respond to physiological stimuli.[20,172]

An important field of application of hydrogels is treatment of diabetes. In a study, pancreatic islets entrapped in alginate were implanted in rats, and sustained release of insulin for 8 weeks was observed.[173] The system released higher insulin concentrations at high glucose levels, thus showed bioresponsiveness. Purified alginate was shown to lead to higher cell survival and better insulin secretion than its impure counterpart.[174] In another study, crosslinked PEG hydrogels were used to entrap murine islets.[175] Cells were shown to survive for 2 weeks and continue to release insulin, and the release rate of insulin was decreased with the increasing crosslinking density.

Hydrogels carrying cells were used in the regeneration of vascular tissue. Implantation of hydrogels encapsulating endothelial cells alone was not successful, because the cells either died or they could not integrate with the existing blood vessels and the smooth muscle cells which are crucial for vascularization.[143] Therefore, cells were entrapped together with certain growth factors that promote angiogenesis. In one such study, endothelial cells, VEGF, and monocyte chemotactic protein-1 (MCP-1) were encapsulated in alginate microparticles which were then transplanted in mice.[176] The system was shown to enhance angiogenesis and increase the density of mature blood vessels containing smooth muscle cells.

Hydrogels carrying cells have also been used in the regeneration of bone. Alginate with adipic acid dihydrazide (AAD) and poly(aldehyde guluronate) (PAG) was used to entrap osteoblasts and then was injected to mice.[177] The hydrogel system promoted mineralized bone formation in 9 weeks. Another study used RGD-modified alginate gels to transplant rat calvarial osteoblasts into mice.[178] Significantly higher regeneration of the bone tissue was observed with the RGD modified gels compared to those not modified. In another study, mesenchymal stem cells and BMP-2 were entrapped in a blend of alginate and chitosan which was injected to mice.[179] New bone formation that histologically resembled trabecular bone was observed. Osteoblast loaded alginate gels were incorporated into calcium phosphate cement to protect the cells during the setting of the cement and the structure was reinforced with chitosan and PLGA fiber meshes.[180] About 80% of the cells were shown to remain alive after a one day culture, and the reinforced cement displayed mechanical properties matching those of porous hydroxyapatite implants and cancellous bone. Chondrocytes and osteoblasts were entrapped in RGD-alginate and then transplanted into mice to induce the formation of epiphysis.[181] The cells were shown to organize into a structure similar to the growth plate.

Hydrogels were also used to entrap cells for cartilage tissue repair. For example, shape-memory alginate macroporous gels were introduced into mice in their dry state via a catheter, followed by *in situ* rehydration with articular chondrocyte suspension.[182] After shape recovery in an hour, the construct conformed to the desired geometry. The original tissue shape was reached after 3 months, and cartilage tissue formation was observed after 6 months. In a study, injectable, biodegradable hydrogels containing OPF carrying rabbit marrow MSCs and gelatin microparticles loaded with transforming growth factor-β1 (TGF-β1) were used to promote chondrocyte differentiation.[183] The entrapped rabbit MSCs were shown to differentiate into chondrocyte-like cells.

In a clinical study, autologous ear chondrocytes entrapped in alginate were injected into the bladder to treat urinary incontinence.[184] The method proved safe and effective; 81% of the patients showed improvement 8 months after a single injection. When collagen was used without cells, repeated injections had to be made to achieve an improvement in the long term.[135] In another study, adipose derived stem cells (ADSCs) were encapsulated in silk fibroin microspheres and injected periurethrally to rats in order to treat severe stress urinary incontinence.[185] The study showed that cell free microspheres were effective in the short term, while those containing cells displayed long term improvement and better tissue regeneration.

3.1.2.2. Hydrogels as tissue engineering scaffolds

Tissue engineering has three major components: cells, scaffolds (cell carriers), and bioactive agents (growth factors). Scaffolds are biodegradable, polymer-based and highly porous structures which enable nutrient and waste transfer, and cell attachment and growth. They are synthetic ECM that house the cells and might have features which help in cell organization and production of the targeted tissue.[186] Since tissues have specific properties, scaffolds are designed to meet these specifications.[134] Cell behavior on the scaffolds can be manipulated with physical and chemical cues created on them using variety of approaches.[20]

Scaffolds can be made of hydrogels which are designed to be porous and bioresorbable, and may be fabricated to release bioactive molecules or present cell receptor binding motifs, in order to induce cells to attach, proliferate, migrate, and differentiate.[2] Some problems related with the use of hydrogel-based scaffolds are low mechanical strength, difficulty in sterilization, and low rate of diffusion.[2,187,188] Various tissues have been produced using hydrogels as scaffolds. Thin or avascular tissues such as skin and bladder were produced and used in the clinical trials. Microperforated HA-based membranes (Hyaff 11) seeded with autologous fibroblasts and keratinocytes were successfully applied in skin regeneration.[189] Bladder shaped collagen scaffolds seeded with urothelial and muscle cells were successfully used in bladder reconstruction.[121]

The main problem of the engineered tissues is transfer of nutrient and oxygen. To solve this problem, angiogenesis is chemically induced with growth factors or mechanically by introducing channels into the scaffolds. An example is that where VEGF and endothelial cells were incorporated in alginate-PLGA gels.[190] A significant increase in the density of the capillaries in the construct was observed after implantation to mice. In another study, endothelial progenitor cells were seeded on RGD modified alginate hydrogels in an attempt to enhance angiogenesis in mice with ischemic hind limb.[191] Increased vasculature density and prevention of the ischemic limb necrosis were reported. Nanopatterned collagen scaffolds seeded with vascular smooth muscle cells were used to tissue engineer blood vessels *in vitro*.[192,193] The scaffolds were shown to have mechanical properties matching the natural blood vessels, and the cells seeded on the scaffolds retained their phenotype after 21 days of culture. PEG scaffolds modified with cell adhesion sequences (KQAGDV and RGDS), enzymatic degradation sites (GGLGPAGGK, collagenase sensitive, or AAAAAAAAAK, elastase sensitive), or growth factors (e.g., TGF-β) have been used to engineer vascular smooth muscle.[194,195] Smooth muscle cells on these hydrogels proliferated and showed enhanced deposition of hydroxyapatite and ECM. In a study hepatocytes were seeded in agarose gels with microchannels.[196] Perfusion through the channels resulted in diffusion of the medium into the gels in a few minutes. However, cell viability was shown to decrease with the increasing distance from the channels. In another study in which human umbilical vein endothelial cells (HUVEC) were seeded on the walls of the microchannels fabricated using the injection molding within a collagen network seeded with HBVPC or HUASMC.[111] After 14 days of culture, HUVECs were shown to form cylindrical lumens and sprouts in the collagen matrix, and protein leakage was recorded in the vicinity of these sprouts. In another study, a leaf tissue with veins was used as a photomask to obtain PDMS molds using soft lithography, and agarose gels with microvascular networks were fabricated using the molds.[197] Hepatocytes were seeded on agarose hydrogels and the microchannels were injected with collagen to obtain perfusable vasculature. Cell viability increased with the presence of microvesicles.

Bioprinting is another method which enables fabrication of vascular structures with hydrogels. For example, smooth muscle cells and fibroblasts were bioprinted into 3D vascular structures of differing size and shapes using agarose rods as molding templates.[198] Agarose rods and cell spheroids were fabricated with RP method, and after building the 3D structure, agarose rods were removed to leave the cell spheroids in the vascular structures. The cells retained their phenotypes after 3 days of culture and they were fused to form vessel-like structures sturdy-enough for handling and further culture. In another study, sacrificial gelatin was printed within crosslinked collagen layers carrying entrapped fibroblasts, and the layers were fused on top of each other to fabricate a multilayer 3D structure.[199] After removal of the gelatin by heat, vascular microchannels were obtained. Perfused channels resisted a pressure of over 100 mm Hg without any cracking of the collagen network, and cells up to 400 μm distance from the microchannels remained viable after 7 days of culture.

Hydrogel scaffolds have also been used for the engineering of the corneal tissue. In an *in vitro* study, stromal keratocytes, epithelial cells and endothelial cells simultaneously seeded on crosslinked hydrogels composed of collagen and chondroitin sulfate were shown to form stratified epithelium and deposit basement membrane components.[28] The structure was pliable and had good resilience. The same group used micropatterned collagen films to align the human stromal keratocytes seeded on them.[200] Higher transparency was achieved with the patterned hydrogels relative to unpatterned ones, due to organized ECM deposition on the films. Another study used a 3D corneal scaffold composed of four micropatterned layers of crosslinked collagen modified with elastin like recombinamer (Col: ELR) to seed human corneal keratocytes.[201] The cell seeded constructs demonstrated higher transparency than the unseeded controls.

Hydrogels have also been used in bone tissue engineering studies. In order to engineer bone tissue, HA-based hydrogel, Hyaff 11, was seeded with bone marrow stromal cells along with basic fibroblast growth factor (bFGF), and then implanted into radial bone defects in rat models.[202] Improved mineralization was observed at the defect site, and better healing was achieved. When bone marrow stromal cells that were differentiated *in vitro* to osteoblasts were seeded in alginate gels and implanted to dogs in an attempt to treat horizontal alveolar bone defects, bone nodules were observed 4 weeks after implantation and mature bone formation was observed after 12 weeks.[203] Hydrogels with composite structures were produced to improve cell attachment on the scaffolds. For example, higher osteosarcoma cell attachment was observed on porous alginate-hydroxyapatite scaffolds compared to pure alginate scaffolds.[204] Micro and macropatterned and crosslinked collagen films carrying calcium phosphate particles were used to differentiate mesenchymal osteoprogenitor cells to osteoblasts.[205] Higher osteogenic activity was observed with the calcium phosphate loaded scaffolds. In another study, MSCs derived from Wharton's jelly and menstrual blood were seeded on the collagen-calcium phosphate composite scaffolds and tested *in vitro* for proliferation and osteogenic differentiation.[206] Cells on the composite scaffolds showed higher proliferation and better osteogenic activity compared to those on calcium phosphate free groups. Sequential delivery of growth factors may be achieved using hydrogel-based tissue engineering scaffolds (Fig. 7). For example, PLGA scaffolds containing poly(4-vinyl pyridine)-alginate microspheres loaded with BMP-2 and BMP-7 were used to engineer bone tissue.[207] The microspheres provided sequential release of the growth factors which led to better osteogenic differentiation relative to those when each growth factor was administered alone. Sequential delivery of BMP-2 and BMP-7 was achieved by encapsulation of BMP-2 and BMP-7 in PLGA and poly(3-hydroxybutyrate-*co*-3-hydroxyvalerate) (PHBV) nanocapsules which were then incorporated into chitosan and chitosan-PEO hydrogels.[55] BMP-2 was released faster than BMP-7, to achieve a high proliferation rate early in the culture duration. Another study showed that human bone marrow stromal cells (HBMSC) seeded onto alginate-VEGF-poly(D,L-lactic acid) (PLA)-BMP-2 scaffolds resulted in the

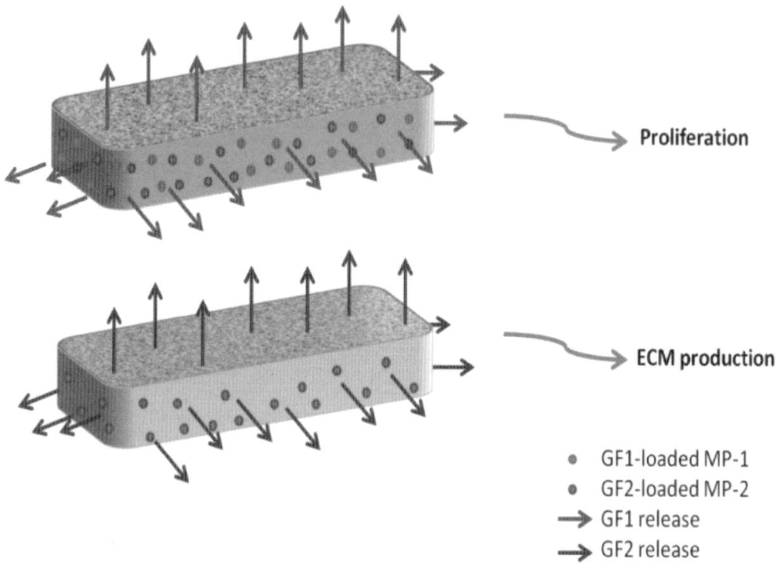

Fig. 7. Sequential release of two growth factors (GF-1 and GF-2) from hydrogel microparticles (MP-1 and MP-2) entrapped in a 3D scaffold. First GF-1 is released to induce proliferation of the cells, and then GF-2 is released to induce ECM production.

repair and regeneration of critical size bone defects in the mouse model.[208] Sequential release of VEGF from alginate and BMP-2 from PLA resulted in the formation of new blood vessels and bone.

Hydrogels have also been used to engineer cartilage tissue. Stiff materials promote chondrocyte proliferation instead of ECM deposition, resulting in less production of cartilage. Softer materials such as HA,[209] collagen,[210,211] alginate,[212] OPF mixed with gelatin microspheres containing TGF-β1, and fibrin-PEG-poly(lactic acid),[213] on the other hand, have been used as scaffolds for chondrocyte seeding. Alginate gels were mixed with chondrocytes and the solution was crosslinked with calcium ions before injection into the molds of facial implants to produce the scaffolds.[212,214] Scaffolds implanted to mice and sheep retained their shape for 30 weeks with mechanical properties matching the native tissue, and the cells demonstrated cartilage phenotype, and produced significant amounts of GAGs and collagen. Collagen microsponges introduced to PLGA sponges seeded with chondrocytes were tested both *in vitro* and *in vivo* for cartilage regeneration.[215] Cells were shown to proliferate and produce collagen type II after 4 weeks of *in vitro* culture. About 8 weeks after subcutaneous implantation of the scaffolds in mice, cells were shown to produce collagen type II and GAGs.

Hydrogels have also been used in the engineering of many other tissues. For skeletal muscle reconstruction, collagen was seeded *in vitro* with skeletal muscle myoblasts to form highly condensed rod-like scaffolds, which were cultured *in vitro* on ring shaped alginate under cyclic stretching conditions.[216] The muscular tissue produced was highly oriented, and led to densely accumulated cells and collagen fibers aligned in the direction of stretching. For adipose tissue engineering, collagen sponges were seeded with preadipocytes and implanted in mice model.[217] Formation of vascularized adipose-like tissue was achieved 8 weeks after implantation. Schwann cells were seeded on alginate hydrogels to create nerve grafts, and the gels were shown *in vitro* to support cell viability and to lead to enhanced neurite growth.[218] Fibrochonrocytes were seeded on alginate coated PLLA-collagen composite to construct meniscus tissue substitute.[219] Constructs containing alginate led to expression of chondrogenic phenotype with high collagen type II production, while those not containing alginate displayed rather a fibroblastic phenotype. In a study, fibroblast seeded PEG microgels with various structures and shapes (beads, cubes, saws, droplets) were used as building blocks to which acoustic waves were applied to assemble them into 3D structures.[220] The study showed that the microgels assembled into 3D structures in seconds, with more than 90% cell viability.

3.2. Hydrogels in controlling stem cell fate for regenerative medicine

The use of stem cells in tissue engineering is appealing because isolation and expansion of mature cells has its problems (invasive isolation, slow expansion),[143] and stem cells can differentiate into a variety of cell lineages and they proliferate much faster.[221] The complex microenvironment in the body (niche), including biochemical factors (receptor ligands, chemokines, cytokines, growth factors), physical factors (organization and mechanical properties of the ECM), and cell–cell interactions contribute to the determination of stem cell fate.[222–224] For example, primary erythroid progenitors require binding to fibronectin in the ECM through $\alpha 4\beta 1$ integrin in order to adhere and proliferate properly.[225] Besides, cells in different tissues require different physical cues to maintain their phenotypes. For example, stem cells in epithelial tissues (e.g., skin and gut) encounter a 2D ECM, while those in other tissues interact with a 3D ECM.[224] Tissue engineering scaffolds should, therefore, be designed to mimic the appropriate organization to guide the stem cells in their differentiation to the desired cells and tissues.

Hydrogels are used in the construction of cellular systems simply because they are highly hydrated and porous, properties which can be manipulated to provide chemical and mechanical signals for the cell to attach, or differentiate.[134] One method used to modify scaffold materials with biochemical factors may be the immobilization of growth factors such as BMP-2 onto hydrogels. Alginate hydrogels with tethered BMP-2 in order to promote *in vivo* bone formation led to bone mineralization in the muscles of the rats, indicating a successful effect.[226] Another study was conducted with the human and porcine MSCs which were seeded on poly(methyl methacrylate)-g-poly(ethylene oxide) carrying tethered EGF to promote cell spreading and prevent apoptosis.[227] Tethered EGF was found to increase cell spreading and survival more than the soluble EGF. Another study showed enhanced osteogenic differentiation of bone marrow derived MSCs after modification of an IPN of RGD-functionalized poly(acrylamide-co-ethylene glycol-acrylic acid) with Sonic hedgehog (Shh) proteins.[228] Immobilized Shh protein had no effect on cell proliferation but it enhanced osteogenic differentiation in the presence of BMP-2 and BMP-7.

Positive contribution of the growth factors on differentiation was shown in the study by Basmanav *et al.*,[207] in which sequential delivery of BMP-2 and BMP-7 from poly(4-vinylpyridine)-alginate microspheres directed the differentiation of rat bone marrow stem cells to osteoblasts *in vitro*. Another study showed the positive effect of alginate gels carrying TGF-β1, dexamethasone, and ascorbate 2-phosphate in the chondrogenic differentiation of the hMSCs.[229] In another study, mesenchymal stem cells and BMP-2 were entrapped in an alginate/chitosan blend which was then injected to mice.[179] Osteogenic tissue formation that histologically resembled trabecular bone was observed. Injectable, biodegradable hydrogel containing OPF entrapped with rabbit marrow MSCs and gelatin microparticles loaded with TGF-β1 were used successfully to induce differentiation of rabbit MSCs into chondrocyte-like cells.[183]

Effect of mechanical cues on the differentiation of stem cells was shown in the study conducted on polyacrylamide scaffolds that were seeded with MSCs.[230] Depending on the stiffness of the scaffolds, MSCs differentiated into bone (with stiff gels), or muscle (with moderately stiff gels), or neural tissue (with soft gels). Another study showed that mechanical cues together with the biochemical factors work in combination to determine the stem cell behavior.[231] It was observed that when seeded on moderately stiff IPN gels, in the presence of FGF-b, adult neural stem cells proliferated, but in the presence of retinoic acid they differentiated into neurons. In the absence of these biochemical factors, however, stiff gels led to differentiation of neural stem cells to astrocytes, whereas softer gels led to differentiation to neurons. Thus, even in the absence of chemical factors, differentiation direction or fate could be altered. Stability of the substrate was also seen to alter the fate of differentiation. Human MSCs on degradable (not photocrosslinked) HA-based hydrogels spread and demonstrated osteogenic differentiation, while those on non-degradable (photocrosslinked) hydrogels remained round in shape and showed adipogenic differentiation.[105] The effect of pattern dimensions on differentiation was studied with hADSCs photoentrapped in alginate/PEG hydrogels.[232] Double-photocrosslinked hydrogels were fabricated with photolithographic method using checkerboard masks of differing dimensions to create the patterns. Stem cells demonstrated highest osteogenic differentiation on gels with pattern dimensions of 200 μm, highest chondrogenic differentiation on gels with pattern dimensions of 100 μm, and highest adipogenic differentiation on gels with dimensions of 50 μm.

4. Conclusion

Hydrogels are widely used in tissue engineering and regenerative medicine applications, because they possess various unique advantages, including stability under different conditions, good oxygen and metabolite permeability, good barrier properties, and avoidance of inflammation and foreign body reaction. They are commonly used as drug delivery devices, cell carriers and as scaffolds. Although they have poor protein and cell adhesiveness, they could be manipulated to have improved protein and cell adhesion, and allow cell proliferation, migration and differentiation. They have been applied successfully in engineering of many tissues including the bone, cartilage, blood vessels, muscle, cornea, bladder, and nerves. They have a great potential to further contribute to tissue engineering and regenerative medicine.

References

1. E. M. Ahmed, Hydrogel: preparation, characterization, and applications. *JAR*, **6**(2), 105–121 (2015).
2. A. S. Hoffman, Hydrogels for biomedical applications. *Adv. Drug Deliv. Rev.*, **54**(1), 3–12 (2002).
3. O. Wichterle and D. Lim D, Hydrophilic gels for biological use. *Nature*, **185**(4706), 117–118 (1960).
4. A. Singh, P. K. Sharma, V. K. Garg, and G. Garg, Hydrogels: A review. *Int. J. Pharm. Sci. Rev. Res.*, **4**(2), 97–105 (2010).
5. O. Okay, General properties of hydrogels. In *Hydrogel Sensors and Actuators*, Eds. G. Gerlach and K. Arndt, Berlin Heidelberg: Springer, pp. 1–14 (2010).
6. K. Deligkaris, T. S. Tadele, W. Olthuis, and A. van den Berg, Hydrogel-based devices for biomedical applications. *Sensor. Actuat. B-Chem.*, **147**(2), 765–774 (2010).
7. S. B. Campbell and T. Hoare, Externally addressable hydrogel nanocomposites for biomedical applications. *Curr. Opin. Chem. Eng.*, **4**, 1–10 (2014).
8. N. A. Peppas, Hydrogels. In *Biomaterials Science: An Introduction to Materials in Medicine*, Ed. B. D. Ratner, A. S. Hoffman, F. J. Schoen and J. E. Lemons, New York: Academic Press, pp. 100–107 (2004).
9. N. A. Peppas, Y. Huang, M. Torres-Lugo, J. H. Ward, and J. Zhang, Physicochemical foundations and structural design of hydrogels in medicine and biology. *Annu. Rev. Biomed. Eng.*, **2**(1), 9–29 (2000).

10. W. E. Hennink and C. F. van Nostrum, Novel crosslinking methods to design hydrogels. *Adv. Drug Deliv. Rev.*, **54**(1), 13–36 (2002).
11. A. S. Hoffman, Hydrogels for biomedical applications. *Adv. Drug Deliv. Rev.*, **64**, 18–23 (2012).
12. J. M. Rosiak and P. Ulański, Synthesis of hydrogels by irradiation of polymers in aqueous solution. *Radiat. Phys. Chem.*, **55**(2), 139–151 (1999).
13. K. H. G. Syed and Saphwan Al-Assaf, Hydrogels: Methods of preparation, characterisation and applications. In *Progress in Molecular and Environmental Bioengineering-From Analysis and Modeling to Technology Applications*, Ed. Angelo Carpi, Croatio: InTech (2011).
14. J. M. Rosiak and F. Yoshii, Hydrogels and their medical applications. *Nucl. Instr. Meth. Phys.*, **151**(1–4), 56–64 (1999).
15. X. Gao, C. He, C. Xiao, X. Zhuang, and X. Chen, Synthesis and characterization of biodegradable pH-sensitive poly(acrylic acid) hydrogels crosslinked by 2-hydroxyethyl methacrylate modified poly(L-glutamic acid). *Mater. Lett.*, **77**, 74–77 (2012).
16. S. S. Vaghani and M. M. Patel, pH-sensitive hydrogels based on semi-interpenetrating network (semi-IPN) of chitosan and polyvinyl pyrrolidone for clarithromycin release. *Drug Dev. Ind. Pharm.*, **37**(10), 1160–1169 (2011).
17. S. Sugiura, K. Sumaru, K. Ohi, K. Hiroki, T. Takagi, and T. Kanamori, Photoresponsive polymer gel microvalves controlled by local light irradiation. *Sensor. Actuat. A-Phys.*, **140**(2), 176–184 (2007).
18. L. A. Wells, S. Furukawa, and H. Sheardown, Photoresponsive PEG-anthracene grafted hyaluronan as a controlled-delivery biomaterial. *Biomacromolecules*, **12**(4), 923–932 (2011).
19. B. V. Slaughter, S. S. Khurshid, O. Z. Fisher, A. Khademhosseini, and N. A. Peppas, Hydrogels in regenerative medicine. *Adv. Mater.*, **21**(32–33), 3307–3329 (2009).
20. J. M. Saul and D. F. Williams, Hydrogels in regenerative medicine. In *Principles of Regenerative Medicine (Second Edition)*, Ed. A. Atala, R. Lanza, J. A. Thomson, and R. Nerem, San Diego: Academic Press, pp. 637–661 (2011).
21. A. S. Sawhney and P. D. Drumheller, Polymer Synthesis. In *Frontiers in Tissue Engineering*, Ed. C. W. Patrick, Jr., A. G. Mikos, L. V. McIntire and R.S. Langer, Pergamon: Oxford, pp. 83–106 (1998).
22. V. M. Correlo, J. M. Oliveira, J. F. Mano, N. M. Neves, and R. L. Reis, Natural origin materials for bone tissue engineering — properties, processing, and performance. In *Principles of Regenerative Medicine (Second Edition)*, ed. A. Atala, R. Lanza, J. A. Thomson, and R. M. Nerem, San Diego: Academic Press, pp. 557–586 (2011).
23. T. J. Wess, Collagen fibril form and function. *Adv. Protein Chem.*, **70**, 341–374 (2005).
24. G. Bou-Gharios and B. de Crombrugghe, Type I collagen structure, synthesis, and regulation. In *Principles of Bone Biology (Third Edition)*, Ed. J. P. Bilezikian, L. G. Raisz and T. J. Martin, San Diego: Academic Press, pp. 285–318 (2008).
25. T. Cawston, Matrix metalloproteinases and TIMPs: properties and implications for the rheumatic diseases. *Mol. Med. Today*, **4**(3), 130–137 (1998).

26. C. H. Lee, A. Singla, and Y. Lee, Biomedical applications of collagen. *Int. J. Pharm.*, **221**(1), 1–22 (2001).
27. W. Liu, K. Merrett, M. Griffith, P. Fagerholm, S. Dravida, B. Heyne *et al.*, Recombinant human collagen for tissue engineered corneal substitutes. *Biomaterials*, **29**(9), 1147–1158 (2008).
28. N. E. Vrana, N. Builles, V. Justin, J. Bednarz, G. Pellegrini, B. Ferrari *et al.*, Development of a reconstructed cornea from collagen-chondroitin sulfate foams and human cell cultures. *Invest. Ophth. Vis. Sci.*, **49**(12), 5325–5331 (2008b).
29. X. Xiao, S. Pan, X. Liu, X. Zhu, C. J. Connon, J. Wu *et al.*, *In vivo* study of the biocompatibility of a novel compressed collagen hydrogel scaffold for artificial corneas. *J. Biomed. Mater. Res. Part A*, **102**(6), 1782–1786 (2014).
30. K. Li, L. Guo, Y. Fan, and X. Zhang, Experimental study on collagen hydrogel scaffolds for cartilage tissue engineering. *Chinese Journal of Reparative and Reconstructive Surgery*, **26**(11), 1356–1361 (2012).
31. X. Chen, F. Zhang, X. He, Y. Xu, Z. Yang, L. Chen *et al.*, Chondrogenic differentiation of umbilical cord-derived mesenchymal stem cells in type I collagen-hydrogel for cartilage engineering. *Injury*, **44**(4), 540–549 (2013).
32. J. M. Ranieri, D. T. Padavan, C. Ghem, D. J. Stewart, M. Ruel, and E. J. Suuronen, Novel physically cross-linked collagen hydrogel for cardiovascular tissue regeneration. *Can. J. Cardiol.*, **29**(10, Supplement), S201 (2013).
33. L. A. Reis, L. L. Y. Chiu, Y. Liang, K. Hyunh, A. Momen, and M. Radisic, A peptide-modified chitosan–collagen hydrogel for cardiac cell culture and delivery. *Acta Biomater.*, **8**(3), 1022–1036 (2012).
34. A. Ndreu Halili, N. Hasirci, and V. Hasirci, A mechanically functional collagen-based construct designed as a meniscus substitute. *J. Biomater. Tissue Eng.*, **3**(2), 173–184 (2013).
35. J. Radhakrishnan, U. M. Krishnan, and S. Sethuraman, Hydrogel based injectable scaffolds for cardiac tissue regeneration. *Biotechnol. Adv.*, **32**(2), 449–461 (2014).
36. X. Xu, A. K. Jha, D. A. Harrington, M. Farach-Carson, and X. Jia, Hyaluronic acid-based hydrogels: From a natural polysaccharide to complex networks. *Soft*, **8**(12), 3280–3294 (2012).
37. J. A. Burdick and G. D. Prestwich, Hyaluronic acid hydrogels for biomedical applications. *Adv. Mater.*, **23**(12), H41–H56 (2011).
38. T. Pouyani, G. S. Harbison, and G. D. Prestwich, Novel hydrogels of hyaluronic acid: synthesis, surface morphology, and solid-state NMR. *J. Am. Chem. Soc.*, **116**(17), 7515–7522 (1994).
39. A. K. Jha, W. Yang, C. B. Kirn-Safran, M. C. Farach-Carson, and X. Jia Perlecan domain I-conjugated, hyaluronic acid-based hydrogel particles for enhanced chondrogenic differentiation via BMP-2 release. *Biomaterials*, **30**(36), 6964–6975 (2009).
40. O. P. Varghese, W. Sun, J. Hilborn, and D. A. Ossipov, *In situ* cross-linkable high molecular weight hyaluronan-bisphosphonate conjugate for localized delivery and cell-specific targeting: A hydrogel linked prodrug approach. *J. Am. Chem. Soc.*, **131**(25), 8781–8783 (2009).

41. A. Fakhari and C. Berkland, Applications and emerging trends of hyaluronic acid in tissue engineering, as a dermal filler and in osteoarthritis treatment. *Acta Biomater.*, **9**(7), 7081–7092 (2013).
42. J. Yeom, S. H. Bhang, B. Kim, M. S. Seo, E. J. Hwang, I. H. Cho *et al.*, Effect of cross-linking reagents for hyaluronic acid hydrogel dermal fillers on tissue augmentation and regeneration. *Bioconjug. Chem.*, **21**(2), 240–247 (2010).
43. H. Tan, C. M. Ramirez, N. Miljkovic, H. Li, J. P. Rubin, and K. G. Marra KG, Thermosensitive injectable hyaluronic acid hydrogel for adipose tissue engineering. *Biomaterials*, **30**(36), 6844–6853 (2009).
44. T. A. Ahmed, E.V. Dare, and M. Hincke, Fibrin: a versatile scaffold for tissue engineering applications. *J. Tissue Eng. Pt. B, Rev.*, **14**(2), 199–215 (2008).
45. B. Petersen, A. Barkun, S. Carpenter, P. Chotiprasidhi, R. Chuttani, W. Silverman *et al.*, Tissue adhesives and fibrin glues: November 2003. *Gastrointest. Endosc.*, **60**(3), 327–333 (2004).
46. G. Dickneite, H. Metzner, T. Pfeifer, M. Kroez, and G. Witzke, A comparison of fibrin sealants in relation to their *in vitro* and *in vivo* properties. *Thromb. Res.*, **112**(1–2), 73–82 (2003).
47. J. W. Choi, B. H. Choi, S. Park, K. S. Pai, T. Z. Li, B. Min *et al.*, Mechanical stimulation by ultrasound enhances chondrogenic differentiation of mesenchymal stem cells in a fibrin-hyaluronic acid hydrogel. *Artif. Organs*, **37**(7), 648–655 (2013).
48. M. N. V. Ravi Kumar, A review of chitin and chitosan applications. *React. Funct. Polym.*, **46**(1), 1–27 (2000).
49. A. Chenite, C. Chaput, D. Wang, C. Combes, M. D. Buschmann, C. D. Hoemann *et al.*, Novel injectable neutral solutions of chitosan form biodegradable gels *in situ*. *Biomaterials*, **21**(21), 2155–2161 (2000).
50. N. Boucard, C. Viton, D. Agay, E. Mari, T. Roger, Y. Chancerelle *et al.*, The use of physical hydrogels of chitosan for skin regeneration following third-degree burns. *Biomaterials*, **28**(24), 3478–3488 (2007).
51. H. Li, A. Wijekoon, and N. Leipzig, Entrapped neural stem cell neuronal differentiation in fluorinated methacrylamide chitosan hydrogels. *Ann. Biomed. Eng.*, **42**(7), 456–469 (2014).
52. M. Dessì, A. Borzacchiello, T. H. A. Mohamed, W. I. Abdel-Fattah, and L. Ambrosio, Novel biomimetic thermosensitive β-tricalcium phosphate/chitosan-based hydrogels for bone tissue engineering. *J. Biomed. Mater. Res. A*, **101**(10), 2984–2993 (2013).
53. C. Isikli, V. Hasirci, and N. Hasirci, Development of porous chitosan-gelatin/hydroxyapatite composite scaffolds for hard tissue-engineering applications. *J. Tissue Eng. Regen. Med.*, **6**(2), 135–143 (2012).
54. I. Saygun, V. Bal, A. Kızıltay, P. Yilgor Huri, F. Avcu, K. Üstün *et al.*, Proliferation and differentiation of mesenchymal stem cells in chitosan scaffolds loaded with nanocapsules containing bone morphogenetic proteins-4, platelet-derived growth factor and insulin-like growth factor 1. *J. Biomater. Tissue Eng.*, **4**, 181–188 (2014).
55. P. Yilgor, K. Tuzlakoglu, R. L. Reis, N. Hasirci, and V. Hasirci, Incorporation of a sequential BMP-2/BMP-7 delivery system into chitosan-based scaffolds for bone tissue engineering. *Biomaterials*, **30**(21), 3551–3559 (2009).

56. R. Jin, L. S. Moreira Teixeira, P. J. Dijkstra, M. Karperien, C. A. van Blitterswijk, Z. Y. Zhong *et al.*, Injectable chitosan-based hydrogels for cartilage tissue engineering. *Biomaterials*, **30**(13), 2544–2551 (2009).
57. S. Ucar, P. Yilgor, V. Hasirci, and N. Hasirci, Chitosan-based wet-spun scaffolds for bioactive agent delivery. *J. Appl. Poly. Sci.*, **130**(5), 3759–3769 (2013).
58. Q. Wang, M. Kong, Y. An, Y. Liu, J. Li, X. Zhou *et al.*, Hydroxybutyl chitosan thermosensitive hydrogel: A potential drug delivery system. *J. Mater. Sci.*, **48**(16), 5614–5623 (2013).
59. A. P. Balgude, X. Yu, A. Szymanski, and R. V. Bellamkonda, Agarose gel stiffness determines rate of DRG neurite extension in 3D cultures. *Biomaterials*, **22**(10), 1077–1084 (2001).
60. H. Ko, C. Sfeir, and P. N. Kumta, Novel synthesis strategies for natural polymer and composite biomaterials as potential scaffolds for tissue engineering. *Philos. Trans. A Math. Phys. Eng. Sci.*, **368**(1917), 1981–1997 (2010).
61. N. T. Khanarian, N. M. Haney, R. A. Burga, and H. H. Lu, A functional agarose-hydroxyapatite scaffold for osteochondral interface regeneration. *Biomaterials*, **33**(21), 5247–5258 (2012).
62. J. A. Puértolas, J. L. Vadillo, S. Sánchez-Salcedo, A. Nieto, E. Gómez-Barrena, and M. Vallet-Regí, Compression behaviour of biphasic calcium phosphate and biphasic calcium phosphate–agarose scaffolds for bone regeneration. *Acta Biomater.*, **7**(2), 841–847 (2011).
63. P. Matricardi, C. Di Meo, T. Coviello, W. E. Hennink, and F. Alhaique, Interpenetrating polymer networks polysaccharide hydrogels for drug delivery and tissue engineering. *Adv. Drug Deliv. Rev.*, **65**(9), 1172–1187 (2013).
64 T. Re'em, Y. Kaminer-Israeli, E. Ruvinov, and S. Cohen, Chondrogenesis of hMSC in affinity-bound TGF-beta scaffolds. *Biomaterials*, **33**(3), 751–761 (2012).
65. E. Ruvinov, J. Leor, and S. Cohen, The promotion of myocardial repair by the sequential delivery of IGF-1 and HGF from an injectable alginate biomaterial in a model of acute myocardial infarction. *Biomaterials*, **32**(2), 565–578 (2011).
66. P. A. Gunatillake and R. Adhikari, Biodegradable synthetic polymers for tissue engineering. *Eur. Cell. Mater.*, **5**, 1–16; discussion 16 (2003).
67. H. Tian, Z. Tang, X. Zhuang, X. Chen, and X. Jing, Biodegradable synthetic polymers: preparation, functionalization and biomedical application. *Prog. Polym. Sci.*, **37**(2), 237–280 (2012).
68. M. C. Hacker and A. G. Mikos, Synthetic polymers. In *Principles of Regenerative Medicine (Second Edition)*, Ed. A. Atala, R. Lanza, J. A. Thomson and R. Nerem, San Diego: Academic Press, pp. 587–622 (2011).
69. K. Merrett, C. M. Griffith, Y. Deslandes, G. Pleizier, and H. Sheardown, Adhesion of corneal epithelial cells to cell adhesion peptide modified pHEMA surfaces. *J. Biomater. Sci. Polym. E.*, **12**(6), 647–671 (2001).
70. A. J. Cadotte and T. B. DeMarse, Poly-HEMA as a drug delivery device for *in vitro* neural networks on micro-electrode arrays. *J. Neural Eng.*, **2**(4), 114 (2005).
71. F. X. Vacanti, PHEMA as a fibrous capsule–resistant breast prosthesis. *Plast. Reconstr. Sur.*, **113**(3), 949–952 (2004).

72. V. P. Bavaresco, C. A. C. Zavaglia, M. C. Reis, and J. R. Gomes, Study on the tribological properties of pHEMA hydrogels for use in artificial articular cartilage. *Wear*, **265**(3), 269–277 (2008).
73. D. Arslantunali, G. Budak, V. Hasirci, Multiwalled CNT-pHEMA composite conduit for peripheralnerve repair. *J. Biomed. Mater. Res. A*, **102**(3), 828–841(2014).
74. T. Tsou, S. Tang, Y. Huang, J. Wu, J. Young, and H. Wang, Poly (2-hydroxyethyl methacrylate) wound dressing containing ciprofloxacin and its drug release studies. *J. Mater. Sci.*, **16**(2), 95–100 (2005).
75. L. Varshney, Role of natural polysaccharides in radiation formation of PVA–hydrogel wound dressing. *Nucl. Instrum. Meth. B*, **255**(2), 343–349 (2007).
76. N. A. Peppas, J. Z. Hilt, A. Khademhosseini, and R. Langer, Hydrogels in biology and medicine: From molecular principles to bionanotechnology. *Adv. Mater.*, **18**(11), 1345–1360 (2006).
77. R. H. Schmedlen, K. S. Masters, and J. L. West, Photocrosslinkable polyvinyl alcohol hydrogels that can be modified with cell adhesion peptides for use in tissue engineering. *Biomaterials*, **23**(22), 4325–4332 (2002).
78. N. E. Vrana, Y. Liu, G. B. McGuinness, and P. A. Cahill, Characterization of poly (vinyl alcohol)/chitosan hydrogels as vascular tissue engineering scaffolds. *Macromol. Symp.*, **269**(1), 106–110 (2008c).
79. M. Kobayashi, J. Toguchida, and M. Oka, Preliminary study of polyvinyl alcohol-hydrogel (PVA-H) artificial meniscus. *Biomaterials*, **24**(4), 639–647 (2003).
80. D. Bezuidenhout, A. Oosthuysen, N. Davies, L. Ahrenstedt, S. Dobner, P. Roberts *et al.*, Covalent incorporation and controlled release of active dexamethasone from injectable polyethylene glycol hydrogels. *J. Biomed. Mater. Res. A*, **101A**(5), 1311–1318 (2013).
81. J. L. Ruan, N. L. Tulloch, V. Muskheli, E. E. Genova, P. D. Mariner, K. S. Anseth *et al.*, An improved cryosection method for polyethylene glycol hydrogels used in tissue engineering. *Tissue Eng. Pt. C Meth.*, **19**(10), 794–801 (2013).
82. A. T. Francisco, P. Y. Hwang, C. G. Jeong, L. Jing, J. Chen, and L. A. Setton, Photocrosslinkable laminin-functionalized polyethylene glycol hydrogel for intervertebral disc regeneration. *Acta Biomater.*, **10**(3), 1102–1111 (2014).
83. F. Yang, J. Wang, J. Hou, H. Guo, and C. Liu, Bone regeneration using cell-mediated responsive degradable PEG-based scaffolds incorporating with rhBMP-2. *Biomaterials*, **34**(5), 1514–1528 (2013).
84. C. Lu, P. Zahedi, A. Forman, and C. Allen, Multi-arm PEG/silica hydrogel for sustained ocular drug delivery. *J. Pharm. Sci.*, **103**(1), 216–226 (2014).
85. I. Strehin, Z. Nahas, K. Arora, T. Nguyen, and J. Elisseeff, A versatile pH sensitive chondroitin sulfate–PEG tissue adhesive and hydrogel. *Biomaterials*, **31**(10), 2788–2797 (2010).
86. W. Wang and A. Wang, Synthesis and swelling properties of pH-sensitive semi-IPN superabsorbent hydrogels based on sodium alginate-g-poly (sodium acrylate) and polyvinylpyrrolidone. *Carbohyd. Polym.*, **80**(4), 1028–1036 (2010).

87. M. Wang, L. Xu, H. Hu, M. Zhai, J. Peng, Y. Nho *et al.*, Radiation synthesis of PVP/ CMC hydrogels as wound dressing. *Nucl. Instrum. Meth. B*, **265**(1), 385–389 (2007).
88. S. W. Suh, J. Y. Shin, J. Kim, C. H. Beak, D. I. Kim *et al.*, Effect of different particles on cell proliferation in polymer scaffolds using a solvent-casting and particulate leaching technique. *ASAIO J.*, **48**(5), 460–464 (2002).
89. M. Verhulsel, M. Vignes, S. Descroix, L. Malaquin, D. M. Vignjevic, and J. Viovy, A review of microfabrication and hydrogel engineering for micro-organs on chips. *Biomaterials*, **35**(6), 1816–1832 (2014).
90. Y. C. Chiu, J. C. Larson, Jr. A. Isom, and E. M. Brey, Generation of porous poly (ethylene glycol) hydrogels by salt leaching. *Tissue Eng. Pt. C Meth.*, **16**(5), 905–912 (2010).
91. N. Annabi, J. W. Nichol, X. Zhong, C. Ji, S. Koshy, A. Khademhosseini *et al.*, Controlling the porosity and microarchitecture of hydrogels for tissue engineering. *J. Tissue Eng. Pt. B Rev.*, **16**(4), 371–383 (2010).
92. C. T. Lim, X. Ren, M. H. Afizah, S. Panjaitan, Z. Yang, Y. We, K. S. Chian, A. G. Mikos, and J. H. P. Hui, Repair of osteochondral defects with rehydrated freeze-dried oligo(poly(ethylene glycol) fumarate) hydrogels seeded with bone marrow mesenchymal stem cells in a porcine model. *Tissue Eng. Pt. A*, **65**, 1–48 (2013).
93. R. Ricciardi, Short time dynamics of solvent molecules and supramolecular organization of poly (vinyl alcohol) hydrogels obtained by freeze/thaw techniques. *Macromolecules*, **38**(15), 6629–6639 (2005).
94. Y. Liu, N. E. Vrana, P. A. Cahill, and G. B. McGuinness, Physically crosslinked composite hydrogels of PVA with natural macromolecules: Structure, mechanical properties, and endothelial cell compatibility. *J. Biomed. Mater. Res. Part B Appl. Biomater.*, **90**(2), 492–502 (2009).
95. B. Kinikoglu, C. Auxenfans, P. Pierrillas, V. Justin, P. Breton, C. Burillon *et al.*, Reconstruction of a full-thickness collagen-based human oral mucosal equivalent. *Biomaterials*, **30**(32), 6418–6425 (2009).
96. V. Keskar, N. W. Marion, J. J. Mao, and R. A. Gemeinhart, *In vitro* evaluation of macroporous hydrogels to facilitate stem cell infiltration, growth, and mineralization. *Tissue Eng. Pt. A*, **15**(7), 1695–1707 (2009).
97. O. R. Davies, A. L. Lewis, M. J. Whitaker, H. Tai, K. M. Shakesheff, and S. M. Howdle, Applications of supercritical CO2 in the fabrication of polymer systems for drug delivery and tissue engineering. *Adv. Drug Deliv. Rev.*, **60**(3), 373–387 (2008).
98. C. Ji, N. Annabi, A. Khademhosseini, and F. Dehghani, Fabrication of porous chitosan scaffolds for soft tissue engineering using dense gas CO_2. *Acta Biomater.*, **7**(4), 1653–1664 (2011).
99. A. Fathi, S. Lee, X. Zhong, N. Hon, P. Valtchev, and F. Dehghani, Fabrication of interpenetrating polymer network to enhance the biological activity of synthetic hydrogels. *Polym.* (*United Kingdom*), **54**(21), 5534–5542 (2013).
100. S. Zhang, X. Liu, S. F. Barreto-Ortiz, Y. Yu, B. P. Ginn, N. A. DeSantis, D. L. Hutton, W. L. Grayson, F. Z. Cui, B. A. Korgel, S. Gerecht, and H. Q. Mao, Creating polymer

hydrogel microfibres with internal alignment via electrical and mechanical stretching. *Biomaterials*, **35**(10), 3243–3251 (2014).
101. S. Selimovic, J. Oh, H. Bae, M. Dokmeci, and A. Khademhosseini, Microscale strategies for generating cell-encapsulating hydrogels. *Polymers*, **4**(3), 1554 (2012).
102. N. Annabi, A. Tamayol, J. A. Uquillas, M. Akbari, L. E. Bertassoni, C. Cha *et al.*, 25th anniversary article: Rational design and applications of hydrogels in regenerative medicine. *Adv. Mater. Res.*, **26**(1), 85–124 (2014).
103. M. E. Helgeson, S. C. Chapin, and P. S. Doyle, Hydrogel microparticles from lithographic processes: Novel materials for fundamental and applied colloid science. *Curr. Opin. Colloid Interface Sci.*, **16**(2), 106–117 (2011).
104. J. W. Nichol, S. T. Koshy, H. Bae, C. M. Hwang, S. Yamanlar, and A. Khademhosseini, Cell-laden microengineered gelatin methacrylate hydrogels. *Biomaterials*, **31**(21), 5536–5544 (2010).
105. S. Khetan and J. A. Burdick, Patterning network structure to spatially control cellular remodeling and stem cell fate within 3-dimensional hydrogels. *Biomaterials*, **31**(32), 8228–8234 (2010).
106. G. M. Whitesides, E. Ostuni, S. Takayama, X. Jiang, and D. E. Ingber, Soft lithography in biology and biochemistry. *Annu. Rev. Biomed. Eng.*, **3**, 335–373 (2001).
107. A. Mujahid, N. Iqbal, and A. Afzal, Bioimprinting strategies: From soft lithography to biomimetic sensors and beyond. *Biotechnol. Adv.*, **31**(8), 1435–1447 (2013).
108. D. Qin, Y. Xia, and G. M. Whitesides, Soft lithography for micro- and nanoscale patterning. *Nat. Protoc.*, **5**(3), 491–502 (2010).
109. J. You, S. Park, D. Shin, D. Patel, V. K. Raghunathan, M. Kim, C. J. Murphy, G. Tae, and A. Revzin, Characterizing the Effects of Heparin Gel Stiffness on Function of Primary Hepatocytes. *Tissue Eng. Part A*, **19**(23 and 24), 2655–2663 (2013).
110. P. Zorlutuna, Z. Rong, P. Vadgama, and V. Hasirci, Influence of nanopatterns on endothelial cell adhesion: Enhanced cell retention under shear stress. *Acta Biomater.*, **5**(7), 2451–2459 (2009b).
111. Y. Zheng, J. Chen, M. Craven, N. W. Choi, S. Totorica, A. Diaz-Santana *et al.*, *In vitro* microvessels for the study of angiogenesis and thrombosis. *Proceedings of the National Academy of Sciences*, **109**(24), 9342–9347 (2012).
112. P. Zorlutuna, J. H. Jeong, H. Kong, and R. Bashir, Stereolithography-based hydrogel microenvironments to examine cellular interactions. *Adv. Funct. Mater.*, **21**(19), 3642–3651 (2011).
113. S. Khetan and J. A. Burdick, Patterning hydrogels in three dimensions towards controlling cellular interactions. *Soft Matter*, **7**(3), 830–838 (2010).
114. V. Chan, P. Zorlutuna, J. H. Jeong, H. Kong, and R. Bashir, Three-dimensional photopatterning of hydrogels using stereolithography for long-term cell encapsulation. *Lab Chip*, **10**(16), 2062–2070 (2010).
115. F. P. W. Melchels, M. A. N. Domingos, T. J. Klein, J. Malda, P. J. Bartolo, and D. W. Hutmacher, Additive manufacturing of tissues and organs. *Prog. Polym. Sci.*, **37**(8), 1104–1079 (2011).

116. T. Billiet, M. Vandenhaute, J. Schelfhout, S. Van Vlierberghe, and P. Dubruel, A review of trends and limitations in hydrogel-rapid prototyping for tissue engineering. *Biomaterials*, **33**(26), 6020–6041 (2012).
117. H. Lin, D. Zhang, P. G. Alexander, G. Yang, J. Tan, A. W. M. Cheng, and R. S. Tuan, Application of visible light-based projection stereolithography for live cell-scaffold fabrication with designed architecture. *Biomaterials*, **34**(2), 331–339 (2013).
118. A. Ovsianikov, S. Mühleder, J. Torgersen, Z. Li, X.-H. Qin, S. Van Vlierberghe, P. Dubruel, W. Holnthoner, H. Redl, R. Liska, and J. Stampfl, Laser photofabrication of cell-containing hydrogel constructs. *Langmuir*, **30**(13), 3787–3794 (2014).
119. J. Torgersen, X. Qin, Z. Li, A. Ovsianikov, R. Liska, and J. Stampfl, Hydrogels for two-photon polymerization: A toolbox for mimicking the extracellular matrix. *Adv. Funct. Mater.*, **23**(36), 4542–4554 (2013).
120. P. Macchiarini, P. Jungebluth, T. Go, M. A. Asnaghi, L. E. Rees, T. A. Cogan *et al.*, Clinical transplantation of a tissue-engineered airway. *Lancet*, **372**(9655), 2023–2030 (2008).
121. A. Atala, S. B. Bauer, S. Soker, J. J. Yoo, and A. B. Retik, Tissue-engineered autologous bladders for patients needing cystoplasty. *The Lancet*, **367**(9518), 1241–1246 (2006).
122. S. Llames, E. Garcia, V. Garcia, M. del Rio, F. Larcher, J. L. Jorcano *et al.*, Clinical results of an autologous engineered skin. *Cell Tissue Bank.*, **7**(1), 47–53 (2006).
123. C. C. Chang, E. D. Boland, S. K. Williams, and J. B. Hoying, Direct-write bioprinting three-dimensional biohybrid systems for future regenerative therapies. *J. Biomed. Mater. Res. B, Appl. Biomater.*, **98**(1), 160–170 (2011).
124. J. Malda, J. Visser, F. P. Melchels, T. Jüngst, W. E. Hennink, W. J. A. Dhert *et al.*, 25th Anniversary Article: Engineering hydrogels for biofabrication. *Adv. Mater.*, **25**(36), 5011–5028 (2013).
125. N. R. Schiele, D. T. Corr, Y. Huang, N. A. Raof, Y. Xie, and D. B. Chrisey, Laser-based direct-write techniques for cell printing. *Biofabrication*, **2**(3), 1–14 (2010).
126. B. Duan, E. Kapetanovic, L. A. Hockaday, and J. T. Butcher, Three-dimensional printed trileaflet valve conduits using biological hydrogels and human valve interstitial cells. *Acta Biomater.*, **10**(5), 1836–1846 (2014).
127. S. Wüst, M. E. Godla, R. Müller, and S. Hofmann, Tunable hydrogel composite with two-step processing in combination with innovative hardware upgrade for cell-based three-dimensional bioprinting. *Acta Biomater.*, **10**(2), 630–640 (2014).
128. S. Tasoglu and U. Demirci, Bioprinting for stem cell research. *Trend. Biotechnol.*, **31**(1), 10–19 (2013).
129. I. Gibas and H. Janik, Review: Synthetic polymer hydrogels for biomedical applications. *Chem. Chem. Tech.*, **4**(4), 297–304 (2010).
130. F. Lim and A. M. Sun, Microentrapped islets as bioartificial endocrine pancreas. *Science*, **210**(4472), 908–910 (1980).
131. J. F. Burke, I. V. Yannas, Jr. W. C. Quinby, C. C. Bondoc, and W. K. Jung, Successful use of a physiologically acceptable artificial skin in the treatment of extensive burn injury. *Ann. Surg.*, **194**(4), 413–428 (1981).

132. R. Langer and J. P. Vacanti, Tissue engineering. *Science*, **260**(5110), 920–926 (1993).
133. A. Atala, Regenerative medicine strategies. *J. Pediatr. Surg.*, **47**(1), 17–28 (2012).
134. J. L. Drury and D. J. Mooney, Hydrogels for tissue engineering: Scaffold design variables and applications. *Biomaterials*, **24**(24), 4337–4351 (2003).
135. A. E. Bent, J. Foote, S. Siegel, G. Faerber, R. Chao and E. A. Gormley, Collagen implant for treating stress urinary incontinence in women with urethral hypermobility. *J. Urol.*, **166**(4), 1354–1357 (2001b).
136. R. R. Dmochowski and R. A. Appell, Injectable agents in the treatment of stress urinary incontinence in women: where are we now? *Urology*, **56**(6, Supplement 1), 32–40 (2000).
137. A. Loebsack, K. Greene, S. Wyatt, C. Culberson, C. Austin, R. Beiler *et al. In vivo* characterization of a porous hydrogel material for use as a tissue bulking agent. *J. Biomed. Mater. Res.*, **57**(4), 575–581 (2001).
138. C. Wiegand, T. Heinze, and U. Hipler, Comparative *in vitro* study on cytotoxicity, antimicrobial activity, and binding capacity for pathophysiological factors in chronic wounds of alginate and silver-containing alginate. *Wound Repair Regen.*, **17**(4), 511–521 (2009).
139. K. A. Smeds and M. W. Grinstaff, Photocrosslinkable polysaccharides for *in situ* hydrogel formation. *J. Biomed. Mater. Res.*, **54**(1), 115–121 (2001).
140. G. Lackgern, N. Wahlin, E. Sköldenberg, and A. Stenberg, Long-term follow up of children treated with dextranomer/hyaluronic acid copolymer for vesicourethral reflux. *J. Urol.*, **166**(5), 1887–1892 (2001).
141. J. H. Ross and R. Kay, Pediatric urinary tract infection and reflux. *Am. Fam. Physician*, **59**(6), 1472–1478, 1485–1486 (1999).
142. B. Balakrishnan, M. Mohanty, P. R. Umashankar, and A. Jayakrishnan, Evaluation of an *in situ* forming hydrogel wound dressing based on oxidized alginate and gelatin. *Biomaterials*, **26**(32), 6335–6342 (2005).
143. K. Y. Lee and D. J. Mooney, Alginate: Properties and biomedical applications. *Prog. Polym. Sci.*, **37**(1), 106–126 (2012).
144. J. Sun and H. Tan, Alginate-based biomaterials for regenerative medicine applications. *Materials*, **6**(4), 1285–1309 (2013).
145. J. S. Boateng, K. H. Matthews, H. N. Stevens, and G. M. Eccleston, Wound healing dressings and drug delivery systems: A review. *J. Pharm. Sci.*, **97**(8), 2892–2923 (2008).
146. H. Bilgic, M. Demiriz, M. Ozler, T. Ide, N. Dogan, S. Gumus *et al.*, Effects of EGF containing gelatin sponges as wound healing scaffolds, *J. Biomater. Tissue Eng.*, **3**(2), 205–211 (2013).
147. W. D. Spotnitz and S. Burks, Hemostats, sealants, and adhesives III: A new update as well as cost and regulatory considerations for components of the surgical toolbox. *Transfusion*, **52**(10), 2243–2255 (2012).
148. K. Ono, Y. Saito, H. Yura, K. Ishikawa, A. Kurita, T. Akaike *et al.*, Photocrosslinkable chitosan as a biological adhesive. *J. Biomed. Mater. Res.*, **49**(2), 289–295 (2000).

149. C. M. Haller, W. Buerzle, A. Kivelio, M. Perrini, C. E. Brubaker, R. J. Gubeli *et al.* Mussel-mimetic tissue adhesive for fetal membrane repair: An *ex vivo* evaluation. *Acta Biomater.*, **8**(12), 4365–4370 (2012).
150. C. E. Brubaker, H. Kissler, L. Wang, D. B. Kaufman, and P. B. Messersmith, Biological performance of mussel-inspired adhesive in extrahepatic islet transplantation. *Biomaterials*, **31**(3), 420–427 (2010).
151. C. E. Brubaker and P. B. Messersmith, Enzymatically degradable mussel-inspired adhesive hydrogel. *Biomacromolecules*, **12**(12), 4326–4334 (2011).
152. M. Mehdizadeh, H. Weng, D. Gyawali, L. Tang, and J. Yang, Injectable citrate-based mussel-inspired tissue bioadhesives with high wet strength for sutureless wound closure. *Biomaterials*, **33**(32), 7972–7983 (2012).
153. A. S. Sawhney, C. P. Pathak, J. J. van Rensburg, R. C. Dunn, and J. A. Hubbell, Optimization of photopolymerizedbioerodible hydrogel properties for adhesion prevention. *J. Biomed. Mater. Res.*, **28**(7), 831–838 (1994).
154. M. Kobayashi, J. Toguchida, and M. Oka, Development of polyvinyl alcohol-hydrogel (PVA-H) shields with a high water content for tendon injury repair. *J. Hand Surg.*, **26**(5), 436–440 (2001).
155. D. L. Elbert and J. A. Hubbell, Reduction of fibrous adhesion formation by a copolymer possessing an affinity for anionic surfaces. *J. Biomed. Mater. Res.*, **42**(1), 55–65 (1998).
156. X. Gao, X. Deng, X. Wei, H. Shi, F. Wang, T. Ye *et al.*, Novel thermosensitive hydrogel for preventing formation of abdominal adhesions. *Int. J. Nanomedicine*, **8**, 2453–2463 (2013).
157. S. E. Sakiyama-Elbert, and J. A. Hubbell, Development of fibrin derivatives for controlled release of heparin-binding growth factors. *J. Control. Release*, **65**(3), 389–402 (2000).
158. K. Y. Lee, M. C. Peters, and D. J. Mooney, Comparison of vascular endothelial growth factor and basic fibroblast growth factor on angiogenesis in SCID mice. *J. Control. Release*, **87**(1), 49–56 (2003).
159. M. C. Peters, B. C. Isenberg, J. A. Rowley, and D. J. Mooney, Release from alginate enhances the biological activity of vascular endothelial growth factor. *J. Biomater. Sci. Polym. E*, **9**(12), 1267–1278 (1998).
160. K. Y. Lee, M. C. Peters, and D. J. Mooney, Controlled drug delivery from polymers by mechanical signals. *Adv. Mater.*, **13**(11), 837–839 (2001a).
161. K. Y. Lee, M. C. Peters, K. W. Anderson, and D. J. Mooney, Controlled growth factor release from synthetic extracellular matrices. *Nature*, **408**(6815), 998–1000 (2000).
162. Y. Tabata, M. Miyao, M. Ozeki, and Y. Ikada, Controlled release of vascular endothelial growth factor by use of collagen hydrogels. *J. Biomater. Sci., Polym. E*, **11**(9), 915–930 (2000).
163. X. Hao, E. A. Silva, A. Mansson-Broberg, K. H. Grinnemo, A. J. Siddiqui, G. Dellgren *et al.*, Angiogenic effects of sequential release of VEGF-A165 and PDGF-BB with alginate hydrogels after myocardial infarction. *Cardiovasc. Res.*, **75**(1), 178–185 (2007).

164. Q. Sun, E. A. Silva, A. Wang, J. C. Fritton, D. J. Mooney, M. B. Schaffler *et al.*, Sustained release of multiple growth factors from injectable polymeric system as a novel therapeutic approach towards angiogenesis. *Pharm. Res.*, **27**(2), 264–271 (2010).
165. B. Balakrishnan, M. Mohanty, A. C. Fernandez, P. V. Mohanan, and A. Jayakrishnan, Evaluation of the effect of incorporation of dibutyryl cyclic adenosine monophosphate in an *in situ*-forming hydrogel wound dressing based on oxidized alginate and gelatin. *Biomaterials*, **27**(8), 1355–1361 (2006).
166. K. Ulubayram, A. N. Cakar, P. Korkusuz, C. Ertan, and N. Hasirci, EGF containing gelatin-based wound dressings. *Biomaterials*, **22**(11), 1345–1356 (2001).
167. S. Dogan, S. Demirer, I. Kepenekci, B. Erkek, A. Kiziltay, N. Hasirci *et al.*, Epidermal growth factor-containing wound closure enhances wound healing in non-diabetic and diabetic rats. *Int. Wound J.*, **6**(2), 107–115 (2009).
168. H. D. Kim, and R. F. Valentini, Retention and activity of BMP 2 in hyaluronic acid based scaffolds *in vitro. J. Biomed. Mater. Res.*, **59**(3), 573–584 (2002).
169. N. Saito, T. Okada, H. Horiuchi, N. Murakami, J. Takahashi, M. Nawata, *et al.*, A biodegradable polymer as a cytokine delivery system for inducing bone formation. *Nat. Biotechnol.*, **19**(4), 332–335 (2001).
170. T. Manaka, A. Suzuki, K. Takayama, Y. Imai, H. Nakamura, and K. Takaoka, Local delivery of siRNA using a biodegradable polymer application to enhance BMP-induced bone formation. *Biomaterials*, **32**(36), 9642–9648 (2011).
171. Y. M. Kolambkar, K. M. Dupont, J. D. Boerckel, N. Huebsch, D. J. Mooney, D. W. Hutmacher *et al.*, An alginate-based hybrid system for growth factor delivery in the functional repair of large bone defects. *Biomaterials*, **32**(1), 65–74 (2011).
172. G. Orive, S. K. Tam, J. L. Pedraz, and J. Hallé, Biocompatibility of alginate–poly-l-lysine microcapsules for cell therapy. *Biomaterials*, **27**(20), 3691–3700 (2006).
173. M. Lamb, R. Storrs, S. Li, O. Liang, K. Laugenour, R. Dorian *et al.*, Function and viability of human islets entrapped in alginate sheets: *In vitro* and *in vivo* culture. *Transplant. Proc.*, **43**(9), 3265–3266 (2011).
174. A. G. Mallett and G. S. Korbutt, Alginate modification improves long-term survival and function of transplanted entrapped islets. *Tissue Eng. Pt. A*, **15**(6), 1301–1309 (2009).
175. L. M. Weber, C. G. Lopez, and K. S. Anseth, Effects of PEG hydrogel crosslinking density on protein diffusion and entrapped islet survival and function. *J. Biomed. Mater. Res. A*, **90**(3), 720–729 (2009).
176. S. M. Jay, B. R. Shepherd, J. W. Andrejecsk, T. R. Kyriakides, J. S. Pober and, W. M. Saltzman, Dual delivery of VEGF and MCP-1 to support endothelial cell transplantation for therapeutic vascularization. *Biomaterials*, **31**(11), 3054–3062 (2010).
177. K. Y. Lee, E. Alsberg, and D. J. Mooney, Degradable and injectable poly (aldehyde guluronate) hydrogels for bone tissue engineering. *J. Biomed. Mater. Res.*, **56**(2), 228–233 (2001b).
178. E. Alsberg, K. W. Anderson, A. Albeiruti, R. T. Franceschi, and D. J. Mooney, Cell-interactive alginate hydrogels for bone tissue engineering. *J. Dnt. Res.*, **80**(11), 2025–2029 (2001).

179. D. Park, B. Choi, S. Zhu, J. Huh, B. Kim, and S. Lee, Injectable bone using chitosan-alginate gel/mesenchymal stem cells/BMP-2 composites. *J. Cranio-Maxill. Surg.*, **33**(1), 50–54 (2005).
180. M. D. Weir, H. H. K. Xu, and C. G. Simon, Strong calcium phosphate cement-chitosan-mesh construct containing cell-encapsulating hydrogel beads for bone tissue engineering. *J. Biomed. Mater. Res. A*, **77**(3), 487–496 (2006).
181. E. Alsberg, K. W. Anderson, A. Albeiruti, J. A. Rowley, and D. J. Mooney, Engineering growing tissues. *Proceedings of the National Academy of Sciences of the United States of America*, **99**(19), 12025–12030 (2002).
182. A. J. Thornton, E. Alsberg, M. Albertelli, and D. J. Mooney, Shape-defining scaffolds for minimally invasive tissue engineering. *Transplantation*, **77**(12), 1798–1803 (2004).
183. H. Park, J. S. Temenoff, Y. Tabata, A. I. Caplan, and A. G. Mikos, Injectable biodegradable hydrogel composites for rabbit marrow mesenchymal stem cell and growth factor delivery for cartilage tissue engineering. *Biomaterials*, **28**(21), 3217–3227 (2007).
184. A. E. Bent, R. T. Tutrone, M. T. McLennan, K. Lloyd, M. J. Kennelly, and G. Badlani, Treatment of intrinsic sphincter deficiency using autologous ear chondrocytes as a bulking agent. *Neurourol. Urodynam.*, **20**(2), 157–165 (2001a).
185. L. B. Shi, H. X. Cai, L. K. Chen, Y. Wu, S. A. Zhu, X. N. Gong *et al.*, Tissue engineered bulking agent with adipose-derived stem cells and silk fibroin microspheres for the treatment of intrinsic urethral sphincter deficiency. *Biomaterials*, **35**(5), 1519–1530 (2014).
186. S. Yang, K. F. Leong, Z. Du, and C. K. Chua, The design of scaffolds for use in tissue engineering. Part I. Traditional factors. *Tissue Eng.*, **7**(6), 679–689 (2001).
187. D. W. Hutmacher, Scaffold design and fabrication technologies for engineering tissues-state of the art and future perspectives. *J. Biomater. Sci. Polym. E*, **12**(1), 107–124 (2001).
188. J. Malda, T. J. Klein, and Z. Upton, The roles of hypoxia in the *in vitro* engineering of tissues. *Tissue Eng.*, **13**(9), 2153–2162 (2007).
189. C. Caravaggi, R. De Giglio, C. Pritelli, M. Sommaria, S. DallaNoce, E. Faglia *et al.*, HYAFF 11-based autologous dermal and epidermal grafts in the treatment of non-infected diabetic plantar and dorsal foot ulcers: A prospective, multicenter, controlled, randomized clinical trial. *Diabetes Care*, **26**(10), 2853–2859 (2003).
190. M. C. Peters, P. J. Polverini, and D. J. Mooney, Engineering vascular networks in porous polymer matrices. *J. Biomed. Mater. Res.*, **60**(4), 668–678 (2002).
191. E. A. Silva, E. S. Kim, H. J. Kong, and D. J. Mooney, Material-based deployment enhances efficacy of endothelial progenitor cells. *Proceedings of the National Academy of Sciences of the United States of America*, **105**(38), 14347–14352 (2008).
192. P. Zorlutuna, A. Elsheikh, and V. Hasirci, Nanopatterning of collagen scaffolds improve the mechanical properties of tissue engineered vascular grafts. *Biomacromolecules*, **10**(4), 814–821 (2009a).
193. P. Zorlutuna, N. Hasirci, and V. Hasirci, Nanopatterned collagen tubes for vascular tissue engineering. *J. Tissue Eng. Regen. Med.*, **2**(6), 373–377 (2008).

194. B. K. Mann, A. S. Gobin, A. T. Tsai, R. H. Schmedlen, and J. L. West, Smooth muscle cell growth in photopolymerized hydrogels with cell adhesive and proteolytically degradable domains: Synthetic ECM analogs for tissue engineering. *Biomaterials*, **22**(22), 3045–3051 (2001a).
195. B. K. Mann, R. H. Schmedlen, and J. L. West, Tethered-TGF-β increases extracellular matrix production of vascular smooth muscle cells. *Biomaterials*, **22**(5), 439–444 (2001b).
196. Y. Ling, J. Rubin, Y. Deng, C. Huang, U. Demirci, J. M. Karp *et al.*, A cell-laden microfluidic hydrogel. *Lab Chip*, **7**(6), 756–762 (2007).
197. J. He, M. Mao, Y. Liu, J. Shao, Z. Jin, and D, Li, Fabrication of nature-inspired microfluidic network for perfusable tissue constructs. *Adv. Healthc. Mater.*, **2**(8), 1108–1113 (2013).
198. C. Norotte, F. S. Marga, L. E. Niklason, and G. Forgacs, Scaffold-free vascular tissue engineering using bioprinting. *Biomaterials*, **30**(30), 5910–5917 (2009).
199. W. Lee, V. Lee, S. Polio, P. Keegan, J. Lee, K. Fischer *et al.*, On-demand three-dimensional freeform fabrication of multi-layered hydrogel scaffold with fluidic channels. *Biotechnol. Bioeng.*, **105**(6), 1178–1186 (2010).
200. N. E. Vrana, N. Builles, M. Hindie, O. Damour, A. Aydinli, and V. Hasirci, Contact guidance enhances the quality of a tissue engineered corneal stroma. *J. Biomed. Mater. Res. Part A*, **84A**(2), 454–463 (2008a).
201. C. Kilic, A. Girotti, J. Rodriguez-Cabello, and V. Hasirci, A collagen-based corneal stroma substitute with micro-designed architecture. *Biomater. Sci.*, **2**(3), 318–329 (2014).
202. G. Lisignoli, M. Fini, G. Giavaresi, N. NicoliAldini, S. Toneguzzi, and A. Facchini, Osteogenesis of large segmental radius defects enhanced by basic fibroblast growth factor activated bone marrow stromal cells grown on non-woven hyaluronic acid-based polymer scaffold. *Biomaterials*, **23**(4), 1043–1051 (2002).
203. Y. Weng, M. Wang, W. Liu, X. Hu, G. Chai, Q. Yan *et al.*, Repair of experimental alveolar bone defects by tissue-engineered bone. *Tissue Eng.*, **12**(6), 1503–1513 (2006).
204. H. Lin and Y. Yeh, Porous alginate/hydroxyapatite composite scaffolds for bone tissue engineering: Preparation, characterization, and *in vitro* studies. *J. Biomed.* Mater. *Res. B: Appl. Biomater.*, **71B**(1), 52–65 (2004).
205. S. Ber, G. Torun Köse, and V. Hasırcı, Bone tissue engineering on patterned collagen films: an *in vitro* study. *Biomaterials*, **26**(14), 1977–1986 (2005).
206. O. Karadas, D. Yucel, H. Kenar, G. Torun Kose, and V. Hasirci, Collagen scaffolds with *in situ*-grown calcium phosphate for osteogenic differentiation of Wharton's jelly and menstrual blood stem cells. *J. Tissue Eng. Regen. Med.*, **8**(7), 534–545 (2012).
207. B. F. Basmanav, G. T. Kose, and V. Hasirci, Sequential growth factor delivery from complexed microspheres for bone tissue engineering. *Biomaterials*, **29**(31), 4195–4204 (2008).
208. J. M. Kanczler, P. J. Ginty, L. White, N. M. Clarke, S. M. Howdle, K. M. Shakesheff *et al.*, The effect of the delivery of vascular endothelial growth factor and bone morphogenic protein-2 to osteoprogenitor cell populations on bone formation. *Biomaterials*, **31**(6), 1242–1250 (2010).

209. B. Grigolo, G. Lisignoli, A. Piacentini, M. Fiorini, P. Gobbi, G. Mazzotti *et al.*, Evidence for redifferentiation of human chondrocytes grown on a hyaluronan-based biomaterial: molecular, immunohistochemical and ultrastructural analysis. *Biomaterials*, **23**(4), 1187–1195 (2002).
210. G. T. Köse, F. Korkusuz, P. Korkusuz, and V. Hasirci, *In vivo* tissue engineering of bone using poly (3-hydroxybutyric acid-co-3-hydroxyvaleric acid) and collagen scaffolds. *Tissue Eng.*, **10**(7–8), 1234–1250 (2004).
211. E. Pfeiffer, S. M. Vickers, E. Frank, A. J. Grodzinsky, and M. Spector, The effects of glycosaminoglycan content on the compressive modulus of cartilage engineered in type II collagen scaffolds. *Osteoarth. Cartil.*, **16**(10), 1237–1244 (2008).
212. S. C. Chang, G. Tobias, A. K. Roy, C. A. Vacanti, and L. J. Bonassar, Tissue engineering of autologous cartilage for craniofacial reconstruction by injection molding. *Plast. Reconstr. Sur.*, **112**(3), 793–799; discussion 800-1 (2003).
213. M. Lind, A. Larsen, C. Clausen, K. Osther, and H. Everland, Cartilage repair with chondrocytes in fibrin hydrogel and MPEG polylactide scaffold: An *in vivo* study in goats. *Knee Surg. Sport. Tr. A.*, **16**(7), 690–698 (2008).
214. S. C. Chang, J. A. Rowley, G. Tobias, N. G. Genes, A. K. Roy, D. J. Mooney *et al.*, Injection molding of chondrocyte/alginate constructs in the shape of facial implants. *J. Biomed. Mater. Res.*, **55**(4), 503–511 (2001).
215. G. Chen, T. Sato, T. Ushida, N. Ochiai, and T. Tateishi, Tissue engineering of cartilage using a hybrid scaffold of synthetic polymer and collagen. *Tissue Eng.*, **10**(3–4), 323–330 (2004).
216. T. Okano and T. Matsuda, Tissue engineered skeletal muscle: preparation of highly dense, highly oriented hybrid muscular tissues. *Cell Transplant.*, **7**(1), 71–82 (1998).
217. D. von Heimburg, S. Zachariah, H. Kühling, I. Heschel, H. Schoof, B. Hafemann *et al.*, Human preadipocytes seeded on freeze-dried collagen scaffolds investigated *in vitro* and *in vivo*. *Biomaterials*, **22**(5), 429–438 (2001).
218. A. Mosahebi, M. Simon, M. Wiberg, and G. Terenghi, A novel use of alginate hydrogel as Schwann cell matrix. *Tissue Eng.*, **7**(5), 525–534 (2001).
219. G. Bahcecioglu, A. Buyuksungur, A. Kiziltay, N. Hasirci, and V. Hasirci, Construction and *in vitro* testing of a multilayered, tissue-engineered meniscus. *J. Bioact. Compat. Pol.*, **29**(3), 235–253 (2014).
220. F. Xu, T. D. Finley, M. Turkaydin, Y. Sung, U. A. Gurkan, A. S. Yavuz *et al.*, The assembly of cell-encapsulating microscale hydrogels using acoustic waves. *Biomaterials*, **32**(31), 7847–7855 (2011).
221. M. F. Pittenger, A. M. Mackay, S. C. Beck, R. K. Jaiswal, R. Douglas, J. D. Mosca *et al.*, Multilineage potential of adult human mesenchymal stem cells. *Science*, **284**(5411), 143–147 (1999).
222. D. E. Discher, D. J. Mooney, and P. W. Zandstra, Growth factors, matrices, and forces combine and control stem cells. *Science*, **324**(5935), 1673–1677 (2009).
223. F. Guilak, D. M. Cohen, B. T. Estes, J. M. Gimble, W. Liedtke, and C. S. Chen, Control of stem cell fate by physical interactions with the extracellular matrix. *Cell Stem Cell*, **5**(1), 17–26 (2009).

224. M. P. Lutolf, P. M. Gilbert, and H. M. Blau, Designing materials to direct stem-cell fate. *Nature*, **462**(7272), 433–441 (2009).
225. S. Eshghi, M. G. Vogelezang, R. O. Hynes, L. G. Griffith, and H. F. Lodish, Alpha4beta1 integrin and erythropoietin mediate temporally distinct steps in erythropoiesis: Integrins in red cell development. *J. Cell Biol.*, **177**(5), 871–880 (2007).
226. Y. Suzuki, M. Tanihara, K. Suzuki, A. Saitou, W. Sufan, and Y. Nishimura, Alginate hydrogel linked with synthetic oligopeptide derived from BMP 2 allows ectopic osteoinduction *in vivo*. *J. Biomed. Mater. Res.*, **50**(3), 405–409 (2000).
227. V. H. Fan, A. Au, K. Tamama, R. Littrell, L. B. Richardson, J. W. Wright *et al.* Tethered epidermal growth factor provides a survival advantage to mesenchymal stem cells. *Stem Cells*, **25**(5), 1241–1251 (2007).
228. J. E. Ho, E. H. Chung, S. Wall, D. V. Schaffer, and K. E. Healy, Immobilized sonic hedgehog N-terminal signaling domain enhances differentiation of bone marrow-derived mesenchymal stem cells. *J. Biomed. Mater. Res. A*, **83**(4), 1200–1208 (2007).
229. H. Ma, S. Hung, S. Lin, Y. Chen, and W. Lo, Chondrogenesis of human mesenchymal stem cells entrapped in alginate beads. *J. Biomed. Mater. Res. A*, **64**(2), 273–281 (2003).
230. A. J. Engler, S. Sen, H. L. Sweeney, and D. E. Discher, Matrix elasticity directs stem cell lineage specification. *Cell*, **126**(4), 677–689 (2006).
231. K. Saha, E. F. Irwin, J. Kozhukh, D. V. Schaffer, and K. E. Healy, Biomimetic interfacial interpenetrating polymer networks control neural stem cell behavior. *J. Biomed. Mater. Res. A*, **81**(1), 240–249 (2007).
232. O. Jeon and E. Alsberg, Regulation of stem cell fate in a three dimensional micropatterned dual crosslinked hydrogel system. *Adv. Funct. Mater.*, **23**(38), 4765–4775 (2013).

Chapter 2

Determining Stem Cell Fate with Hydrogels

Aylin Acun[*,‡], **Andreana Panzo**[†] **and Pinar Zorlutuna**[*,†,§]
[*]*Bioengineering Graduate Program, University of Notre Dame, Notre Dame, IN 46556, USA*
[†]*Aerospace and Mechanical Engineering Department, University of Notre Dame, Notre Dame, IN 46556, USA*
[‡]*aacun@nd.edu, apanzo@nd.edu*
[§]*pinar.zorlutuna.1@nd.edu*

Cellular microenvironment is crucial for basic cell functions including growth, migration, differentiation, and maturation. These microenvironmental factors are mainly composed of (1) other cells, (2) soluble factors coming from nearby cells or from the blood, and (3) the extracellular matrix (ECM). In cell culture, it is relatively easy to include the first two factors into the culture environment through co-cultures and addition of soluble factors to the culture media while incorporating the ECM is more challenging. Hydrogels are fully hydrated polymeric materials that resemble the ECM, and therefore have been used for providing cells a biomimetic environment for applications ranging from 3D cell cultures for studying cell biology to fabricating engineered tissues for regenerative medicine or disease modeling.

Stem cells are cells with the capacity to give rise to different cell types while keeping their self-renewal capability. Whether a stem cell self-renews or differentiates to a specific cell type, as well as which cell type it differentiates into, or in other words the stem cell fate, is closely regulated through above-mentioned microenvironmental factors. As such, hydrogels have been used in determining stem cell fate through their tunable physical and chemical properties, as well as incorporating biological cues in a spatiotemporally controlled manner. Getting stem cells in contact with different biological factors by incorporating them in the hydrogel structure has proven to be effective in guiding distinct stem cell fates. Physical properties of the hydrogels such as stiffness or geometry (i.e., confining cells in patterns of defined shape and size) were also shown to have control over stem cell behavior. Similarly, the chemistry and composition of hydrogels can guide stem cell fate. In this chapter, the basic features and types of stem cells as well as the interactions of stem cells with hydrogels to control or direct the stem cell behavior are explained.

1. Stem Cells

Stem cells are cells with a unique ability to differentiate into more than one cell type while maintaining their self-regenerative properties.[1] Unlike terminally differentiated cells the regenerative potential of stem cells compensate for the cell death due to natural causes or damage, and help maintain the tissues' functionality over time.[2] The decision whether a stem cell will have a differentiated or undifferentiated phenotype is regulated during the stem cell division. A stem cell can give rise to two differentiated or two undifferentiated cells as well as one differentiated and one undifferentiated cell depending on the demand.[1] The maintenance of stemness by giving rise to other stem cells upon division is referred to as self-renewal. This process is regulated by the interactions of stem cells with each other, with the ECM surrounding them, and with other surrounding cells. These interactions constitute a well-controlled microenvironment referred as a "stem cell niche".[1,3] Keeping a constant ratio of differentiated to undifferentiated cells under normal conditions is of upmost importance since the excess of either phenotype would cause unfavorable effects on the tissues. For example, the increase in stem cell number to abnormal levels can cause accumulation of mutations in the cells, which in turn may lead to tumor formation.[3] On the other hand, in the case of excessive differentiation, the stem cell reserve may be depleted and the regenerative potential of the tissue would decrease.

Although most stem cells have several characteristics in common, there are some factors that help us distinguish between the different types of stem cells. Not all stem cells have the same potency level, which is defined by the number of different cell types that a stem cell can give rise to. Totipotent cells like the zygote, can give rise to a whole organism whereas, pluripotent stem cells can differentiate into all cell types found in the body except the extra-embryonic structures. Some stem cells like mesenchymal stem cells can give rise to a limited number of cell types and therefore are called multipotent stem cells. The cells that can give rise only to one specific cell type are unipotent and are still accepted as stem cells.[4] Stem cells can be classified as embryonic stem cells (ESCs), adult stem cells (ASCs), and induced pluripotent stem cells (iPSCs) according to their level of potency and their origin (Fig. 1).[5]

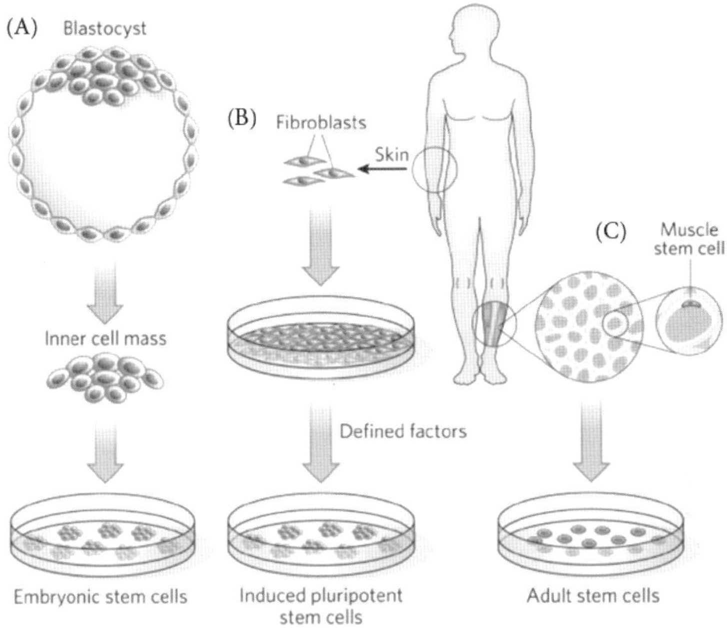

Fig. 1. Schematic representation of the sources of different stem cell types.[5]

1.1. ASCs

ASCs are multipotent stem cells that are stored in specific locations in the body. These cells are no longer pluripotent since they are more specialized compared to ESCs.[6] Being multipotent, they can give rise to the cell types found in the tissue where their niches are located.[1,7] ASCs were discovered when the bone marrow stem cells and hematopoietic stem cells were identified in bone marrow. This finding led to discovery of many other sources of ASCs. The major stem cell niches in the body are dental pulp, adipose tissue, heart, pancreas, intestine, brain, and liver.[1,2]

Different types of ASCs have been isolated and used for tissue engineering and regenerative medicine purposes. One of the most commonly used type of ASCs is the mesenchymal stem cells. They are the stromal cells that can be isolated from several sources such as bone marrow, umbilical cord, adipose tissue, and menstrual blood.[8] These cells have the potential to be differentiated into osteocytes, chondrocytes, skeletal and smooth muscle myocytes, and adipocytes.[8,9] Isolation of these cells from bone marrow can be done from the patients undergoing a joint replacement surgery.[10] There are more readily accessible stem cell sources like adipose tissue[11] or peripheral blood.[12] Adipose tissue is a relatively new and preferable source as the discarded fat tissue from liposuction procedures are used for isolation of stem cells.[13]

1.2. ESCs

The single cell zygote is formed by fertilization of an oocyte with a sperm. After several divisions a liquid filled structure called a blastocyst, which comprise of an outer cell lining and an inner cell mass is formed. The cells of this inner cell mass can be isolated and cultured *in vitro*, and they are called ESCs.[14]

ESCs can be cultured to proliferate stably in their undifferentiated state, while maintaining their broad differentiation potential.[1,15] This makes them a favorable cell source for tissue engineering especially for target tissues where primary cell isolation is limited, such as neurons or heart muscle. Therefore, these cells have been used extensively for tissue engineering applications. Different lines of ESCs have been successfully differentiated into bone,[16,17] cardiac,[18,19] liver,[20,21] pancreatic,[22] and neural[23,24] tissues. However, the ethical concerns related to human ESCs limit their usage and their clinical applications.

1.3. iPSCs

A recent discovery by Takahashi and Yamanaka introduced a new and promising type of stem cell to the field.[25,26] These pluripotent cells are called iPSCs and they share many characteristics with ESCs, except the ethical concerns that the ESCs have.[27] A totally differentiated cell from an animal[28] or a human[25] can regain its pluripotency by genetic reprogramming. The first demonstration of this was performed on mouse embryonic fibroblasts. In this study, the cells were reprogrammed by retroviral infection to activate the expression of four key transcription factors that were shown to regulate the pluripotency in ESCs, namely, Oct4, Sox2, Klf4, and c-Myc.[26,29] The following year the same group successfully produced human iPSCs.[25]

Some disadvantages of this method were reported as introducing mutations during viral infection,[30] potential risk of teratoma formation,[30,31] low efficiency about 1%, and heterogenous population of wild type and induced cells.[32,33] Soon after, alternatives to avoid these disadvantages were proposed. The reprogramming of the cells was successfully done using chemical methods,[34] episomal vectors,[35] plasmids,[36] transposons,[37] and nanoparticles.[38]

The potential of iPSCs has been tested via differentiating them towards several different cell lineages. iPSCs were successfully differentiated to neural,[39,40] liver,[41] chondroprogenitor,[42,43] and heart[44–46] cells.

iPSCs are promising for tissue engineering applications as, theoretically, the somatic cells from a patient can be isolated and his/her own iPSCs can be produced. Then these cells can be differentiated to the intended cell types and regenerate a specific target tissue. Since the patient's own cells will be used, the risk of immune rejection will be minimized. This technology can be used for constructing tissue models and for personalized medicine applications. Instead of using the constructs for implantation, they can be used as models to test drugs to assess toxicity or efficiency. It also provides a good platform to study the basic mechanisms of cell differentiation or tissue/organ development.

2. Stem Cells in Tissue Engineering and Regenerative Medicine

Tissue engineering and regenerative medicine aims to regenerate the lost function of a tissue or an organ. Primary cells obtained from patients are not the ideal candidates for this purpose for several reasons, including the requirement of a second surgical procedure, donor site morbidity, problems with culturing the primary cells for an extended period of time, and low cell numbers that can be obtained compared to the defect size when the damage is extensive.

On the other hand, allogenic and xenogeneic sources of cells present problems such as donor scarcity, interspecies differences, the risk of disease transmission, and immune rejection. Even if the organs/tissues match, there is a high risk of rejection and re-hospitalization. Furthermore, the patients have to take immunosuppressive drugs for the rest of their lives. According to U.S. Organ Procurement and Transplantation Network (OPTN) and the Scientific Registry of Transplant Recipients (SRTR) Annual Data Report 2011, released by U.S. Department of Health and Human Services on December 2012, more than 70% of patients who received a pancreas transplant were re-hospitalized after 1 year and 40% of the transplants were rejected at the end of 60 months (OPTN/SRTR 2011 Annual Data Report). The similar results were recorded for intestine transplants as more than 50% of intestine transplants were rejected 36 months after the transplantation. In the case of heart transplants 11.6% of the patients waiting for a heart transplant had died since there were no suitable donor organs in 2008. Among the ones that received the transplants, 50.9% rejected the transplant by the 5th year of the procedure (OPTN/SRTR 2011 Annual Data Report). Therefore, there is a need for tissue and organ substitutes that can be tailor-made for patients using their own cells in a timely and cost effective manner. With further understanding of how to regulate and control the stem cell fate it could be possible to make use of the regenerative potential of stem cells to engineer fully functional tissue and organ substitutes.

3. Effect of Biochemical and Physical Properties of Hydrogels on Stem Cell Fate

It is crucial to provide the proper cues to direct stem cell differentiation into the cells that comprise a targeted tissue. In order to achieve healthy and functional differentiated cells, these cues should be introduced at the correct time, dose, and location in

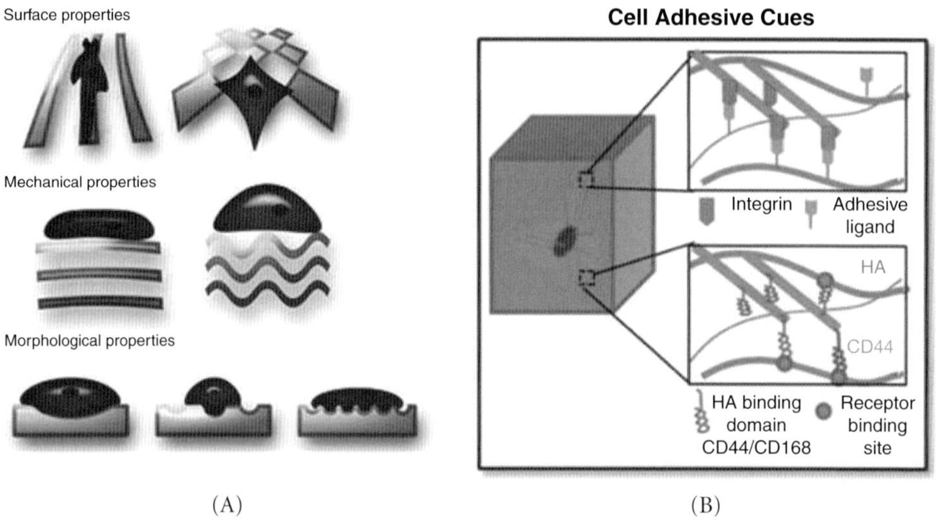

Fig. 2. Schematic representation of (A) ECM-based microenvironmental factors that can be used to guide stem cell fate by altering hydrogel properties[29] and (B) how stem cells interact with cell adhesive motifs in hydrogels.[47]

a biomimetic manner.[48] Hydrogels provide a biomimetic environment for stem cells to attach, proliferate, and differentiate,[5,49] and therefore have been used to study how the physical and biochemical cues affect the stem cell behavior. Several types of hydrogels with different properties were used to investigate the relationship between these cues on stem cell fate commitment.[5,50,51] As such, hydrogels have also been used to direct stem cell fate *in vitro* and *in vivo* mostly for tissue engineering applications. Control of stem cell fate can be achieved by spatiotemporal distribution of bioactive factors, by controlling the hydrogel composition and geometry, and by tuning mechanical properties of the hydrogels (Fig. 2).[52,53]

3.1. Effect of incorporating bioactive factors to hydrogels

Soluble and ECM bound bioactive molecules such as growth factors, cytokines, and other paracrine or autocrine factors are among the best studied factors that control the stem cell fate. Such bioactive factors have been incorporated into 2D or 3D culture systems by the help of hydrogels. Hydrogels not only provide control over spatial distribution of these factors but can also provide temporal control, thus they have shown to be effective carriers for bioactive molecules.[54,55] By incorporating the required biological cues into the hydrogel structures, the fate of stem cells can be controlled in a biomimetic manner. In this section, stem cell–hydrogel interactions are reviewed focusing on types of bioactive molecules used to direct the stem cell fate as well as various approaches to integrate these bioactive cues into the hydrogels.

3.1.1. *Modification of hydrogel surfaces with bioactive molecules*

Hydrogels have been used as 2D substrates for determining stem cell fate since they provide a more biomimetic surface for cell attachment when compared to traditional culture substrates (Fig. 3).

In 2D applications, bioactive factors are usually immobilized on the surface of the hydrogels to create spatially controlled bioactive factor distributions or bioactive factor gradients as cues to direct stem cell behavior.[56,57] In one such study, growth factors were immobilized on a polyacrylamide-based hydrogel surface using inkjet printing and the effect of growth factor gradients on stem cell differentiation was tested and compared to soluble growth factor presence (Fig. 3(A)).[58] Results showed that neural stem cells (NSCs) were differentiated into astrocytes and smooth muscle cells when cultured on hydrogels functionalized with ciliary neurotrophic factor (CNTF), and fetal bovine serum, respectively. When cultured on fibroblast growth factor 2 (FGF-2) immobilized hydrogels, NSCs remained undifferentiated. These results were similar to the control groups where the growth factors were introduced through the culture media as soluble factors. When the cells were cultured on CNTF gradients, higher number of cells differentiated into astrocytes on the higher CNTF concentration regions of the gradient compared to the cells on the lower CNTF concentrations. In another study, differentiation of endothelial progenitor cells to endothelial cells was achieved by seeding the cells on heparin and CD34 functionalized hyaluronic acid (HA) hydrogels.[59] Cell

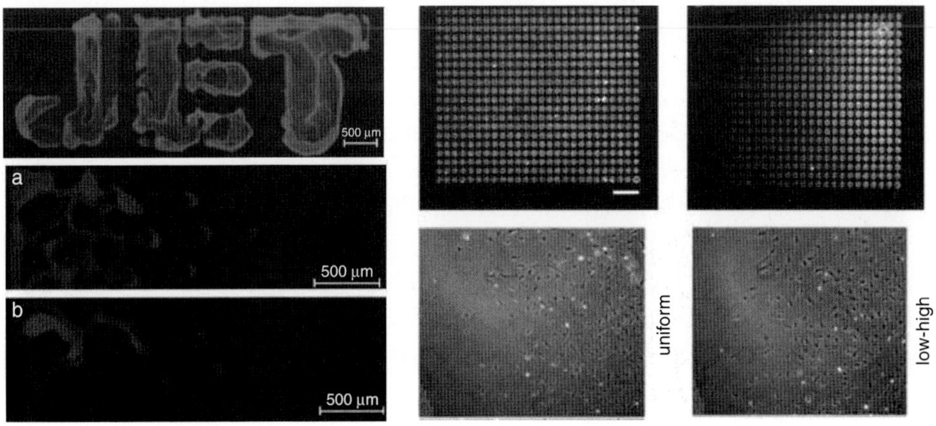

(A) (B)

Fig. 3. Patterning biochemical factors on hydrogel substrates for directing stem cell fate. (A) Inkjet printing of transferrin on polyacrylamide based hydrogels. Transferrin gradient printed on hydrogels before (a) and after (b) 22 h incubation in culture conditions.[58] (B) Uniform (left) and high to low concentration gradient (right) of inkjet printed fibrin gels with HB-EGF. Mouse mesenchymal stem cells were shown to migrate in both uniform and gradient patterns of HB-EGF.[60]

attachment, spreading, and differentiation were higher in the functionalized hydrogels compared to their non-functionalized counterparts. Immobilizing growth factors at specific locations on ECM-like materials without altering their potency has the advantage of directing the fate of the stem cells to multiple different lineages simultaneously. This is not possible when the growth factors or other bioactive factors are introduced as soluble agents since they would affect all the cells exposed to the liquid environment through diffusion.

In addition to controlling their differentiation, immobilization of bioactive molecules on hydrogel surfaces can be used for guiding stem cell migration.[82] Fibrin gel surfaces were patterned with heparin-binding epidermal growth factor-like growth factor (HB-EGF) using an inkjet printer to form an array of low to high or high to low concentrations of the growth factor (Fig. 3(B)). It was shown that the presence of the growth factor was a more important parameter than its gradient to guide stem cell migration.

Another way to create gradients of bioactive factors is to use microfluidics techniques.[61,62] In one such study, Villa-Diaz *et al.* used a microfluidic device to keep ESCs undifferentiated (Fig. 4).[63] The cell culture chamber was coated with ECM molecules to provide attachment for the cells. The design allowed for spatiotemporal control over the different solutions introduced to the cell culture chamber of the device. Half of the culture chamber was perfused with trypsin-EDTA, while the other half was perfused with culture media. The cells and the ECM coating were washed away from the trypsin-EDTA treated half. Cell proliferation was observed to remain limited on the side with the intact ECM coating. The highly controllable environment provided by microfluidics enable researchers to study the specific factors affecting stem cell behavior such as the effect of ECM-based hydrogel surface on stem cell attachment and proliferation as exemplified above. In another study a micro-bioreactor system was used to control the growth conditions of ESCs.[64] Under highly controlled flow or static conditions ESCs were cultured on collagen gels. The two different configurations of the micro-bioreactor, low and high shear conditions were created using the same flow rate. Under high shear conditions ESC differentiated into smooth muscle cells, however, differentiation was not induced under low shear or static conditions. Micro-scale devices provide highly controlled environments to study stem cell behavior, including cell attachment, proliferation and differentiation, and thus can be used to direct stem cells in combination with hydrogels.

3.1.2. *3D incorporation of bioactive molecules to stem cell-laden hydrogels*

In order to provide a more biomimetic and 3D environment, stem cells can be encapsulated within the hydrogels instead of seeding on their surfaces.

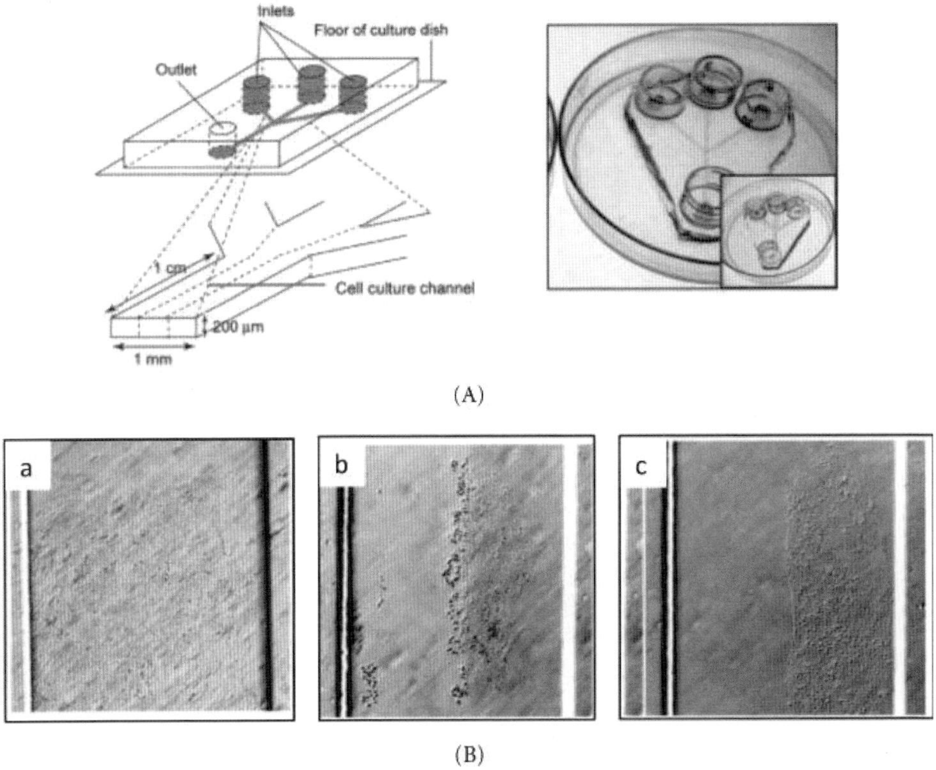

Fig. 4. Combination of hydrogel substrates with microfluidic techniques can be used to control stem cell fate. (A) Microfluidic device schematic (left) and photograph (right) for ESC culture.[63] (B) ESC in the microfluidic device seeded on ECM coating (a), after treating left half with trypsin-EDTA to remove cells and ECM coating (b), and 48 h after the treatment indicating the selective growth of ESC on ECM coated surface (c).[63]

Earlier studies on the effect of bioactive factors on hydrogel-stem cell interactions in 3D had mostly focused on providing biochemical cues exogenously or by directly incorporating these cues to cell-laden hydrogels by mixing the hydrogel polymer solution with the bioactive factors (Fig. 5(A)). For example, osteogenic differentiation of poly(ethylene glycol) (PEG)-encapsulated human mesenchymal stem cells (hMSCs) was examined in the presence of exogenous growth factors and compared to cells cultured on 2D cell culture substrates.[65] When cultured in osteogenic media, PEG-encapsulated hMSC showed increased expression of osteogenic lineage-specific cell markers and showed increased mineralization within one week. In another study, Zhao *et al.* encapsulated human umbilical cord mesenchymal stem cells to fabricate cell-laden hydrogel beads and showed that they can be

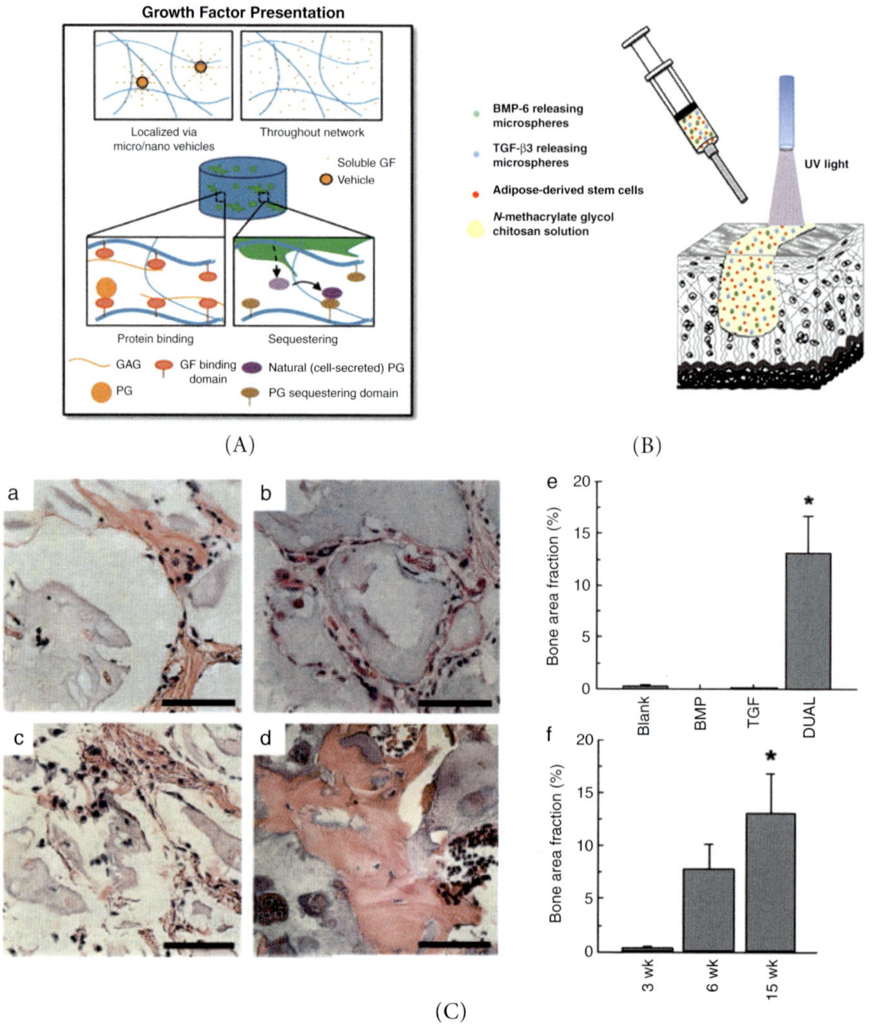

Fig. 5. Effect of presence of bioactive factors on stem cells cultured in 3D using hydrogels. (A) Schematic representation of growth factor incorporation to hydrogels.[47] (B) Schematic representation of encapsulating adipose-derived stem cells with BMP-6 and TGF-β3 releasing microbeads in chitosan-based hydrogel.[66] (C) Bone formation (pink) in RGD-conjugated alginate hydrogels (a) without growth factors, (b) with only BMP-2, (c) only TGF-β3, and (d) with co-delivery of BMP-2 and TGF-β3. (e) Greater bone area fraction was observed when two growth factors were delivered simultaneously. (f) Bone formation has increased in time in the hydrogels with dual growth factor delivery.[67]

differentiated into bone cells when cultured in osteogenic media containing dexamethasone, β-glycerophosphate, ascorbic acid, and 1,25-dihydroxyvitamin.[68] Alginate hydrogel–cell mixture was made into beads by extrusion from a syringe. These cell-encapsulated beads were then mixed with calcium phosphate cement functionalized with chitosan to improve the mechanical properties. It was shown that the cement–hydrogel bead mix was suitable for injection using low forces, which in turn helped obtaining high cell viability. It was observed that the cells were viable and differentiated to osteogenic cells by the osteogenic media they were cultured in. When injected, cells showed high mineralization and expression of bone specific cell markers like alkaline phosphatase, osteocalcin, and collagen type I.[68] Lei and Schaffer showed that human pluripotent stem cells can be successfully differentiated into different cell lineages using the thermoresponsive hydrogel poly(N-isopropylacrylamide)-co-poly(ethylene glycol) (PNIPAAm-PEG) and exogenously introduced small molecules.[69] Via the use of respective inductive media components, the stem cells encapsulated in thermoresponsive hydrogels were differentiated into dopaminergic neurons and cardiomyocytes. It was also shown that proliferation of the encapsulated neuronal cells was significantly higher than their 2D counterparts. This sets an example of how hydrogels provide a suitable environment for stem cells to proliferate and differentiate when the correct bioactive molecules are supplied exogenously.

Growth factors can be incorporated into the cell-laden hydrogels by directly mixing them with the cell-hydrogel solution to mimic the native developmental environment of stem cells. With a broader understanding of the embryonic development, introduction of multiple growth factors simultaneously or subsequently in order to mimic the different developmental stages has become important. In an attempt to apply this approach to induce bone tissue formation, Simmons *et al.* have successfully delivered bone morphogenetic protein-2 (BMP-2) and transforming growth factor β3 (TGF-β3) individually or simultaneously to see the effect of dual delivery on bone formation of rat bone marrow stromal cells.[67] Cells were encapsulated in arginine–glycine–aspartate (RGD) peptide-functionalized alginate hydrogels mixed with the growth factors and implanted ectopically to mice. Bone formation was significantly higher in the samples with dual growth factor incorporated cell-laden hydrogels compared to their single growth factor containing or no growth factor containing counterparts[67] (Fig. 5(C)). Similarly, delivery of bone morphogenic protein 6 (BMP-6) and TGF-β3 simultaneously was shown to direct human adipose-derived stem cells (hASCs) to differentiate into chondrogenic lineage (Fig. 5(B)).[66]

Incorporating growth factors through direct physical encapsulation in hydrogels has shown to cause loss of activity or inefficient release due to growth factors' interaction with the hydrogels.[47,70] Therefore, to have a better controlled-release profile, growth factors were incorporated into the hydrogels after encapsulating them in microcarriers (Fig. 5(B)).[66,71,72] Using this technique to guide chondrogenic differentiation, Park *et al.* have developed a novel hydrogel construct to encapsulate rabbit bone marrow mesenchymal stem cells in oligo(poly(ethylene glycol) fumarate) (OPF), a biodegradable and injectable hydrogel composite.[72] Gelatin microbeads releasing transforming growth factor-β1 (TGF-β1) were incorporated to the hydrogel system to enhance the chondrogenic differentiation. The effect of TGF-β1 presence was shown to differentiate the cells to chondrogenic linage better when it was incorporated in microbeads. In another study, Ferreira *et al.* incorporated vascular endothelial growth factor (VEGF) releasing microbeads to an RGD peptide functionalized, dextran-based hydrogel along with ESCs to direct vascular differentiation. When cultured under attachment preventing conditions, ESCs can form cell aggregates called embryoid bodies (EBs). EBs were used as controls and the results indicated that the vascular differentiation of the cells cultured in the hydrogel-VEGF microparticle system was higher compared to the EBs. Even though the cells did not fully differentiate to vascular lineage, the VEGF receptor expression was improved.[71] In another study, chondrogenic differentiation of hASCs was promoted through incorporation of BMP-6 and TGF-β3 containing microspheres to methacrylated chitosan hydrogels.[66] The hASCs expressed cartilage specific proteins when they were encapsulated with a mixture of microspheres carrying the two growth factors.

In addition to incorporating bioactive molecules through physical means either in the form of direct entrapment of the factors within the hydrogel bulk or adding them after encapsulating within microcarriers, it is also possible to incorporate bioactive factors through chemical conjugation to the hydrogel structure.

One example to direct stem cell fate by conjugating bioactive molecules to the hydrogel structure was proposed by Benoit *et al.*[73] Heparin, a bioactive factor known to be interacting with many different proteins including the ones regulating the attachment and differentiation of mesenchymal stem cells to osteogenic linage, was tested for its effects on hMSC differentiation. Methacrylated heparin was synthesized and conjugated to PEG dimethacrylate (PEGDM) via crosslinking. The resulting composite was used for encapsulating hMSCs and osteogenic differentiation was higher in heparin-conjugated samples as increased alkaline phosphatase and osteopontin expression suggested. The mechanism behind the osteogenic differentiation was further assessed and a strong correlation between differentiation and heparin–BMP-2 and heparin–fibronectin interactions was observed.[73] In another study, neural stem/progenitor cells (NSPCs) were encapsulated in platelet derived growth factor AA (PDGF-AA) immobilized agarose hydrogels. The effect of incorporating PDGF-AA through immobilization Versus as a soluble factor was assessed. The results showed that PDGF-AA specifically directed the NSPCs to differentiate into oligodendrocytes. Also, immobilized growth factor had a better effect on enhancing differentiation. In a similar study, pro-neural rat interferon-g (rIFN-g) conjugated hydrogels were shown to direct the NSPC differentiation towards neuronal lineage.[74] Biotinylated rIFN-g was incorporated to methacrylated and thiolated chitosan conjugated with maleimide–strepatavidin. The growth factor was incorporated to the hydrogel homogenously via the affinity binding of biotin–strepatavidin. The growth factor conjugated hydrogel was compared to its counterparts that lack the growth factor in presence or absence of the soluble growth factor in the culture media. The results demonstrated that immobilizing the growth factor to the hydrogel structure increased the number of neuronal committed cells when compared to the cell-laden hydrogels with and without the addition of soluble growth factor. In another example VEGF was chemically conjugated to agarose hydrogels.[75] The effect of immobilized Versus soluble VEGF on differentiating ESCs towards blood progenitor cells was tested. The gene expression levels of blood progenitor specific proteins in encapsulated cells showed that immobilized VEGF enhanced the differentiation of ESCs better than introducing VEGF as a soluble factor, proving the importance of mimicking the native environment of stem cells.[76]

In addition to growth factors, ECM molecule-derived motifs have been conjugated to hydrogels in an attempt to control stem cell differentiation.[77] Hwang and coworkers showed that the RGD functionalized PEG diacrylate (PEGDA) hydrogels enhanced the chondrogenic differentiation of ESC-derived mesenchymal stem cells compared to collagen type I, HA incorporated or pristine PEGDA.[78] The photocrosslinked PEGDA hydrogels with the RGD functionalization, 1:1 mixtures of collagen type I:PEGDA or HA:PEGDA, and non-functionalized PEGDA were tested. The chondrogenic inductive media containing TGF-β1 directed the differentiation of the cells to chondrogenic linage only in the PEGDA gels with RGD functionalization. The cells encapsulated in the collagen, HA conjugated or non-functionalized PEGDA gels did not show any specific cell proliferation or matrix production. In another study, Nguyen *et al.* have differentiated bone marrow stem cells into the three different zones of cartilage by manipulating the hydrogel structure. The superficial, transitional, and deep zones of articular cartilage were reconstituted by encapsulating bone marrow stem cells in PEG hydrogels functionalized with chondroitin sulfate and matrix metalloproteinase-sensitive peptides (PEG:CS:MMP-pep), PEG hydrogels functionalized with only chondroitin sulfate (PEG:CS), and PEG hydrogels functionalized with HA (PEG:HA), respectively (Fig. 6(A)). The three different compositions of PEG hydrogels led marrow cells to have varying levels of collagen type II and proteoglycan expression, resembling the different zones found in the cartilage (Fig. 6(B)). In addition to zone specific ECM production, zone specific compressive modulus was observed.[79] The construct was prepared by encapsulating the marrow stem cells in corresponding hydrogels sequentially, starting from deep zone. It was shown that the sequential crosslinking was successful in preparing a 3D scaffold that can derive a single population of stem cells into three different cell types.

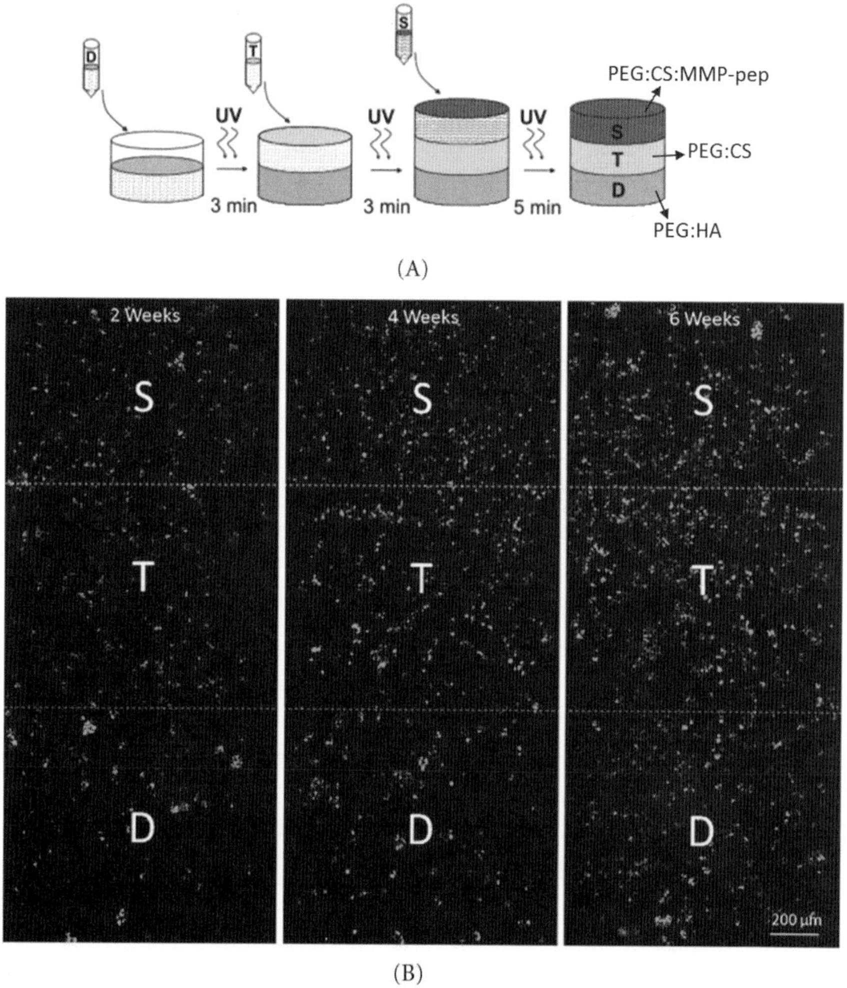

Fig. 6. Effect of composition of hydrogels on directed 3D differentiation of stem cells. (A) Schematic representation of preparation of the 3D hydrogel scaffold for zonal chondrogenic differentiation. (B) Immunostaining for Collagen types II (green) and X (red) at different time points in different zones, showing the zonal increase in type II collagen deposition.[79]

Incorporating ECM molecule-derived motifs was also shown to regulate metabolic responses of stem cells.[80] hASC were shown to have different levels of reactive oxygen species (ROS) production and mitochondrial activity depending on the cell attachment cues provided with the hydrogel scaffolds. PEG hydrogels functionalized with RGD or GRD short peptides were used to encapsulate hASC. The hydrogels incorporated with RGD peptide showed lower ROS production and a better mitochondrial activity compared to GRD peptide incorporated hydrogels. As these findings suggest various signaling pathways of stem cells can be altered by functionalizing hydrogels with short peptides that mimic ECM molecules. As a result, stem cell differentiation into different tissue-specific cell types[73,81] or evoking different metabolic responses[94] can be achieved by tailoring hydrogel scaffolds.

In addition to growth factors and ECM molecules or ECM motifs, chemical modification of hydrogels was also shown to be effective in directing stem cell differentiation into different cell lineages. Adipogenic and osteogenic differentiation of hMSC was guided by incorporating two different functional molecules to PEG hydrogels.[81,82] The hydrogels with t-butyl functionalization were shown to induce osteogenic differentiation, whereas the phosphate functionalized hydrogels induced adipogenic differentiation, both in the presence or absence of growth factors.

Hydrogels were also used to control cell–cell interactions to guide stem cell behavior.[102] Sugiura *et al.* constructed a temporally controlled active co-culture of mouse ESCs with human hepatocellular carcinoma (HepG2) cells encapsulated in PEG and calcium alginate hydrogels, respectively (Fig. 7). The mESC-laden PEG was micropatterned so that islands of ESCs were surrounded by HepG2 encapsulated in calcium-crosslinked alginate. The alginate part of the co-culture construct was removed after 4 days of culture by making use of the reversible calcium crosslinking. The ESCs started to differentiate into cardiomyocytes after 4 days of co-culture of these cells compared to 16 days of single culture and had higher number of beating cells compared to their single cultured counterparts. Also continuous co-culture was compared to their temporary co-culture. The ESCs in temporary co-cultures were shown to have a higher number of beating colonies than the continuous co-cultures. It was concluded that the soluble factors secreted by HepG2 cells were important in enhancing cardiac differentiation of ESCs at an early stage even though there was no direct cell–cell contact. Furthermore, removal of HepG2 cells helped ESCs to have better access to nutrients and less accumulation of metabolic waste, improving the differentiation of ESCs to cardiac lineage.

Defining the localization of proteins in 3D provides critical control over the cell behavior. 3D micropatterning techniques have been used to control the spatial distribution of growth factors which can in turn be used for controlling differentiation of the cells encapsulated in these specific locations of the 3D constructs. Wylie and coworkers used multiphoton chemical patterning technique to immobilize two different proteins that promote differentiation to neuronal lineage within 3D hydrogels.[83] First, the binding proteins were immobilized into predetermined locations in the hydrogel. Then the hydrogels were immersed in a solution of the growth factor-binding protein pair conjugate to create the 3D patterns of the two growth factors. The growth factors were immobilized using the affinity of barnase–barnstar and streptavidin–biotin (Fig. 8(A)). The neuronal precursor cells were

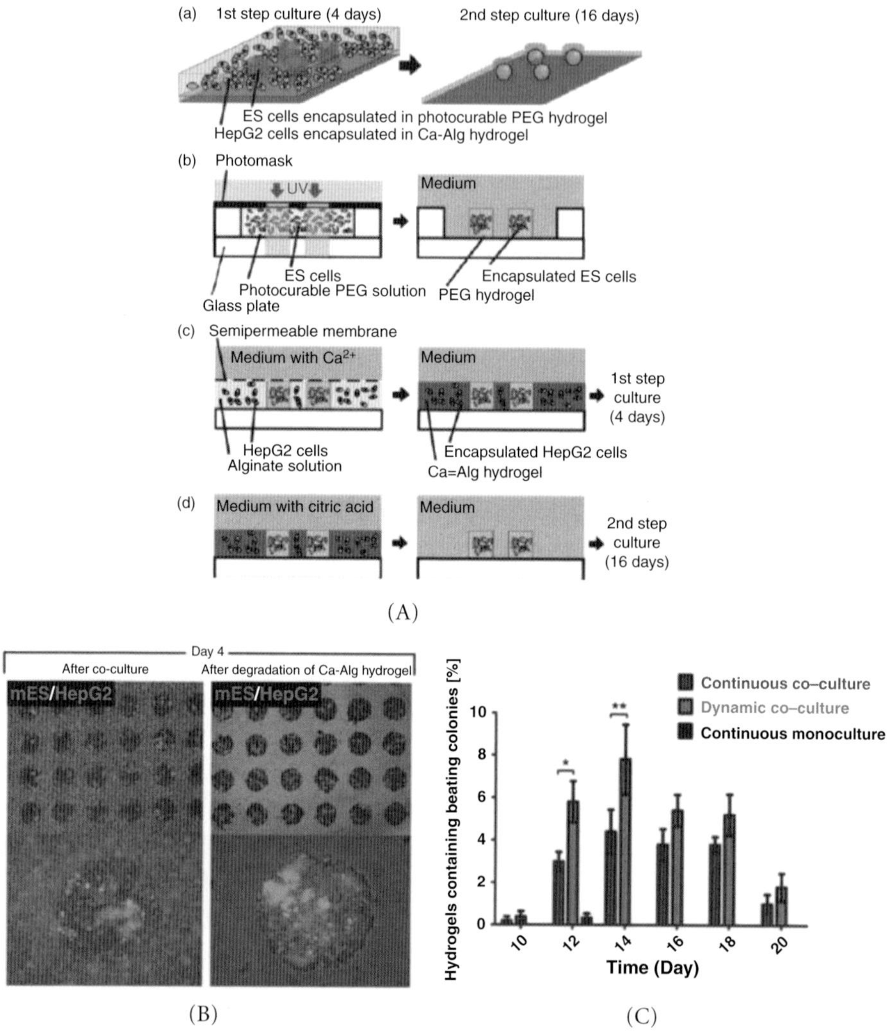

Fig. 7. Spatiotemporally controlled stem cell culture to control stem cell fate. (A) Schematic representation of the dynamic 3D micropatterned co-culture. (a) In the 1st step culture, ES cells in photocurable-PEG are cultured with HepG2 cells in a Ca–Alg hydrogel for 4 days. In the 2nd step culture, the ES cells are cultured without the HepG2 cells, which are removed after the Ca–Alg hydrogel is degraded for 16 days. (b) Encapsulation of ES cells in the photocurable PEG hydrogel. (c) Encapsulation of HepG2 cells in the Ca–Alg hydrogel. (d) Degradation of the Ca–Alg hydrogel and removal of HepG2 cells after the 1st step culture for 4 days. (B) mES and HepG2 cells before (left) and after (right) the removal of HepG2 cells. (C) Beating colony percent of different co-culture conditions at different time points.[17]

successfully patterned and the respective signaling pathways of these two growth factors were shown to be activated, resulting in differentiation to neural cells. In another study, image guided photopatterning was used to pattern specific biomolecules into 3D hydrogels to resemble native ultrastructure of tissues.[84] The confocal micrographs of different tissue sections were used to recapitulate the same structure into the hydrogel using multiphoton laser scanning lithography. To show the precision of this method, the complex structure of NSC niche was recapitulated using two different hydrogels. PEG hydrogels were functionalized with RGDS (labeled red) and IKVAV (labeled green) to recapitulate the vessel structure and the neural progenitor cells, respectively (Fig. 8(B)).

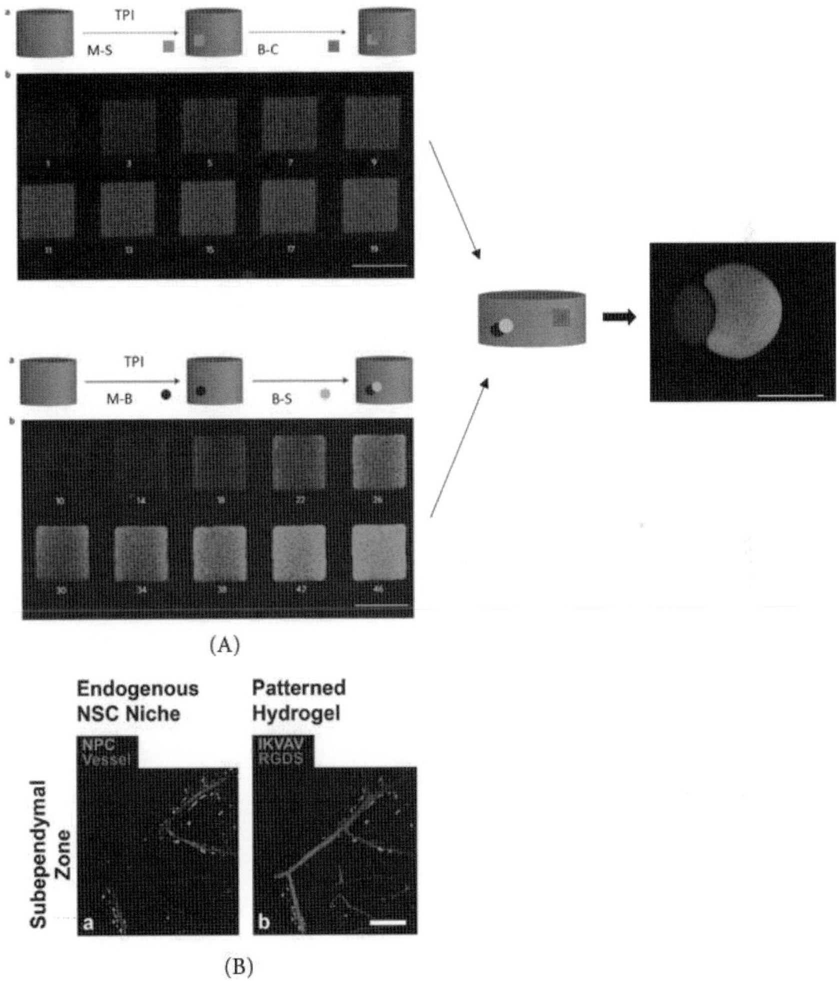

Fig. 8. Two-photon photopatterning to direct stem cell fate. (A) 3D localization of growth factors was achieved by two-photon irradiation (TPI). Malamide–streptavidin (M–S) and biotin–CNTF (B–C) couple was used to immobilize CNTF (red), and malamide–barnase (M–B) and barstar-sonic hedgehog (SHH) (B–S) (green) couple was used to immobilize SHH. The gradients of the two growth factors in response to different irradiation times, and the confocal micrograph of the co-localized growth factors in defined circular areas.[83] (B) (a) Mouse subependymal zone NSC niche immunostaining, vasculature (red), and neural progenitor cells (green). (b) PEG hydrogel patterned with PEG-RGDS (red) recapitulates the vessel structure and PEG-IKVAV (green) recapitulates the neural progenitor cells.[84]

Another important aspect of stem cell culture is to keep the stem cells undifferentiated and enhance their viability. To prolong the undifferentiated state of ESCs Gerecht *et al.* used HA hydrogels.[85] The ESC encapsulated in HA hydrogels shown to be capable of sustaining self-renewal unlike the ones encapsulated in a dextranbased hydrogel. In another study a PEG-based hydrogel was incorporated with 4 specific integrins to promote self-renewal of the ESCs.[86] When integrins a5b1 plus avb5, a6b1, and a9b1 were all incorporated into PEG hydrogels, encapsulated ESCs showed a prolonged expression of pluripotency genes, indicating that PEG hydrogels can be used as synthetic, 3D platforms to enhance stem cell self-renewal.

Incorporating 3D hydrogel cultures into microfluidic systems is very promising in terms of testing the effect of different doses of various biological factors on stem cell fate owing to the highly controllable microenvironment provided by the microfluidic systems. A very recent study demonstrated how microfluidic systems can provide a controlled environment for NSCs encapsulated in collagen gels, by exposing them to steady concentrations and gradients of specific growth factors.[87] The blood flow was recapitulated using endothelial cell seeded channels on both sides of the NSC culture chamber (Figs. 9(A) and 9(B)). The whole system was

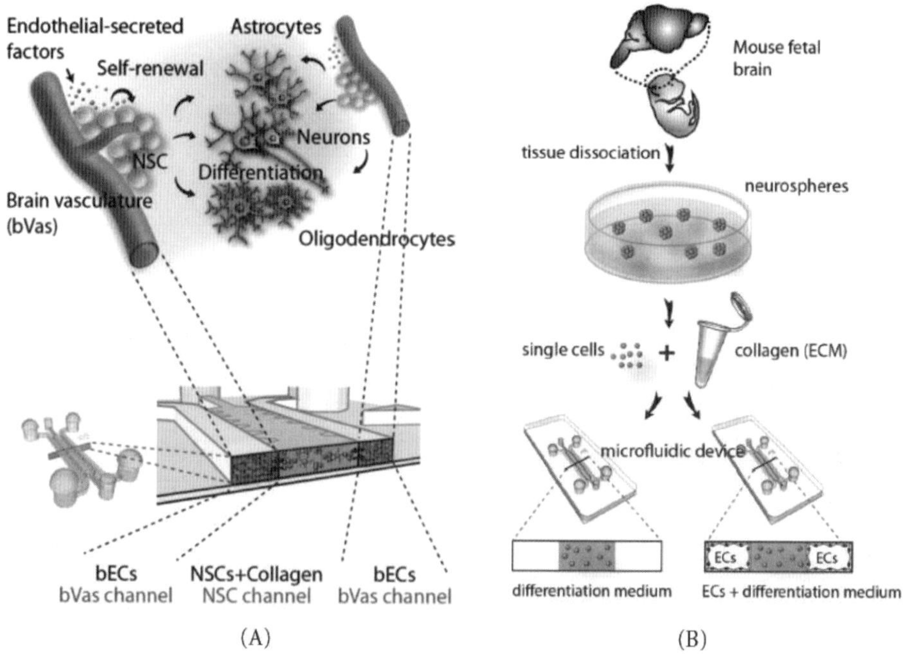

Fig. 9. Using microfluidics and hydrogel-based systems to create *in vitro* stem cell niches. (A) The schematic of recapitulating NSC niche using a microfluidic system. (B) Schematic of the preparation of the 3D co-culture system.[87]

designed to resemble the NSC niche. NSCs were shown to differentiate into astrocytes in the presence of the pseudo-vasculature system showing the effect of endothelial cell secreted bioactive molecules on NSC differentiation.

Hydrogels are good platforms for manipulating stem cell fate as they provide a 2D or 3D environment for cell attachment and growth, and they can be engineered with various bioactive cues such as growth factors, proteins or ECM-mimicking short peptides. In addition, the physical and mechanical properties of hydrogels can be tailor-made to direct stem cells to a specific fate. Factors such as hydrogel composition, stiffness, and geometric structure were shown to be important in determining stem cell behavior.

3.2. Effect of physical properties of hydrogels

It is known that stem cells respond to mechanical properties of the environment they reside in but the exact biophysical mechanism or the extent of its influence on stem cell differentiation is not well understood.[88] Various tools and methods have been used to investigate how physical factors control stem cell function, one of which is hydrogels. Hydrogels can be designed to match a variety of physical parameters that can be seen *in vivo*. By manipulating the degradability and strength of the hydrogel, the biophysical effects of the microenvironment on stem cell fate can be better understood.

Integrins are transmembrane cell surface receptors, which mediate cell attachment to the ECM.[89] Integrins tend to aggregate and interact with other proteins to form supermolecular structures called focal adhesion proteins or focal adhesion complexes. Focal adhesion complexes are drawn to the area of activity where the ECM–cell interaction occurs. On the interior side of the cells, these adhesion complexes are connected to cytoskeleton through which multiple cell functions can be regulated, most well-known of which are cell spreading and migration. The assembly and disassembly of these focal adhesions are dependent on both biochemical signals and actin–myosin contractions.[90]

External biophysical cues can be sensed by the cell and transduced to achieve a specific cell function mostly through these focal adhesion complexes (Fig. 10).[91–95] One of the most studied external physical factors that can be sensed by the stem cells is the ECM elastic modulus. During cell adhesion, the resistive forces from the ECM are balanced by the cytoskeletal contractile forces exhibited by the cell. The process where cells convert these biophysical cues into an intracellular response is called mechanotransduction. One well-understood mechanotransduction pathway is the RhoA/ROCK mediated cytoskeleton tension.[91,96] RhoA is responsible for cellular signaling and cytoskeletal organization and its effector, ROCK, stimulates cytoskeleton tension. Multiple studies showed that

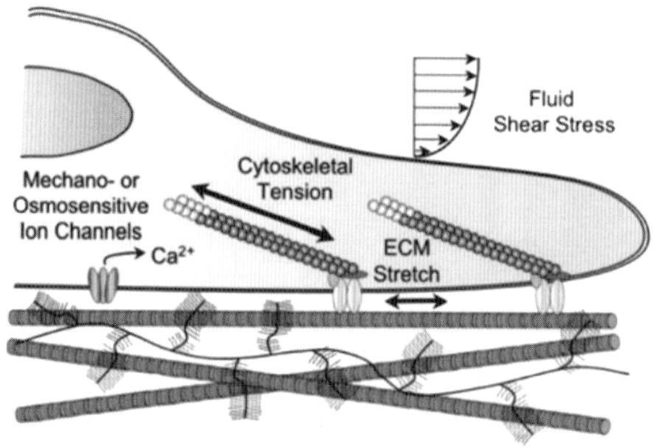

Fig. 10. The biophysical signals within the stem cell niche such as shear stress, ECM stretch and rigidity regulate stem cell fate. The cell internalizes these stimuli with mechanosensors such as cytoskeleton, mechanosensitive ion channels, and focal adhesion.[97]

commitment to a particular stem cell fate attained using soluble bioactive factors can be overridden by manipulating the RhoA/ROCK pathway, demonstrating that the biophysical sensing process is upstream of biochemical cell fate commitment. For example, activating RhoA to create a sense of high tension in the cell results in osteogenesis in the presence of adipogenic medium.[98] Similar results were observed even for non-spreading cells where the cells were forced to remain rounded within a hydrogel. When CD44 interaction was blocked, tension mediated signaling was inhibited with effects similar to the restrictive hydrogel. These results stress the importance of biophysical forces in stem cell fate commitment and suggest that understanding the interaction between cell and microenvironmental biophysical factors can ultimately be used to control the stem cell fate. Hydrogels with tunable physical properties that are in the range of native tissue ECM are ideal candidates to further understand how these physical factors affect stem cell fate, with an ultimate goal of controlling stem cell fate commitment.

3.2.1. *Manipulating hydrogel rigidity to influence stem cell fate*

The stiffness of the hydrogels is tunable and dependent on crosslinking density. The crosslinking density of a hydrogel can be altered by varying the weight percentage or molecular weight of the macromere.[99] The structure of the macromer is also an important factor, and crosslinking density will differ for hydrogels made with linear macromers than those made with their branched counterparts.[100] A more complete description of the physical properties of hydrogels and the variables related to tailoring and characterizing them are explained in Volume 1 of this three-volume set.

3.2.1.1. Effect of surface stiffness of hydrogels on stem cell fate

The rigidity of the substrate on which the stem cells are cultured has been shown to provide biophysical cues that can direct stem cell fate.[101,102] Stem cells match their own stiffness to that of the substrate they have attached to, which has direct effect on gene expression.[103] It was shown by Engler *et al.* that the substrate stiffness can directly influence stem cell phenotype.[101] In this study, polyacrylamide gels were coated with a collagen layer that MSC attached to. The stiffness of the hydrogel was adjusted based on the degree of chemical crosslinking. When the stiffness of the gel matched that of a specific native tissue, the MSCs differentiated to the corresponding cell types (Fig. 11). For instance when the stiffness of the gel matched the brain tissue the phenotype of the MSCs was neuronal. Similarly, the cells followed a myoblast lineage when the stiffness resembled that of striated muscle and osteoblast lineage when the stiffness of the matrix mimicked that of osteoid (the precursor to bone). A similar stiffness dependent differentiation trend was achieved with gelatin hydroxyphenyl-propionic acid (GtneHPA).[104] The hydrogel directly supported hMSCs attachment, so no cell adhesive coating was needed. In this study, stiffer hydrogels were also shown to increase the rate of proliferation and cell spreading.

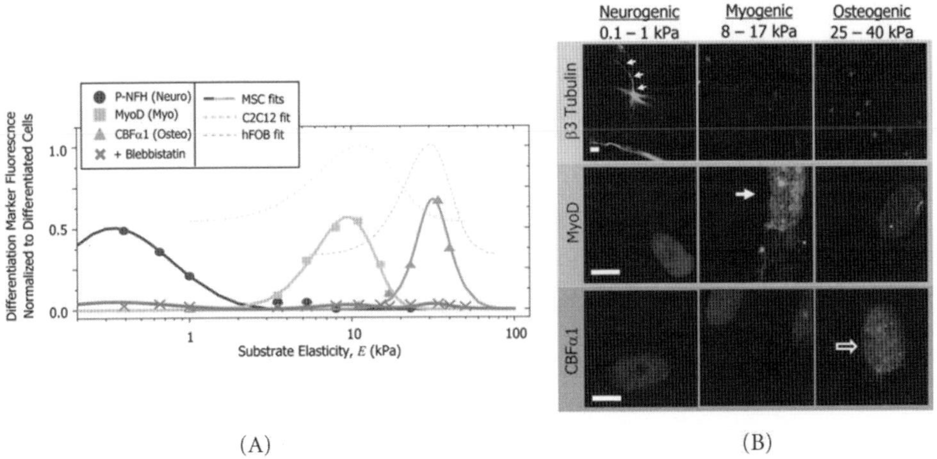

Fig. 11. Effect of matrix stiffness on stem cell fate. (A) The elasticity range for the following soft tissues: brain, fat, striated muscle, stiff cartilage and pre-calcified bone. The level of neurogenic, myogenic and osteogenic differentiation marker expressions seen from hMSCs of differing matrix elasticity correspond to *in vivo* tissue elasticity. (B) Varying expression levels of different cell markers illustrate how MSC fate is stiffness dependent. The arrows point to the cells with the highest expression of either $\beta 3$ tubulin, a neurogenic marker, MyoD1, a muscle transcription factor or CBFα1, an osteoblast transcription factor.[101]

3.2.1.2. Effect of stiffness on stem cell fate in 3D hydrogels

The critical effect of hydrogel stiffness on stem cell fate on 2D hydrogel surfaces led to the question of whether stem cells would respond similarly in 3D systems that are more physiologically relevant. To better understand how stem cells respond to 3D changes in physical microenvironments, the mechanical properties of the hydrogel can be altered to respond to enzymes secreted by the cells or to an external stimulus like light. To manipulate the strength of the hydrogel using light, either cleaving photochemical groups can be used to decrease the density of hydrogel crosslinks (hydrogel degradation),[105] or additional crosslinking chains between polymers can be formed to increase the strength of the hydrogel.[106] The dynamic nature of these hydrogels can be used to closely ressemble *in vivo* conditions.

In a study by Huebsch *et al.* the stiffness of mouse MSC (mMSC) encapsulated hydrogels was varied between 2.5 and 110 kPa to match the range of stiffness seen in a range of different tissues. They used RGD modified agarose[107] and PEGDM[108] to prevent enzymatic degradation by the encapsulated cells.[109,110] The hydrogel stiffness influenced stem cell fate in a similar manner to what had previously observed for stem cells grown on 2D surfaces; mMSC became adipogenic in the stiffness range of 2.5 to 5 kPa mMSC and osteogenic in the stiffness in the range of 11 to 30 kPa (Fig. 12).

Effect of stiffness on cells encapsulated within hydrogels was also studied using covalently crosslinked methacrylated HA (MeHA).[98] Unlike the previous example the mMSCs encapsulated in the hydrogel underwent adipogenesis for all stiffness levels ranging from 4 to 91 kPa (Fig. 12(B)). These results show the importance of

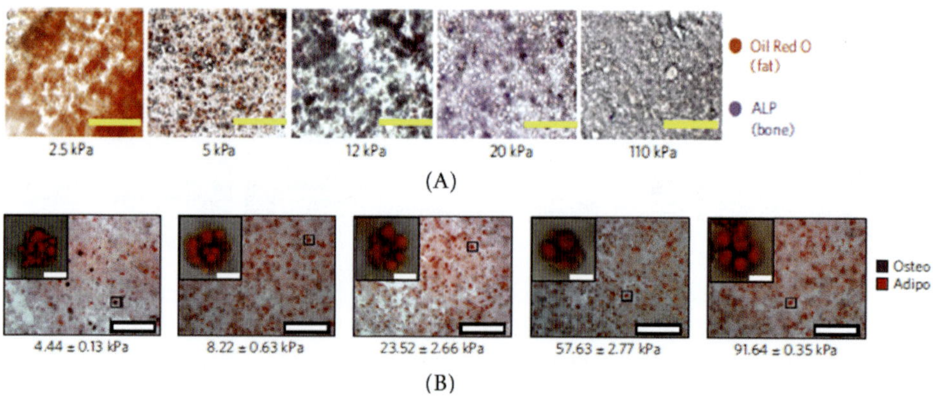

Fig. 12. Controlling stem cell fate through hydrogel stiffness. Stained encapsulated mMSCs with ALP osteogenic biomarker and Oil Red O adipogenic biomarker, (A) RGD-modified agarose with different stiffness and fate choices seen with the biomarkers,[88] (B) fate choice of mMSCs in photopolymerized MeHA hydrogels with a range of stiffness.[98]

understanding the distinct interaction between different hydrogel types and stem cells on determining stem cell fate.

For covalently crosslinked hydrogels, degradation-mediated cellular traction is believed to direct stem cell fate.[98] This idea was tested using a two-step crosslinked HA, which is degradable after initial crosslinking (referred to as "the permissive state"), but became restrictive and non-degradable after a secondary crosslinking with UV exposure. The mMSCs encapsulated in permissive state hydrogels became osteogenic with high traction and spreading, while the mMSCs in the restrictive hydrogel became adipogenic with low traction and spreading. The cell shape was shown to be less influential on stem fate, and osteogenic cells became adipogenic after the secondary crosslinking while the stretched cellular morphology remained unchanged.

Photocrosslinkable hydrogels can be altered to create stiffness variations based on the degree of light exposure. This method can be used to examine how stem cells respond to mechanical differences and how these differences affect their fate. Also hydrogels can be designed to crosslink both chemically and upon light exposure (i.e., photocrosslinking). A dual-crosslinkable system was used to pattern restrictive and permissive microenvironments within the HA hydrogel in a precise manner[112] (Fig. 13(A)). MSCs were first encapsulated with degradable peptide bonds and then non-degradable kinetic chains were formed with a UV induced radical reaction. Remodeling occurred within the permissive primary crosslinked hydrogel and cells within the microenvironment led to an osteogenic fate. The opposite was true for cells within restrictive secondary crosslinked area, they remained rounded with an adipogenic fate (Fig. 13(B)).

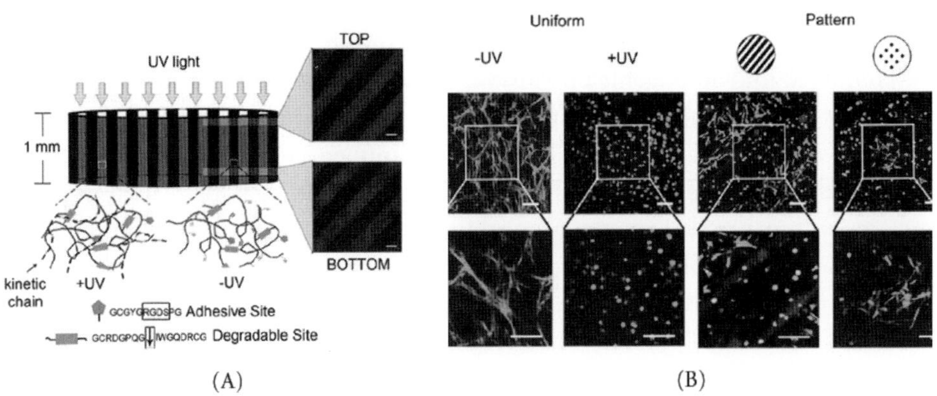

Fig. 13. Controlling stem cell fate through hydrogel degradability. (A) Photopatterning of HA hydrogels with high resolution photomask, (B) remodeling by encapsulated hMSCs in permissive and restrictive hydrogel.[112]

Hydrogel can also be engineered to respond to enzymes that are released by the cells through addition of enzymatic degradation sensitive sequences such as GGLGPAGGK (degraded by collagenase),[113] AAP (Nva) (degraded by elastase),[114] and GPQGILGQ (degraded by collagenase and various matrix metalloproteinases).[115] In one such study, thiol-ene photopolymerized PEG peptide hydrogel was used to culture human MSCs.[116] The four arm PEG macromer was functionalized with cysteine flanked peptides, MMP-degradable peptide sequence (KCGPQG↓WGQCK) and a cell adhesive RGD pendant (CGRGDS) which resulted in a degradable polymer that enhances cell attachment. In this study, the ratio of degradable and non-degradable linkers was varied to determine the effect on the hMSCs. It was shown that differentiation was enhanced in hydrogels that allowed degradation by cell-secreted enzymes.

3.2.2. *Effect of hydrogel microstructures on stem cell fate*

Micro- and nanostructured patterns also control many properties of the cells, including cell orientation, elongation, growth, and migration.[117] Accordingly, researchers showed that the geometric shape in which stem cells are seeded on or confined to within a hydrogel can influence the cell phenotype.

In one such study, molds were used to create wrinkle patterns on the surface of the hydrogel that hMSCs were cultured on.[111] The wrinkle morphology and size were shown to affect both stem cell shape and differentiation. The process used partially polymerized poly(2-hydroxyethyl methacrylate) (PHEMA) hydrogels and PDMS molds with lamellar and hexagonal surface pattern. When the mold was pressed to the surface of the hydrogel and exposed to light the pattern was replicated (Fig. 14). Cells on lamellar pattern spread taking the shape of the pattern. These cells exhibited high aspect ratio, and differentiated into osteogenic phenotype. Cells on the hexagonal patterns remained rounded with low spreading and differentiated into adipogenic phenotype. The results showed that surface patterns that affect cell morphology and spreading can also direct stem cell fate.

Stem cell differentiation has shown to be influenced by size of the stem cell aggregates (also called EBs) they form. Hydrogels can be used to create non-sticky surfaces with controlled geometries and therefore have been used to confine ESCs to obtain size-controlled stem cell aggregates.[118] For example, microwells made from cell repellent PEG hydrogels were used to create uniform sized EBs with diameters of 150, 300, and 450 μm (Fig. 15(A)).[119] Smaller EBs (150 μm) had a higher expression of WNT5a and enhanced endothelial differentiation, while the larger EBs (450 μm) had a higher expression of WNT11 and differentiated into cardiomyocytes as shown using immunostaining against endothelial and cardiac specific markers (Fig. 15(B)). After day 5, the EBs were replated in tissue culture

(A)

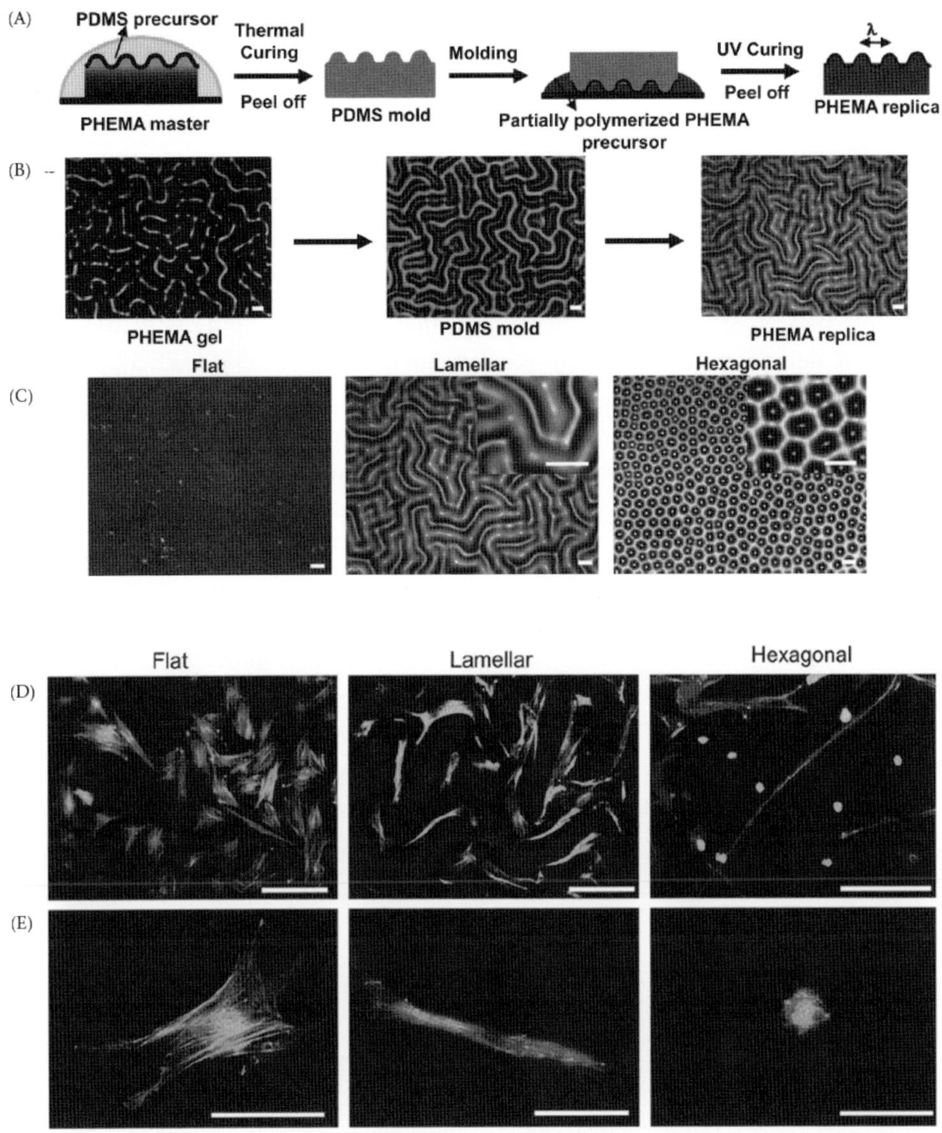

Fig. 14. Controlling stem cell fate through micropatterned hydrogel surfaces. (A) Schematic of fabrication process used for micropatterning the hydrogel surfaces, (B) optical microscopy images of the original gradient-PHEMA gel with swelling surface patterns, the PDMS mold and the PHEMA replica, (C) surface topography of different patterns: Flat, lamellar, and hexagonal, (D–E) fluorescent microscopy images (staining: blue — nuclei, green — actin, and red — hydrogel) of hMSCs seeded on flat, lamellar and hexagonal patterned PHEMA. Scale bar is 100 μm and 25 μm for (D) and (E).[111]

plates, and the beating frequency of the cells was recorded for the varying sized EBs until day 15. Beating frequency was higher for the larger EBs throughout the 15 days. The vessel sprouting activity of the EBs was also evaluated after being transferred to Matrigel coated substrates on day 5 and cultured with endothelial cell growth medium for 11 days. Vessels sprouted after day 6 and sprouting activity was higher in smaller sized EBs (Fig. 15(C)).

A dual-crosslinkable system was used to investigate how geometry and size of the hydrogel regions that cells are encapsulated in affect the stem cell fate.[55] Multiple pattern sizes were tested (0, 25, 50, 100, and 200 μm) with the methacrylated alginate and 8-arm poly(ethylene glycol) amine (OMA/PEG) hydrogel. To assess the effect of size on hASC differentiation level, the photoencapsulated cells were cultured for 4 weeks in osteogenic, chondrogenic or adipogenic differentiation media and characterized for different phenotypic markers. The samples

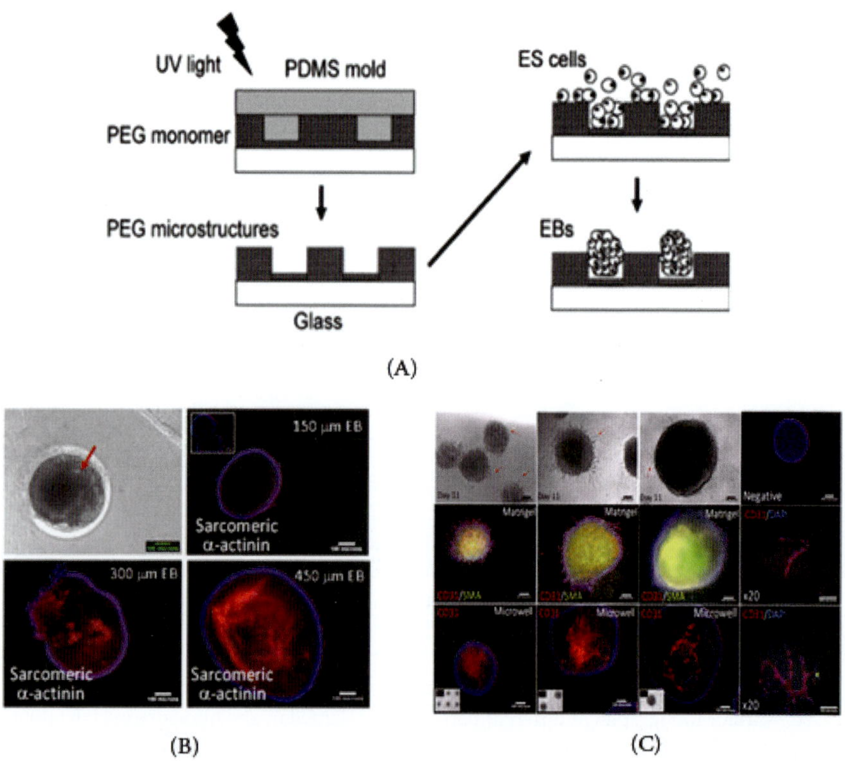

Fig. 15. Controlling stem cell aggregate size using micropatterned hydrogels. (A) Procedure used to create identically sized EBs with PEG hydrogel microwells,[118] (B) Immunocytochemical characterization with sarcomeric α-actinin for cardiomyogenic differentiation, (C) Phase contrast and immunocytochemical characterization with CD31 (red) and SMA (green) for endothelial cell differentiation.[119]

cultured in osteogenic differentiation media showed higher levels of osteocalcin and calcium deposition with the increased size of the micropatterns. The same size dependent results were paralleled when the encapsulated gels were cultured in chondrogenic conditions; the condrogenic markers, glycosaminoglycan (GAG) and chondroadherin (CHAD) content increased with increased size of the micropatterns. Fatty acid binding protein-4 (FABP-4) levels were not significantly different among the varying sized patterns which indicated that 3D adipogenic differentiation was less influenced by pattern size.

4. Conclusion

Stem cells hold promise for future clinical applications like artificial tissue and organ transplants or personalized medicine, owing to their self-regenerative capacity. Therefore, the mechanisms to direct stem cell fate are under intensive research. Hydrogels, providing the stem cell niche-like environments on 2D or 3D, have been one of the best candidates for guiding the stem cell fate. The cell–cell and cell–ECM interactions observed *in vivo* can be recapitulated *in vitro* using hydrogels for tissue engineering and regenerative medicine applications. Stem cell fate can be guided in response to introduction of different types and doses of bioactive molecules like ECM proteins, growth factors and small peptides. Moreover, the physical properties of the 2D or 3D cell environments such as, shape, size, stiffness, and degradability are critical to determine stem cell fate. Given the unique properties of hydrogels to properly mimic physiological conditions, when combined with stem cells, they make promising platforms for studying cell behavior and offer a way towards constructing fully functional organ models or ready-to-transplant organs.

References

1. E. Fuchs, T. Tumbar, *et al.*, Socializing with the neighbors: Stem cells and their niche. *Cell*, **116**(6), 769–778 (2004).
2. L. Li and T. Xie, Stem cell niche: Structure and function. *Annu. Rev. Cell Dev. Biol.*, **21,** 605–631 (2005).
3. S. J. Morrison, N. M. Shah, *et al.*, Regulatory mechanisms in stem cell biology. *Cell*, **88**(3), 287–298 (1997).
4. S. Mitalipov and D. Wolf, Totipotency, pluripotency and nuclear reprogramming. *Adv. Biochem. Eng. Biotechnol.*, **114,** 185–199 (2009).
5. M. P. Lutolf, P. M. Gilbert, *et al.*, Designing materials to direct stem-cell fate. *Nature*, **462**(7272), 433–441 (2009).
6. M. F. Pittenger, A. M. Mackay, *et al.*, Multilineage potential of adult human mesenchymal stem cells. *Science*, **284**(5411), 143–147 (1999).

7. L. A. Boyer, T. I. Lee, *et al.*, Core transcriptional regulatory circuitry in human embryonic stem cells. *Cell*, **122**(6), 947–956 (2005).
8. D. C. Ding, W. C. Shyu, *et al.*, Mesenchymal stem cells. *Cell Transplant.*, **20**(1), 5–14 (2011).
9. A. I. Caplan, Mesenchymal stem cells. *J. Orthopaed. Res.*, **9**(5), 641–650 (1991).
10. R. H. Lee, B. Kim, *et al.*, Characterization and expression analysis of mesenchymal stem cells from human bone marrow and adipose tissue. *Cell Physiol. Biochem.*, **14**(4–6), 311–324 (2004).
11. J. Fraser, M. Zhu, *et al.*, Adipose-derived stem cells. In *Mesenchymal Stem Cells*, Eds. D. Prockop, B. Bunnell, and D. Phinney, Humana Press, *Vol.* 449, pp. 59–67 (2008).
12. C. Cutler and J. H. Antin, Peripheral blood stem cells for allogeneic transplantation: A review. *Stem Cells,* **19**(2), 108–117 (2001).
13. L. Aust, B. Devlin, *et al.*, Yield of human adipose-derived adult stem cells from liposuction aspirates. *Cytotherapy*, **6**(1), 7–14 (2004).
14. H. J. Rippon and A. E. Bishop, Embryonic stem cells. *Cell Prolif.*, **37**(1), 23–34 (2004).
15. T. Vazin and W. J. Freed, Human embryonic stem cells: Derivation, culture, and differentiation: A review. *Restorative Neurology and Neuroscience*, **28**(4), 589–603 (2010).
16. D. Marolt, I. M. Campos, *et al.*, Engineering bone tissue from human embryonic stem cells. *Proc. Natl. Acad. Sci. USA,* **109**(22), 8705–8709 (2012).
17. S. Sugiura, J. M. Cha, *et al.*, Dynamic three-dimensional micropatterned cell co-cultures within photocurable and chemically degradable hydrogels. *J. Tissue Eng. Regen. M.* (2013).
18. O. Caspi, I. Huber, *et al.*, Transplantation of human embryonic stem cell-derived cardiomyocytes improves myocardial performance in infarcted rat hearts. *J. Am. Coll. Cardiol.*, **50**(19), 1884–1893 (2007).
19. O. Caspi, A. Lesman, *et al.*, Tissue engineering of vascularized cardiac muscle from human embryonic stem cells. *Circ. Res.*, **100**(2), 263–272 (2007).
20. T. Touboul, N. R. Hannan, *et al.*, Generation of functional hepatocytes from human embryonic stem cells under chemically defined conditions that recapitulate liver development. *Hepatology*, **51**(5), 1754–1765 (2010).
21. D. H. Woo, S. K. Kim, *et al.*, Direct and indirect contribution of human embryonic stem cell-derived hepatocyte-like cells to liver repair in mice. *Gastroenterology*, **142**(3), 602–611 (2012).
22. O. G. Kelly, M. Y. Chan, *et al.*, Cell-surface markers for the isolation of pancreatic cell types derived from human embryonic stem cells. *Nat. Biotechnol.,* **29**(8), 750–756 (2011).
23. W. Li, W. Sun, *et al.*, Rapid induction and long-term self-renewal of primitive neural precursors from human embryonic stem cells by small molecule inhibitors. *Proc. Natl. Acad. Sci. USA*, **108**(20), 8299–8304 (2011).
24. Y. Li, H. Wang, *et al.*, Global transcriptional and translational repression in human-embryonic-stem-cell-derived Rett syndrome neurons. *Cell Stem Cell*, **13**(4), 446–458 (2013).

25. K. Takahashi, K. Tanabe, *et al.*, Induction of pluripotent stem cells from adult human fibroblasts by defined factors. *Cell*, **131**(5), 861–872 (2007).
26. K. Takahashi and S. Yamanaka, Induction of pluripotent stem cells from mouse embryonic and adult fibroblast cultures by defined factors. *Cell*, **126**(4), 663–676 (2006).
27. N. Manzar, B. Manzar, *et al.*, The ethical dilemma of embryonic stem cell research. *Science and Engineering Ethics*, **19**(1), 97–106 (2013).
28. J. Liao, C. Cui, *et al.*, Generation of induced pluripotent stem cell lines from adult rat cells. *Cell Stem Cell*, **4**(1), 11–15 (2009).
29. S. Martino, F. D'Angelo, *et al.*, Stem cell–biomaterial interactions for regenerative medicine. *Biotechnol. Adv.*, **30**(1), 338–351 (2012).
30. M. Li, M. Chen, *et al.*, How far are induced pluripotent stem cells from the clinic? *Ageing Res. Rev.*, **9**(3), 257–264 (2010).
31. U. Ben-David and N. Benvenisty, The tumorigenicity of human embryonic and induced pluripotent stem cells. *Nat. Rev. Cancer*, **11**(4), 268–277 (2011).
32. C. Feng, Y.-D. Jia, *et al.*, Pluripotency of induced pluripotent stem cells. *Genomics, Proteomics & Bioinformatics*, **11**(5), 299–303 (2013).
33. S. Yamanaka, Induced pluripotent stem cells: Past, present, and future. *Cell Stem Cell*, **10**(6), 678–684 (2012).
34. T. Lin, R. Ambasudhan, *et al.*, A chemical platform for improved induction of human iPSCs. *Nat. Methods*, **6**(11), 805–808 (2009).
35. K. Hu and I. Slukvin, Generation of transgene-free iPSC lines from human normal and neoplastic blood cells using episomal vectors. *Methods Mol. Biol.*, **997**, 163–176 (2013).
36. Q. E. Zhang, S. P. Liu, *et al.*, Efficient reprogramming of human cord blood CD34(+) cells for formation of induced pluripotent stem cells with non-integrating plasmid system. *Zhongguo Shi Yan Xue Ye Xue Za Zhi*, **21**(3), 728–734 (2013).
37. W. A. Kues, D. Herrmann, *et al.*, Derivation and characterization of sleeping beauty transposon-mediated porcine induced pluripotent stem cells. *Stem Cells Dev.*, **22**(1), 124–135 (2013).
38. X. Cao, W. Deng, *et al.*, Non-viral co-delivery of the four Yamanaka factors for generation of human induced pluripotent stem cells via calcium phosphate nanocomposite particles. *Adv. Funct. Mater.*, **23**(43), 5403–5411 (2013).
39. R. Dolmetsch and D. H. Geschwind, The human brain in a dish: The promise of iPSC-derived neurons. *Cell*, **145**(6), 831–834 (2011).
40. C.-C. Huang, C.-K. Liao, *et al.*, A strategy for fabrication of a three-dimensional tissue construct containing uniformly distributed embryoid body-derived cells as a cardiac patch. *Biomaterials*, **31**(24), 6218–6227 (2010).
41. T. Takebe, K. Sekine, *et al.*, Vascularized and functional human liver from an iPSC-derived organ bud transplant. *Nature*, **499**(7459), 481–484 (2013).
42. R. M. Guzzo, J. Gibson, *et al.*, Efficient differentiation of human iPSC-derived mesenchymal stem cells to chondroprogenitor cells. *J. Cell Biochem.*, **114**(2), 480–490 (2013).

43. C. Qu, K. A. Puttonen, *et al.*, Chondrogenic differentiation of human pluripotent stem cells in chondrocyte co-culture. *Int. J. Biochem. Cell Biol.*, **45**(8), 1802–1812 (2013).
44. T. Y. Lu, B. Lin, *et al.*, Repopulation of decellularized mouse heart with human induced pluripotent stem cell-derived cardiovascular progenitor cells. *Nat. Commun.*, **4**, 2307 (2013).
45. S. G. Priori, Induced pluripotent stem cell-derived cardiomyocytes and long QT syndrome: Is personalized medicine ready for prime time? *Circ. Res.*, **109**(8), 822–824 (2011).
46. T. Yu, S. Miyagawa, *et al.*, *In vivo* differentiation of induced pluripotent stem cell-derived cardiomyocytes. *Circ. J.*, **77**(5), 1297–1306 (2013).
47. M. Guvendiren and J. A. Burdick, Engineering synthetic hydrogel microenvironments to instruct stem cells. *Curr. Opin. Biotechnol.*, **24**(5), 841–846 (2013).
48. E. Dawson, G. Mapili, *et al.*, Biomaterials for stem cell differentiation. *Adv. Drug Deliv. Rev.*, **60**(2), 215–228 (2008).
49. K. C. Rustad, V. W. Wong, *et al.*, Enhancement of mesenchymal stem cell angiogenic capacity and stemness by a biomimetic hydrogel scaffold. *Biomaterials*, **33**(1), 80–90 (2012).
50. O. Jeon and E. Alsberg, Regulation of stem cell fate in a three-dimensional micropatterned dual-crosslinked hydrogel system. *Adv. Funct. Mater.*, **23**(38), 4765–4775 (2013).
51. J. Lee, A. A. Abdeen, *et al.*, Directing stem cell fate on hydrogel substrates by controlling cell geometry, matrix mechanics and adhesion ligand composition. *Biomaterials*, **34**(33), 8140–8148 (2013).
52. S. M. Dellatore, A. S. Garcia, *et al.*, Mimicking stem cell niches to increase stem cell expansion. *Curr. Opin. Biotechnol.*, **19**(5), 534–540 (2008).
53. K. Saha, J. F. Pollock, *et al.*, Designing synthetic materials to control stem cell phenotype. *Curr. Opin. Chem. Biol.*, **11**(4), 381–387 (2007).
54. N. Annabi, A. Tamayol, *et al.*, 25th anniversary article: Rational design and applications of hydrogels in regenerative medicine. *Adv. Mater.*, **26**(1), 85–124 (2014).
55. J. L. Drury and D. J. Mooney, Hydrogels for tissue engineering: Scaffold design variables and applications. *Biomaterials*, **24**(24), 4337–4351 (2003).
56. K. Alberti, R. E. Davey, *et al.*, Functional immobilization of signaling proteins enables control of stem cell fate. *Nat. Meth.*, **5**(7), 645–650 (2008).
57. S. J. Jhaveri, M. R. Hynd, *et al.*, Release of nerve growth factor from HEMA hydrogel-coated substrates and its effect on the differentiation of neural cells. *Biomacromolecules*, **10**(1), 174–183 (2008).
58. S. Ilkhanizadeh, A. I. Teixeira, *et al.*, Inkjet printing of macromolecules on hydrogels to steer neural stem cell differentiation. *Biomaterials*, **28**(27), 3936–3943 (2007).
59. G. Camci-Unal, J. W. Nichol, *et al.*, Hydrogel surfaces to promote attachment and spreading of endothelial progenitor cells. *J. Tissue Eng. Regen. M.*, **7**(5), 337–347 (2013).

60. E. D. Miller, K. Li, *et al.*, Spatially directed guidance of stem cell population migration by immobilized patterns of growth factors. *Biomaterials*, **32**(11), 2775–2785 (2011).
61. B. G. Chung, L. A. Flanagan, *et al.*, Human neural stem cell growth and differentiation in a gradient-generating microfluidic device. *Lab Chip*, **5**(4), 401–406 (2005).
62. R. Gomez-Sjoberg, A. A. Leyrat, *et al.*, Versatile, fully automated, microfluidic cell culture system. *Anal. Chem.*, **79**(22), 8557–8563 (2007).
63. L. G. Villa-Diaz, Y. S. Torisawa, *et al.*, Microfluidic culture of single human embryonic stem cell colonies. *Lab Chip*, **9**(12), 1749–1755 (2009).
64. E. Cimetta, E. Figallo, *et al.*, Micro-bioreactor arrays for controlling cellular environments: Design principles for human embryonic stem cell applications. *Methods*, **47**(2), 81–89 (2009).
65. C. R. Nuttelman, M. C. Tripodi, *et al.*, *In vitro* osteogenic differentiation of human mesenchymal stem cells photoencapsulated in PEG hydrogels. *J. Biomed. Mater. Res. A*, **68A**(4), 773–782 (2004).
66. A. Sukarto, C. Yu, *et al.*, Co-delivery of adipose-derived stem cells and growth factor-loaded microspheres in RGD-grafted N-methacrylate glycol chitosan gels for focal chondral repair. *Biomacromolecules*, **13**(8), 2490–2502 (2012).
67. C. A. Simmons, E. Alsberg, *et al.*, Dual growth factor delivery and controlled scaffold degradation enhance *in vivo* bone formation by transplanted bone marrow stromal cells. *Bone*, **35**(2), 562–569 (2004).
68. L. Zhao, M. D. Weir, *et al.*, An injectable calcium phosphate-alginate hydrogel-umbilical cord mesenchymal stem cell paste for bone tissue engineering. *Biomaterials*, **31**(25), 6502–6510 (2010).
69. Y. Lei and D. V. Schaffer, A fully defined and scalable 3D culture system for human pluripotent stem cell expansion and differentiation. *Proceedings of the National Academy of Sciences* (2013).
70. E. Jabbari, Bioconjugation of hydrogels for tissue engineering. *Curr. Opin. Biotechnol.*, **22**(5), 655–660 (2011).
71. L. S. Ferreira, S. Gerecht, *et al.*, Bioactive hydrogel scaffolds for controllable vascular differentiation of human embryonic stem cells. *Biomaterials*, **28**(17), 2706–2717 (2007).
72. H. Park, J. S. Temenoff, *et al.*, Injectable biodegradable hydrogel composites for rabbit marrow mesenchymal stem cell and growth factor delivery for cartilage tissue engineering. *Biomaterials*, **28**(21), 3217–3227 (2007).
73. D. S. Benoit, A. R. Durney, *et al.*, The effect of heparin-functionalized PEG hydrogels on three-dimensional human mesenchymal stem cell osteogenic differentiation. *Biomaterials*, **28**(1), 66–77 (2007).
74. N. D. Leipzig, R. G. Wylie, *et al.*, Differentiation of neural stem cells in three-dimensional growth factor-immobilized chitosan hydrogel scaffolds. *Biomaterials*, **32**(1), 57–64 (2011).
75. N. Rahman, K. A. Purpura, *et al.*, The use of vascular endothelial growth factor functionalized agarose to guide pluripotent stem cell aggregates toward blood progenitor cells. *Biomaterials*, **31**(32), 8262–8270 (2010).

76. Y. Aizawa, N. Leipzig, *et al.*, The effect of immobilized platelet derived growth factor AA on neural stem/progenitor cell differentiation on cell-adhesive hydrogels. *Biomaterials*, **29**(35), 4676–4683 (2008).
77. C. N. Salinas and K. S. Anseth, The influence of the RGD peptide motif and its contextual presentation in PEG gels on human mesenchymal stem cell viability. *J.Tissue Eng. Regen. M.*, **2**(5), 296–304 (2008).
78. N. S. Hwang, S. Varghese, *et al.*, Chondrogenic differentiation of human embryonic stem cell-derived cells in arginine-glycine-aspartate-modified hydrogels. *Tissue Eng.*, **12**(9), 2695–2706 (2006).
79. L. H. Nguyen, A. K. Kudva, *et al.*, Engineering articular cartilage with spatially-varying matrix composition and mechanical properties from a single stem cell population using a multi-layered hydrogel. *Biomaterials*, **32**(29), 6946–6952 (2011).
80. B. Reid, J. M. Afzal, *et al.*, Enhanced tissue production through redox control in stem cell-laden hydrogels. *Tissue Eng. Part A*, **19**(17–18), 2014–2023 (2013).
81. D. S. Benoit, M. P. Schwartz, *et al.*, Small functional groups for controlled differentiation of hydrogel-encapsulated human mesenchymal stem cells. *Nat. Mater.*, **7**(10), 816–823 (2008).
83. R. G. Wylie, S. Ahsan, *et al.*, Spatially controlled simultaneous patterning of multiple growth factors in three-dimensional hydrogels. *Nat. Mater.*, **10**(10), 799–806 (2011).
84. J. C. Culver, J. C. Hoffmann, *et al.*, Three-dimensional biomimetic patterning in hydrogels to guide cellular organization. *Adv. Mater.*, **24**(17), 2344–2348 (2012).
85. S. Gerecht, J. A. Burdick, *et al.*, Hyaluronic acid hydrogel for controlled self-renewal and differentiation of human embryonic stem cells. *Proc. Natl. Acad. Sci. USA*, **104**(27), 11298–11303 (2007).
86. S. T. Lee, J. I. Yun, *et al.*, Engineering integrin signaling for promoting embryonic stem cell self-renewal in a precisely defined niche. *Biomaterials*, **31**(6), 1219–1226 (2010).
87. Y. Shin, K. Yang, *et al.*, Reconstituting vascular microenvironment of neural stem cell niche in three-dimensional extracellular matrix. *Adv. Healthc. Mater*, **3**(9), 1457–1464 (2014).
88. N. Huebsch, P. R. Arany, *et al.*, Harnessing traction-mediated manipulation of the cell/matrix interface to control stem-cell fate. *Nat. Mater.*, **9**(6), 518–526 (2010).
89. D. Boettiger, Mechanical control of integrin-mediated adhesion and signaling. *Curr. Opin. Cell Biol.*, **24**(5), 592–599 (2012).
90. K. Burridge and M. Chrzanowska-Wodnicka, Focal adhesions, contractility, and signaling. *Annu. Rev. Cell Dev. Biol.*, **12,** 463–518 (1996).
91. C. S. Chen, Mechanotransduction a field pulling together? *J. Cell Sci.*, **121**(Pt 20), 3285–3292 (2008).
92. B. Geiger, J. P. Spatz, *et al.*, Environmental sensing through focal adhesions. *Nat. Rev. Mol. Cell Biol.*, **10**(1), 21–33 (2009).
93. A. W. Orr, B. P. Helmke, *et al.*, Mechanisms of mechanotransduction. *Developmental Cell*, **10**(1), 11–20 (2006).
94. M. A. Schwartz and D. W. DeSimone, Cell adhesion receptors in mechanotransduction. *Curr. Opin. Cell Biol.*, **20**(5), 551–556 (2008).

95. V. Vogel and M. Sheetz, Local force and geometry sensing regulate cell functions. *Nat. Rev. Mol. Cell Biol.*, **7**(4), 265–275 (2006).
96. D. E. Ingber, Cellular mechanotransduction: Putting all the pieces together again. *Faseb. J.*, **20**(7), 811–827 (2006).
97. F. Guilak, D. M. Cohen, *et al.*, Control of stem cell fate by physical interactions with the extracellular matrix. *Cell Stem Cell*, **5**(1), 17–26 (2009).
98. S. Khetan, M. Guvendiren, *et al.*, Degradation-mediated cellular traction directs stem cell fate in covalently crosslinked three-dimensional hydrogels. *Nat. Mater.*, **12**(5), 458–465 (2013).
99. S. J. Bryant and K. S. Anseth, Hydrogel properties influence ECM production by chondrocytes photoencapsulated in poly(ethylene glycol) hydrogels. *J. Biomed. Mater. Res.*, **59**(1), 63–72 (2002).
100. J. Zhu, Bioactive modification of poly(ethylene glycol) hydrogels for tissue engineering. *Biomaterials*, **31**(17), 4639–4656 (2010).
101. A. Engler, S. Sen, *et al.*, Matrix elasticity directs stem cell lineage specification. *Cell*, **126**(4), 677–689 (2006).
102. H. J. Kong, T. R. Polte, *et al.*, FRET measurements of cell-traction forces and nano-scale clustering of adhesion ligands varied by substrate stiffness. *Proc. Natl. Acad. Sci. U.S.A.*, **102**(12), 4300–4305 (2005).
103. F. Chowdhury, S. Na, *et al.*, Material properties of the cell dictate stress-induced spreading and differentiation in embryonic stem cells. *Nat. Mater.*, **9**(1), 82–88 (2010).
104. L. S. Wang, J. Boulaire, *et al.*, The role of stiffness of gelatin-hydroxyphenylpropionic acid hydrogels formed by enzyme-mediated crosslinking on the differentiation of human mesenchymal stem cell. *Biomaterials*, **31**(33), 8608–8616 (2010).
105. A. M. Kloxin, A. M. Kasko, *et al.*, Photodegradable hydrogels for dynamic tuning of physical and chemical properties. *Science*, **324**(5923), 59–63 (2009).
106. M. Guvendiren and J. A. Burdick, Stiffening hydrogels to probe short- and long-term cellular responses to dynamic mechanics. *Nat. Commun.*, **3**, 792 (2012).
107. J. T. Connelly, A. J. Garcia, *et al.*, Interactions between integrin ligand density and cytoskeletal integrity regulate BMSC chondrogenesis. *J. Cell Physiol.*, **217**(1), 145–154 (2008).

107a. M. Tang, W. Chen, *et al.*, Human embryonic stem cell encapsulation in alginate microbeads in macroporous calcium phosphate cement for bone tissue engineering. *Acta Biomater*, **8**(9), 3436–3445 (2012).

108. S. A. Bencherif, A. Srinivasan, *et al.*, Influence of the degree of methacrylation on hyaluronic acid hydrogels properties. *Biomaterials*, **29**(12), 1739–1749 (2008).
109. E. Alsberg, H. J. Kong, *et al.*, Regulating bone formation via controlled scaffold degradation. *J. Dent. Res.*, **82**(11), 903–908 (2003).
110. M. P. Lutolf, F. E. Weber, *et al.*, Repair of bone defects using synthetic mimetics of collagenous extracellular matrices. *Nat. Biotechnol.*, **21**(5), 513–518 (2003).
111. M. Guvendiren and J. A. Burdick, The control of stem cell morphology and differentiation by hydrogel surface wrinkles. *Biomaterials*, **31**(25), 6511–6518 (2010).

112. S. Khetan and J. A. Burdick, Patterning network structure to spatially control cellular remodeling and stem cell fate within 3-dimensional hydrogels. *Biomaterials*, **31**(32), 8228–8234 (2010).
113. B. K. Mann, A. S. Gobin, *et al.*, Smooth muscle cell growth in photopolymerized hydrogels with cell adhesive and proteolytically degradable domains: Synthetic ECM analogs for tissue engineering. *Biomaterials*, **22**(22), 3045–3051 (2001).
114. A. A. Aimetti, M. W. Tibbitt, *et al.*, Human neutrophil elastase responsive delivery from poly(ethylene glycol) hydrogels. *Biomacromolecules*, **10**(6), 1484–1489 (2009).
115. H. Nagase and G. B. Fields, Human matrix metalloproteinase specificity studies using collagen sequence-based synthetic peptides. *Biopolymers*, **40**(4), 399–416 (1996).
116. S. B. Anderson, C. C. Lin, *et al.*, The performance of human mesenchymal stem cells encapsulated in cell-degradable polymer-peptide hydrogels. *Biomaterials*, **32**(14), 3564–3574 (2011).
117. R. G. Flemming, C. J. Murphy, *et al.*, Effects of synthetic micro- and nano-structured surfaces on cell behavior. *Biomaterials*, **20**(6), 573–588 (1999).
118. J. M. Karp, J. Yeh, *et al.*, Controlling size, shape and homogeneity of embryoid bodies using poly(ethylene glycol) microwells. *Lab Chip*, **7**(6), 786–794 (2007).
119. Y.-S. Hwang, B. G. Chung, *et al.*, Microwell-mediated control of embryoid body size regulates embryonic stem cell fate via differential expression of WNT5a and WNT11. *Proceedings of the National Academy of Sciences* (2009).

Chapter 3

Applications of Hydrogels in 3D Functional Tissue Models

Tamer Çırak*,†, Tayfun Vural‡, Doğa Kavaz§ and Emir Baki Denkbaş‡,¶,‖,**

**Hacettepe University, Division of Nanotechnology and Nanomedicine, Beytepe, 06800 Ankara, Turkey*

†Aksaray University, Faculty of Arts & Science, Department of Biology, 68100 Aksaray, Turkey

‡Hacettepe University, Department of Chemistry, Biochemistry Division, Beytepe, 06800 Ankara, Turkey

§Cyprus International University, Department of Bioengineering, 98258, Northern Cyprus via Mersin 10, Turkey

¶Hacettepe University, Advanced Technologies Research and Applications Center (HUNITEK), Beytepe, 06800 Ankara, Turkey

‖denkbas@hacettepe.edu.tr

In vitro studies should be generated with very similar features of *in vivo* for obtaining realistic results on cell cytotoxicity, therapeutic efficacy, cell apoptosis and necrosis. *In vitro* studies have the advantages on mass production of pharmaceutical industry since they provide less expensive, faster, and higher throughput screening compared to *in vivo* studies. Additionally wide species of human cells like cardiac cells, nerve cells, stem cells etc. can be used *in vitro* studies. However, cells do not have surrounding tissues and there is lack of blood supply and native nutrient supply. So, cells proliferate on a two-dimensional (2D) surface instead of native three-dimensional (3D) environment. Compared to traditional cell culture methods, in 3D models cells can communicate and proliferate in all directions. 3D tissue engineering models closely mimic complex structure of a

** Corresponding author.

tissue. By use of these models, more realistic results were observed in lots of cell culture studies like controlled drug delivery to cells in tissue, anticancer drug cytotoxicity on a particular cancer tissue, or control of stem cell fate.

1. Introduction

Preclinical drug tests are an important step of drug development as well as discovery of drug itself and clinical development research. Preclinical drug development is a combination of *in vitro* and *in vivo* laboratory testing which involve pharmacological and toxicological evaluations, and drug delivery mechanisms. Typically, three phases clinical studies are asked to meet the regulatory requirements; Phase-I studies are performed to evaluate pharmacological and general adverse effects; Phase-II studies are performed to evaluate the effectiveness and safety of the drug for a particular disease indication; Phase-III studies are performed to evaluate the full efficiency and safety of the drug in long-term drug usage. From Phase I to III, the number of human subjects and number of control group is enlarged to improve the statistical significance of toxicity and pharmacological data and to suppress the effects of life-style, nutritional, genetic and environmental factors.

Animal testing is a bridge between the laboratory tests and clinical studies, to predict the effects of the drug on pharmacological and toxicological results on human. However, human adverse effect prediction success rate is very low (10 to 70%) in animal toxicology tests and it is suitable for some specific applications like gastrointestinal, cardiovascular and hematological toxicities. Nevertheless it is generally insufficient in prediction for the side effects on other systems toxicities. Therefore, more sophisticated animal tests or better understanding of the all other environmental parameters should be developed and used for fully acceptable preclinical drug development.[1]

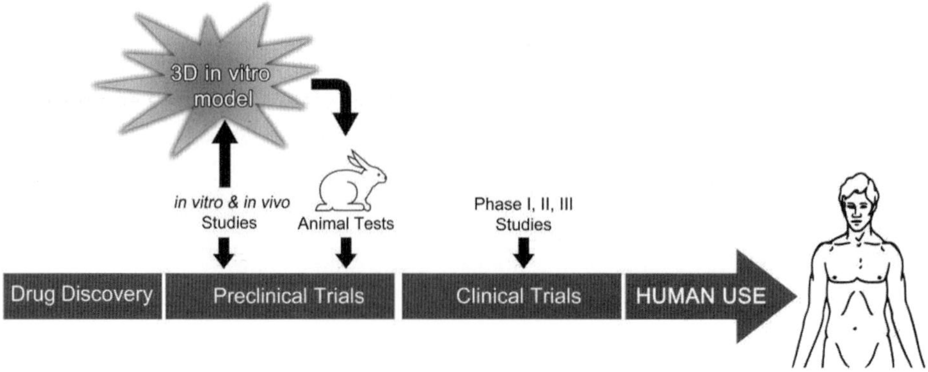

Fig. 1. Importance of 3D in vitro models on drug research.

In vivo studies have importance also on cancer related studies as well as many other clinical studies. However, to be a guide for *in vivo* studies, *in vitro* systems with the simulation abilities of real-like clinical conditions are important because of the ethical problems and experimental difficulties of animal trials. A complex evaluation of drug targeting efficiency into organ or tumor sites, therapeutic activity, cytotoxicity and biocompatibility is a must for patient application of drug delivery system on clinical therapies. In general, therapeutic activities of anti-cancer drugs are investigated *in vitro* in traditional cell culture flasks by culturing two-dimensional (2D) monolayer cells instead of costly and difficult *in vivo* studies. Because of the morphological variances of tumor cells obtained in 2D cell culture, poor cell–cell and cell–extracellular matrix (ECM) interactions, *in vivo* drug activity in tumor microenvironment could not be understood correctly. Therefore, to close the gap between *in vivo* and *in vitro* drug activity studies, some approaches are developed to culture tumor cells in three-dimensional (3D) architecture.

Owen *et al.* prepared hyaluronic acid based hydrogels to create 3D microenvironments for cell culture and related future applications. In their study, they evaluate the mechanical, structural, chemical, and biological properties of proposed 3D hydrogels. Especially, porosity and pore size distribution is examined in order to mimic the ECM conditions. As seen in this work too, a small change in the 3D architecture change all the mechanical properties and fluid mechanics, and so this affects all the mechanism in cell's life cycle.[2]

Albrecht *et al.* designed a biocompatible and stimuli-responsive hydrogel based 3D chondrocyte cell culture structure to form a real-like tissue model. In this case, they used a photo-polymerizable polyethylene glycol (PEG) hydrogel as a scaffold to seed the dispersed chondrocyte cells. Dielectrophoretic forces are used for cell manipulation and trapping. The proposed thin hydrogel structure is facilitating nutrient transport, which is also capable of transport any kind of bio-receptors for future screening applications like drug release studies, toxicological evaluations, gene expression studies and histological investigations (Fig. 2).[3]

Ling *et al.* designed a cell-laden agarose microfluidic hydrogel system, a candidate tool for generation of biomimetic synthetic vasculature for tissue engineering, diagnostics and drug screening applications. Proposed systems diffusion performance is investigated in terms of varying channel dimensions and shapes. Seeded cells are homogeneously distributed in hydrogel and cell culture media is circulated through the channels, which nutrients and wastes are able to diffuse in and out. As a result, the viability of the cells embedded in hydrogel is depending on the distance to microfluidic channels, which indicates the significance of a perfused network of micro-channels for circulating nutrients and oxygen to support cell viability in large hydrogels (Fig. 3).[4]

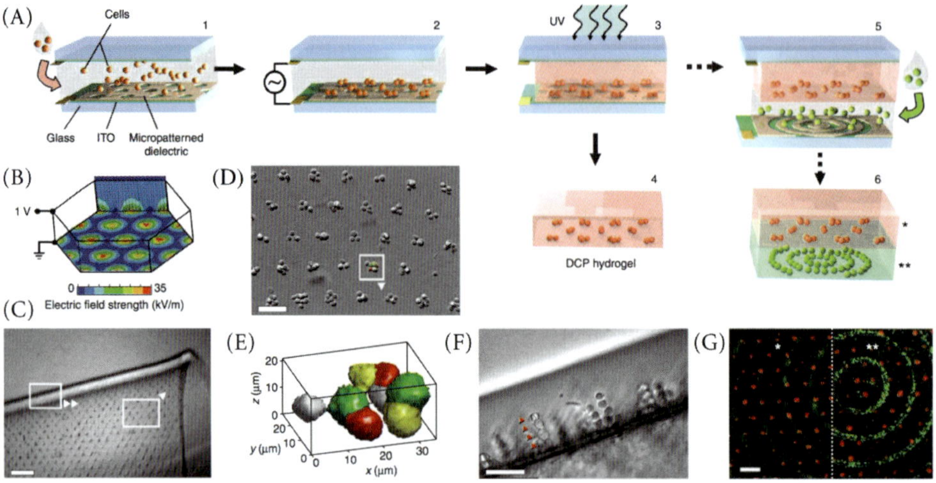

Fig. 2. Fabrication method and examples of DCP hydrogels. (A) Cells in prepolymer solution are introduced into the transparent chamber (1) and localized via dielectrophoretic forces to micropatterned gaps in the 1.8-mm-thick dielectric layer upon application of the a.c. chamber bias (2). Next UV light exposure polymerizes the hydrogel (3), embedding cells in a stable microorganization. The 'DCP hydrogel' can then be removed and cultured (4) or incorporated into multilayer constructs by expanding the chamber height and repeating steps 1–3; each layer may contain distinct cell organization, cell type and hydrogel formulation (5, 6). (B) Electric field strength is high above circular gaps in the dielectric layer arranged in a hexagonal array (100-mm spacing), as shown by finite element modeling (CFD Research Corp.). (C–F) The 100-mm-thick DCP hydrogels contain a microarray of embedded fibroblast clusters. In (C), the free-floating hydrogel is shown folded; boxes indicate orientation of panels (D) and (F). In (E), the 3D shape of an 8-cell cluster from (D) is rendered from confocal data (MATLAB; pseudocolored to depict individual cells). Cross-sectional view of linear columns 4–5 cells long (arrowheads) demonstrates cluster shape versatility (F). (G) A bilayered hydrogel (200 mm total thickness) contains distinct fluorescently labeled fibroblasts in a cluster array above (*) and in concentric rings below (**), as depicted in a (step 6). The same microscope field is shown focused on the upper or lower cell pattern, as delineated by the dotted line. Scale bars: (C) — 250 mm; (d,g) — 100 mm; (F) — 50 mm (Figure reprinted with permission from Ref. 3).

Ng *et al.* prepared a 3D cell-matrix tissue culture in a perfusable chamber in order to evaluate nanoparticle vehicle penetration and transport *in situ. In situ* observation of the cells embedded in matrigel filled into the chamber is performed by confocal microscope imaging of fluorescent-tagged nanoparticles with various diameters. Also, this 3D hydrogel system allows harvesting cells after culture for further biological and cellular studies such as flow-cytometry in order to investigate for applications like gene and drug delivery research.[5]

Toh *et al.* designed a microfluidic channel based 3D cell perfusion-culture system. HepG2 and MCF7 carcinoma cell lines, hepatocytes, and bone marrow

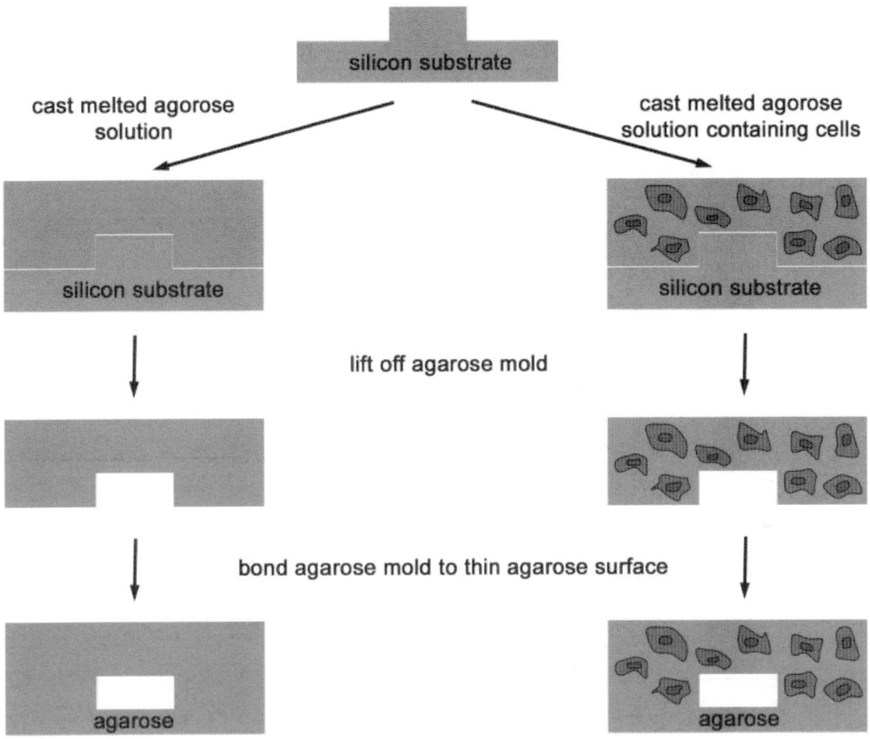

Fig. 3. Schematic of the fabrication of agarose microfluidic devices with (right) and without (left) embedded cells (Figure reproduced with permission from Ref. 4).

mesenchymal stem cells are cultured to demonstrate the biological versatility of this method. High-resolution optical and confocal imaging of the cells is possible through the transparent cover of 3D microfluidic channel-based cell culture system and this could be a tool for evaluation *in situ* assays using fluorescence reporters or immune-labeling for probing cellular events.[6]

2. Drug Delivery Systems

In vitro cell culture methods come to the forefront with the easy and economical ways on understanding of complex cellular mechanisms. Traditional 2D monolayer cell culture systems are used as standard in wide range of *in vitro* model applications for drug delivery, pharmacology, cancer research studies and many more. However, these cultures are differing from *in vivo* conditions because of 2D nature and they are losing their functions rapidly due to the lack of the ability of 3D environment formations. *In vivo* conditions, cells are living in a 3D microenvironment and they are in interaction with ECM molecules.[7,8]

In recent years, a new idea, "3D cell culture model" is developed to overcome the problems of preclinical drug research, which is able to simulate the cell–cell, cell–ECM and cell–target organ interactions.

Yip *et al.* developed a tumor model to use in anti-cancer drug test studies by using a new 3D multicellular heterospheroid collagen hydrogel culture, which is capable of simulating the *in vivo* tumor microenvironment better. In this study, they work on the modeling of 3D structure, multi-cellularity, and ECM barrier features, which are three important properties related to drug resistance abilities of cancer towards the anti-cancer drugs. In order to prepare controlled and uniform heterospheroids hanging drop method, liver carcinoma and stromal fibroblast co-culture are used. Following, heterospheroids are encapsulated in collagen gels. It is claimed that the obtained 3D liver cancer model is simulating the *in vivo* ECM conditions real-like (Fig. 4).[9]

In the study of Shin *et al.*, they prepared *in vitro* 3D tumor model by implementing the hydrogel template strategy to uniform sized spheroids inside the hydrogel scaffolds including micro wells. The goal of this study is to investigate the effects of anti-cancer drug delivery systems in better real-like conditions by simulating solid tumor and microenvironment *in vivo* conditions. Multicellular tumor spheroids

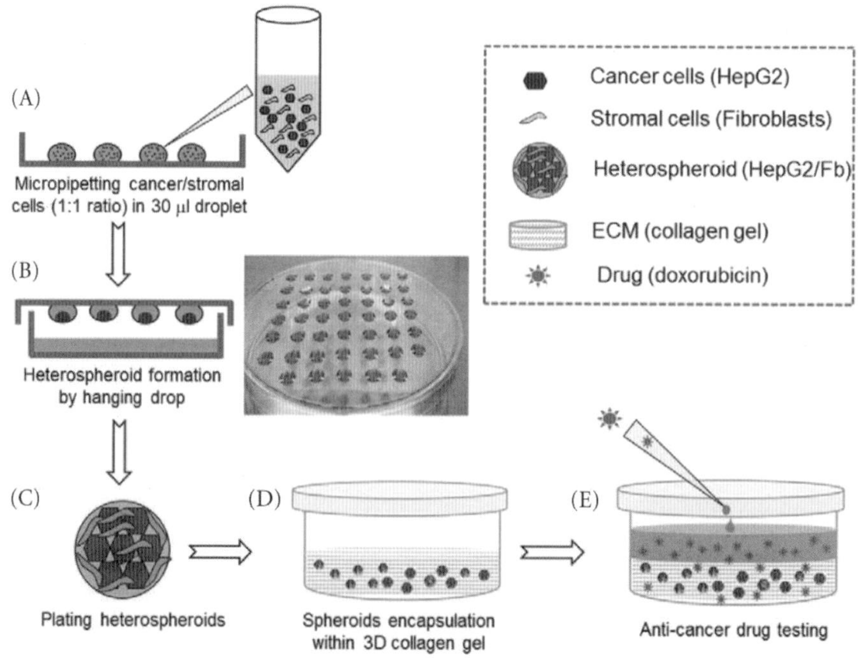

Fig. 4. Schematic diagram on the formation of heterospheroids by hanging drop method and 3D collagen gel culture for anti-cancer drug testing (Figure reprinted with permission from Ref. 9).

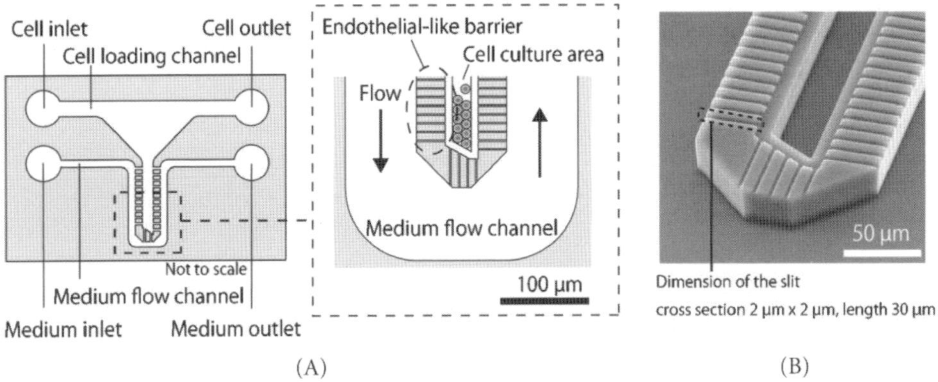

Fig. 5. (A) Design of a microfluidic device mimicking the structure of a hepatic cord. The device has four terminal ports and consists of flow channels, a cell culture area, and an endothelial-like barrier. The width of the cell culture area is designed to be 37 μm, which enables to align hepatocytes in two lines. Culture medium is perfused in the flow channel. (B) An image of the device structure by scanning electron microscope (Figure reprinted with permission from Ref. 10).

(MTSs) cultured in hydrogel scaffold are used for identifying the effects of culture conditions on the drug effectiveness. In order to simulate the dynamic *in vivo* micro-environment, free MTS, which are released from scaffold, are transferred into microfluidic channels. This work is a good example of how the *in vitro* models with the ability of simulating the biological factors like cell–cell, cell–ECM, and dynamic environment interactions, could help to develop the drug delivery systems.[8]

Nakao *et al.* designed a liver tissue simulating microfluidic cell culture model with a hepatic cord-like structure. In this model, hepatocytes were aligned linearly in two lines to form a functional bile canaliculi structure as mimicking the arrangement of the cells in the hepatic cord *in vivo*. The proposed device could be used as an alternative technique to animal testing to evaluate the drug release kinetics and drug induced toxicity on liver tissue (Fig. 5).[10]

3. Toxicological Evaluations

The preclinical testing methods for drug toxicity evaluations by using 2D monolayer cell cultures and animal models are incapable systems to predict the toxicity responses of human patients. For instance, in these preclinical tests, nephrotoxicity based failure rate is found only about 7%. However, in clinical cases, nearly half of the severe acute renal failure is depending on drug-induced nephrotoxicity. The 3D cell culture approaches are used also in toxicological evaluations of drugs similarly used in drug release studies. Desrochers *et al.* cultured the NKi-2 cell lines (immortalized human renal cortical epithelial cells) in a polycarbonate and rat-tail collagen based matrigel 3D hydrogel complex in order to biomimic the renal tissue system and evaluate the drug-induced nephrotoxicity. This bioengineered 3D kidney model is used comparatively with 2D cell culture to investigate acute and chronic toxicities induced by Cisplatin, Gentamicin and Doxorubicin. The 3D model shows significant results based on drug induced toxicity, while 2D cell culture does not give any conclusive result. Even this 3D method is still incapable of recapitulating the complexity of the *in vivo* environment; drug tests using this 3D model could enhance the predictability of drug-induced toxicity responses.[11]

Lan *et al.* developed *in vitro* 3D drug screening model to evaluate hepatotoxicity and drug activity of target drug components. In this study, they prepared a high-density 3D liver cell culture combined with monolayer grown target breast cancer cell line (MCF-7) in a static environment. Rate of drug conversion by HepG2 cells are evaluated and by using commercial drugs 50% cytotoxicity tests for HepG2 cells are performed. By using the model presented in this study, hepatotoxicity and effects of drug on target cell became to be predictable by evaluation of drug dosage concentrations.[12]

In the study of Sung *et al.*, a microfluidic device combined with 3D hydrogel cell culture is prepared in order to evaluate cytotoxicity of anti-cancer drugs by forming multiple-organ interactions. Colon cancer cells (HCT-116) are encapsulated in a 3D matrigel scaffold and cultured with cells to form a tumor compartment. Hepatoma cells (HepG2/C3A) are encapsulated in matrigel scaffold and cultured to form a liver-simulating compartment. Myoblasts (Kasumi-1) are encapsulated in alginate scaffold and cultured to form marrow compartment. These 3D scaffolds are connected together with blood circulation simulating channels. The circulation of drug in these different compartments is designed to simulate pharmacokinetic profile of drug in human body. In contrast to previous works on Tegafur, which are performed with traditional 96 well plate cell cultures, in this study cytotoxic effects are observed. Tagafur and 5-FU applied cell line's cytotoxic results are compatible with clinical studies. These findings show that, prepared 3D *in vitro* model gives more effective and reliable results for evaluation of anticancer drug's metabolism depended toxicological effects in comparison with standard 2D *in vitro* models.[13]

Chen *et al.* designed a microfluidic 3D tumor cell culture model for *in vitro* tumor spheroid formation and anticancer agent screening. *In situ* encapsulation of breast tumor cells was performed along the micro-channels by gelation of alginate. Following the cell seeding, MTSs were formed. An anticancer agent, doxorubicin were injected to the tumor spheroids in several concentrations to investigate the effect of doxorubicin dosage on cell viability. Formation of multi-cellular aggregates on 3D cell cultures is an advantage in comparison with traditional monolayer cultures. In future, this microfluidic device could be an alternative tool to investigate drug related toxicities *in vitro*, by integrating some biological and chemical extracellular components (Fig. 6).[14]

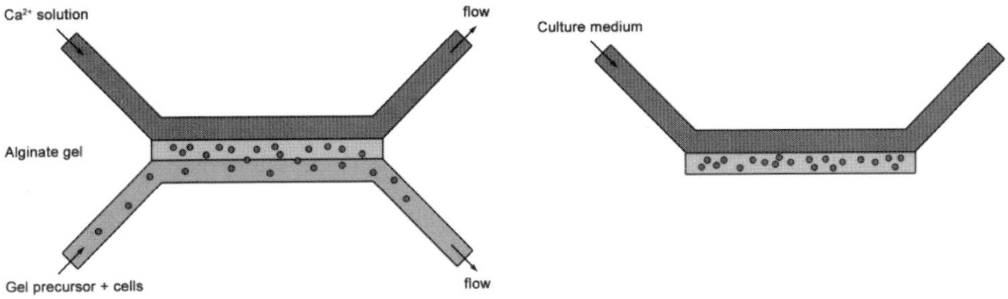

(A) Gel formation and cell entrapment

(B) Cell culture within hydrogel

Fig. 6. Top view of microfluidic device during alginate gel formation and cell entrapment using parallel, laminar flows. (A) Hydrogel forms at the interface between the flows of the gel precursor and Ca^{2+} ions. (B) The diffusion of Ca^{2+} to the opposite channel wall forms a solid band of alginate hydrogel, with cells embedded throughout. The flow is then switched to culture medium for long-term perfusion (Figure reproduced with permission from Ref. 14).

4. Cardiac Process

Myocardial Infarction (MI) causes permanent damage on myocardium in one-third of an hour; leads to areas of inadequate blood supply resulting billions of cell death and causes heart failure.[15,16] In an event of heart failure, heart transplantation is required, but there are not many heart donors available and as well the compatibility is another issue.[17] So new strategies like engineered cardiac tissues are being developed for treatment of damaged heart tissues.[18] Many studies have shown prospecting results using natural and synthetic hydrogels,[19,20] which support cell attachment, cytokine incorporation etc.[21] Furthermore, hydrogels are used for indication of contraction.[22] The studies direct to biocompatible, biodegradable, injectable, localized at injection area and mechanically resistant to heart beating hydrogel models.[23,24]

Cell encapsulated hydrogels build more durable and effective cardiac tissues, which have an impact of repairing damaged area. Injectable hydrogels are useful for mechanical resistance, transportation of bioactive molecules and renovator cells. F. Rask *et al.* developed cell protective peptide functionalized (QHREDGS) photo-cross linkable chitosan hydrogels. These hydrogels caused limited applicability because of the UV penetration depth.[25] Thus, Reis *et al.* used ionic gelation method between chitosan and collagen in order to synthesize cross linker free blend hydrogels. Photo cross linkable hydrogels were prepared in 1–2 min however cross linker free blend hydrogels took place for 30 min in an incubator at 37°C for gelation.[26]

Functional groups on biomaterials allow conjugation with a variety of molecules (ligands, drugs, growth factors, etc.) and improve the efficiency of hydrogel-based cell transplantation treatment.[27] Amine groups on chitosan biomaterials can help covalently attachment of QHREDGS molecules to provide a suitable hardness and mechanical resistance for heart beating, cell adhesion and viability.[26] In addition, covalently attached VEGF thermo-sensitive polyester hydrogels were shown to provide controlled release of VEGF at the accumulated infarct area.[28]

Recently, the nanostructure studies lead to stimulation of biostructures and cells by magnetic field for the purpose of developing tissue constructs. Xu *et al.*, used magnetic field on magnetic nanoparticles embedded cell encapsulated GELMA and PEG hydrogels to fabricate arc- and dome-shaped tissue constructs. Using magnetic field, these hydrogels with different cell types accumulated into one group creating 3D tissue construct.[29]

Akiyama *et al.* reported a new technique to design ring shaped tissue model by applying magnetic field. In this technique, a polycarbonated cylinder is positioned at the center of a well, afterwards mixture of cardiomyocyte cell suspension and collagen precursor solution is added into the well that leads to formation of cardiomyocyte layer, ends up with cardiomyocyte ring tissue model. In this regard, highly concentrated and homogenously aligned cells, aided self-generated contraction and susceptibility of electrical excitement.[30]

Carbon nanotubes have been used as multifunctional sensors for drugs or reactive oxygen species (ROS) inside hydrogels. In this manner, Barone *et al.* crosslinked gels with apo-glucose oxidase to sense glucose that may play a key role as controlled insulin releasing system to cure diabetics.[31] In another study, Shin *et al.*, showed that electrophysiological and mechanical behaviors of hydrogels were totally changed by embedded amount of carbon nanotubes. In the same manner, compressibility, cell adhesion, cell viability, and cell alignment were also affected. Also, gelatin polymer ensured a long time survival of cardiac cells by means of the porous structure. In order to prevent damage on cardiac tissue construct, carbon nanotubes served as free radical scavengers.[32]

5. Regenerative Medicine

Self-induced or donor induced cell transplantation used in classical surgeries has a number of major restrictions, failure to find a suitable donor tissue and experience immune system problems are some of them.[33] Tissue engineering that integrates engineering sciences with life sciences for creation of functional cell units is a relatively recent area.[34] Development of suitable replacement cell units for recruitment of disease or injury damages, have the potential to develop groundbreaking innovations in medicine. One of the benefits of tissue regeneration is finding solutions for self-induced or donor induced cell transplantation. In addition, by checking the current status of the case *in vitro* culture medium, allows modeling *in vivo*.[35] To create an *in vitro* model which is very similar to *in vivo* model by using tissue engineering, normal or abnormal tissue growth factors can also be detected. There are still problems on creating *in vitro* model to be solved. Thus new solutions can be found for difficulties during gene regulation by using other biological fields.[36]

Nowadays for tissue regeneration, transplantation studies are concentrated on constructing 3D structures. Living cells that can be taken from a donor, stem or progenitor, is the initial step for creating optimal cell units. 3D structures, allow cells to grow and to make the cells communicate with each other. Arranging the chemical and mechanical properties of these provide imitation of *in vivo* environment.[37] These 3D structures of hydrogels can be made of synthetic or natural materials.[38] Chemical inducer to allow cell growth and differentiation is used; and the bioactive signals are needed in 3D structures to enhance the interaction with the environment. These are the basics that lie under the ideas of tissue engineering to obtain suitable cell units for replacement of failure tissues.[39]

Furthermore, studies on 3D structures are not only focused on tissue regeneration but also lead to drug efficiency, toxicology, tissue development and morphogenesis.[40] Although *in vivo* studies, *in vitro* studies provide both financial and ethical aspects on these studies and mechanical studies can be performed on easy and controllable systems.[1] 2D culture systems do not allow interaction of cells that cause cell migration to other tissues as well as the lack of 3D environment result in major differences in gene expression. These evidences show that 3D cell culture systems have advantage over 2D culture systems for tissue regeneration.[33] The tissue regeneration studies on *in vitro* culture system are formed models of embryo, kidney, cardiac, skin and bone etc. In pharmaceutical field, human skin models, for electrophysiological evaluation of cardiac tissue models are used in order to simplify *in vivo* studies.[8] In addition, tumor models are used to identify factors influencing initiation and progression of the tumor and the effects of chemotherapeutics agents on the tumor therapy.[41] Also, by using obtained physiological and pathological findings in *in vitro* models, fertility and pregnancy can be increased.[42]

Elasticity plays a crucial role for the cardiac cell regeneration, in this purpose high elastic Methacrylated tropoelastic (MeTro) hydrogels have been generated using microfabrication techniques.[43] Also microfabrication techniques are used to produce high resolution micro channels on the surface of hydrogels, thus these hydrogels are used to enhance the functionality of structures, cellular tendency and elongation.[44] Methacrylated tropoelastin hydrogels were crosslinked under UV light in less than a minute. Tropoelastin is a derivative of elastin which has been demonstrated by crosslinking, sustains the properties of remarkable rigidity, elasticity, continuity and cell interaction. Densely multi-layered high elastic 3D cardiac tissue structures were unified by encapsulating cardiomyocytes into microfabricated MeTro gels. In this manner, benefits in simulation of elastic tissue models, have highlighted the use of meTro gels.

Mechanical properties of manufactured meTro hydrogels can be easily adjusted in different hardness for each particular usage. Consequently variations in elastic modulus and ultimate strength values, which are based on meTro concentration and methacrylation degree, were observed. Cardiomyocyte encapsulating meTro Gels increased cell adhesion, proliferation, and elongation that showed similar mechanical characteristics with natural myocardium because of molecular structure of tropoelastin. Furthermore, *in vitro* cardiomyocyte contraction movements were shown with an increase in contraction capability by using electrical impulses.[43]

To address cellular alignment and controllability in cellular manner on 3D structures using microfabrication techniques, Aubin *et al.* encapsulated different cell types such as fibroblast, myoblast, endothelial cells, and cardiac stem cells within GelMA Hydrogels. In this approach, in order to obtain aligned tissues, internal potential of cells were used to control alignment by retaining cells within micro fabricated gelatin hydrogels.[45]

By the use of adjustable mechanical properties of hydrogels, necessary for differentiation of cells, appropriate physiological environments can be prepared.[19] Most of the body tissues are made up from collagen. Because of its biocompatible, biodegradable and flexible properties, it can be tuned through the application. With this respect, cardiomyocytes isolated from undeveloped tissues, can differentiate in collagen into structures with similar morphological characteristics like cardiac muscle bundles.[46] Zimmermann *et al.*, produced large scale, round shaped collagen/matrigels that even kept its shape constant when subjected to mechanical stress. Gel indicated electrical integration and demonstrated heart contraction even after 28-day culture.[47]

Recently, mesenchymal stem cells are widely used in regenerative medicine and particularly for cartilage regeneration; they differentiate into chondrocytes.[48,49] In this approach, Silva *et al.*, fabricated silk hydrogels, which are elastic and have instantaneous degradation ability in protease XIV existence, can be directly applied for cartilage regeneration based on their cellular potential. This study demonstrated that the usage of ionic liquids increased effectiveness and dissolution ratio of hydrogels. Silk hydrogels provided almost natural environment to human adipose stem cells so that they kept cellular activity up to 28 days.[50]

The other application of hydrogels in regenerative medicine is controlled drug delivery systems.[51] Previous reports have shown that the use of growth factors and cytokines, helped reducing mechanical blockage of damaged heart as well as they supported growth and differentiation of cells by increasing bioactivity.[52] In the study of Takada *et al.*, negative groups on the surface of heparin and heparin sulfate formed a complex structure with the amino acids of BMP-2, generating an increase in the bioactivity of BMP-2 in bone generation.[53] On the other hand, Xu *et al.*, fabricated heparin attached hyaluronic acid hydrogels to show supplementary effect of heparin on bone regeneration and controlled release of BMP-2.[54] Recently, Cao *et al.*, produced rhBMP-2 loaded 2-N, 6-O-sulfated chitosan nanoparticles included photo cross linkable gelatin hydrogel for *in vitro* sustained release and bioactivity of rhBMP-2. Produced hydrogels have also been shown to be useful for controlled delivery carrier in *in vivo* bone regeneration.[55]

6. Cancer

Currently, *in vitro* 3D cancer models have an invaluable role in cancer biology, providing important benefits. Cell–cell and cell–matrix interactions can be understood by *in vitro* 3D tissue culture models in cancer biology.[56] Various types of these models reveal novel strategies to understand the formation of cancer, homeostasis, cellular differentiation, and hypoxia. Recent developments in 3D culture and tissue engineering techniques head towards complex heterogen 3D tumor models. Nowadays, numerous research groups focus on cell proliferation, migration, invasion, and differentiation of cancer cells in 3D culture.

Agata Nyga *et al.* improved a 3D *in vitro* biomimetic colorectal cancer model using colorectal cancer cells and connective tissue cells. Firstly, they manufactured an artificial cancer mass (ACM) by using collagen type-I hydrogel seeded with cancer cells in the hydrogel. Then partial compression was applied to increase density of hydrogel and cells. Thus, they showed that hydrogel density can be adjusted mimicking matrix density. Also fibroblast and endothelial cells were used for mimicking the natural tumor stroma. Hypoxia within the tumor was also demonstrated, and confirmation of VEGF expression by cancer cells was considered as a marker of the initiation of angiogenesis.[57]

Hypoxia results from an imbalance between oxygen supply and demand. Ashley A. Jaeger *et al.* developed a bioreactor system by using silicone hydrogel material that provides controllable O_2 delivery to cancer cells in 3D culture, mimicking *in vivo* O_2 diffusion from vasculature. In this study, silicone micropillars act as vessels that oxygenate spheroid-forming epithelial cancer cells.[58] Another novel method mimicking O_2 delivery to cancer cells in 3D culture uses paper-supported gels. In this study, a hydrogel containing suspended cells was added to a multi-layer paper stack. Layers of paper create a 3D structure that allow to delivery of O_2 and nutrients. Method was applied to several cell line with several dyes (Fig. 7).[59]

Recent findings have shown that cell-based immunotherapy can be very effective against metastasis. Chunming Wang *et al.* fabricated a dual-layer hydrogel/microsphere (GS) composite system by encapsulated murine macrophages.

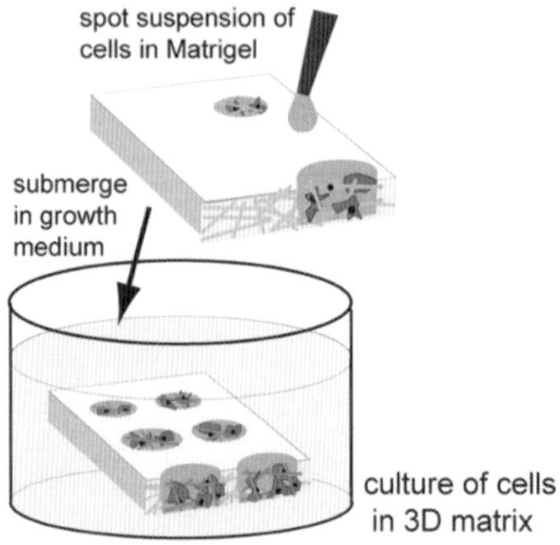

Fig. 7. Generation of 3D fiber-supported cell contained hydrogels with each layer composed of chromatography or filter paper. And incubation of cells in 3D matrix in a growth medium (Figure reprinted with permission from Ref. 59).

Composite material indicates some advantageous features like injectability and favorable permeability to achieve genuine localization and physical immobilization of the executing immunocytes–macrophages. According to the study, *in vitro* investigations on the composite system present effective elimination of carcinoma cells as well as high safety without gene alteration, cell leakage or unwanted proliferation.[60]

3D *in vitro* hydrogel models have been extensively used to probe cancer cell invasion and migration.[61–63] Liu *et al.* constructed a 3D structure which includes lowlands and highlands. They showed metastatic behaviors of LNCaP and PC-3 cancer cells by measuring vertical ascent of cancer cells from the lowlands to the highlands. They also used Fisher's equation to describe contact inhibition in cancer cells that affects both proliferation and migration.[61] Matrigel is known as a good model matrix by scientist. Hyaluronan is an essential and very important polysaccharide produced by many organism. Also it has been in clinical use for over 30 years with different applications like cell delivery, cell recovery, drug delivery, tissue repair. David *et al.* synthesized hyaluronan hydrogel and determined invasiveness of tumor cells in this hydrogel and compared that with matrigel. Hyaluronan showed better features of the natural tumor microenvironment because of the hyaluronan-binding sites.[62]

Cancer cells can spread to another tissue that is known as micrometastasis and causes to secondary tumors. The International Union Against Cancer (IUAC) defined tumor deposits between 0.2 and 2 mm as micrometastases. Single tumor cells or small clusters of cells smaller than 0.2 mm are defined as Isolated Tumor Cells (ITC). Understanding of metastasis is crucial to develop new therapeutic strategies. Transendothelial migration of highly metastatic breast cancer cells within a 3D *in vitro* bone-like microenvironment was studied by Bersini *et al.* They developed a microfluidic 3D *in vitro* model to analyze the extravasion of highly metastatic breast cancer cells into an osteo-cell conditioned microenvironment.[63] Also microfluidic systems can mimic the tumor microenvironment more realistically. Therefore, this type of systems provide easy and high resolution imaging of tumor cell migration and extravasation.[64] Alginate, a naturally occurring anionic polysaccharide gels have been used to generate *in vitro* 3D models. By using this hydrogel that have large scalability and reproducibility, not only metastatic ability of cancer cells but also screening drug compounds for hepatotoxicity and drug efficacy can be investigated.[12,65]

Natural matrix materials like collagen, gelatin, matrigel (Bioscience), chitosan, alginate etc. are used frequently as 3D cell culture material and shows similar properties to natural ECM. Besides chemical nature, mechanical properties of the hydrogel influence behaviors of cells. Metastatic spread is regulated by mechanical properties of the tissue microenvironment. By adjusting crosslinking ratio and/or changing concentration of hydrogel, matrix stiffness and mechanical force can be changed that affects cell growth, survival, and differentiation.[66,67]

7. Stem Cells

Last decades, stem cell based therapies have great interest as an alternative treatment for various diseases such as skin wound, infarcted heart, and brain. Also numerous studies have suggested that they have potential therapeutic value. The most important property of stem cells is the ability to differentiate into various cell types therefore stem cells can help to repair and regenerate damaged tissues.

The increasing evidence has supported the breakthrough role of human pluripotent stem cells (HPSCs) that also include embryonic and bone marrow derived stem cells, in various biomedical applications as cell replacement therapies, tissue and organ regenerations, wound healing and autoimmune diseases treatments.[68] Under *in vitro* conditions, prolonged self-renewing activity is the fundamental purpose of utilization of pluripotent stem cells clinically and scientifically. So far, 2D system in plastic culture dish with liquid medium has been used however it is clear that providing the factors for homeostasis of cells as *in vivo* condition is deprived. To optimize the culture system, administration of cellular component, non-cellular environment, ECM proteins and various substances to culture environment are the obvious requirements. A 3D system mainly composed of synthetic polymer and peptide motif that include all the factors for cell homeostasis can be prepared. This artificial system provides an appropriate area for expandable cells by decreasing mechanical stress and stimulate cell signal for cellular activity. To compare the 2D with 3D system, 3D system mimics the *in vivo* environment, and therefore gives an alternative way for self-renewal of stem cells.

The principal method for the delivering of stem cells and growth factors is the direct injection to the wound bed. But this method comes out with poor engraftment and poor efficacy because of proteolytic degradation a few hours after injection. That prevents the beneficial effects of the expensive stem cell therapy. To overcome these problems, using a scaffold derived from natural and synthetic origin for the delivery of stem cells is a promising way. For instance, 3D hydrogel scaffold have been used recently for their hydrophilic, biodegradable and injectable properties that make them convenient and qualified for drug and stem cell delivery. Hydrogels can encapsulate stem cells by this way so that they can be protected from outer periphery and also provide microenvironment to mimic the *in vivo* conditions such as stimulation of cellular proliferation, differentiation, and regeneration.

One of the key factors in stem cell niche study is the interactions between different cell types. Other cell types in microenvironment of the stem cells can affect the behavior of the stem cell by direct contact or via secreted proteins or signals coming from the cells which lead to differentiation, expansion and self-renewal. Cell–cell interactions have been studied by developing *in vitro* models including mixed cells cultured in 2D. In these models, cell distribution and density cannot be controlled. In addition to this, several studies have suggested that, chondrocytes tend to de-differentiate to fibroblasts.[69] Therefore, a new generation co-culture models which better mimic microenvironment are more preferred to cell–cell interactions studies. Biomimetic hydrogels are used to obtain a 3D microenvironment which is a direct differentiation in culture system, unlike conventional *in vitro* co-culture models. For instance, formation of cartilage cells is dependent on the distribution of the cells and intracellular distance and these parameters can be controlled by the hydrogels.[70]

Traumatic brain injury occurs when an external force causes the destruction of brain cells. Many brain injuries may lead to tissue loss of cerebral parenchyma which leads to cavities formation because of the primary destruction and secondary injuries such as ischemia and inflammation. Although there are numerous studies upon central nervous system regeneration, brain damage is almost unrecoverable because of the poor regenerative capability of neural tissue.[71]

In the present day, tissue engineering approach seems the best way for repairing damaged neural tissue. For the treatment of injured neural tissue in brain, a suitable 3D matrix or scaffold is necessary to use. However, some of the endogenous progenitor neural cells in the adult brain has limited ability to generate new functional neurons. Neural stem cells have the potential of unlimited self-renewal and the ability to differentiate into neurons, astrocytes, and oligodendrocytes which can be used as a source for neural regeneration. In addition, with the use of functionalized hydrogels neuronal differentiation of stem cells, neurite proliferation, and outgrowth can be further increased.[72]

Hematopoietic stem cells (HSCS) are a kind of clinically relevant stem cells that are administered to patients having hematological disorders such as leukemia.[73] HSCSs cannot be cultured *in vitro* for a period of time without differentiation, however they can be able to replenish the blood system and self-renew. Therefore, it is important to mimic native environment *in vitro.* To mimic the physical and biological parameters as 3D structure, the adhesive ECM, soluble factors and the stromal cell compartment of the HSC niche, an easy approach has been developed. The hydrogel scaffold mentioned here is a direct approach to create an analog that mimic the bone marrow. The scaffold mimics the 3D structure of related bone region, biofunctionalized with RGD presents adhesive area as ECM does in nature. MSCs are crucial because of their cytokine production properties therefore the cytokine-supplemented medium provides soluble factors. On the other hand, it is easy to produce a scaffold, thereof it presents a convenient and easy way to apply the application in every laboratory.[74]

Stem cells differentiation into specialized cell types is managed by the micro environment. The constructions that affect the micro environment are, ECM,[75] matrix elasticity,[76,77] substrate topography[77,78] and direct cell–cell contact.[79] To sum up, 3D nature of the ECM plays an important role in the regulation of cell behavior.

According to the study presented by Pirouz *et al.*, a cell-loaded 3D gel micro array platform has been used to observe the response of differentiation of human mesenchymal stem cells with different ECM and growth factor combinations. Gel microarray platform, is composed of automated printing strategy with miniaturized cell-laden hydrogel. The hydrogels composed of GE and hMSCs along with ECM proteins was exposed to osteogenic bone morphogenic proteins (BMPs). The microarray system allowed rapid, high throughput and cost efficient screening of stem cell-matrix interactions inside various 3D microenvironment. This microarray platform indicates the evaluations of the osteogenic potential of HMSCs in 96 distinctive 3D niches. By these results, cellular responses and appropriate matrix components of the stem cell niches can be predicted for various tissue engineering applications (Fig. 8).[80]

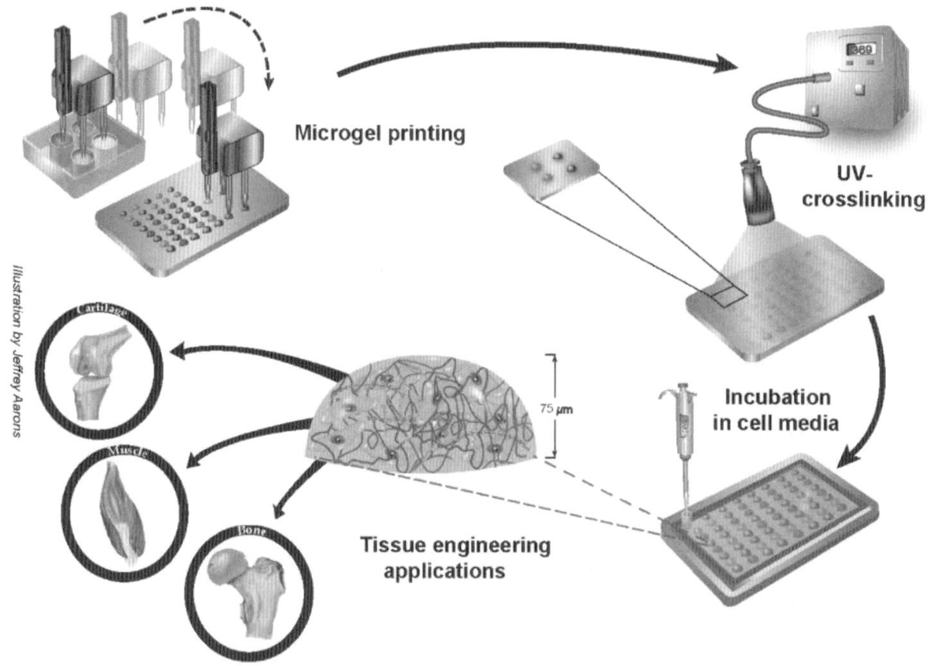

Fig. 8. Schematic demonstration of fabrication of 3D hMSC-laden microgels with a robotic microarray system. The solution consists of hMSCs, gelatin methacrylate (GE)-based prepolymer solution and various ECM proteins. In the next step, printed solution was exposed to UV light for 15 s. Formed cell-laden gel constructs were placed inside sealed chambers for incubation (Figure reprinted with permission from Ref. 80).

8. Conclusions

A new study area of hydrogels has shown a remarkable potential to develop 3D functional tissue models. These structures create 3D native environments for cells to proliferate and differentiate. Various models have been developed for different purposes like anticancer drug delivery in a cancer tissue or the control of stem cell fate. Developing 3D cancer tissue models aided analyzing the formation of cancer and toxicological effect of an anticancer drug on a cancer tissue much more accurately. Furthermore, stem cell differentiation can be controlled by using 3D hydrogel model. Promotion of cardiac, fibroblast, myoblast, endothelial cells etc. can be used to generate different types of tissues for regenerative medicine applications. In this chapter, the new point of view on the study of hydrogels as 3D functional tissue models is summarized. This new approach has improved the studies on drug discovery, therapeutic approaches and regenerative medicine as mentioned. However along with advanced science and technology, novel and innovative models for tissue engineering, regenerative medicine, stem cell applications and drug discovery are being developed every day. The challenge in the future could be summarized as "How to create a model that entirely mimic tissue?"

References

1. R. Dixit and U. A. Boelsterli, Healthy animals and animal models of human disease(s) in safety assessment of human pharmaceuticals, including therapeutic antibodies. *Drug Discov. Today*, **12**(7–8), 336–342 (2007).
2. S. C. Owen, S. A. Fisher, R. Y. Tam, C. M. Nimmo, and M. S. Shoichet, Hyaluronic acid click hydrogels emulate the extracellular matrix. *Langmuir*, **29**(24), 7393–7400 (2013).
3. D. R. Albrecht, G. H. Underhill, T. B. Wassermann, R. L. Sah, and S. N. Bhatia, Probing the role of multicellular organization in three-dimensional microenvironments. *Nat. Methods*, **3**(5), 369–375 (2006).
4. Y. Ling, J. Rubin, Y. Deng, C. Huang, U. Demirci, J. M. Karp, and A. Khademhosseini, A cell-laden microfluidic hydrogel. *Lab Chip*, **7**(6), 756–762 (2007).
5. C. P. Ng and S. H. Pun, A perfusable 3D cell-matrix tissue culture chamber for *in situ* evaluation of nanoparticle vehicle penetration and transport. *Biotechnol. Bioeng.*, **99**(6), 1490–1501 (2008).
6. Y. C. Toh, C. Zhang, J. Zhang, Y. M. Khong, S. Chang, V. D. Samper, D. van Noort, D. W. Hutmacher, and H. R. Yu, A novel 3D mammalian cell perfusion-culture system in microfluidic channels. *Lab Chip*, **7**(3), 302–309 (2007).
7. S. C. Wen and M. S. Shoichet, Design of three-dimensional biomimetic scaffolds. *J. Biomed. Mater. Res. A*, **94A**(4), 1321–1331 (2010).
8. C. S. Shin, B. Kwak, B. Han, and K. Park, Development of an *in vitro* 3D tumor model to study therapeutic efficiency of an anticancer drug. *Mol. Pharm.*, **10**(6), 2167–2175 (2013).
9. D. Yip and C. H. Cho, A multicellular 3D heterospheroid model of liver tumor and stromal cells in collagen gel for anti-cancer drug testing. *Biochem. Bioph. Res. Co.*, **433**(3), 327–332 (2013).
10. Y. Nakao, H. Kimura, Y. Sakai, and T. Fujii, Bile canaliculi formation by aligning rat primary hepatocytes in a microfluidic device. *Biomicrofluidics*, **5**, 2 (2011).
11. T. M. DesRochers, L. Suter, A. Roth, and D. L. Kaplan, Bioengineered 3D human kidney tissue, a platform for the determination of nephrotoxicity. *Plos One*, **8**, 3 (2013).
12. S. F. Lan and B. Starly, Alginate based 3D hydrogels as an *in vitro* co-culture model platform for the toxicity screening of new chemical entities. *Toxicol. Appl. Pharm.*, **256**(1), 62–72 (2011).
13. J. H. Sung and M. L. Shuler, A micro cell culture analog (mu CCA) with 3-D hydrogel culture of multiple cell lines to assess metabolism-dependent cytotoxicity of anti-cancer drugs. *Lab Chip*, **9**(10), 1385–1394 (2009).
14. M. C. W. Chen, M. Gupta, and K. C. Cheung, Alginate-based microfluidic system for tumor spheroid formation and anticancer agent screening. *Biomed. Microdevices*, **12**(4), 647–654 (2010).

15. V. L. Roger, A. S. Go, D. M. Lloyd-Jones, R. J. Adams, J. D. Berry, T. M. Brown, M. R. Carnethon, S. Dai, G. de Simone, E. S. Ford, C. S. Fox, H. J. Fullerton, C. Gillespie, K. J. Greenlund, S. M. Hailpern, J. A. Heit, P. M. Ho, V. J. Howard, B. M. Kissela, S. J. Kittner, D. T. Lackland, J. H. Lichtman, L. D. Lisabeth, D. M. Makuc, G. M. Marcus, A. Marelli, D. B. Matchar, M. M. McDermott, J. B. Meigs, C. S. Moy, D. Mozaffarian, M. E. Mussolino, G. Nichol, N. P. Paynter, W. D. Rosamond, P. D. Sorlie, R. S. Stafford, T. N. Turan, M. B. Turner, N. D. Wong, and J. Wylie-Rosett, American Heart Association Statistics, C., Stroke Statistics, S., Heart disease and stroke statistics — 2011 update: A report from the American Heart Association. *Circulation*, **123**(4), e18–e209 (2011).
16. E. Braunwald, Congestive heart failure: A half century perspective. *Eur. Heart J.*, **22**(10), 825–836 (2001).
17. A. A. Rane and K. L. Christman, Biomaterials for the treatment of myocardial infarction: A 5-year update. *J. Am. Coll. Cardiol.*, **58**(25), 2615–2629 (2011).
18. C. E. Murry, L. J. Field, and P. Menasche, Cell-based cardiac repair: Reflections at the 10-year point. *Circulation*, **112**(20), 3174–3183 (2005).
19. T. Dvir, B. P. Timko, D. S. Kohane, and R. Langer, Nanotechnological strategies for engineering complex tissues. *Nat. Nanotechnol.*, **6**(1), 13–22 (2011).
20. A. P. Nowak, V. Breedveld, L. Pakstis, B. Ozbas, D. J. Pine, D. Pochan, and T. J. Deming, Rapidly recovering hydrogel scaffolds from self-assembling diblock copolypeptide amphiphiles. *Nature*, **417**(6887), 424–428 (2002).
21. D. Fan, Z. Yin, R. Cheong, F. Q. Zhu, R. C. Cammarata, C. L. Chien, and A. Levchenko, Subcellular-resolution delivery of a cytokine through precisely manipulated nanowires. *Nat. Nanotechnol.*, **5**(7), 545–551 (2010).
22. K. Shapira-Schweitzer and D. Seliktar, Matrix stiffness affects spontaneous contraction of cardiomyocytes cultured within a PEGylated fibrinogen biomaterial. *Acta Biomater.*, **3**(1), 33–41 (2007).
23. L. E. Freed, F. Guilak, X. E. Guo, M. L. Gray, R. Tranquillo, J. W. Holmes, M. Radisic, M. V. Sefton, D. Kaplan, and G. Vunjak-Novakovic, Advanced tools for tissue engineering: Scaffolds, bioreactors, and signaling. *Tissue Eng.*, **12**(12), 3285–3305 (2006).
24. O. Jeon, C. Powell, S. M. Ahmed, and E. Alsberg, Biodegradable, photocrosslinked alginate hydrogels with independently tailorable physical properties and cell adhesivity. *Tissue Eng. A*, **16**(9), 2915–2925 (2010).
25. F. Rask, A. Mihic, L. Reis, S. M. Dallabrida, N. S. Ismail, K. Sider, C. A. Simmons, M. A. Rupnick, R. D. Weisel, R. K. Li, and M. Radisic, Hydrogels modified with QHREDGS peptide support cardiomyocyte survival *in vitro* and after sub-cutaneous implantation. *Soft Matter*, **6**(20), 5089–5099 (2010).
26. L. A. Reis, L. L. Chiu, Y. Liang, K. Hyunh, A. Momen, and M. Radisic, A peptide-modified chitosan-collagen hydrogel for cardiac cell culture and delivery. *Acta Biomater.*, **8**(3), 1022–1036 (2012).
27. C. A. Deforest, E. A. Sims, and K. S. Anseth, Peptide-functionalized click hydrogels with independently tunable mechanics and chemical functionality for 3D cell culture. *Chem. Mater.*, **22**(16), 4783–4790 (2010).

28. J. Wu, F. Zeng, X. P. Huang, J. C. Chung, F. Konecny, R. D. Weisel, and R. K. Li, Infarct stabilization and cardiac repair with a VEGF-conjugated, injectable hydrogel. *Biomaterials*, **32**(2), 579–586 (2011).
29. F. Xu, C. A. Wu, V. Rengarajan, T. D. Finley, H. O. Keles, Y. Sung, B. Li, U. A. Gurkan, and U. Demirci, Three-dimensional magnetic assembly of microscale hydrogels. *Adv. Mater.*, **23**(37), 4254–4260 (2011).
30. H. Akiyama, A. Ito, M. Sato, Y. Kawabe, and M. Kamihira, Construction of cardiac tissue rings using a magnetic tissue fabrication technique. *Int. J. Mol. Sci.*, **11**(8), 2910–2920 (2010).
31. P. W. Barone, H. Yoon, R. Ortiz-Garcia, J. Zhang, J. H. Ahn, J. H. Kim, and M. S. Strano, Modulation of single-walled carbon nanotube photoluminescence by hydrogel swelling. *ACS Nano*, **3**(12), 3869–3877 (2009).
32. S. R. Shin, S. M. Jung, M. Zalabany, K. Kim, P. Zorlutuna, S. B. Kim, M. Nikkhah, M. Khabiry, M. Azize, J. Kong, K. T. Wan, T. Palacios, M. R. Dokmeci, H. Bae, X. S. Tang, and A. Khademhosseini, Carbon-nanotube-embedded hydrogel sheets for engineering cardiac constructs and bioactuators. *ACS Nano*, **7**(3), 2369–2380 (2013).
33. B. Subramanian, D. Rudym, C. Cannizzaro, R. Perrone, J. Zhou, and D. L. Kaplan, Tissue-engineered three-dimensional *in vitro* models for normal and diseased kidney. *Tissue Eng. A*, **16**(9), 2821–2831 (2010).
34. F. Berthiaume, T. J. Maguire, and M. L. Yarmush, Tissue engineering and regenerative medicine: History, progress, and challenges. *Ann. Rev. Chem. Biomol. Eng.*, **2**(2), 403–430 (2011).
35. R. Langer, Selected advances in drug delivery and tissue engineering. *J. Control. Release*, **62**(1–2), 7–11 (1999).
36. H. Kim, Y. Phung, and M. Ho, Changes in global gene expression associated with 3D structure of tumors: An *ex vivo* matrix-free mesothelioma spheroid model. *Plos One*, **7**, 6 (2012).
37. N. Bursac, M. Papadaki, R. J. Cohen, F. J. Schoen, S. R. Eisenberg, R. Carrier, G. Vunjak-Novakovic, and L. E. Freed, Cardiac muscle tissue engineering: Toward an *in vitro* model for electrophysiological studies. *Am. J. Physiol.-Heart C.*, **277**(2), H433–H444 (1999).
38. N. A. Peppas, J. Z. Hilt, A. Khademhosseini, and R. Langer, Hydrogels in biology and medicine: From molecular principles to bionanotechnology. *Adv. Mater.*, **18**(11), 1345–1360 (2006).
39. Z. Y. Ye, Y. Zhou, H. B. Cai, and W. S. Tan, Myocardial regeneration: Roles of stem cells and hydrogels. *Adv. Drug Deliv. Rev.*, **63**(8), 688–697 (2011).
40. H. Naderi, M. M. Matin, and A. R. Bahrami, Review paper: Critical issues in tissue engineering: Biomaterials, cell sources, angiogenesis, and drug delivery systems. *J. Biomater. Appl.*, **26**(4), 383–417 (2011).
41. S. Talukdar, M. Mandal, D. W. Hutmacher, P. J. Russell, C. Soekmadji, and S. C. Kundu, Engineered silk fibroin protein 3D matrices for *in vitro* tumor model. *Biomaterials*, **32**(8), 2149–2159 (2011).

42. M. Xu, P. K. Kreeger, L. D. Shea, and T. K. Woodruff, Tissue-engineered follicles produce live, fertile offspring. *Tissue Eng.*, **12**(10), 2739–2746 (2006).
43. N. Annabi, K. Tsang, S. M. Mithieux, M. Nikkhah, A. Ameri, A. Khademhosseini, and A. S. Weiss, Highly elastic micropatterned hydrogel for engineering functional cardiac tissue. *Adv. Funct. Mater.*, **23**(39), 4950–4959 (2013).
44. C. A. Bashur, L. Venkataraman, and A. Ramamurthi, Tissue engineering and regenerative strategies to replicate biocomplexity of vascular elastic matrix assembly. *Tissue Eng. B*, **18**(3), 203–217 (2012).
45. H. Aubin, J. W. Nichol, C. B. Hutson, H. Bae, A. L. Sieminski, D. M. Cropek, P. Akhyari, and A. Khademhosseini, Directed 3D cell alignment and elongation in microengineered hydrogels. *Biomaterials*, **31**(27), 6941–6951 (2010).
46. M. Cortes-Morichetti, G. Frati, O. Schussler, J. P. D. Van Huyen, E. Lauret, J. A. Genovese, A. F. Carpentier, and J. C. Chachques, Association between a cell-seeded collagen matrix and cellular cardiomyoplasty for myocardial support and regeneration. *Tissue Eng.*, **13**(11), 2681–2687 (2007).
47. W. H. Zimmermann, I. Melnychenko, G. Wasmeier, M. Didie, H. Naito, U. Nixdorff, A. Hess, L. Budinsky, K. Brune, B. Michaelis, S. Dhein, A. Schwoerer, H. Ehmke, and T. Eschenhagen, Engineered heart tissue grafts improve systolic and diastolic function in infarcted rat hearts. *Nat. Med.*, **12**(4), 452–458 (2006).
48. A. I. Caplan, Adult mesenchymal stem cells for tissue engineering versus regenerative medicine. *J. Cell. Physiol.*, **213**(2), 341–347 (2007).
49. C. Csaki, P. R. A. Schneider, and M. Shakibaei, Mesenchymal stem cells as a potential pool for cartilage tissue engineering. *Ann. Anat.*, **190**(5), 395–412 (2008).
50. S. S. Silva, E. G. Popa, M. E. Gomes, M. B. Oliveira, S. Nayak, B. Subia, J. F. Mano, S. C. Kundu, and R. L. Reis, Silk hydrogels from non-mulberry and mulberry silkworm cocoons processed with ionic liquids. *Acta Biomater.*, **9**(11), 8972–8982 (2013).
51. G. W. Ashley, J. Henise, R. Reid, and D. V. Santi, Hydrogel drug delivery system with predictable and tunable drug release and degradation rates. *Proc. Natl. Acad. Sci. USA*, **110**(6), 2318–2323 (2013).
52. T. Simon-Yarza, F. R. Formiga, E. Tamayo, B. Pelacho, F. Prosper, and M. J. Blanco-Prieto, Vascular endothelial growth factor-delivery systems for cardiac repair: An overview. *Theranostics*, **2**(6), 541–552 (2012).
53. T. Takada, T. Katagiri, M. Ifuku, N. Morimura, M. Kobayashi, K. Hasegawa, A. Ogamo, and R. Kamijo, Sulfated polysaccharides enhance the biological activities of bone morphogenetic proteins. *J. Biol. Chem.*, **278**(44), 43229–43235 (2003).
54. X. Xu, A. K. Jha, R. L. Duncan, and X. Q. Jia, Heparin-decorated, hyaluronic acid-based hydrogel particles for the controlled release of bone morphogenetic protein 2. *Acta Biomater.*, **7**(8), 3050–3059 (2011).
55. L. Y. Cao, J. A. Werkmeister, J. Wang, V. Glattauer, K. M. McLean, and C. S. Liu, Bone regeneration using photocrosslinked hydrogel incorporating rhBMP-2 loaded 2-N, 6-O-sulfated chitosan nanoparticles. *Biomaterials*, **35**(9), 2730–2742 (2014).
56. J. B. Kim, Three-dimensional tissue culture models in cancer biology. *Semin. Cancer Biol.*, **15**(5), 365–377 (2005).

57. A. Nyga, M. Loizidou, M. Emberton, and U. Cheema, A novel tissue engineered three-dimensional in vitro colorectal cancer model. *Acta Biomater.*, **9**(8), 7917–7926 (2013).
58. A. A. Jaeger, C. K. Das, N. Y. Morgan, R. H. Pursley, P. G. McQueen, M. D. Hall, T. J. Pohida, and M. M. Gottesman, Microfabricated polymeric vessel mimetics for 3-D cancer cell culture. *Biomaterials*, **34**(33), 8301–8313 (2013).
59. R. Derda, A. Laromaine, A. Mammoto, S. K. Y. Tang, T. Mammoto, D. E. Ingber, and G. M. Whitesides, Paper-supported 3D cell culture for tissue-based bioassays. *Proc. Natl. Acad. Sci. U.S.A.*, **106**(44), 18457–18462 (2009).
60. C. M. Wang, G. N. Adrianus, N. Sheng, S. Toh, Y. H. Gong, and D. A. Wang, *In vitro* performance of an injectable hydrogel/microsphere based immunocyte delivery system for localised anti-tumour activity. *Biomaterials*, **30**(36), 6986–6995 (2009).
61. L. Y. Liu, B. Sun, J. N. Pedersen, K. M. A. Yong, R. H. Getzenberg, H. A. Stone, and R. H. Austin, Probing the invasiveness of prostate cancer cells in a 3D microfabricated landscape. *Proc. Natl. Acad. Sci. USA*, **108**(17), 6853–6856 (2011).
62. L. David, V. Dulong, D. Le Cerf, C. Chauzy, V. Norris, B. Delpech, M. Lamacz, and J. P. Vannier, Reticulated hyaluronan hydrogels: A model for examining cancer cell invasion in 3D. *Matrix Biol.*, **23**(3), 183–193 (2004).
63. S. Bersini, J. S. Jeon, G. Dubini, C. Arrigoni, S. Chung, J. L. Charest, M. Moretti, and R. D. Kamm, A microfluidic 3D *in vitro* model for specificity of breast cancer metastasis to bone. *Biomaterials*, **35**(8), 2454–2461 (2014).
64. J. S. Jeon, I. K. Zervantonakis, S. Chung, R. D. Kamm, and J. L. Charest, *In vitro* model of tumor cell extravasation. *Plos One*, **8**, 2 (2013).
65. X. X. Xu, C. Liu, Y. Liu, N. Li, X. Guo, S. J. Wang, G. W. Sun, W. Wang, and X. J. Ma, Encapsulated human hepatocellular carcinoma cells by alginate gel beads as an *in vitro* metastasis model. *Exp. Cell Res.*, **319**(14), 2135–2144 (2013).
66. T. Z. Zhang, J. S. Chen, Q. Y. Zhang, J. Dou, and N. Gu, Poly(ethylene glycol)-cross linked poly(methyl vinyl ether-co-maleic acid)hydrogels for three-dimensional human ovarian cancer cell culture. *Colloid. Surface A*, **422**, 81–89 (2013).
67. Y. Liang, J. Jeong, R. J. DeVolder, C. Cha, F. Wang, Y. W. Tong, and H. Kong, A cell-instructive hydrogel to regulate malignancy of 3D tumor spheroids with matrix rigidity. *Biomaterials*, **32**(35), 9308–9315 (2011).
68. V. Tabar and L. Studer, Pluripotent stem cells in regenerative medicine: Challenges and recent progress. *Nat. Rev. Genet.*, **15**(2), 82–92 (2014).
69. E. M. Darling and K. A. Athanasiou, Rapid phenotypic changes in passaged articular chondrocyte subpopulations. *J. Orthopaed. Res.*, **23**(2), 425–432 (2005).
70. J. H. Lai, G. Kajiyama, R. L. Smith, W. Maloney, and F. Yang, Stem cells catalyze cartilage formation by neonatal articular chondrocytes in 3D biomimetic hydrogels. *Scientific Reports*, **3**, 3553 (2013).
71. R. J. Nudo, Recovery after brain injury: Mechanisms and principles. *Front. Hum. Neurosci.*, **7**, 887 (2013).
72. T. Y. Cheng, M. H. Chen, W. H. Chang, M. Y. Huang, and T. W. Wang, Neural stem cells encapsulated in a functionalized self-assembling peptide hydrogel for brain tissue engineering. *Biomaterials*, **34**(8), 2005–2016 (2013).

73. P. Kaimakis, M. Crisan, and E. Dzierzak, The biochemistry of hematopoietic stem cell development. *BBA — Gen. Subjects*, **1830**(2), 2395–2403 (2013).
74. A. Raic, L. Rodling, H. Kalbacher, and C. Lee-Thedieck, Biomimetic macroporous PEG hydrogels as 3D scaffolds for the multiplication of human hematopoietic stem and progenitor cells. *Biomaterials*, **35**(3), 929–940 (2014).
75. M. Schuldiner, O. Yanuka, J. Itskovitz-Eldor, D. A. Melton, and N. Benvenisty, Effects of eight growth factors on the differentiation of cells derived from human embryonic stem cells. *Proc. Natl. Acad. Sci. USA*, **97**(21), 11307–11312 (2000).
76. D. E. Discher, Matrix elasticity directs stem cell lineage — Soluble factors that limit osteogenesis. *Bone*, **44**(2), S205–S206 (2009).
77. G. M. Harris, M. E. Piroli, and E. Jabbarzadeh, Deconstructing the effects of matrix elasticity and geometry in mesenchymal stem cell lineage commitment. *Adv. Funct. Mater.*, **24**(16), 2396–2403 (2014).
78. L. D. Markert, J. Lovmand, M. Foss, R. H. Lauridsen, M. Lovmand, E. M. Fuchtbauer, A. Fuchtbauer, K. Wertz, F. Besenbacher, F. S. Pedersen, and M. Duch, Identification of distinct topographical surface microstructures favoring either undifferentiated expansion or differentiation of murine embryonic stem cells. *Stem Cells Dev.*, **18**(9), 1331–1342 (2009).
79. J. Tang, R. Peng, and J. D. Ding, The regulation of stem cell differentiation by cell–cell contact on micropatterned material surfaces. *Biomaterials*, **31**(9), 2470–2476 (2010).
80. A. Dolatshahi-Pirouz, M. Nikkhah, A. K. Gaharwar, B. Hashmi, E. Guermani, H. Aliabadi, G. Camci-Unal, T. Ferrante, M. Foss, D. E. Ingber, and A. Khademhosseini, A combinatorial cell-laden gel microarray for inducing osteogenic differentiation of human mesenchymal stem cells. *Scientific Reports*, **4**, 3896 (2014).

Chapter 4

Engineering Regenerative Dextran Hydrogels for Acute Skin Wound Healing

Jie Cheng*, Ying Jin*, David M. Owens†,‡, Jeremy J. Mao* and Guoming Sun*,§

**Center for Craniofacial Regeneration, Columbia University Medical Center, 630 West 168th Street, NY 10032, USA*

†Department of Dermatology, Columbia University, College of Physicians and Surgeons, NY 10032, USA

‡Department of Pathology and Cell Biology, Columbia University, College of Physicians and Surgeons, NY 10032, USA

§gsun@sunogel.com

As the largest organ in the body, the skin and its appendages provide a protective barrier against physical, chemical and biological pathogens and also prevent dehydration to support and maintain human health. However, deep skin injuries are hard to heal. Wound healing is one of the most intricate biological processes, which requires the coordinated efforts of various cell lineages, matrix and signaling molecules to work delicately at their hierarchical levels so as to achieve perfect regeneration. Full-thickness skin wounding commonly leads to non-functioning scar formation with substantial loss of skin appendages. Cutaneous scar formation can cause severe cosmetic, functional, and psychological problems. Despite tremendous efforts to attenuate scarring, the regeneration of perfect skin remains a challenge and highly prized goal, and has thus become a major aim in wound healing. This chapter introduces the mechanism and factors that influence the wound healing outcome, especially scarless wound healing. We will discuss recent approaches to treat acute wound healing, especially hydrogel therapy.

1. Introduction

Skin is the largest organ in the body and conducts a lot of essential functions including physical protection, systemic and innate immunity, sensation and temperature regulation. Multiple cells and matrix form distinct and interdependent compartments that regulate skin behavior throughout life. However, the loss of skin integrity upon diverse injuries commonly leads to non-functional fibrotic scarring with substantial loss of skin appendages.[1] Cutaneous scar formation frequently brings about cosmetic, functional and psychological problems, and subsequent treatments cause even more economic burden.[2] Furthermore, chronic wounds that are compromised by such diseases as ischemia, diabetes mellitus, and venous stasis are a major clinical challenge largely due to the lack of efficient treatments.[3] Therefore, it is of paramount importance to uncover the molecular mechanisms underlying skin wound healing and develop the novel treatment therapy to prevent abnormal healing. In recent years, more attention and interests have been paid to translational research that bridges the gaps between basic research and wound healing treatments. The development of biologically active scaffolds has been increasingly recognized as a novel and effective therapeutic strategy to enhance wound healing.[4,5]

Wound healing is one of the most intricate biological processes that require the coordinated efforts of various cell lineages to work delicately at their hierarchical levels. Past decades have witnessed great advances in unveiling the mechanism of normal and abnormal skin wound healing and provided valuable insight into intrinsic restorative pathways in this complex organ.[5] The normal mammalian responses to skin injuries occur in the three sequential overlapping but distinct stages: inflammation, proliferation (new tissue formation) and tissue remodeling. Once injuries occur, many cell types including immune cells, endothelial cells, keratinocytes and fibroblasts become activated and synchronized to restore the tissue integrity and hemostasis.[1,6] Meanwhile, various intracellular and intercellular pathways are also activated and coordinated to orchestrate cell proliferation, differentiation and migration.[7]

During the cutaneous wound healing process, hemostasis serves as the initiating step. Immediately after the clot is formed, multiple inflammatory cell subsets influx into the injury sites to clear invading bacteria and cellular debris. Neutrophils infiltrate the wounding site first and soon are replaced by macrophages which are further activated to mediate the transition into the proliferative phase. Macrophages migrate into wound at 48 to 96 hours after injury and become predominant before fibroblasts are activated. Accumulating evidence has indicated that macrophages play key roles during wound healing by performing the debridement, sterilizing the wound and ending the initial inflammatory response, and they also secrete many cytokines and growth factors.[8] Macrophages produce both inflammatory and angiogenic growth factors, such as TNF-α, VEGF, PDGF, bFGF, and TGF-β, which mediate the later healing process such as angiogenesis, re-epithelization and matrix biosynthesis.[9] Macrophages also work closely with fibroblasts during matrix remodeling. However, the detailed and precise role of macrophages during wound healing remains incompletely understood and further study should be carried out to unveil their mechanisms both *in vitro* and *in vivo.*

In addition to macrophages, various stem cells (e.g., stem cells from dermal papilla) are also involved in skin wound healing. Multiple mesenchymal stem cell (MSC) populations have been confirmed to contribute to skin homeostasis and repair.[10,11] Epithelial stem cells that originate from the interfollicular epidermis, sebaceous glands, and bulge region of hair follicles, and dermal stem cells that are derived from either dermal papilla or perivascular regions also contribute cellular input into healing wounds.[12–14] In addition, although their identities and roles remain controversial, multi-lineage stem cells from subcutaneous adipose tissue and body circulation have also been found to involve in skin wound healing.[15] These stem cells are involved in all three stages of wound healing at different degrees.[16] Upon injuries, these stem cells are activated by chemotactic signals, and differentiated into hierarchic cell types that are essential for wound healing and thus restoring the tissue integrity.[17] On the other hand, they are capable of attenuating inflammatory responses to facilitate wound healing beyond the inflammatory phase. MSCs promote wound healing primarily through their strong paracrine functions.[18] MSCs secrete many growth factors vital for tissue repair. These growth factors bind to corresponding receptors in cell surfaces of epithelial, endothelial, keratinocytes and fibroblasts, and then induced a series of cellular responses including cell proliferation, migration and functional gene expression. MSC-conditioned medium significantly enhances the migration of macrophages, keratinocytes and endothelial cells and proliferation of keratinocytes and endothelial cells.[19] Interestingly, MSCs release VEGF and HGF and maintain the delicate balance between pro-scarring TGF-β1 and anti-scarring TGF-β3 so as to counter scar formation.[20] The administration of MSCs to either acute or diabetic wounds improved wound closure by accelerating re-epithelialization, increasing granulation tissue formation and enhancing angiogenesis.[15,21] Moreover, the transplantation of MSCs to the burn wounds attenuated inflammatory cell infiltration and improved the formation of new vessels and granulation tissue.[22] Collectively, these findings have demonstrated that MSCs have beneficial impacts during all phases of wound repair.

Genome-wide expression profiling has identified many genes and signaling pathways that are critical for regulating wound healing process and outcomes.[23,24] The most prominent mediators for cutaneous wound repair are the TGF-β and Wnt pathways. Once injured, many TGF-β ligands are secreted from cell lineages such as macrophages and fibroblasts. They then bind to TGF-β receptors, resulting in signaling activation and subsequent gene expression.[25] A recent study has revealed that TGF-β1 is a scar-promoting factor, while TGF-β3 is anti-scarring during wound repair.[26] Treatment of wounds with topical TGF-β1 or its activation improved fibroblast proliferation, extracellular matrix (ECM) deposition, angiogenesis, phenotypic transition of fibroblasts to myofibroblast and wound contraction, thus promoting skin healing.[27–30] Genetically modified animal models deficient in TGF-β signaling (Smad 3 knockout, Tgfb1−/− Scid−/− and TGF-βRII deletion mice) show diverse defects in skin healing evidenced by fewer fibroblast, reduced matrix production, decreased granulation and delayed healing.[31,32] These data clearly demonstrated that TGF-β signaling is indispensable and crucial during cutaneous repair. However, excessive TGF-β1 activity leads to excessive and disorganized collagen synthesis and scar formation.[33] In fact, scarless embryonic wounds have less TGF-β1 but much more TGF-β3 levels than the scar-forming adult wounds. Local application of TGF-β3 has been shown to reduce scarring through effects on matrix production, cell movement and inflammatory mediators in multiple rodent wound model.[34,35] Thus, the family of TGF-β signaling like TGF-β1 and TGF-β3 may be exploited therapeutically to enhance wound repair and reduce scar formation.

In addition to TGF-β signaling, the Wnt pathway is intimately implicated in skin wound healing. Various components of canonical Wnt/β-catenin signaling are expressed in diverse compartments during wound healing. Cutaneous wound healing studies have revealed that β-catenin is a key regulator for re-epitheliazation and fibroblast behaviors.[36,37] Its levels and transcriptional activity are elevated in dermal fibroblasts at the proliferative phase, but they subsequently drop to baseline.[38,39] Meanwhile, the level and activity of β-catenin contributes to different skin wound phenotype. It has been found that the high β-catenin leads to an enlarged, hypercellular dermal compartment while low level results in a smaller and less cellular one,[36] which can lead to excessive fibrosis and scar formation.[40] In addition, the cell fate of MSCs is tightly regulated by the Wnt/β-catenin pathway, and their biological behaviors are modulated in a similar way during healing. Ito and colleagues have reported that inhibition of Wnt signaling after re-epithelialization completely abrogates wound-induced neofolliculogenesis, whereas overexpression of its ligands in the epidermis remarkably increases the number of regenerated hair follicles.[41] This finding has led to therapeutic manipulation of Wnt signaling, and gives rise to a more complete skin wound healing.[42]

2. Scar Formation During Wound Healing

Although fetal skin can completely regenerate skin devoid of scarring, adults lose this capacity and frequently heal with non-functional scarring after deep wound injuries.[33] Currently, research efforts have focused on unveiling the mechanisms underlying the scarless wound healing. Significant differences in skin structure, inflammatory response, cellular reactions, stem cell function and gene profiles of fetal scarless healing and postnatal repair have been identified and shed mechanistic insights into the scarless skin repair and regeneration.[33,43,44] These findings will undoubtedly help the development of scarless wound treatment. Perfect skin regeneration has recently become a major aim in wound healing.[45,46]

In the mammalian fetus, scarless wound healing is age-dependent. Scar formation begins approximately 24 weeks of gestation in humans and embryonic day 18.5 in mice. Although the scarless repair capacity was initially attributed to the specific intrauterine environment, recent studies have demonstrated that skin regenerative capacity is intrinsic to fetal skin itself.[44,47,48] The delicate structure and composition of ECM plays an important role in cell adhesion, differentiation, and proliferation during wound healing.[49] The fetal ECM is optimized to facilitate cellular migration and proliferation, providing an environment that is conducive toward scarless wound repair.[50] In addition, there are phenotypic differences between the collagen content and crosslinking patterns in fetal scarless and postnatal scarring wounds. Collagen type III is rapidly deposited in a fine reticular network in fetal wounds, while the ratio of type I to type III collagen in postnatal wounds increases with type I collagen predominance.[51,52] Recent study showed that, the skin regenerating ECM was dominated by collagen type III, while the scar formation ECM was rich in collagen type I, which antagonizes skin appendage regeneration.[53]

Although inflammation and angiogenesis are indispensable for a normal wound healing, robust inflammatory responses and angiogenesis also contribute to a scarring phenotype.[54] Inflammation has significant impact on the outcome of wound healing.[55] Fewer neutrophils and macrophages are observed in fetal compared to adult skin wounds,[56] and fetal wounds heal rapidly without prominent scar formation. However, adult skin injury triggers inflammatory cascade and stimulates cytokine release and chemotaxis of neutrophils and macrophages, which gives rise to significant scar formation. Moreover, the inflammatory cytokines that are important for chemotaxis and inflammatory activation, including interleukin-6 and -8 are significantly lower and last for a shorter time at the fetal wounds when compared with adult wounds.[57,58] This is generally correlated to the fact that adults have a more mature immune system than the developing fetus. Meanwhile, the introduction of inflammation and the addition of IL-6 into normal scarless wounds significantly increased scarring collagen deposition,[31] while wounds treated with IL-10 expression vector reduced inflammation and yielded scarless healing.[59,60] Additionally, mast cells migrate to the wound and release pro-inflammatory growth factors that also enhance fibrosis synthesis and scar formation. Mast cells in scarless wounds do not degranulate and are fewer and less mature at embryonic day 15 (E15), but granulate and lead to fibrotic scar formation at E18. Further study revealed that the injection of mast cell lysates into E15 wounds disrupted scarless healing. However, mast cell-deficient Kit(W/W-v) mice healed with significantly smaller scars at E18, suggesting that mast cells enhance scar formation and mediate the transition from scarless to fibrotic healing.[61] Collectively, therapeutic reduction of inflammation has beneficial effects to reduce scar formation and promote scarless regeneration upon injuries.

Angiogenesis is a complex factor that impacts wound healing outcome. Although angiogenesis is critical for deep wound healing, robust angiogenesis compromises cutaneous wound healing and leads to scar formation.[62] Recent studies have uncovered that scarless fetal wounds had lower levels of vascular endothelial growth factor (VEGF) and were less vascularized than fibrotic fetal wounds. VEGF expression quickly elevates during scarless compared with scarring repair and fluctuates with angiogenesis or vascular regression, which likely contributes to the more rapid scarless fetal repair rate.[63] Adding exogenous VEGF converts the scarless phenotype into a scar-forming phenotype, whereas neutralization of VEGF reduced vascularity and decreased scar formation in adult wounds.[64] These findings suggested that angiogenic factors like VEGF have a strong influence on scar tissue formation and might become potential targets for skin regenerative repair.

Notably, wound healing outcome differs from one tissue to another.[65] Oral mucosa is generally known to heal without scar formation, and is characterized by faster resolution of inflammation and control of myofibroblast action compared to skin wounds in a pig model.[66,67] Interestingly, there is no significant difference in molecular composition and clinical and histological scar scales between pig and human oral mucosal wounds,[67] which makes pig a good model for scarless wound healing. However, the number of macrophages, mast cells and other immunopositive cells was significantly lower in the oral mucosal wounds than that in the skin wounds, which is consistent with the aforementioned discussion about the inflammation effect on wound healing. Scarred skin encompassed a significantly increased number of type I procollagen immunopositive cells and increased fibronectin content, while the oral mucosal wounds had a prolonged accumulation of tenascin-C. Although the number of myofibroblasts was significantly higher in oral mucosal wounds, the oral mucosal healing showed significantly less contraction as compared with the skin wounds. This scarless wound healing model in oral mucosa provides us a valuable way to further look into the nature of wound healing and regeneration and ultimately lead to novel therapeutic strategies to prevent fibrotic scars.[68]

3. Wound Healing Treatment

In addition to scar formation, other clinical conditions can result from perturbed wound healing in adults. Delayed or insufficient healing can lead to debilitating chronic ulcers whereas excessive healing can lead to keloid formation. Although tremendous advances in multidisciplinary wound care have greatly improved clinical outcomes, abnormal healing inevitably brings heavy social, economic and biomedical burden each year. There is an urgent and unmet need for novel strategies to both prevent and treat these detrimental healing outcomes. At present, many therapeutic options are available in the clinic to treat the acute wounds, such as wound dressing, topical antimicrobials, local application of PDGF-BB and engineered skin substitute.[69–71] Although these treatments partially achieved successful and therapeutic efficacies, they all have substantial limitations and disadvantages such as immunological reaction, low efficiency and high price. Ideally, the ultimate goal is to completely restore the morphology and function of injured skin via a process that is balanced between sufficient but not excessive regeneration and remodeling.[4]

The aim of tissue engineering is to repair and regenerate diseased or damaged tissues or organs so as to restore their functions. In wound care, current research focuses largely on the use of tissue engineering in dermal tissue regeneration. A great deal of interest and attention has been paid to exploiting the regenerative potential of stem cells and biomaterials.[72,73] The synergistic efforts of biomedical engineers, material scientists and biologists have yielded promising therapies for wounds management and explosion of commercial products for wound care.[74] Recent advances in stem cell biology have facilitated the isolation and characterization of diverse cell subpopulation with stem cell properties in skin. In parallel, strides in biomaterials science have fabricated lots of bioactive scaffolds like hydrogels which make it possible to combine stem cells with scaffolds as complex to promote wound healing.[75]

Wound dressing on the wound surface represents an efficient way to improve healing by providing physical protection against microorganism deposition, wound dehydration and external secondary injuries.[70] They also directly interact with the wound beds and accelerate the healing via the release of beneficial bioactive molecules, and/or by maintaining the moist environment suitable for effective wound healing.[76] Various formulations of composites containing synthetic and/or biological agents, such as foams, films and hydrogels, have been developed to maintain the favorable environmental conditions, and/or to deliver bioactive chemical compounds, and hence enhance the healing process.

4. Engineering Hydrogels for Wound Healing

Hydrophilic polymeric hydrogels were introduced into biomedicine over 50 years ago and have found a broad range of pharmaceutical and biomedical applications.[77] Because of their structural similarity to natural ECM, hydrogel can be engineered as an instructive template to direct cells to assemble into functional tissues. Therefore,

Table 1. Hydrogel scaffold for wound healing.

Polymers	Polymer source	Applications	Results	Refs.
Hyaluronic Acid	Polysaccharide	Incorporation of bioactive molecules and/or stem cells	Enhanced vascularization or wound healing	84, 85
Dextran	Polysaccharide	Supports vascular and tissue regeneration	Complete wound healing	79, 86
Alginate	Polysaccharide	Promotion of granulation and capillary formation, wound closure, moisture regulation, and hemostasis	Repair infected skin wounds	87, 88
Chitosan	Polysaccharide	Promoting cell proliferation and neovascularization induce granulation tissue formation	Accelerate wound healing; healed third-degree burn	89–92
Pullulan	Polysaccharide	Improve MSC survival, engraftment and differentiation in high-oxidative-stress environments	Enhanced angiogenesis and enhanced wound healing	93, 94
PHEMA-PMPC-PHEMA	Triblock copolymer	Thermoresponsive biodegradable copolymer gels	Potentially used as wound dressings	95
Oligopeptide	Protein	Topical treatment of simulated uranium wounds, burn wounds	Reduced the toxicity of uranyl oxide; promoted wound healing	96, 97

hydrogels are extensively studied for tissue engineering.[78] Also, hydrogels have been popularly applied as bioresponsive scaffolds, drug delivery vehicles, and cell culture platforms.[78] As Table 1 indicates, many hydrogel scaffolds have been examined to enhance wound healing, and some hydrogels have showed promising results in diverse animal wound healing models.[75,79,80] These sophisticated hydrogels are biocompatible with the host and promote wound healing by functional interacting with different kinds of cells involved in healing. They have capabilities in maintaining moist environment, providing adequate gaseous exchange, absorbing fluids, and preventing contamination during healing.[81–83] Luo *et al.* developed a bletilla striata polysaccharide (BSP) hydrogel by oxidation and tested the hydrogel for cutaneous wound healing. Attributed to attenuated TNF-α and elevated EGF, the BSP hydrogel achieved much better healing results than controls. Thus proving that this new BSP hydrogel is able to control the inflammatory responses and accelerate wound closure and has potential application in wound healing.[82]

Beyond their utilities as scaffolds, hydrogels can also deliver cytokines and growth factors, antibiotics and cells. Ito and colleague demonstrated that dexamethasone could be slowly released from an injectable hydrogel to reduce inflammatory cytokines (TNF-α, IL-6) production, and caused less inflammation when used in skin wounds.[85] Obara *et al.* found that application of the chitosan hydrogel containing fibroblast growth factor-2 (FGF-2) significantly induced wound contraction and accelerated wound closure and addition of FGF-2 in the chitosan hydrogel further accelerated wound closure by advanced granulation tissue formation, capillary formation and epithelialization.[98] Recently, Rustad *et al.* reported that the wounds treated with pullulan-collagen hydrogels that were encapsulated with MSCs showed significantly accelerated healing by enhanced angiogenesis.[93] In this study, they revealed that hydrogels induced MSC secretion of angiogenic cytokines and expression of transcription factors associated with maintenance of pluripotency and self-renewal (Oct4, Sox2, Klf4). In contrast, the direct injection of MSCs did not significantly improve wound closure. This study clearly demonstrated that this biomimetic hydrogel provided a functional niche capable of augmenting MSC regenerative potential to enhance wound healing. Meanwhile, more and more studies have showed the great potential of hydrogel for wound healing.[75]

4.1. General requirements for biomimetic scaffolds

The delicate structure and composition of the ECM are vital for complete wound healing. African spiny mice are capable of skin autotomy and develop more porous ECM during skin regeneration than non-skin regenerating mice.[49] The slow ECM deposition allows different molecules to communicate and coordinate to restore the full skin structure. Designing and tailoring their properties for wound healing are of increasingly interest. The design of naturally inspired scaffolds takes into consideration such properties as structural characteristics, mechanical strength, and biodegradation.

4.1.1. *Biocompatibility*

Biocompatibility is the most basic requirement for all tissue-engineered scaffolds. However, this requirement is critically important for wound healing. As discussed aforementioned, excessive inflammatory cells result in fibrosis tissue formation in adults. Consequently, a scaffold that causes no or less immune response is especially desirable for wound healing. Polyester-based polymers that are widely studied for tissue engineering generally degrade into a large amount of acidic group that may accelerate immunological responses,[100] and are not ideal scaffold for wound healing. Thus, while designing wound healing scaffolds, such factors as limiting the number of ester group should be taken into consideration, which may reduce inflammatory response, and help achieve complete wound healing. Though some collagen-based scaffolds have been accepted for clinical use, increased inflammation has been observed in clinical research.[100] Recently, Zhang *et al.* demonstrated that ultra-low-fouling zwitterionic hydrogels can resist macrophage deposition,[101] which shows great potential as ideal wound healing scaffolds. However, other than the inflammatory response, an ideal wound dressing should protect the wound from bacterial infection, control evaporative water loss and prevent dehydration, allow diffusion of oxygen and carbon dioxide, absorb wound exudate, and provide cells an environment similar to the native tissue to maintain cell phenotype and organization to promote complete skin regeneration.

4.1.2. *Porous structure*

Cells are the key players in tissue engineering, and a porous scaffold that can maintain cell viability and facilitate cell proliferation is thus critical for them to form functional tissues. In general, a porous scaffold should have interconnected channels to transport chemicals, nutrients, and metabolic wastes. Cell and nutrition transport into the large pore scaffolds occurs relatively easier than into smaller pores, and self-assemble into new tissues. Neither our previous results[102] nor published results with denser crosslinking[103] showed any distinct vessel formation within the hydrogel, unless they are degraded. Burdick *et al.* demonstrated that the sequential crosslinking of HA-based hydrogel has significant effect on cell migration. Further crosslinking reduced pore size and retarded cell spreading in the hydrogel.[104] Apparently, designing macroporous interconnected hydrogel scaffolds that can facilitate nutrient transport and cell infiltration will be a prerequisite for deep cutaneous wound healing. As discussed above, the inflammatory cells contribute to the scar tissue formation. The porous scaffold may also help diffuse the excessive inflammatory cells that contribute to scar formation. However, the fabrication of precise porous scaffolds for tissue engineering remains a significant challenge.[105] Although 3D printing has been extensively investigated to build porous scaffolds,[106] it is still a challenge to build macroporous hydrogel scaffolds.

4.1.3. *Degradation*

Hydrogel serves as a temporal scaffold and eventually degrades, and degradation helps create porous structures. The direct fabrication of functional macroporous hydrogel scaffold still remains a significant challenge, but degradation will break down crosslinks and give rise to loose network structures. Tailorable degradation is thus critical in designing wound healing scaffolds. Kloxin *et al.* demonstrated that a hydrogel with dense structure could not support cell migration.[107] Without porous structure or degradation, no cell can infiltrate or spread within the hydrogel, while further photo degradation allowed the encapsulated cells to spread and migrate. Phelps *et al.* also showed that a matrix metalloproteinase (MMP)-degradable hydrogel promoted vascularization.[103] Burdick group recently demonstrated that the degradation via MMP group even mediates cellular traction that directs stem cell fate in the three-dimensional network.[108] Furthermore, degradation is also one of the major release mechanisms, i.e., diffusion and degradation. For wound healing, scaffolds provide more than support for cell fate and assembly — they may also act as carriers for various biological molecules that promote wound repair. An appropriate degradation rate is essential for continuous growth factor release, which is indispensable for tissue growth. Thus, the degradation property of scaffolds is crucial.

4.1.4. *Mechanical stimulation*

Body movement can cause significant deformation, such as stretch, compression, or torsion, and these deformations will give rise to a wide change of mechanical forces.[110] However, the effect of mechanical force that regulates cell–matrix interactions is often overlooked during wound repair.[110] All phases of the cutaneous wound healing are influenced by mechanical forces. Focal adhesion components, such focal adhesion kinase (FAK),[111] could be activated after cutaneous wounding and enhanced by mechanical forces, leading to more inflammation and fibrosis synthesis. Guvendiren *et al.* recently demonstrated that the stiffness hydrogels mediate cell behaviors and differentiation.[112] Not only can mechanical cues regulate cell fate, they can also modulate the response of cells to their microenvironment environment.[113] When applied to scaffolds, these mechanical forces can be transferred to the cells and further change tissue development. Meanwhile, the mechanical force can also influence the release profiles.[114] Additionally, appropriate mechanical strength also enables the hydrogel to fill the wound region and prevent contraction, which thus reduce the distortion and scar formation.

4.1.5. *Angiogensis*

Angiogenesis and neovascularization are essential for the outcomes of the deep wound healing.[115,116] The newly formed vasculature facilitates oxygen and nutrient transport to the injured tissue,[117] which enhances and maintains the viability and functionality of newly formed tissues, thus governing the wound healing outcomes. Consequently, the design of hydrogel for wound healing has to take angiogenesis into consideration. Thus far, there are three general strategies to enhance vascularization of tissue-engineered scaffolds,[118] which include the incorporation of regulatory factors (e.g., VEGF) that induce the growth of vasculature from surrounding tissues or by cell homing,[119,120] the *in vitro* vascularization via cell engineering[121,122] and the *in vivo* prevascularization.[118,119,123–125] All these strategies are achieved with the support of an efficient hydrogel scaffold. As discussed in aforementioned scar formation, the effect of cutaneous wound healing could be compromised by robust angiogenesis, which could lead to fibrotic tissue formation during wound healing. The delicate design of angiogenesis is extremely important to achieve regenerative healing.

4.2. Engineering regenerative dextran hydrogel

Although many reports have shown promising results in improving wound healing, there is still a long way to achieve the ideal regenerative healing with perfect skin regenerations. Instead of regenerative mechanism, the current treatments follow repair mechanism and usually yielded the fibrotic scar-like structures without simultaneous regeneration of skin appendages.[5,126] Therefore, ideal restoration of skin integrity and function requires regenerative strategy to accomplish more favorable healing. Recently, our studies and others have focused on tailing the properties of chemically modified dextran hydrogel to improve cutaneous wound healing. A series of *in vitro* and *in vivo* experiments findings from our team have demonstrated that dextran hydrogel scaffolds alone, without additional growth factors or cells, significantly promoted remarkable neovascularization and skin regeneration in third-degree burn wounds.[80,102,127]

Dextran is a bacterially derived non-toxic hydrophilic polysaccharide. Its biocompatibility is well known and dextran has been extensively explored for biomedical and pharmaceutical applications.[129–131] The chemically reactive hydroxyl groups allowed dextran to be modified with desirable functionalities,[132] and developed into hydrogels with specific characteristics. Ours and other teams have reported that various functional groups including allyl isocyanate (Dex-AI), ethylamine (Dex-AE), chloroacetic acid (Dex-AC), or maleic-anhydride (Dex-AM) were successfully incorporated into dextrans and produced different dextran-based biodegradable hybrid hydrogels with varied physical and biological

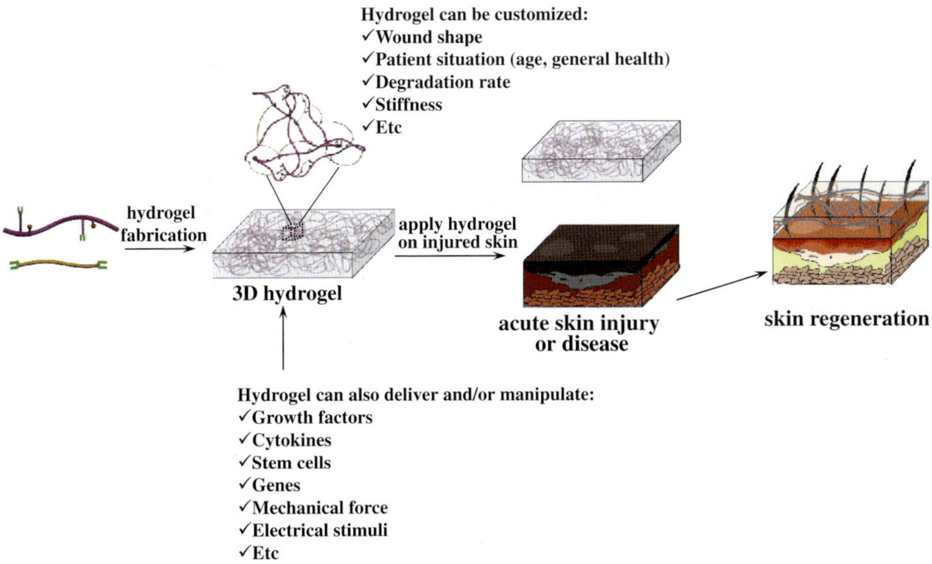

Fig. 1. The design of dextran hydrogels as wound healing scaffold for complete skin regeneration. The dextran hydrogels can be customized and manipulated to deliver stimulating factors. Ideally, the hydrogel will create an embryonic environment that helps regenerate complete skin structures. Reprinted with permission of the National Academy of Sciences, USA.[128]

properties.[102,133,134] Intriguingly, the incorporation of amine groups into dextran gives rise to hydrogels with better biocompatible and release properties.

The modifications of dextran backbone allow the development of many favorable hydrogels with specific characteristics that enhance their utilization in various biomedical applications.[135,136] As illustrated in Fig. 1, the dextran hydrogels can be engineered to administer or delivery many components and stimuli (e.g., stimuli, growth factors) that are critical for complete wound healing or skin regeneration. Moreover, the hydrogel can be customized on the basis of patient condition (e.g., age, general health), wound shape, etc., so as to develop more efficient and personalized wound care products. Most wound healing scaffolds, however, are not customized to treat these wounds, which prevents them from achieving desirable outcomes. After the customized scaffolds are applied, they will activate, induce, and maintain an environment ideally similar to the embryonic phenotype, thus regenerating the whole skin structures to achieve ideal wound healing.

Toward this end, we further remodeled the Dex-AE-based hydrogel, which has better biocompatibility than other dextran precursors,[102] by reducing crosslinking density via decreased degree of substitution of crosslinking groups, leading to improved hydrogel properties such as reduced rigidity, increased swelling, increased VEGF release capability, and facilitated rapid hydrogel disintegration and tissue ingrowth. The expedited biodegradation allowed the more endothelial cell homing, resulting in faster vascularization. Moreover, we also found that the synergistic effect of multiple angiogenic growth factors was established as indicated by more and larger blood vessels in treatment with co-immobilization of multiple factors than treatment with individual one. However, no synergistic effects were established unless VEGF was involved. This finding demonstrates that rapid, efficient, and functional neovascularization can be achieved by precisely manipulating hydrogel scaffold properties and immobilizing defined angiogenic growth factors.[127]

The question that whether such hydrogel scaffolds alone or in combination with cells or peptides have the capabilities to induce tissue regeneration during wound repair still remains largely unexplored. To evaluate the hydrogel potential as wound healing scaffold, we examined our dextran hydrogel in a burn injury model that follows current clinical practice (Fig. 2). These wounds were covered with either dextran hydrogels or Integra (state-of-the-art product used in clinic), and non-treatment as control followed coverage with Duoderm dressings. The burn site and surrounding tissue were collected at different time intervals and immunohistochemical stains for wound responses and skin regeneration were performed. During three-week observation, we found that dextran hydrogel promoted dermal regeneration with complete skin appendages. By day 21, burn wounds treated with hydrogel developed a mature epithelial structure with hair follicles and sebaceous glands. After 5 weeks of treatment, the hydrogel scaffolds promoted new hair growth and epidermal morphology and thickness similar to normal mouse skin. Our studies clearly demonstrate that dextran hydrogel alone promotes rapid neovascularation and complete skin regeneration, thus holding great potential to serve as a unique device for superior treatment of dermal wounds in clinical settings.[128]

However, little is still known about how dextran scaffold coordinates cell lineages to regenerate complete dermal structures, and these are foreseeable barriers to improve our design and efficiency of hydrogel scaffolds for deep dermal wound healing. Despite a myriad of biological approaches geared towards improvement of dermal regeneration, a clinically translational therapy is still far from being available and still remains a significant challenge. Our study provides a rare opportunity to understand how dextran hydrogel reconstructs an embryonic skin niche for intact skin regeneration, and identify the appropriate differentiation lineages that are designated to skin regeneration under dextran-based hydrogel treatment. On the basis of our findings, more efficient and personalized hydrogel scaffolds may be developed to further translational medicine.

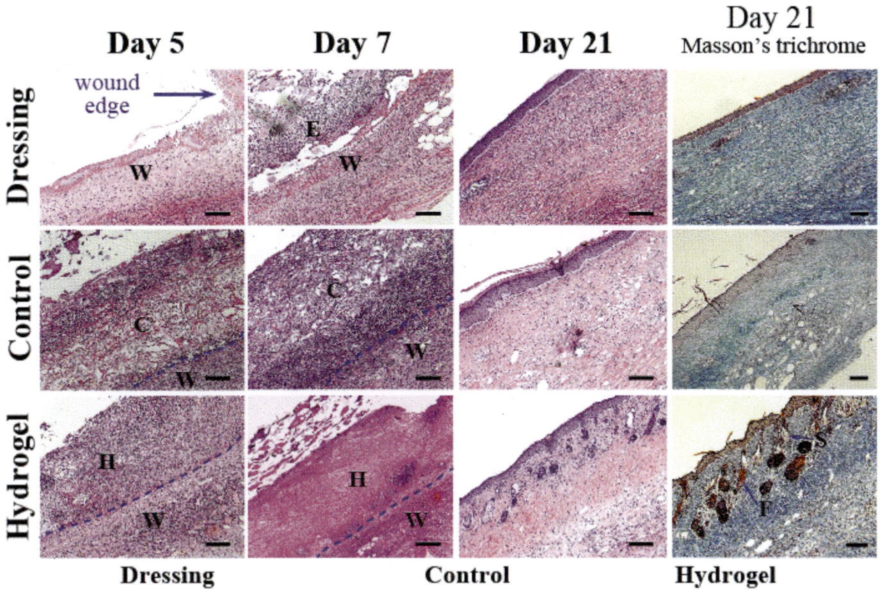

(A)

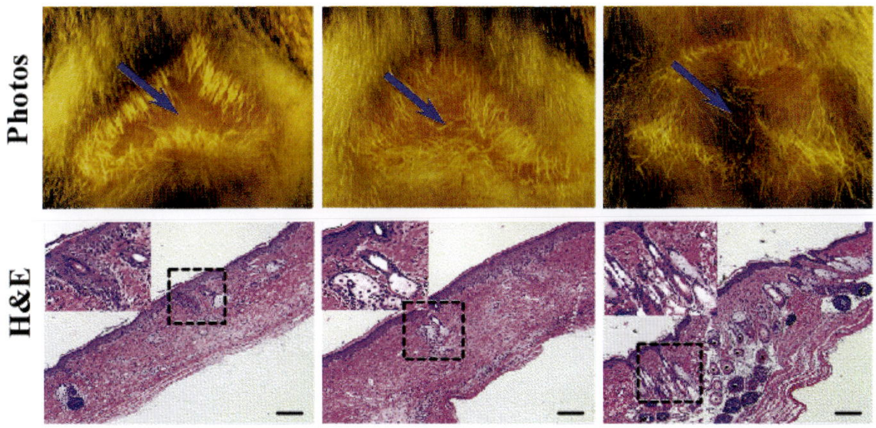

(B)

Fig. 2. Dextran hydrogel as a therapeutic modality promotes complete skin regeneration. (A) Representative images of histological sections at time intervals show that dextran hydrogel promoted wound healing with complete skin appendage regeneration. Masson's trichrome staining indicates distinct collagen structures formed in dermal layer by day 21. (B) Photo and H&E staining of newly regenerated skin treated with hydrogel after 35 days. W = Wound area, H = Hydrogel scaffold, C = Control scaffold, E = Eschar, F = Follicle, and S = sebaceous gland. Scale bars = 100 μm. Reprinted with permission of the National Academy of Sciences, USA.[128]

5. Future Perspectives

The development of tissue-engineered skin containing hair follicles and associated structures, including adipose tissue, would provide new treatment options for many acute and chronic wounds. However, due to the complexity and variety in wound healing mechanisms, many challenges must be overcome before it become useful in clinical practice. It will require multidisciplinary collaboration among biologists, materials scientists, engineers and physicians.

To regenerate perfect skin structures, future development should focus on the reconstruction of the embryonic skin niche for intact skin regeneration, and on understanding how to induce, activate, or maintain signaling molecules and pathways to facilitate stem cells to form complete skin structures. The delicate structure and composition of the ECM are vital for complete wound healing. Some basic questions about hydrogel scaffolds have to be answered before significant advances in wound healing therapy can be made. The function of hydrogels has yet to be fully uncovered. Much is still unknown about how hydrogel scaffolds coordinate cell lineages to regenerate complete skin structures. From the clinical practice point of view, the dream of perfect scaffold still remains out of reach. More work should be done to screen the optimal hydrogel precursors for skin regeneration, though some scaffolds showed great potential in treating acute wounds. The potential of dextran is still underestimated, and more dextran scaffolds may be further explored for better wound treatment. With appropriate manipulation and administration of multiple factors, it is possible to treat cutaneous injuries.

6. Conclusion

Wound healing is one of the most complex biological processes characterized by sequential and delicate coordination of multiple cells, molecular regulators and pathways as well as scaffold synthesis to achieve skin regeneration. Once disturbed, this biological event might lead to unfavorable outcomes with functional and cosmetic defects. Tremendous progress in skin biology and biomaterial science offers the possibility to make the repair transformed into a regenerative process, which holds great potential and values in clinical wound healing management. Hydrogel scaffolds can be engineered to administer many factors that facilitate skin regeneration. Moreover, hydrogel can also be customized according to patients' situation, which makes them an ideal scaffold to treat cutaneous injuries. Dextran hydrogels show great promise as acute wound healing scaffolds, but their potential has not been fully exploited yet. More investigation should be pursued to optimize and expand their applications.

References

1. G. Broughton, J. E. Janis, and C. E. Attinger, The basic science of wound healing. *Plast. Reconstr. Surg.*, **117**, 12S–34S (2006).
2. C. Tziotzios, C. Profyris, and J. Sterling, Cutaneous scarring: Pathophysiology, molecular mechanisms, and scar reduction therapeutics Part II. Strategies to reduce scar formation after dermatologic procedures. *J. Am. Acad. Dermatol.*, **66**, 13–24; quiz 25–16 (2012).
3. K. N. Francis-Goforth, A. H. Harken, and J. D. Saba, Normalization of diabetic wound healing. *Surgery*, **147**, 446–449 (2010).
4. P. Murawala, E. M. Tanaka, and J. D. Currie, Regeneration: the ultimate example of wound healing. *Semin. Cell Dev. Biol.*, **23**, 954–962 (2012).
5. G. C. Gurtner, S. Werner, Y. Barrandon, and M. T. Longaker, Wound repair and regeneration. *Nature*, **453**, 314–321 (2008).
6. A. J. Singer and R. A. Clark, Cutaneous wound healing. *New Engl. J. Med.*, **341**, 738–746 (1999).
7. K. A. Bielefeld, S. Amini-Nik, and B. A. Alman, Cutaneous wound healing: Recruiting developmental pathways for regeneration. *Cell Mol. Life Sci.*, **70**, 2059–2081 (2013).
8. R. Mirza, L. A. DiPietro, and T. J. Koh, Selective and specific macrophage ablation is detrimental to wound healing in mice. *Am. J. Pathol.*, **175**, 2454–2462 (2009).
9. P. Martin and S. J. Leibovich, Inflammatory cells during wound repair: The good, the bad and the ugly. *Trends Cell Biol.*, **15**, 599–607 (2005).
10. W. M. Jackson, L. J. Nesti, and R. S. Tuan, Concise review: Clinical translation of wound healing therapies based on mesenchymal stem cells. *Stem Cells Transl. Med.*, **1**, 44–50 (2012).
11. G. Solanas and S. A. Benitah, Regenerating the skin: A task for the heterogeneous stem cell pool and surrounding niche. *Nat. Rev. Mol. Cell Biol.*, **14**, 737–748 (2013).
12. H. J. Snippert, A. Haegebarth, M. Kasper, V. Jaks, J. H. van Es, N. Barker, M. van de Wetering, M. van den Born, H. Begthel, R. G. Vries, D. E. Stange, R. Toftgard, and H. Clevers, Lgr6 marks stem cells in the hair follicle that generate all cell lineages of the skin. *Science*, **327**, 1385–1389 (2010).
13. N. D. Evans, R. O. Oreffo, E. Healy, P. J. Thurner, and Y. H. Man, Epithelial mechanobiology, skin wound healing, and the stem cell niche. *J. Mech. Behav. Biomed.*, **28**, 397–409 (2013).
14. E. N. Arwert, E. Hoste, and F. M. Watt, Epithelial stem cells, wound healing and cancer. *Nat. Rev. Cancer*, **12**, 170–180 (2012).
15. M. K. Maharlooei, M. Bagheri, Z. Solhjou, B. M. Jahromi, M. Akrami, L. Rohani, A. Monabati, A. Noorafshan, and G. R. Omrani, Adipose tissue derived mesenchymal stem cell (AD-MSC) promotes skin wound healing in diabetic rats. *Diabetes Res. Clin. Pract.*, **93**, 228–234 (2011).
16. S. Maxson, E. A. Lopez, D. Yoo, A. Danilkovitch-Miagkova, and M. A. Leroux, Concise review: Role of mesenchymal stem cells in wound repair. *Stem Cells Transl. Med.*, **1**, 142–149 (2012).

17. A. Nuschke, Activity of mesenchymal stem cells in therapies for chronic skin wound healing. *Organogenesis*, **10**, 29–37 (2013).
18. L. Chen, E. E. Tredget, P. Y. Wu, and Y. Wu, Paracrine factors of mesenchymal stem cells recruit macrophages and endothelial lineage cells and enhance wound healing. *PLoS One*, **3**, e1886 (2008).
19. P. Jayaraman, P. Nathan, P. Vasanthan, S. Musa, and V. Govindasamy, Stem cells conditioned medium: A new approach to skin wound healing management. *Cell Biol. Int.*, **37**, 1122–1128 (2013).
20. W. M. Jackson, L. J. Nesti, and R. S. Tuan, Mesenchymal stem cell therapy for attenuation of scar formation during wound healing. *Stem Cell Res. Ther.*, **3**, 20 (2012).
21. K. Ma, S. Liao, L. He, J. Lu, S. Ramakrishna, and C. K. Chan, Effects of nanofiber/stem cell composite on wound healing in acute full-thickness skin wounds. *Tissue Eng. A*, **17**, 1413–1424 (2011).
22. E. Bey, M. Prat, P. Duhamel, M. Benderitter, M. Brachet, F. Trompier, P. Battaglini, I. Ernou, L. Boutin, M. Gourven, F. Tissedre, S. Crea, C. A. Mansour, T. de Revel, H. Carsin, P. Gourmelon, and J. J. Lataillade, Emerging therapy for improving wound repair of severe radiation burns using local bone marrow-derived stem cell administrations. *Wound Repair Regen.*, **18**, 50–58 (2010).
23. A. K. Muller, M. Meyer, and S. Werner, The roles of receptor tyrosine kinases and their ligands in the wound repair process. *Semin. Cell Dev. Biol.*, **23**, 963–970 (2012).
24. K. Deonarine, M. C. Panelli, M. E. Stashower, P. Jin, K. Smith, H. B. Slade, C. Norwood, E. Wang, F. M. Marincola, and D. F. Stroncek, Gene expression profiling of cutaneous wound healing. *J. Transl. Med.*, **5**, 11 (2007).
25. J. Massague, TGFbeta signalling in context. *Nat. Rev. Mol. Cell Biol.*, **13**, 616–630 (2012).
26. J. W. Penn, A. O. Grobbelaar, and K. J. Rolfe, The role of the TGF-beta family in wound healing, burns and scarring: A review. *Int. J. Burns Trauma*, **2**, 18–28 (2012).
27. C. K. Joo and Y. Seomun, Matrix metalloproteinase (MMP) and TGF beta 1-stimulated cell migration in skin and cornea wound healing. *Cell Adh. Migr.*, **2**, 252–253 (2008).
28. J. Liu, Y. Wang, Q. Pan, Y. Su, Z. Zhang, J. Han, X. Zhu, C. Tang, and D. Hu, Wnt/beta-catenin pathway forms a negative feedback loop during TGF-beta1 induced human normal skin fibroblast-to-myofibroblast transition. *J. Dermatol. Sci.*, **65**, 38–49 (2012).
29. S. A. Ham, H. J. Kim, H. J. Kim, E. S. Kang, S. Y. Eun, G. H. Kim, M. H. Park, I. S. Woo, H. J. Kim, K. C. Chang, J. H. Lee, and H. G. Seo, PPARdelta promotes wound healing by up-regulating TGF-beta1-dependent or -independent expression of extracellular matrix proteins. *J. Cell. Mol. Med.*, **14**, 1747–1759 (2010).
30. P. A. Puolakkainen, M. J. Reed, W. R. Gombotz, D. R. Twardzik, I. B. Abrass, and H. E. Sage, Acceleration of wound healing in aged rats by topical application of transforming growth factor-beta(1). *Wound Repair Regen.*, **3**, 330–339 (1995).
31. G. S. Ashcroft, X. Yang, A. B. Glick, M. Weinstein, J. L. Letterio, D. E. Mizel, M. Anzano, T. Greenwell-Wild, S. M. Wahl, C. Deng, and A. B. Roberts, Mice lacking Smad3 show

accelerated wound healing and an impaired local inflammatory response. *Nat. Cell Biol.*, **1**, 260–266 (1999).
32. M. J. Crowe, T. Doetschman, and D. G. Greenhalgh, Delayed wound healing in immunodeficient TGF-beta 1 knockout mice. *J. Invest. Dermatol.*, **115**, 3–11 (2000).
33. B. J. Larson, M. T. Longaker, and H. P. Lorenz, Scarless fetal wound healing: A basic science review. *Plast. Reconstr. Surg.*, **126**, 1172–1180 (2010).
34. N. L. Occleston, S. O'Kane, H. G. Laverty, M. Cooper, D. Fairlamb, T. Mason, J. A. Bush, and M. W. Ferguson, Discovery and development of avotermin (recombinant human transforming growth factor beta 3): A new class of prophylactic therapeutic for the improvement of scarring. *Wound Repair Regen.*, **19** (Suppl. 1), s38–s48 (2011).
35. S. Ohno, S. Hirano, S. Kanemaru, I. Tateya, Y. Kitani, T. Kojima, T. Nakamura, and J. Ito, Prevention of buccal mucosa scarring with transforming growth factor beta3. *Laryngoscope*, **121**, 1404–1409 (2011).
36. S. S. Cheon, Q. Wei, A. Gurung, A. Youn, T. Bright, R. Poon, H. Whetstone, A. Guha, and B. A. Alman, Beta-catenin regulates wound size and mediates the effect of TGF-beta in cutaneous healing. *FASEB J.*, **20**, 692–701 (2006).
37. S. S. Cheon, A. Y. Cheah, S. Turley, P. Nadesan, R. Poon, H. Clevers, and B. A. Alman, beta-Catenin stabilization dysregulates mesenchymal cell proliferation, motility, and invasiveness and causes aggressive fibromatosis and hyperplastic cutaneous wounds. *Proc. Natl. Acad. Sci. USA*, **99**, 6973–6978 (2002).
38. M. B. Labus, C. M. Stirk, W. D. Thompson, and W. T. Melvin, Expression of Wnt genes in early wound healing. *Wound Repair Regen.*, **6**, 58–64 (1998).
39. T. Okuse, T. Chiba, I. Katsuumi, and K. Imai, Differential expression and localization of WNTs in an animal model of skin wound healing. *Wound Repair Regen.*, **13**, 491–497 (2005).
40. S. Cheon, R. Poon, C. Yu, M. Khoury, R. Shenker, J. Fish, and B. A. Alman, Prolonged beta-catenin stabilization and tcf-dependent transcriptional activation in hyperplastic cutaneous wounds. *Lab Invest.*, **85**, 416–425 (2005).
41. M. Ito, Z. Yang, T. Andl, C. Cui, N. Kim, S. E. Millar, and G. Cotsarelis, Wnt-dependent de novo hair follicle regeneration in adult mouse skin after wounding. *Nature*, **447**, 316–320 (2007).
42. J. L. Whyte, A. A. Smith, B. Liu, W. R. Manzano, N. D. Evans, G. R. Dhamdhere, M. Y. Fang, H. Y. Chang, A. E. Oro, and J. A. Helms, Augmenting endogenous wnt signaling improves skin wound healing. *PLoS One*, **8**, e76883 (2013).
43. A. S. Colwell, M. T. Longaker, and H. Peter Lorenz, Identification of differentially regulated genes in fetal wounds during regenerative repair. *Wound Repair Regen.*, **16**, 450–459 (2008).
44. D. D. Lo, A. S. Zimmermann, A. Nauta, M. T. Longaker, and H. P. Lorenz, Scarless fetal skin wound healing update. *Birth Defects Res. Pt. C*, **96**, 237–247 (2012).
45. M. C. Y. Heng, Wound healing in adult skin: Aiming for perfect regeneration. *Int. J. Dermatol.*, **50**, 1058–1066 (2011).
46. P. Martin, Wound healing — aiming for perfect skin regeneration. *Science*, **276**, 75–81 (1997).

47. T. A. Wilgus, Regenerative healing in fetal skin: A review of the literature. *Ostomy Wound Manage*, **53**, 16–31 (2007).
48. M. T. Longaker, D. J. Whitby, M. W. Ferguson, H. P. Lorenz, M. R. Harrison, and N. S. Adzick, Adult skin wounds in the fetal environment heal with scar formation. *Ann. Surg.*, **219**, 65–72 (1994).
49. A. W. Seifert, S. G. Kiama, M. G. Seifert, J. R. Goheen, T. M. Palmer, and M. Maden, Skin shedding and tissue regeneration in African spiny mice (Acomys). *Nature*, **489**, 561–565 (2012).
50. C. Soo, S. R. Beanes, F. Y. Hu, X. Zhang, C. Dang, G. Chang, Y. Wang, I. Nishimura, E. Freymiller, M. T. Longaker, H. P. Lorenz, and K. Ting, Ontogenetic transition in fetal wound transforming growth factor-beta regulation correlates with collagen organization. *Am. J. Pathol.*, **163**, 2459–2476 (2003).
51. A. S. Colwell, T. M. Krummel, M. T. Longaker, and H. P. Lorenz, An *in vivo* mouse excisional wound model of scarless healing. *Plast. Reconstr. Surg.*, **117**, 2292–2296 (2006).
52. L. Cuttle, M. Nataatmadja, J. F. Fraser, M. Kempf, R. M. Kimble, and M. T. Hayes, Collagen in the scarless fetal skin wound: Detection with picrosirius-polarization. *Wound Repair Regen.*, **13**, 198–204 (2005).
53. A. Satoh, A. Makanae, A. Hirata, and Y. Satou, Blastema induction in aneurogenic state and Prrx-1 regulation by MMPs and FGFs in Ambystoma mexicanum limb regeneration. *Develop. Biol.*, **355**, 263–274 (2011).
54. N. A. Coolen, K. C. Schouten, B. K. Boekema, E. Middelkoop, and M. M. Ulrich, Wound healing in a fetal, adult, and scar tissue model: A comparative study. *Wound Repair Regen.*, **18**, 291–301 (2010).
55. M. H. Kim, W. Liu, D. L. Borjesson, F. R. Curry, L. S. Miller, A. L. Cheung, F. T. Liu, R. R. Isseroff, and S. I. Simon, Dynamics of neutrophil infiltration during cutaneous wound healing and infection using fluorescence imaging. *J. Invest. Dermatol.*, **128**, 1812–1820 (2008).
56. O. O. Olutoye, D. R. Yager, I. K. Cohen, and R. F. Diegelmann, Lower cytokine release by fetal porcine platelets: A possible explanation for reduced inflammation after fetal wounding. *J. Pediatr. Surg.*, **31**, 91–95 (1996).
57. K. W. Liechty, N. S. Adzick, and T. M. Crombleholme, Diminished interleukin 6 (IL-6) production during scarless human fetal wound repair. *Cytokine*, **12**, 671–676 (2000).
58. K. W. Liechty, T. M. Crombleholme, D. L. Cass, B. Martin, and N. S. Adzick, Diminished interleukin-8 (IL-8) production in the fetal wound healing response. *J. Surg. Res.*, **77**, 80–84 (1998).
59. A. Gordon, E. D. Kozin, S. G. Keswani, S. S. Vaikunth, A. B. Katz, P. W. Zoltick, M. Favata, A. P. Radu, L. J. Soslowsky, M. Herlyn, and T. M. Crombleholme, Permissive environment in postnatal wounds induced by adenoviral-mediated overexpression of the anti-inflammatory cytokine interleukin-10 prevents scar formation. *Wound Repair Regen.*, **16**, 70–79 (2008).
60. K. W. Liechty, H. B. Kim, N. S. Adzick, and T. M. Crombleholme, Fetal wound repair results in scar formation in interleukin-10-deficient mice in a syngeneic murine model of scarless fetal wound repair. *J. Pediatr. Surg.*, **35**, 866–872 (2000).

61. B. C. Wulff, A. E. Parent, M. A. Meleski, L. A. DiPietro, M. E. Schrementi, and T. A. Wilgus, Mast cells contribute to scar formation during fetal wound healing. *J. Invest. Dermatol.*, **132**, 458–465 (2012).
62. S. A. Eming, B. Brachvogel, T. Odorisio, and M. Koch, Regulation of angiogenesis: Wound healing as a model. *Prog. Histochem. Cytoc.*, **42**, 115–170 (2007).
63. A. S. Colwell, S. R. Beanes, C. Soo, C. Dang, K. Ting, M. T. Longaker, J. B. Atkinson, and H. P. Lorenz, Increased angiogenesis and expression of vascular endothelial growth factor during scarless repair. *Plast. Reconstr. Surg.*, **115**, 204–212 (2005).
64. T. A. Wilgus, A. M. Ferreira, T. M. Oberyszyn, V. K. Bergdall, and L. A. Dipietro, Regulation of scar formation by vascular endothelial growth factor. *Lab. Invest.*, **88**, 579–590 (2008).
65. H. Larjava, C. Wiebe, C. Gallant-Behm, D. A. Hart, J. Heino, and L. Hakkinen, Exploring scarless healing of oral soft tissues. *J. Can. Dent. Assoc.*, **77**, b18 (2011).
66. K. Mak, A. Manji, C. Gallant-Behm, C. Wiebe, D. A. Hart, H. Larjava, and L. Hakkinen, Scarless healing of oral mucosa is characterized by faster resolution of inflammation and control of myofibroblast action compared to skin wounds in the red Duroc pig model. *J. Dermatol. Sci.*, **56**, 168–180 (2009).
67. J. W. Wong, C. Gallant-Behm, C. Wiebe, K. Mak, D. A. Hart, H. Larjava, and L. Hakkinen, Wound healing in oral mucosa results in reduced scar formation as compared with skin: Evidence from the red Duroc pig model and humans. *Wound Repair Regen.*, **17**, 717–729 (2009).
68. J. E. Glim, M. van Egmond, F. B. Niessen, V. Everts, and R. H. Beelen, Detrimental dermal wound healing: What can we learn from the oral mucosa? *Wound Repair Regen.*, **21**, 648–660 (2013).
69. T. Lazic and V. Falanga, Bioengineered skin constructs and their use in wound healing. *Plast. Reconstr. Surg.*, **127**(Suppl. 1), 75S–90S (2011).
70. K. Fan, J. Tang, J. Escandon, and R. S. Kirsner, State of the art in topical wound-healing products. *Plast. Reconstr. Surg.*, **127**(Suppl. 1), 44S–59S (2011).
71. J. E. Mogford, B. Tawil, S. Jia, and T. A. Mustoe, Fibrin sealant combined with fibroblasts and platelet-derived growth factor enhance wound healing in excisional wounds. *Wound Repair Regen.*, **17**, 405–410 (2009).
72. V. W. Wong and G. C. Gurtner, Tissue engineering for the management of chronic wounds: current concepts and future perspectives. *Exp. Dermatol.*, **21**, 729–734 (2012).
73. M. Chen, M. Przyborowski, and F. Berthiaume, Stem cells for skin tissue engineering and wound healing. *Crit. Rev. Biomed. Eng.*, **37**, 399–421 (2009).
74. A. D. Metcalfe and M. W. Ferguson, Harnessing wound healing and regeneration for tissue engineering. *Biochem. Soc. Trans.*, **33**, 413–417 (2005).
75. C. Gong, Q. Wu, Y. Wang, D. Zhang, F. Luo, X. Zhao, Y. Wei, and Z. Qian, A biodegradable hydrogel system containing curcumin encapsulated in micelles for cutaneous wound healing. *Biomaterials*, **34**, 6377–6387 (2013).
76. G. Chaby, P. Senet, M. Vaneau, P. Martel, J. C. Guillaume, S. Meaume, L. Teot, C. Debure, A. Dompmartin, H. Bachelet, H. Carsin, V. Matz, J. L. Richard, J. M. Rochet, N. Sales-Aussias, A. Zagnoli, C. Denis, B. Guillot, and O. Chosidow, Dressings for acute and chronic wounds: A systematic review. *Arch. Dermatol.*, **143**, 1297–1304 (2007).

77. C. M. Kirschner and K. S. Anseth, Hydrogels in healthcare: From static to dynamic material microenvironments. *Acta Mater.*, **61**, 931–944 (2013).
78. J. L. Drury and D. J. Mooney, Hydrogels for tissue engineering: Scaffold design variables and applications. *Biomaterials*, **24**, 4337–4351 (2003).
79. G. Sun, X. Zhang, Y. I. Shen, R. Sebastian, L. E. Dickinson, K. Fox-Talbot, M. Reinblatt, C. Steenbergen, J. W. Harmon, and S. Gerecht, Dextran hydrogel scaffolds enhance angiogenic responses and promote complete skin regeneration during burn wound healing. *Proc. Natl. Acad. Sci. USA*, **108**, 20976–20981 (2011).
80. B. Singh, S. Sharma, and A. Dhiman, Design of antibiotic containing hydrogel wound dressings: Biomedical properties and histological study of wound healing. *Int. J. Pharm.*, **457**, 82–91 (2013).
81. N. Lei, C. Gong, Z. Qian, F. Luo, C. Wang, H. Wang, and Y. Wei, Therapeutic application of injectable thermosensitive hydrogel in preventing local breast cancer recurrence and improving incision wound healing in a mouse model. *Nanoscale*, **4**, 5686–5693 (2012).
82. Y. Luo, H. Diao, S. Xia, L. Dong, J. Chen, and J. Zhang, A physiologically active polysaccharide hydrogel promotes wound healing. *J. Biomed. Mater. Res. A*, **94**, 193–204 (2010).
83. X. Huang, Y. Zhang, X. Zhang, L. Xu, X. Chen, and S. Wei, Influence of radiation crosslinked carboxymethyl-chitosan/gelatin hydrogel on cutaneous wound healing. *Mater. Sci. Eng. C Mater. Biol. Appl.*, **33**, 4816–4824 (2013).
84. D. Hanjaya-Putra, Y.-I. Shen, A. Wilson, K. Fox-Talbot, S. Khetan, J. A. Burdick, C. Steenbergen, and S. Gerecht, Integration and regression of implanted engineered human vascular networks during deep wound healing. *Stem Cells Translational Medicine*, **2**, 297–306 (2013).
85. T. Ito, I. P. Fraser, Y. Yeo, C. B. Highley, E. Bellas, and D. S. Kohane, Anti-inflammatory function of an *in situ* cross-linkable conjugate hydrogel of hyaluronic acid and dexamethasone. *Biomaterials*, **28**, 1778–1786 (2007).
86. M. P. Ribeiro, P. I. Morgado, S. P. Miguel, P. Coutinho, and I. J. Correia, Dextran-based hydrogel containing chitosan microparticles loaded with growth factors to be used in wound healing. *Mater. Sci. Eng. C Mater. Biol. Appl.*, **33**, 2958–2966 (2013).
87. H. Xu, X.-D. Yuan, B.-D. Shen, J. Han, Q.-Y. Lv, L. Dai, M.-G. Lin, C. Yu, J.-X. Bai, and H.-L. Yuan, Development of poly(N-isopropylacrylamide)/alginate copolymer hydrogel-grafted fabrics embedding of berberine nanosuspension for the infected wound treatment. *J. Biomater. Appl.* **28**, 1376–1385 (2014).
88. X. Li, S. Chen, B. Zhang, M. Li, K. Diao, Z. Zhang, J. Li, Y. Xu, X. Wang, and H. Chen, *In situ* injectable nano-composite hydrogel composed of curcumin, N,O-carboxymethyl chitosan and oxidized alginate for wound healing application. *Int. J. Pharm.*, **437**, 110–119 (2012).
89. D. Poranki, W. Whitener, S. Howse, T. Mesen, E. Howse, J. Burnell, O. Greengauz-Roberts, J. Molnar, and M. Van Dyke, Evaluation of skin regeneration after burns *in vivo* and rescue of cells after thermal stress *in vitro* following treatment with a keratin biomaterial. *J. Biomater. Appl.* **29**, 26–35 (2013).

90. S.-H. Chen, C.-T. Tsao, C.-H. Chang, Y.-T. Lai, M.-F. Wu, C.-N. Chuang, H.-C. Chou, C.-K. Wang, and K.-H. Hsieh, Assessment of reinforced poly(ethylene glycol) chitosan hydrogels as dressings in a mouse skin wound defect model. *Mater. Sci. Eng. C*, **33**, 2584–2594 (2013).
91. N. Boucard, C. Viton, D. Agay, E. Mari, T. Roger, Y. Chancerelle, and A. Domard, The use of physical hydrogels of chitosan for skin regeneration following third-degree burns. *Biomaterials*, **28**, 3478–3488 (2007).
92. T. Kiyozumi, Y. Kanatani, M. Ishihara, D. Saitoh, J. Shimizu, H. Yura, S. Suzuki, Y. Okada, and M. Kikuchi, The effect of chitosan hydrogel containing DMEM/F12 medium on full-thickness skin defects after deep dermal burn. *Burns*, **33**, 642–648 (2007).
93. K. C. Rustad, V. W. Wong, M. Sorkin, J. P. Glotzbach, M. R. Major, J. Rajadas, M. T. Longaker, and G. C. Gurtner, Enhancement of mesenchymal stem cell angiogenic capacity and stemness by a biomimetic hydrogel scaffold. *Biomaterials*, **33**, 80–90 (2012).
94. V. W. Wong, K. C. Rustad, J. P. Glotzbach, M. Sorkin, M. Inayathullah, M. R. Major, M. T. Longaker, J. Rajadas, and G. C. Gurtner, Pullulan hydrogels improve mesenchymal stem cell delivery into high-oxidative-stress wounds. *Macromol. Biosci.*, **11**, 1458–1466 (2011).
95. J. Madsen, S. P. Armes, K. Bertal, H. Lomas, S. MacNeil, and A. L. Lewis, Biocompatible wound dressings based on chemically degradable triblock copolymer hydrogels. *Biomacromolecules*, **9**, 2265–2275 (2008).
96. Z. Yang, K. Xu, L. Wang, H. Gu, H. Wei, M. Zhang, and B. Xu, Self-assembly of small molecules affords multifunctional supramolecular hydrogels for topically treating simulated uranium wounds. *Chem. Commun.*, **21**, 4414–4416 (2005).
97. H. Meng, L. Chen, Z. Ye, S. Wang, and X. Zhao, The effect of a self-assembling peptide nanofiber scaffold (peptide) when used as a wound dressing for the treatment of deep second degree burns in rats. *J. Biomed. Mater. Res. B*, **89B**, 379–391 (2009).
98. K. Obara, M. Ishihara, M. Fujita, Y. Kanatani, H. Hattori, T. Matsui, B. Takase, Y. Ozeki, S. Nakamura, T. Ishizuka, S. Tominaga, S. Hiroi, T. Kawai, and T. Maehara, Acceleration of wound healing in healing-impaired db/db mice with a photocrosslinkable chitosan hydrogel containing fibroblast growth factor-2. *Wound Repair Regen.*, **13**, 390–397 (2005).
99. M. E. Gomes and R. L. Reis, Biodegradable polymers and composites in biomedical applications: from catgut to tissue engineering. Part 1 Available systems and their properties. *Int. Mater. Rev.*, **49**, 261–273 (2004).
100. P. Muangman, H. Deubner, S. Honari, D. M. Heimbach, L. H. Engrav, M. B. Klein, and N. S. Gibran, Correlation of clinical outcome of integra application with microbiologic and pathological biopsies. *Journal of Trauma and Acute Care Surgery*, **61**, 1212–1217 (2006).
101. L. Zhang, Z. Cao, T. Bai, L. Carr, J.-R. Ella-Menye, C. Irvin, B. D. Ratner, and S. Jiang, Zwitterionic hydrogels implanted in mice resist the foreign-body reaction. *Nat. Biotech.*, **31**, 553–556 (2013).

102. G. Sun, Y. I. Shen, C. C. Ho, S. Kusuma, and S. Gerecht, Functional groups affect physical and biological properties of dextran-based hydrogels. *J. Biomed. Mater. Res. A*, **93**, 1080–1090 (2010).
103. E. A. Phelps, N. Landazuri, P. M. Thule, W. R. Taylor, and A. J. Garcia, Bioartificial matrices for therapeutic vascularization. *Proc. Natl. Acad. Sci. U. S.A.*, **107**, 3323–3328 (2010).
104. S. Khetan, J. S. Katz, and J. A. Burdick, Sequential crosslinking to control cellular spreading in 3-dimensional hydrogels. *Soft Matter*, **5**, 1601–1606 (2009).
105. S. J. Hollister, Porous scaffold design for tissue engineering. *Nat. Mater.*, **4**, 518–524 (2005).
106. K. Pataky, T. Braschler, A. Negro, P. Renaud, M. P. Lutolf, and J. Brugger, Microdrop printing of hydrogel bioinks into 3D tissue-like geometries. *Adv. Mater.*, **24**, 391–396 (2012).
107. A. M. Kloxin, A. M. Kasko, C. N. Salinas, and K. S. Anseth, Photodegradable hydrogels for dynamic tuning of physical and chemical properties. *Science*, **324**, 59–63 (2009).
108. S. Khetan, M. Guvendiren, W. R. Legant, D. M. Cohen, C. S. Chen, and J. A. Burdick, Degradation-mediated cellular traction directs stem cell fate in covalently crosslinked three-dimensional hydrogels. *Nat. Mater.*, **12**, 458–465 (2013).
109. G. Sun, S. Kusuma, and S. Gerecht, The integrated role of biomaterials and stem cells in vascular regeneration. In *Biomaterials as Stem Cell Niche*, Ed. K. Roy, Heidelberg, Berlin: Springer, pp. 195–223 (2010).
110. R. Agha, R. Ogawa, G. Pietramaggiori, and D. P. Orgill, A review of the role of mechanical forces in cutaneous wound healing. *J. Surg. Res.*, **171**, 700–708 (2011).
111. V. W. Wong, K. C. Rustad, S. Akaishi, M. Sorkin, J. P. Glotzbach, M. Januszyk, E. R. Nelson, K. Levi, J. Paterno, I. N. Vial, A. A. Kuang, M. T. Longaker, and C. C. Gurtner, Focal adhesion kinase links mechanical force to skin fibrosis via inflammatory signaling. *Nat. Med.*, **18**, 148–152 (2012).
112. M. Guvendiren and J. A. Burdick, Stiffening hydrogels to probe short- and long-term cellular responses to dynamic mechanics. *Nat. Commun.*, **3**, 792 (2012).
113. G. Kasper, N. Dankert, J. Tuischer, M. Hoeft, T. Gaber, J. D. Glaeser, D. Zander, M. Tschirschmann, M. Thompson, G. Matziolis, and G. N. Duda, Mesenchymal stem cells regulate angiogenesis according to their mechanical environment. *Stem Cells*, **25**, 903–910 (2007).
114. K. Y. Lee, M. C. Peters, K. W. Anderson, and D. J. Mooney, Controlled growth factor release from synthetic extracellular matrices. *Nature*, **408**, 998–1000 (2000).
115. E. E. Tredget, The basis of fibrosis and following thermal injury wound healing disorders. *J. Trauma-Injury Infect. Crit. Care*, **62**, S69–S69 (2007).
116. M. G. Tonnesen, X. Feng, and R. A. F. Clark, Angiogenesis in wound healing. *J. Investig. Dermatol. Symp. Proc.*, **5**, 40–46 (2000).
117. J. Li, Y.-P. Zhang, and R. S. Kirsner, Angiogenesis in wound repair: Angiogenic growth factors and the extracellular matrix. *Microsc. Res. Techniq.*, **60**, 107–114 (2003).
118. B. V. Slaughter, S. S. Khurshid, O. Z. Fisher, A. Khademhosseini, and N. A. Peppas, Hydrogels in regenerative medicine. *Adv. Mater.*, **21**, 3307–3329 (2009).

119. T. P. Richardson, M. C. Peters, A. B. Ennett, and D. J. Mooney, Polymeric system for dual growth factor delivery. *Nat. Biotech.*, **19**, 1029–1034 (2001).
120. A. M. Porter, C. M. Klinge, and A. S. Gobin, Covalently grafted VEGF165 in hydrogel models upregulates the cellular pathways associated with angiogenesis. *Am. J. Physiol.*, **301**, C1086–C1092 (2011).
121. J. E. Saik, D. J. Gould, A. H. Keswani, M. E. Dickinson, and J. L. West, Biomimetic hydrogels with immobilized EphrinA1 for therapeutic angiogenesis. *Biomacromolecules*, **12**, 2715–2722 (2011).
122. D. Hanjaya-Putra, V. Bose, Y.-I. Shen, J. Yee, S. Khetan, K. Fox-Talbot, C. Steenbergen, J. A. Burdick, and S. Gerecht, Controlled activation of morphogenesis to generate a functional human microvasculature in a synthetic matrix. *Blood* (2011).
123. A. G. Mikos, G. Sarakinos, M. D. Lyman, D. E. Ingber, J. P. Vacanti, and R. Langer, Prevascularization of porous biodegradable polymers. *Biotechnol. Bioeng.*, **42**, 716–723 (1993).
124. T. Dvir, A. Kedem, E. Ruvinov, O. Levy, I. Freeman, N. Landa, R. Holbova, M. S. Feinberg, S. Dror, Y. Etzion, J. Leor, and S. Cohen, Prevascularization of cardiac patch on the omentum improves its therapeutic outcome. *Proc. Natl. Acad. Sci. USA*, **106**, 14990–14995 (2009).
125. E. A. Phelps and A. J. Garcia, Update on therapeutic vascularization strategies. *Regen. Med.*, **4**, 65–80 (2008).
126. V. W. Wong, G. C. Gurtner, and M. T. Longaker, Wound healing: A paradigm for regeneration. *Mayo Clin. Proc.*, **88**, 1022–1031 (2013).
127. G. Sun, Y. I. Shen, S. Kusuma, K. Fox-Talbot, C. J. Steenbergen, and S. Gerecht, Functional neovascularization of biodegradable dextran hydrogels with multiple angiogenic growth factors. *Biomaterials*, **32**, 95–106 (2011).
128. G. Sun, X. Zhang, Y.-I. Shen, R. Sebastian, L. E. Dickinson, K. Fox-Talbot, M. Reinblatt, C. Steenbergen, J. W. Harmon, and S. Gerecht, Dextran hydrogel scaffolds enhance angiogenic responses and promote complete skin regeneration during burn wound healing. *Proc. Natl. Acad. Sci. USA*, **108**, 20976–20981 (2011).
129. G. Sun and C.-C. Chu, Synthesis, characterization of biodegradable dextran-allyl isocyanate-ethylamine/polyethylene glycol-diacrylate hydrogels and their *in vitro* release of albumin. *Carbohyd. Polym.*, **65**, 273–287 (2006).
130. S. Mitra, U. Gaur, P. C. Ghosh, and A. N. Maitra, Tumour targeted delivery of encapsulated dextran–doxorubicin conjugate using chitosan nanoparticles as carrier. *J. Contr. Rel.*, **74**, 317–323 (2001).
131. L. S. Ferreira, S. Gerecht, J. Fuller, H. F. Shieh, G. Vunjak-Novakovic, and R. Langer, Bioactive hydrogel scaffolds for controllable vascular differentiation of human embryonic stem cells. *Biomaterials*, **28**, 2706–2717 (2007).
132. G. Sun and J. J. Mao, Engineering dextran-based scaffolds for drug delivery and tissue repair. *Nanomedicine*, **7**, 1771–1784 (2012).
133. L. Ferreira, M. M. Figueiredo, M. H. Gil, and M. A. Ramos, Structural analysis of dextran-based hydrogels obtained chemoenzymatically. *J. Biomed. Mater. Res. B*, **77**, 55–64 (2006).

134. J. M. Jukes, L. J. van der Aa, C. Hiemstra, T. van Veen, P. J. Dijkstra, Z. Zhong, J. Feijen, van C. A. Blitterswijk, and J. de Boer, A newly developed chemically crosslinked dextran-poly(ethylene glycol) hydrogel for cartilage tissue engineering. *Tissue Eng. A*, **16**, 565–573 (2010).
135. S. R. Van Tomme and W. E. Hennink, Biodegradable dextran hydrogels for protein delivery applications. *Expert Rev. Med. Dev.*, **4**, 147–164 (2007).
136. C. Demirbilek and C. O. Dinc, Synthesis of diethylaminoethyl dextran hydrogel and its heavy metal ion adsorption characteristics. *Carbohyd. Polym.*, **90**, 1159–1167 (2012).

Chapter 5

Application of Hydrogels in Ocular Tissue Engineering

Vipuil Kishore[*,†,**], **Yunus Alapan**[‡], **Ranjani Iyer**[†], **Ryan Mclay**[*] **and Umut A. Gurkan**[‡,§,¶,‖,††]

[*]*Department of Chemical Engineering, Florida Institute of Technology, Melbourne, FL 32901, USA*

[†]*Department of Biomedical Engineering, Florida Institute of Technology, Melbourne, FL 32901, USA*

[‡]*Case Biomanufacturing and Microfabrication Laboratory, Mechanical and Aerospace Engineering Department, Case Western Reserve University, Cleveland, OH 44106, USA*

[§]*Department of Orthopedics, Case Western Reserve University, Cleveland, OH 44106, USA*

[¶]*Advanced Platform Technology Center, Louis Stokes Cleveland Veterans Affairs Medical Center, Cleveland, OH 44106, USA*

[‖]*Biomedical Engineering Department, Case Western Reserve University, Cleveland, OH 44106, USA*

[**]*vkishore@fit.edu*

[††]*umut@case.edu*

Corneal disease is a leading cause of blindness worldwide. Transplantation of donor cornea is the preferred treatment for corneal disease. However, these are limited in supply, especially in developing countries. Therefore, there is a need for alternative treatment methods to meet the rising demand for corneal replacements. Over the past two decades, much work has been done in this realm. In this chapter, we first describe the significance of the problem associated with corneal disease, the current treatment options available and their related limitations. Following this, studies in the literature that focus on the development of alternative treatment methodologies including decellularization of xenogenic corneal tissue, development of hydrogel based polymeric bioscaffolds and microengineering are discussed in detail. Finally, the chapter is concluded with potential future directions that need to be pursued for the realization of a fully functional bioartificial cornea as an alternative to donor cornea for corneal repair and regeneration.

1. Introduction

Cornea is a transparent, sparsely cellular and avascular organ that forms the outermost part of the eye. It mainly comprises of three functional layers: (i) corneal epithelium, (ii) corneal stroma, and (iii) corneal endothelium. The corneal epithelium acts as a barrier to protect the stromal layer from pathogen invasion and maintain the tear/air interface. The stromal layer forms 90% of the cornea and is responsible for its transparency and refractive index. The corneal endothelium layer comprises of approximately 400,000 endothelial cells that help conserve the transparency of the stromal layer by maintaining the fluid balance within the stroma. The three layers together regulate the function of the cornea.

Corneal disease is the second major cause for blindness and has rendered more than 10 million people blind across the world.[1] The most significant factors that lead to corneal disease include infection (trachoma), injury (ocular trauma) and corneal ulcerations. Keratoplasty (transplantation of donor cornea) is the current gold standard for the treatment of corneal disease. In the United States alone, approximately 72,000 corneal transplants are performed each year.[2] However, donor corneas are associated with a rejection rate of 18% due to immunological complications.[3] Further, in developing countries, donor corneas are limited in supply.[4–6] Application of synthetic polymer-based artificial corneas (keratoprosthesis) to restore corneal function is limited due to serious complications such as glaucoma and other inflammatory/immunological outcomes.

Over the past two decades, amniotic membrane derived from the inner wall of the fetal placenta, has been successfully applied in the clinic for ocular surface reconstruction.[7–9] Since the amniotic tissue is usually discarded after birth, it is easily accessible with little ethical concern. The use of the amniotic membrane as a graft for corneal repair has been reported to promote re-epithelialization, reduce inflammation and inhibit vascularization.[10] These key elements make the amniotic membrane a viable source for ocular surface reconstruction. However, reliable use of amniotic membrane is associated with limitations that include suboptimal transparency,[11] poor biomechanical strength[12] and donor-to-donor variation in tissue quality.[13] Therefore, development of bioengineered cornea by either using decellularized xenogenic tissues[14] or polymeric matrices as scaffolds is being widely explored for corneal repair and regeneration.

Preservation of the biochemical composition and tissue structure of the native cornea by employing optimal decellularization methods is a promising approach to generate a biological scaffold material for corneal tissue engineering applications. Decellularized tissues present a natural microenvironment to the cells populating it and thereby modulate various facets of the host response including cell adhesion, orientation, proliferation, vascularization, and inflammation. Application of decellularized cornea for ocular applications is being widely studied both as an acellular matrix[15,16] relying on the host cells post implantation, and as a scaffold material seeded with cells[17,18] prior to implantation. Porcine corneas are most commonly decellularized for corneal applications because they are easily available and they match the size and refractive power of the human cornea.[19]

Efforts have also focused on the development of tissue-engineered constructs using hydrogels and biopolymeric materials for the regeneration of the native cornea. An ideal tissue engineered construct must closely mimic the native cornea to be able to perform its critical functions. Ruberti and Zeiske have defined the three design considerations that must be met for an artificial corneal construct to be functional: (i) protection, (ii) transmission, and (iii) refraction.[20] To be able to meet the aforementioned design criteria, the tissue engineered construct must have the following essential characteristics: (i) promote the migration and proliferation of corneal epithelial cells for the formation of a functional corneal epithelium and protect the intraocular contents from pathogenic invasion, (ii) possess tensile strength (3.8 MPa)[110] and viscoelastic properties comparable to the native cornea to withstand the intraocular pressure and thereby prevent rupture, (iii) mimic the nanoscale fibrillar organization of the corneal stroma to exhibit high degree of transparency (>90%) and refractive index, (iv) support the maintenance of functional keratocytes to conserve transparency and synthesize *de novo* collagenous tissue, and (v) match the swelling ratio of the native cornea to be able to retain and distribute the water content and thereby prevent the formation of an optical haze.

This chapter is a comprehensive summary of the various studies in the literature that focus on the development of decellularized corneas or polymeric hydrogels as substrates for corneal tissue engineering applications. Further, some of the key studies that employ stem cells for corneal regeneration have been highlighted. Finally, studies that employ microengineering to pattern polymeric substrates and mimic the highly organized structure of native cornea have been described.

2. Decellularization of Biological Tissues for Corneal Repair

Decellularization is a process by which cellular and nuclear material is removed from the tissue via chemical, physical, and biological means whilst retaining the extracellular matrix (ECM) of tissue.[14,22,111] The ECM is a biological scaffold with an intact tissue microstructure comprising of proteins arranged in a unique architecture specific to the tissue. By presenting a substrate that is akin to the native tissue, decellularized scaffolds modulate the host tissue response by influencing cell adhesion, cell–cell signaling, and cell proliferation. Further, the tissue specific architecture has been shown to guide cell differentiation and maintain the phenotype of the differentiated cells to enable the regeneration of a functional tissue.[21]

Unlike other tissues in the body, cornea has the specific requirement of being optically clear and transparent. Therefore, optimal decellularization of the tissue must not only result in the removal of the cellular material and antigens but also protect the transparency of the tissue. Further, the decellularized tissue must support the maintenance and growth of corneal cells.[24] While the most effective decellularization method that can be reliably used for corneal applications is yet to be determined, there are a plethora of methods in use today that have shown promise in the development of decellularized corneal scaffolds.[14,22,111] Most researchers use a combination of methods to facilitate complete decellularization while maintaining the key attributes of the tissue to enable their applicability for a functional corneal.[22] Currently, a great deal of interest has been placed on decellularization of porcine corneas as they are easily available and closely mimic the ECM of the native human cornea.[19]

2.1. Chemical methods

Acids and bases are commonly employed for decellularization of tissues.[22,30] Treatment with acids, such as peracetic acid, and bases, such as ammonium hydroxide, results in the digestion of the cell cytoplasmic contents and disruption of nucleic acids within the cell. However, these are associated with limitations that include altering the tissue microstructure and weakening the mechanical properties of the tissue.

Detergents are another class of chemicals also used for decellularization of tissues.[22,30] They are classified into three basic categories (non-ionic, ionic and zwitterionic) based on the polar head group (charge). Non-ionic detergents (uncharged head groups) such as Triton X-100 are mild and non-denaturing. They work by disrupting lipid–lipid and lipid–protein interactions, while leaving protein–protein interactions intact[22] and thus removing the biological information (antigens, etc.) that could illicit an immune response while preserving the unique protein–protein architecture of the tissue. However, non-ionic detergents have limited ability to fully decellularize the tissue.[16] Ionic detergents (charged head groups) such as sodium dodecyl sulfate (SDS) are much stronger than Triton X-100 resulting in more efficient solubilization of cytoplasmic and nuclear membranes.[22] However, they have a tendency to denature proteins and reduce the transparency of the tissue.[25] Zwiiterionic detergents have a net charge of zero and exhibit properties of both ionic and non-ionic detergents. While zwitterionic reagents such as CHAPS (3-((3-cholamidopropyl)dimethylammonio)-1-propanesulfonate) protect the native state of proteins, they are not commonly used because they are not as effective in removing nuclear information and cell fragments which can lead to immunological rejection after implantation.[16]

Other chemicals commonly used for decellularization include alcohols, hypotonic and hypertonic solutions, and chelating agents. Alcohols work via two mechanisms: dehydration of the tissue and removal of the lipids (solvation). Since alcohols are typically used as fixatives, their use for decellularization may result in altering the ultrastructure of the corneal tissue.[22] Hypotonic and hypertonic solutions rely on osmotic shock resulting from high differences in ion concentration to disrupt the cell membrane and lyse cells. Most commonly, deionized water or sodium chloride solutions have been used for cell lysis.[22] Chelating agents such as EDTA bind to metallic ions (divalent cations) and disrupt cell adhesion causing dislocation of cells.[22,26]

Most of the aforementioned chemicals have been used either by themselves or in combination for effective decellularization of corneal tissue. Findings from some of the key studies have been elucidated here. Choi *et al.*, employed a combination of Triton X-100 and ammonium hydroxide to produce decellularized human corneal stroma scaffolds with intact ECM proteins and biomechanical properties comparable to normal cornea.[27] Further, they reported that culture of human corneal endothelial cells on decellularized human corneal stromas resulted in the regeneration of the corneal endothelium with potential for transplantation. Du *et al.*, developed a full thickness acellular porcine matrix for corneal tissue engineering applications by decellularizing porcine cornea via immersion in 0.5 wt% sodium dodecyl sulfate.[16] They reported that the process resulted in efficient decellularization while preserving the major structural components and strength of the cornea suggesting that acellular porcine matrix can be used for corneal transplantation and tissue engineering applications. Gonzalez-Andrades *et al.* compared two different chemical methods (NaCl and SDS) to identify the most optimal method of decellularization.[24] The results indicated that while treatment with 0.1% SDS treatment resulted in high level of fibril disorganization and poor optical behavior of the corneas, 1.5 M NaCl preserves the histologic structure, composition, and optical behavior of porcine corneas. Other research groups have also reported a similar negative outcome including loosening of collagen fibrils and removal of glycosaminoglycans (GAGs) with the use of SDS as a decellularization agent.[28,29] GAGs are highly polar and attract water which is essential for the health of the eye. Removal of GAGs during decellularization is not optimal as it may cause loss of hydration in the corneal stroma resulting in corneal haze.

Ponce Marquez *et al.*, studied the effect of three different chemical methods to decellularize bovine corneas.[30] The three methods included: (1) treatment with 1%

SDS followed by 75% ethanol, (2) treatment with 75% ethanol followed by trypsin-EDTA, and (3) treatment with varying ratios of peracetic acid and ethanol mixtures. Histological analyses revealed that while treatment with peracetic acid and ethanol mixtures (method 3) resulted in only partial decellularization, no nuclei or cell debris were found when the corneas were treated with methods 1 and 2. Further, corneas treated with method 2 were reported to maintain their transparency. Keratocytes cultured on corneas treated with method 2 were observed to be more metabolically active and produce more collagen compared to method 1. Mechanical characterization results indicated that the Young's modulus of corneas treated with method 1 was significantly higher compared to method 2; however, the mechanical properties of corneas treated with method 2 were comparable to native cornea. Glycerol has also been used in the decellularization of cornea.[14,31] It has been shown to maintain and even restore corneal transparency after edema (over swelling due to high water content).

2.2. Physical methods

Physical methods most commonly employed for decellularization of tissues include ultrahigh hydrostatic pressure (UHP), freeze-thaw cycling, lyophilization, and mechanical agitation. UHP works via physical disruption of the cell membrane such that the cellular contents are exposed and washed out as debris. A simple and benign cultured medium wash may be used to remove the compromised cellular material out of the ECM of the tissue, thus making this method non-cytotoxic unlike the detergents.[32] Sasaki *et al.*, have shown that while treatment of porcine corneas with detergents Triton X-100 and SDS resulted in decrease in GAG content and denaturation of the superstructure of the cornea, UHP treatment had no such effect and maintained the corneal structure.[29] Further, more efficient removal of cell components was observed with the UHP method compared to detergents. However, the corneas treated with both detergents and UHP were extremely cloudy. Upon treatment with glycerol, the transparency was recovered for corneas treated with UHP but the corneas treated with detergents were found to be irreversibly cloudy.[29] The mechanical properties of corneas decellularized using UHP were not altered and comparable to native cornea.

Proulx *et al.*, have reported the reconstruction of a healthy corneal endothelium using devitalized human corneas.[33] Native human corneas were decellularized by using three consecutive freeze-thaw cycles that resulted in cell lysis via the formation of ice crystals within the cellular membrane. Feline corneal endothelial cells cultured on denuded Descemet's membrane of devitalized human corneas maintain endothelial morphology and form a high cell density monolayer. Further, the healthy corneal endothelium expressed functional proteins such as ZO-1, Na^+/

K^+-APTase and NA^+/HCO_3^- cotransporter, thus demonstrating potential to be used as a replacement for diseased endothelium. Corneal stroma from pig, rabbit and human source was decellularized by freezing using liquid nitrogen followed by thawing rapidly at 37°C.[34] While the rabbit corneas became opaque, pig and human corneas were found to be transparent after three freeze thaw cycles. Further, the collagen fibrils were severely damaged in rabbit corneas Versus well preserved collagen fibrils in pig and human corneas. These findings suggest that techniques for successful decellularization are not necessarily transferable across species. Further, it is important to note that the effect of freeze-thaw method on the mechanical properties of cornea were not reported in these studies. Fu *et al.*, used a combination of Triton X-100 and freeze drying to reconstruct tissue engineered cornea from porcine acellular matrix.[35] The decellularized scaffold was seeded with keratocytes followed by epithelial and endothelial cells on either side of the scaffold to form a viable biological corneal equivalent for corneal tissue engineering applications. Recently, porcine corneas were decellularized using a combination of freeze-thaw cycles and pepsin treatment to generate optically transparent and mechanically robust hydrogels for corneal stroma regeneration.[112]

Lyophilization is another common method to fabricate acellular scaffolds. It differs from freeze-thaw cycling in that the process itself produces pores in the tissue which are useful for cell seeding and distribution within the decellularized tissue. The diameter of the pores resulting from the freezing process can be controlled by varying the temperature: smaller ice crystals form at lower freezing temperatures while larger ice crystals form when there is a larger temperature gradient during freezing. Xiao *et al.*, employed the lyophilization method to fabricate porous acellular porcine corneal stroma scaffolds and reported that keratocytes remain viable and proliferate well on these scaffolds.[15] Light transmittance of these scaffolds after transplantation in rabbit eyes was found to be comparable to native cornea. Lee *et al.*, have shown that lyophilized acellular porcine cornea have better survival compared to acellular porcine cornea in rat model.[36] In a follow-up study, Lee *et al.*, showed that lyophilized acellular porcine cornea repopulated with cells showed better optical transparency and higher expression of corneal markers suggesting that recellularization of scaffolds prior to transplantation may be beneficial for ocular surface reconstruction.[36]

Mechanical agitation is generally used in combination with other methods for more effective decellularization. The use of mechanical agitation alone is largely ineffective as it results in removal of cells only from the surface of the tissue. Additionally, the mechanical agitation greatly affects the integrity of the unique architecture of the tissue itself.[26] Therefore, it is not a method that should be used in whole-layer tissues' decellularization, where maintaining structure and fibril orientation is of utmost importance, such as in corneal tissue scaffolding.

2.3. Biological methods

Biological methods for decellularization entail the use of enzymatic agents. Trypsin is commonly used as a pretreatment during the decellularization process and it works via hydrolytic degradation and breakdown of proteins into smaller amino acids resulting in disruption of the ECM.[26,31] However, prolonged exposure to trypsin can destroy the cornea's unique ECM including the removal of collagen and GAG which affects the transparency and strength of the cornea. Therefore, the decellularization of corneal tissue via trypsin needs to be carried out judiciously to preserve the unique architecture of the cornea.[14]

Nuclease is a type of enzyme that can cleave phosphodiester bonds which make up the backbone of DNA and RNA. There are two primary categories of nucleases: endonuclease and exonuclease. Endonucleases are enzymes that catalyze the hydrolysis of bonds between nucleic acids in the interior of DNA or RNA chains. Exonucleases are enzymes that catalyze the hydrolysis of nucleotides at the ends of DNA or RNA chains. One of the limitations of using nucleases for decellularization is some amounts of the enzyme remains in the tissue even after the decellularization process which could elicit an immune response that could result in the rejection of a corneal scaffold.[26] Additionally, nucleases also tend to require additional treatment by a different decellularization method for complete removal of cellular and nuclear content.

Gonzalez-Andrades *et al.*, employed Dispase to remove the corneal epithelium and endothelium layers prior to employing chemical means to fully decellularize porcine cornea.[24] The phospholipase class of enzymes is characterized by the hydrolysis of phospholipids into fatty acids. Wu *et al.*, employed phospholipase A2 to decellularize porcine corneal scaffolds. Unlike other enzymatic decellularizing agents, phospholipase A2 preserved the collagen and fibril architecture of the tissue without breaking down the proteoglycans. Further, phospholipase A2 did not significantly affect the transparency of the decellularized cornea and the mechanical strength was also maintained.

Aprotinin has been used together with EDTA and SDS for efficient elimination of corneal cells from porcine corneas.[17] The resultant decellularized porcine cornea supported the growth of corneal epithelial, stromal and endothelial cells and thus demonstrating promise to be used for corneal tissue engineering applications.

3. Hydrogels and Polymeric Matrices for Corneal Repair

While application of xenografts (porcine and bovine corneas) post decellularization for corneal repair is meritorious, they are associated with several limitations that include immune response and risk of disease transmission. Additionally, a fully efficient decellularization process that completely removes the cellular material and also maintains the microstructure and transparency of the corneal tissue is challenging and yet to be realized. Tissue engineering using natural or synthetic biomaterials with or without cells has been widely studied as a promising approach for the regeneration of cornea. Efforts in the corneal tissue engineering realm have centered on two different strategies: 1) development of thin sheet like matrices for the regeneration of specific layers (i.e., epithelium or endothelium) of the cornea, 2) fabrication of multilayered scaffolds for the development of hemi-corneas (stroma + epithelium) or full-thickness tissue engineered cornea. Various materials, both natural and synthetic, have been used for the development of a corneal scaffold. This section presents a summary of some of the key studies on the developement of hydrogels and biopolymeric materials for corneal tissue engineering applications.

3.1. Collagen hydrogels

Type-I collagen is the primary component of native cornea and hence is of significant interest as a biomaterial for corneal tissue engineering applications. In 1999, Griffith *et al.*, developed functional human corneal equivalents using a collagen–chondroitin sulfate substrate.[37] In this study, human cells isolated from the individual layers of the cornea were immortalized, well-characterized and then seeded on the substrate. Stromal cells were seeded within the substrate and the eptihelial cells and endothelial cells were seeded on the top and bottom of the substrate, respectively. The resultant corneal equivalent resembled the human cornea in morphology, histology, and transparency. Further, the corneal equivalents were also comparable to the human cornea in terms of the stromal swelling, physiological endothelium activity, and response to chemicals. However, despite the positive outcome *in vitro*, these scaffolds lacked the mechanical robustness required to withstand surgery and implantation procedure. In a follow-up study, Griffith *et al.*, fabricated collagen-based composites by blending collagen and a synthetic acrylamide-based polymer [poly(N-isopropylacrylamide)] (pNIPAAm). To assess the feasibility of these composites in ocular surface reconstruction, a 3 mm circular wound was made in the central cornea area of Japanese white rabbits and the collagen-pNIPAAm scaffolds were implanted.[38,39] The results showed regrowth of epithelium within three days for 87.5% of the rabbits (21 out of 24). While rapid epithelialization was observed on the composite scaffold, histological findings revealed that epithelial differentiation overlaying the polymer was abnormal as indicated by disarrayed cell stratification on the scaffold. To improve this outcome, Li *et al.*, synthesized a hydrogel by blending collagen and a copolymer poly(N-isopropylacrylamide-coacrylicacid-coacryloxysuccinimide) (TERP).[40] A laminin adhesion pentapeptide motif (YIGSR) was then grafted onto the hydrogel (TERP 5). Upon implantation of TERP 5 in pig corneas, rapid recruitment of the host corneal epithelial and stromal cells was observed together with functional innervation within the implant. This mechanically robust and suturable scaffold was the first physiologically functional tissue substitute reported for corneal applications.

A simple crosslinked collagen tissue substitute for corneal implantation [41] was synthesized in contact lens molds by mixing porcine collagen (pH adjusted to 5.5) and aqueous solutions of 1-ethyl-3-(3-dimethylaminopropyl)carbodiimide (EDC) and N-hydroxysuccinimide (NHS) followed by curing at 21°C for 24 h and 37°C for 24 h. The optical transparency of the crosslinked collagen matrices was reported to be superior to the human cornea. Further, *in vivo* studies in rabbit and porcine models showed that 23 out of 24 crosslinked collagen matrices remained transparent six months post implantation and showed stable host-graft integration. Further, the crosslinked collagen matrices promote corneal cell regeneration and innervation. Together, these findings suggest that stable implantable collagen matrix substitutes can be synthesized via simple crosslinking using water soluble carbodiimides.

Since the use of xenogenic sources of collagen (bovine, porcine etc.) is associated with several limitations including disease transmission and many immunological and allergic reactions, recombinant human collagen is a viable source for the synthesis of collagen hydrogels for corneal repair. Comparative studies to assess the efficacy of recombinant human type I collagen (RHC I) and recombinant human type III collagen (RHC III) for use as corneal substitutes revealed that both RHC I and RHC III had similar optical clarity to the human cornea.[42,43] However, RHC III hydrogels were found to be optically and mechanically superior to RHC I hydrogels. Twelve months post implantation in pig corneas, both RHC I and RHC III hydrogels showed good integration with the host tissue. Confocal microscopic images showed regeneration of corneal cells and nerve and also the retention of optical clarity by the hydrogels.[42] Re-innervation of these RHC substitutes one year post implantation in a pig model was found to be comparable to porcine allograft and biosynthetic porcine collagen based grafts.[44]

The promising outcome of the RHC III hydrogels, as described above, led to the first Phase I clinical study. EDC-NHS crosslinked RHC III corneal substitutes were implanted into 10 patients with vision loss.[45] Six to seven months post implantation, the RHC III corneal substitutes were found to be well-integrated accompanied by regeneration of corneal epithelium, stroma and nerves. No evidence of neovascularization, inflammation or rejection was observed. Remodeling and seamless integration of the RHC III collagen substitutes with the host tissue was witnessed as seen in the prior animal studies. A follow-up study 24 months post-implantation confirmed that the RHC III corneal substitutes remain integrated to the host tissue with no evidence of infection, inflammation or vascularization.[46] Visual acuity test revealed that the vision of 6 out of 10 patients improved, two remained unchanged and two deteriorated when compared with values prior to implantation of the RHC III collagen substitutes.[46] A recent assessment four years post-implantation revealed that the RHC III collagen substitutes outperformed donor human corneas in many realms.[47] While one out of nine patients implanted with donor human cornea had a rejection episode, the regenerated neocorneas for patients implanted with RHC III corneal substitutes were found to be stably integrated with no episodes of rejection. Further, unlike donor human corneas, corneal cell and nerve regeneration in RHC III corneal substitute was comparable to healthy human corneas. However, visual acuity of RHC III corneal substitutes was lower (20/54) compared to donor human corneas (20/36) suggesting that further optimization of the corneal substitutes is needed to improve vision. This Phase I clinical study demonstrates that RHC III corneal substitutes have significant potential to be used for the treatment of corneal blindness in the future.

Collagen matrices have been formed as gels, foams and sponges for corneal applications. A tissue-engineered collagen sponge matrix has been shown to maintain the phenotype of human corneal cells and support the co-culture of different corneal cell types.[48] The transparency of the collagen sponge matrix was five times greater than conventional collagen gels but only 50–60% of the native cornea. Highly porous collagen-based foams were developed as biodegradable scaffolds using lyophilization followed by EDC-NHS crosslinking.[49] Human corneal keratocytes penetrated and populated the scaffold over one month and synthesized corneal ECM composed of collagen types I, V, and VI. Reconstructed full-thickness corneal substitute was developed using porous collagen-chondroitin sulfate scaffold seeded with stromal keratocytes followed by epithelial and endothelial cells on the top and bottom of the scaffold, respectively.[50] Results showed that keratocytes populate and synthesize *de novo* ECM on the scaffold. Further, the epithelial cells form a stratified epithelium and the endothelial cells remain viable suggesting that collagen-chondroitin sulfate scaffolds are good substrates for artificial cornea. Duan and Sheardown developed a dendrimer (polypropyleneimine octaamine) based crosslinking approach of concentrated collagen solutions and reported that the transparency of dendrimer crosslinked gels was significantly greater compared to glutaraldehyde and EDC crosslinked gels.[51] Further, dendrimer crosslinked collagen gels were significantly stiffer (5 MPa) than EDC crosslinked gels (0.1 MPa) and comparable to human cornea (up to[3–13] MPa) thereby demonstrating promise to be used for corneal tissue engineering applications. However, the suture strength of dendrimer crosslinked collagen gels was significantly lower than the native cornea.

Pure collagen-based hydrogels are mechanically weak and degrade rapidly. While crosslinking improves the strength and stiffness of collagen hydrogels to some extent, additional processing methods have been developed and employed along with crosslinking to further enhance the mechanical properties and biodegradability of the hydrogels for corneal applications.[52–55] One of the primary reasons for the weak mechanical strength of traditional collagen hydrogels is the high water content. Plastic compression of traditional collagen gels has been reported to remove excess water and form dense and mechanically strong collagen matrices. Mi *et al.*, have reported that plastically compressed collagen scaffolds are transparent with fiber diameter and orientation comparable to native cornea.[52] Further, photochemical crosslinking of plastically compressed collagen gels using riboflavin significantly improved the mechanical properties and suturability of the scaffold without altering the structural and optical properties.[53] In a follow-up study, Xiao *et al.* implanted the plastically compressed collagen gels in rabbit corneas and reported that the biocompatibility of the semi-transparent gels was comparable to the amniotic membrane.[54] Levis *et al.*, developed plastically compressed collagen constructs incorporated with human limbal fibroblasts as a substrate for the expansion of human limbal epithelial cells.[55] The fibroblasts remained viable within the collagen constructs post plastic compression and the epithelial cells expanded well on the surface of the constructs. Further, the epithelial cells formed a multilayered epithelium with polygonal morphology similar to that of the native central cornea suggesting that plastically compressed collagen constructs may be used for limbal epithelial cell transplantation. Levis *et al.*, have also shown that plastically compressed collagen maintain the cobblestone morphology and support the expansion of corneal endothelial cells.[56] Additionally, the ease of handling during transplantation make plastically compressed collagen a promising carrier for corneal endothelial cell transplantation.

Incorporation of natural or synthetic polymers within the collagen network is another method that is widely employed to improve the mechanical properties and biodegradability of collagen hydrogels. A collagen-phospholipid corneal substitute was developed using a combination of EDC-NHS crosslinked collagen and poly(ethylene glycol) diacrylate crosslinked 2-methacryloyloxyethyl phosphorylcholine (MPC).[57,58] The resultant hybrid hydrogel was mechanically superior and more stable compared to hydrogels made from individual components. Further, the optical transparency and cellular compatibility of these hybrid hydrogels was comparable to traditional collagen hydrogels. Implantation of these hybrid hydrogels into the corneas of minipigs resulted in regeneration of the corneal epithelium and stroma together with nerve ingrowth suggesting that these hydrogels can be used as corneal substitutes in clinical applications. Rafat *et al.*, synthesized hybrid polymer network scaffolds using collagen and chitosan and crosslinked the scaffold with EDC-NHS along with poly (ethylene glycol) dibutyraldehyde (PEG DBA).[59] The resultant scaffolds were found to be transparent, optimally strong and biocompatible. Silica-collagen biohybrid materials have shown to be transparent and exhibit better mechanical properties than collagen crosslinked scaffolds.[60]

The high level of transparency and robust mechanical properties of native cornea are attributed to the orthogonal arrangement of aligned collagen fibrils within the corneal stroma. Development of bioengineered corneas that mimics such structural organization would greatly enhance their potential to be used in corneal regeneration applications. Torbet *et al.*, gelled neutralized type I collagen solution in the presence of a horizontal magnetic field of 7 Tesla to assemble collagen molecules and form oriented collagen fibrils within the gel.[61] A series of gelation–rotation–gelation cycles was employed to produce a scaffold of orthogonal lamellae composed of aligned collagen fibrils. Keratocytes seeded onto the scaffold composed of three orthogonal lamellae were found to rapidly proliferate, penetrate the scaffold and align along the direction of the collagen fibrils. Incorporation of proteoglycans improved the transparency of the scaffold and enhanced the proliferation of keratocytes. In a follow-up study, human keratocytes were seeded onto magnetically aligned collagen scaffolds and cultured for 30 days to form the reconstructed stroma.[62] Human limbal stem cell-derived epithelial cells seeded on the surface of the reconstructed stroma exhibited a well-differentiated epithelium consistent with the epithelial structure of the native cornea. Additionally, in a preliminary study, the scaffolds were implanted into five rabbits to assess the *in vivo* response.[62] Four of the five rabbits lost their scaffolds despite extensive suturing. However, the retained scaffold showed good outcomes in case of re-epithelialization, regaining transparency and minimal neo-vascularization. Further work is required in terms of crosslinking of these scaffolds and suturing for better scaffold retention.

3.2. Non-collagenous hydrogels

Biomaterials for the development of hydrogels and polymeric matrices for corneal applications are not limited to collagen alone. Other materials that include gelatin, keratin, chitosan and silk have also been used. Some of the key studies that focus on the development of non-collagenous matrices for corneal regeneration are elucidated here.

3.2.1. *Keratin*

Human hair keratin in nanoparticulate form was mixed with a softening agent (glycerol) and subjected to a curing process to develop a transferable, stable, and transparent keratin film for corneal applications.[63] The transparency and stiffness of keratin films were reported to be significantly higher than amniotic membrane (currently in use for ocular surface reconstruction). Further, growth behavior of human corneal epithelial cells was comparable between keratin films and amniotic membrane. Based on these findings, the authors suggest that keratin films are promising candidates for ocular surface reconstruction. In a follow-up study,

Borrelli *et al.*, employed different sterilization protocols to assess the effect of sterilization on the optical transparency and biomechanical properties of keratin films.[64] The sterilization process showed no significant differences in both optical transparency and biomechanical properties of keratin films. However, suturing of keratin films onto enucleated porcine eyes was found to be more challenging than amniotic membrane due to poor elasticity leading to concerns of irritation and neovascularization post implantation *in vivo*. Therefore, the authors suggest that additional work to improve the elasticity of the material is needed for the use of keratin films in ocular surface reconstruction.

3.2.2. *Chitosan*

A blend of chitosan, gelatin and chondroitin sulfate was used to synthesize optically transparent membranes with a tensile strength of 1.48 MPa and elongation at break of 45.64%.[65] These membranes support the growth of corneal endothelial cells and may be used as carriers for corneal endothelial cell transplantation. Immobilization of hyaluronic acid onto chitosan membranes was reported to maintain the transparency (99%) and improve the hydrophilicity and growth of corneal epithelial cells suggesting that these membranes are promising candidates for corneal regeneration.[66] The transparency of ultrathin chitosan films (~50 μm thick) crosslinked with diepoxy-polyethylene glycol (PEG) and cystamine was reported to be on par with human cornea.[67] Further, the films were biodegradable and supported the growth of corneal endothelial cells. Optimization of PEG concentration for crosslinking resulted in mechanically robust films that possessed excellent characteristics for physical manipulation and implantation demonstrated by an *ex vivo* ovine surgical model. The smooth topology together with excellent optical and biomechanical properties suggest that ultrathin chitosan-PEG films are promising candidates to be used for corneal transplantation procedures.[67] Grolik *et al.*, reported that the genipin crosslinked chitosan-collagen hydrogels were significantly stronger (46.93 MPa) than amniotic membrane (2.3 MPa).[68] Further, growth of corneal epithelial cells on genipin crosslinked chitosan-collagen hydrogels was comparable to amniotic membrane. However, the hydrogels exhibited a bluish–brown color due to genipin crosslinking. The authors suggest that the bluish–brown color of the hydrogels should not pose a problem as the hydrogels would be resorbed post implantation and instead can be used as an indicator of resorption of the hydrogels.

3.2.3. *Silk Fibroin*

Several studies have reported the merit of using silk fibroin as a biomaterial for ocular tissue reconstruction.[69] Silk fibroin membranes have been reported to be transparent, easy to handle and support the growth of human limbal epithelial cells.[70] Further, the morphology and phenotype of corneal epithelial cells on the silk fibroin membrane has shown to be comparable to the amniotic membrane.[71,72] Madden *et al.*, were the first to report that primary human corneal endothelial cells exhibit a polygonal morphology and grow to confluency on collagen coated silk fibroin substrates thus demonstrating that these substrates may be used as tissue constructs for the repair of the corneal endothelium.[73] A dual layered silk fibroin scaffold consisting of a thin epithelial cell substrate layered onto a 3D stromal network was developed to provide an intact stem cell niche for long-term corneolimbal tissue regeneration.[74] Human limbal epithelial cells and limbal mesenchymal stromal cells maintain a similar phenotype (corneal and stromal, respectively) on separate layers and on the dual layered scaffold thus demonstrating the feasibility of the 3D scaffold for corneolimbal tissue regeneration applications.[74] Silk fibroin membranes functionalized with chitosan were reported to be biocompatible and support the growth of keratocytes.[75] While the mechanical properties of the silk membranes improved considerably with the addition of chitosan (1.3 MPa), they were still weaker compared to native cornea (3.8 MPa). Further, the transparency of silk-chitosan membranes (62%) was also lower than native cornea (>85%). Recent studies that focus on the development patterned silk substrates for corneal regeneration are elucidated in the microengineering corneal tissue section.[76,77]

3.2.4. *Other materials*

A full thickness biological substitute of the rabbit cornea was developed using a sequential culturing technique of the endothelial, stromal and epithelial cells on porous cell culture inserts.[78] The endothelial cultured at the bottom of the insert were layered with keratocytes entrapped in a human fibrin and 0.1% agarose gel and the epithelial cells were grown on the surface of the gel. Microscopic evaluation revealed that the corneal epithelial cells form a stratified epithelium, keratocytes proliferate rapidly and endothelial cells exhibit a pattern similar to the corneal endothelium. These fibrin-agarose matrices mimic the native cornea and hence may be used as an *in vitro* model to understand the various biological mechanisms underlying the function of the cornea.

Due to the lack of fibrous structure, mechanical properties of most hydrogels are suboptimal. Tonsomboon *et al.*, significantly improved the mechanical properties of alginate hydrogels by reinforcing the hydrogels with electrospun gelatin nanofibers (450 KPa from 78 KPa).[79] Carbodiimide crosslinking further improved the stiffness of gelatin nanofiber reinforced alginate hydrogels to ~800 KPa but at the expense of the transparency of the hydrogel. These mechanically robust nanofiber reinforced hydrogels are promising materials for corneal transplantation.

4. Stem Cells in Corneal Regeneration

Regenerative medicine through stem cell transplantation is one of the promising treatments in ocular diseases.

4.1. Stem cells in corneal regeneration

Stem cells have been used in regenerative medicine since it was demonstrated that a laboratory grown human epidermis could be transplanted onto burnt patients to reconstitute a functional epidermis.[80,81] There are many different sources of stem cells used for corneal regeneration. Mesenchymal stem cells from different sources and limbal stem cells from the eye itself are the two main stem cell types used for cellular therapy.

4.2. Limbal stem cells

Limbal stem cells are found in the limbal corner of the cornea. Due to their capacity for self-renewal and proliferation, their role is accepted as to maintain the integrity of corneal epithelium. Even if there is no specific marker for the limbal stem cells, they can be distinguished by high nucleus to cytoplasm ratio, proliferative potential and capacity for self-renewal.[82] Injuries and several diseases can cause depletion of these stem cells which is called limbal stem cell deficiency. Deficiency of limbal stem cells causes visual loss and severe discomfort. To treat these deficiency, limbal stem cells have been used for more than 10 years.[83]

Ex vivo expansion of human LSCs for the treatment of LSCD was first described in 1997. Pellegrini *et al.* used limbal stem cells to restore the surface of the cornea of the patients who had limbal stem cell deficiency syndrome. In their study, biopsies from the limbus part of the eye were taken. The cells were plated on to 3T3-J2 cells. The cells were then mounted on a graft and implanted into the eyes of the patients. It was found that the limbal stem cells and patients' visions were restored.[83] At first, the original approach for *ex vivo* expansion of limbal stem cells were to seed the cells obtained from limbus of the healthy eye on 3T3 fibroblast cells. In time, different strategies were developed such as growing the cells on human amniotic membrane. Amniotic membrane can be used either alone or with 3T3 fibroblast cells. In one of the first studies using human amniotic membrane, 13 patients with severe limbal deficiency were treated. Healthy limbal tissue samples were taken from the donors. Tissue cells were then cultured on amniotic membrane. After a certain confluence, amniotic membrane with cells was transplanted into the eye of the patient and another amniotic membrane was used to cover the cells. At end of the surgical experiments, 6 of the 13 patients recovered.[84]

Use of thermo-responsive cell culture plates for culturing limbal stem cells is another novel technique which was firstly used by Nishida *et al.* In this study, human and rabbit limbal stem cells were co-cultured with 3T3 fibroblast cells on temperature sensitive culture plates. Cells were harvested from temperature responsive dishes after 2 week just by reducing the temperature to 200°C without using any proteases or EDTA. In this technique, multilayered corneal epithelial sheet were obtained intact simply by adjusting the temperature. The cell sheets were then transplanted to the eye directly without using any carrier and also reconstruction of the cornea of the rabbits were found very high.[85] In another study, fibrin cultured limbal stem cells were used in patients with limbal stem cell deficiency. Transplanted fibrin-cultured limbal stem cells were successful in 14 of 18 patients.[86] In summary, although there is still a lot of way to go and the studies that have been done so far are still experimental studies, it has been shown that limbal stem cells have the potential for treating limbal stem cell deficiency and treating the corneal dysfunctions.

4.3. Mesenchymal stem cells

Mesenchymal stem cells are multipotent progenitor cells found in many tissues including bone marrow, adipose tissue, heart, dental pulp, peripheral blood, cord blood and limbal stroma of the eye.[87–93] These cells can differentiate to form lineages of mesenchymal tissue.[94] Recently, researchers have started to use mesenchymal stem cells for corneal reconstruction. Mesenchymal stem cells can be administered by either intravenous injection or local administration. For corneal reconstruction, local administration is used. Mesenchymal stem cells can be administered to cornea directly or by carriers.[95–97] In a recent study done by Liu *et al.*, 10^4 mesenchymal stem cells was directly transplanted to mouse cornea which resulted formation of keratinocyte like cell formation.[95] In another *in vivo* study, healing of chemically burned rat cornea was accelerated with subconjuctival injection of mesenchymal stem cells.[98] In another study, human umbilical cord derived mesenchymal stem cells were used to treat thin and cloudy corneas of lumican null mouse. Intrastromal injection of mesenchymal stem cells resulted more transparent and thick cornea in mice.[99] In another *in vivo* study, adipose derived mesenchymal stem cells were used to treat post-traumatic persistent sterile corneal epithelial defect. It was observed that healing process started in 10 days after topical application of mesenchymal stem cells and after 1 month, complete healing was observed.[100] In mesenchymal stem cell therapies, carriers and fibrin gels can be also used to deliver the cells other than the direct injection of mesenchymal stem cells. Ma *et al.* showed that chemically burned rat cornea reconstructed by using bone marrow derived mesenchymal stem cells grown on amnion membrane. In this study, mesenchymal stem cells were isolated from human bone marrow. Cells then were seeded on amnion membrane. After seven days of injury, the damaged corneal surfaces were carefully keratectomized. An amnion membrane with cells was sutured to cornea and a blank amnion membrane put on to cover them. It was found that cornea was reconstructed in 4 week time.[101] In another study, rat corneal injury induced by alkali treatment was treated by rat bone marrow derived mesenchymal stem cells. Stem cells were co-cultured with corneal stromal cells and then they were seeded on amniotic membrane. After keratectomization of the cornea, grafts were transplanted. The transplanted cells reconstructed the chemically burnt rat cornea.[96] Fibrin gels with mesenchymal stem cells are also used in corneal reconstruction. In a study, cornea of rabbits damaged by NaOH and the damage was reconstructed by transplantation of rabbit bone marrow derived mesenchymal stem cells seeded in fibrin gel.[97]

5. Microengineering Corneal Tissue

Corneal tissue exhibits a highly organized compositional layer structure along with specific nanotopographical features at each layer. An intricate relationship exists between the complex topographical structures of the corneal tissue and biophysical characteristics of the comprised cell types. Since these topographical structures are at the micro-nano scale levels, it is vital to approach corneal tissue engineering at the same order of scale. Although current state of the art microengineering methods have grown to be essential tools in engineering of other tissues, their applications are limited in corneal tissue engineering. Studies that focused on developing microengineering based approaches for corneal applications (Fig. 1) are described in this section.

One of the earliest studies is conducted by Teixeira *et al.*,[102] where they investigated the morphological and biophysical behavior of human corneal epithelial cells cultured on substrates with nano grooves and ridges with various spatial configuration. Topographically patterned substrate dimensions were ranging from as small as 70 nm to 2.1 μm with feature pitch between 400 nm to 4 μm, whereas the groove depths used were 150 nm and 600 nm. Elongated and aligned epithelial cells were observed on substrates with ridge and groove features whereas substrates with smooth surfaces resulted in spherical cell morphologies. Furthermore, focal adhesion and microfilament formation was not observed regularly, but when it did, alignment of these structures were in accordance with the alignment of the cell body.

In a follow-up study, Karuri *et al.*[103] studied the effect of surface topography on adhesion strength of corneal epithelial cells by using shear flow in a parallel flow chamber. Adhesion strength of cells cultured on smooth surfaces, and nano, and microscale grooved surfaces, with varying pitch size between 400 nm to 4 μm, were compared. Increase in adhesion strength was observed with decreasing feature dimensions, which indicates cells exhibit a greater adherence to substrate topographies at the nanoscale that is also the native biological scale.

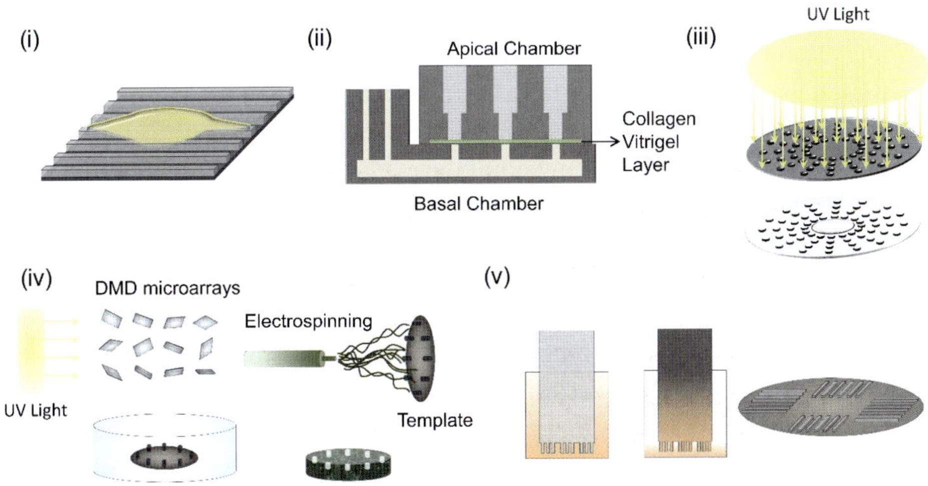

Fig. 1. Microengineering approaches in corneal tissue engineering. (i) Micro ridges/grooves for cell elongation and alignment.[76,77,102,103,105,106] (ii) Organ-on-a-chip devices for co-culture of different cell layers to mimic compositional structure of cornea.[107] (iii) Microfabrication of a 3D hydrogel construct for artificial cornea consisting of strong central optic part and resilient lateral part using photolithography.[104] (iv) Fabrication of biodegradable scaffold for limbus using stereolithography with digital microarrays and electrospinning.[108] (v) Microfabrication of limbal crypts from collagen hydrogel for artificial limbus via hydrophilic porous absorbers (HPAS) with microridge patterned topography.[109]

Myung *et al.*[104] designed and fabricated a 3D hydrogel construct for an artificial cornea, consisting of a mechanically strong and rich in water central optic part and a resilient and microperforated lateral skirt part by using photolithography. Poly(ethylene glycol)/poly(acrylicacid) (PEG/PAA) was utilized for the central optic part, whereas poly(hydroxyethyl acrylate) (PHEA) hydrogel was used for the peripheral skirt. Both parts were functionalized with 5-azidonitrobenzoyloxy N-hydroxysuccinimide to covalently bind the collagen type 1 for cell adhesion. Furthermore, adhesion of endothelial cells and fibroblast cells onto the central and peripheral part, respectively, was demonstrated to further support the cellular compatibility of the implant.

Crabb *et al.*,[105] studied effects of collagen processing methods (isolation and crosslinking) and film topography on optical and biomechanical properties of collagen film-based stroma. To micropattern collagen films, a dispersion of soluble and insoluble collagen was cast onto holographic diffraction grating replicas with 1 μm and 2 μm groove spacing and 1 μm depth. On microgrooved substrates flattened stellate cells were observed in a tightly packed form, on the other hand, cells on the smooth substrates were spindle shaped and in a formation presenting greater gaps between cells. After a week in culture, cells were aligned along the grooves with 2 μm spacing, whereas no alignment was noted for the grooves with 1 μm spacing. There were no significant differences between smooth and microgroove substrates, but transparency decreased after cell culture for both substrates. Ultimate tensile strength was similar for both smooth and micropatterned substrates with or without cells at any point during the 3 week culture period.

In a similar study,[76] microgrooves were utilized for cell alignment on silk protein films. Silk films used in the study were 2 μm thick as the corneal collagen lamellae and 0.5 μm to 5 μm pores were introduced into the film to facilitate translamellar nutrient diffusion and cell–cell interactions. Furthermore, corneal fibroblast proliferation, alignment and corneal ECM expression were observed in 2D and 3D cultures using micropatterned silk protein films. In a follow-up study, Gil *et al.*[77] investigated the effect of micropatterned grooves with depth and spacing ranging from 37 to 342 nm and 445 to 3,580 nm, respectively. Greater alignment of human corneal fibroblasts was observed with deeper and narrower microgrooves. However, there was not any significant difference in DNA content between cells seeded on different substrates after 4 day in culture.

In an organ-on-a-chip study, Puleo *et al.*[107] fabricated a co-culture bilayer microfluidic device with collagen vitrigel between the layers as a barrier. The device consisted of two separate culture chambers that were bonded with vacuum with vitrified collagen between. First, epithelial cells were seeded into apical culture wells and after culture of an epithelium layer, collagen vitrigel, which structurally supported epithelium initially, was sacrificially degraded through injection of a collagenase etching media. Afterwards, keratocytes, as stromal layer, were seeded through the other layer onto the epithelium. Since vacuum provides a reversible sealing, micro corneal patches were able to be removed from the chip for further analysis or use. After viability analysis and culture of bilayered corneal micropatches, sacrificial etching was found to be biocompatible. Furthermore, results of laser-induced fluorescence epithelial permeability assays and histological examinations validated the development of bilayer corneal tissue micropatch.

Ortega *et al.*[108] combined stereolithography with electrospinning to produce a biodegradable scaffold that incorporates artificial stem cell niches imitating the palisades of Vogt in the limbus, for cell delivery application in corneal regeneration. A digital micromirror device was utilized in the stereolithography part, to fabricate a template for the scaffold from a photocurable polymer PEG diacrylate. Afterwards, a biodegradable membrane of poly(lactic-co-glycolic acid) 50:50 was deposited on this template to obtain the scaffold with specific microstructure features. To demonstrate the functionality of the scaffold for corneal regeneration, the authors cultured rabbit limbal fibroblasts and rabbit limbal epithelial cells on the electrospun scaffold.

In a similar study, Levis *et al.*[109] designed and fabricated bioengineered limbal crypts from collagen hydrogel to mimic the native limbal stem cell niche. Bioengineered limbal crypts were produced by incubating HPAs with ridged base topography on top of the collagen hydrogels with or without human limbal fibroblast cells. During 15 min incubation of HPA, water content of hydrogels was depleted by capillary wicking. After removal of HPA, topography of the thin collagen substrate resembles the native limbal crypts. Next, human limbal cells were seeded on the collagen construct and formed several layers. The results demonstrated that cells lining the limbal crypts showed expression of stem cell marker p63α.

In a recent study, Kilic *et al.*[106] produced different combinations of stacked 3D collagen and/or collagen–elastin like recombinamer (Col:ELR) with or without ridge and groove patterns as a substitute for native corneal tissue. The effect of several parameters, such as ridge and groove patterns, scaffold material (collagen or Col:ELR), and various levels of dehydrothermal treatment for crosslinking, on cell alignment, proliferation, and transparency were investigated. The results showed that dehydrothermal treatment at 150°C produced the best results in film fabrication with respect to durability and transparency. Moreover, both collagen and Col:ELR showed sufficient support for cell adhesion and proliferation and cells seeded on micropatterned films showed elongation as soon as culture day 1. Also, transparency of micropatterned films with seeded keratocytes showed improvement over the course of the study. It was shown that while seeding keratoctyes increased the transparency, seeding 3T3 fibroblasts decreased the transparency. Overall, after the 30 days of incubation period, transparency of the Col:ELR films were better than the collagen films and were close to the transparency of native cornea.

6. Conclusions and Future Directions

As summarized in this chapter, researchers over the past two decades have focused on the development of bioartificial corneas to be used as an alternative to donor corneas for the treatment of corneal disease. Efforts have mainly focused on hydrogel based approaches for the development of bioscaffolds mimicking the physiochemical properties of the native cornea. Although much work has been done, a fully tissue engineered cornea has not yet been realized. While many studies in the literature report novel fabrication methodologies and processing protocols to synthesize materials that mimic the physical (transparency, stability etc.), mechanical (strength and stiffness) and biological (cellular compatibility) properties of native cornea, very few studies have focused on mimicking the lamellae structure of the corneal stroma. Each lamellae of the corneal stroma is composed of highly aligned collagen fibrils and the fibrils of adjacent lamellae are arranged in an orthogonal manner. This organized structure of the corneal stroma imparts a high level of transparency and robust biomechanical properties to the native cornea. Development of functional bioscaffolds that mimic the orthogonal lamellae structure of the native corneal stroma will help gain significant headway towards realizing a bioartificial cornea for corneal tissue engineering applications.

Further, limited work has been done in the direction of stem cell based approaches for the development of a fully tissue engineered cornea. Since the corneal tissue is multicellular, understanding cell differentiation pathways will be a key step in devising regenerative approaches using different cell sources such as corneal limbal stem cells, mesenchymal stem cells and induced pluripotent stem cells for corneal regeneration. In conclusion, employing a combinatory approach of developing functional bioscaffolds that mimic the native cornea and fostering stem cell based regenerative therapies will help expedite the development of bioartificial corneas that can be used in the clinic for corneal repair and regeneration.

References

1. J. P. Whitcher, M. Srinivasan, and M. P. Upadhyay, Corneal blindness: A global perspective. *Bull. World Health. Organ.*, **79**(3), 214–221 (2001).
2. Eye Bank Association of America, Annual Report (2013).
3. R. W. Thompson, Jr. *et al.*, Long-term graft survival after penetrating keratoplasty. *Ophthalmology*, **110**(7), 1396–1402 (2003).
4. J. Shimazaki *et al.*, Efficacy and safety of international donor sharing: A single-center, case-controlled study on corneal transplantation. *Transplantation*, **78**(2), 216–220 (2004).
5. K. Y. Cao *et al.*, Demographics of corneal transplantation in Canada in 2004. *Can. J. Ophthalmol.*, **41**(6), 688–692 (2006).
6. A. P. Barboza *et al.*, Project of cornea donation in the hospital complex of Santa Casa de Porto Alegre, Brazil. *Transplant. Proc.*, **39**(2), 341–343 (2007).
7. H. S. Dua *et al.*, The amniotic membrane in ophthalmology. *Surv. Ophthalmol.*, **49**(1), 51–77 (2004).
8. J. Shimazaki, N. Shinozaki, and K. Tsubota, Transplantation of amniotic membrane and limbal autograft for patients with recurrent pterygium associated with symblepharon. *Br. J. Ophthalmol.*, **82**(3), 235–240 (1998).
9. S. C. Tseng, P. Prabhasawat, and S. H. Lee, Amniotic membrane transplantation for conjunctival surface reconstruction. *Am. J. Ophthalmol.*, **124**(6), 765–774 (1997).
10. S. C. Tseng *et al.*, How does amniotic membrane work? *Ocul. Surf.*, **2**(3), 177–187 (2004).
11. C. J. Connon *et al.*, The variation in transparency of amniotic membrane used in ocular surface regeneration. *Br. J. Ophthalmol.*, **94**(8), 1057–1061 (2010).
12. A. J. Shortt *et al.*, *Ex vivo* expansion and transplantation of limbal epithelial stem cells. *Ophthalmology*, **115**(11), 1989–1997 (2008).

13. H. S. Dua *et al.*, Variations in amniotic membrane: Relevance for clinical applications. *Br. J. Ophthalmol.*, **94**(8), 963–964 (2010).
14. L. Sidney, L. E. S. Wilson, S. E. Dunphy, J. B. Rose, and A. Hopkinson, Keeping an Eye on Decellularized Corneas: A Review of Methods, Characterization and Applications. *J. Funct. Biomat.*, **4**, 114–161 (2013).
15. J. Xiao *et al.*, Construction of the recellularized corneal stroma using porous acellular corneal scaffold. *Biomaterials*, **32**(29), 6962–6971 (2011).
16. L. Du, and X. Wu, Development and characterization of a full-thickness acellular porcine cornea matrix for tissue engineering. *Artif. Organs*, **35**(7), 691–705 (2011).
17. E. Yoeruek *et al.*, Decellularization of porcine corneas and repopulation with human corneal cells for tissue-engineered xenografts. *Acta. Ophthalmol*, **90**(2), e125–e131 (2012).
18. H. Luo *et al.*, Construction of tissue-engineered cornea composed of amniotic epithelial cells and acellular porcine cornea for treating corneal alkali burn. *Biomaterials*, **34**(28), 6748–6759 (2013).
19. H. J. Choi *et al.*, Efficacy of pig-to-rhesus lamellar corneal xenotransplantation. *Invest. Ophthalmol. Vis. Sci.*, **52**(9), 6643–6650 (2011).
20. Ruberti, J. W. and J. D. Zieske, Prelude to corneal tissue engineering — gaining control of collagen organization. *Prog. Retin. Eye Res.*, **27**(5), 549–577 (2008).
21. N. K. Monika Abedin, Diverse evolutionary paths to cell adhesion. *Trends. Cell biol.*, **20**(12), 734–742 (2010).
22. T. W. Gilbert, T. L. Sellaro, and S. F. Badylak, Decellularization of tissues and organs. *Biomaterials*, **27**(19), 3675–3683 (2006).
23. S. Amano *et al.*, Decellularizing corneal stroma using N_2 gas. *Mol. Vis.*, **14**, 878–882 (2008).
24. M. Gonzalez- Andrades *et al.*, Generation of bioengineered corneas with decellularized xenografts and human keratocytes. *Invest. Ophthalmol. Vis. Sci.*, **52**(1), 215–222 (2011).
25. A. M. Seddon, P. Curnow, and P. J. Booth, Membrane proteins, lipids and detergents: Not just a soap opera. *Biochim. Biophys. Acta.*, **1666**(1–2), 105–117 (2004).
26. P. M. Crapo, T. W. Gilbert, and S. F. Badylak, An overview of tissue and whole organ decellularization processes. *Biomaterials*, **32**(12), 3233–3243 (2011).
27. J. S. Choi, J. K. Williams, M. Greven, K. A. Walter, P. W. Laber, G. Khang, and S. Soker, Bioengineering endothelialized neo-corneas using donor-derived corneal endothelial cells and decellularized corneal stroma. *Biomaterials*, **31**(26), 6738–6745 (2010).
28. K. Pang, L. Du, and X. Wu, A rabbit anterior cornea replacement derived from acellular porcine cornea matrix, epithelial cells and keratocytes. *Biomaterials*, **31**(28), 7257–7265 (2010).
29. S. Sasaki *et al.*, *In vivo* evaluation of a novel scaffold for artificial corneas prepared by using ultrahigh hydrostatic pressure to decellularize porcine corneas. *Mol. Vis.*, **15**, 2022–2028 (2009).
30. S. Ponce Marquez *et al.*, Decellularization of bovine corneas for tissue engineering applications. *Acta. Biomater.*, **5**(6), 1839–1847 (2009).

31. M. A. Shafiq *et al.*, Decellularized human cornea for reconstructing the corneal epithelium and anterior stroma. *Tissue Eng. Pt. C,* **18**(5), 340–348 (2012).
32. S. Funamoto *et al.*, The use of high-hydrostatic pressure treatment to decellularize blood vessels. *Biomaterials,* **31**(13), 3590–3595 (2010).
33. S. Proulx *et al.*, Transplantation of a tissue-engineered corneal endothelium reconstructed on a devitalized carrier in the feline model. *Invest. Ophthalmol. Vis. Sci.,* **50**(6), 2686–2694 (2009).
34. J. Y. Oh *et al.*, Comparative observation of freeze-thaw-induced damage in pig, rabbit, and human corneal stroma. *Vet. Ophthalmol.,* **12** (Suppl. 1), 50–56 (2009).
35. Y. Fu *et al.*, Reconstruction of a tissue-engineered cornea with porcine corneal acellular matrix as the scaffold. *Cells. Tissues. Organs.,* **191**(3), 193–202 (2010).
36. J. K. Lee *et al.*, The effect of lyophilization on graft acceptance in experimental xenotransplantation using porcine cornea. *Artif. Organs.,* **34**(1), 37–45 (2010).
37. M. Griffith *et al.*, Functional human corneal equivalents constructed from cell lines. *Science,* **286**(5447), 2169–2172 (1999).
38. M. Griffith *et al.*, Artificial human corneas: Scaffolds for transplantation and host regeneration. *Cornea,* **21**(Suppl. 7), S54–S61 (2002).
39. S. Shimmura *et al.*, Collagen-poly(N-isopropylacrylamide)-based membranes for corneal stroma scaffolds. *Cornea,* **22**(Suppl. 7), S81–S88 (2003).
40. F. Li *et al.*, Cellular and nerve regeneration within a biosynthetic extracellular matrix for corneal transplantation. *Proc. Natl. Acad. Sci. USA,* **100**(26), 15346–15351 (2003).
41. Y. Liu *et al.*, A simple, cross-linked collagen tissue substitute for corneal implantation. *Invest. Ophthalmol. Vis. Sci.,* **47**(5), 1869–1875 (2006).
42. K. Merrett *et al.*, Tissue-engineered recombinant human collagen-based corneal substitutes for implantation: Performance of type I versus type III collagen. *Invest. Ophthalmol. Vis. Sci.,* **49**(9), 3887–3894 (2008).
43. W. Liu *et al.*, Recombinant human collagen for tissue engineered corneal substitutes. *Biomaterials,* **29**(9), 1147–1158 (2008).
44. N. Lagali *et al.*, Innervation of tissue-engineered recombinant human collagen-based corneal substitutes: A comparative *in vivo* confocal microscopy study. *Invest. Ophthalmol. Vis. Sci.,* **49**(9), 3895–3902 (2008).
45. P. Fagerholm *et al.*, Corneal regeneration following implantation of a biomimetic tissue-engineered substitute. *Clin. Transl. Sci.,* **2**(2), 162–164 (2009).
46. P. Fagerholm *et al.*, A biosynthetic alternative to human donor tissue for inducing corneal regeneration: 24-month follow-up of a phase 1 clinical study. *Sci. Transl. Med.,* **2**(46), 46–61 (2010).
47. P. Fagerholm *et al.*, Stable corneal regeneration four years after implantation of a cell-free recombinant human collagen scaffold. *Biomaterials,* **35**(8), 2420–2427 (2014).
48. E. J. Orwin, and A. Hubel, *In vitro* culture characteristics of corneal epithelial, endothelial, and keratocyte cells in a native collagen matrix. *Tissue. Eng.,* **6**(4), 307–319 (2000).

49. N. E. Vrana *et al.*, EDC/NHS cross-linked collagen foams as scaffolds for artificial corneal stroma. *J. Biomater. Sci. Polym. Ed.*, **18**(12), 1527–1545 (2007).
50. N. E. Vrana *et al.*, Development of a reconstructed cornea from collagen-chondroitin sulfate foams and human cell cultures. *Invest. Ophthalmol. Vis. Sci.*, **49**(12), 5325–5331 (2008).
51. X. Duan, and H. Sheardown, Dendrimer crosslinked collagen as a corneal tissue engineering scaffold: Mechanical properties and corneal epithelial cell interactions. *Biomaterials*, **27**(26), 4608–4617 (2006).
52. S. Mi *et al.*, Plastic compression of a collagen gel forms a much improved scaffold for ocular surface tissue engineering over conventional collagen gels. *J. Biomed. Mater. Res.* A, **95**(2), 447–453 (2010).
53. S. Mi *et al.*, Photochemical cross-linking of plastically compressed collagen gel produces an optimal scaffold for corneal tissue engineering. *J. Biomed. Mater. Res.* A, **99**(1), 1–8 (2011).
54. X. Xiao *et al.*, *In vivo* study of the biocompatibility of a novel compressed collagen hydrogel scaffold for artificial corneas. *J. Biomed. Mater. Res.* A, **102**(6), 1782–1787 (2014).
55. H. J. Levis, R. A. Brown, and J. T. Daniels, Plastic compressed collagen as a biomimetic substrate for human limbal epithelial cell culture. *Biomaterials*, **31**(30), 7726–7737 (2010).
56. H. J. Levis *et al.*, Plastic compressed collagen as a novel carrier for expanded human corneal endothelial cells for transplantation. *PLoS One*, **7**(11), e50993 (2012).
57. W. Liu *et al.*, Collagen-phosphorylcholine interpenetrating network hydrogels as corneal substitutes. *Biomaterials*, **30**(8), 1551–1559 (2009).
58. C. R. McLaughlin *et al.*, Regeneration of functional nerves within full thickness collagen-phosphorylcholine corneal substitute implants in guinea pigs. *Biomaterials*, **31**(10), 2770–2778 (2010).
59. M. Rafat *et al.*, PEG-stabilized carbodiimide crosslinked collagen-chitosan hydrogels for corneal tissue engineering. *Biomaterials*, **29**(29), 3960–3972 (2008).
60. M. D. DiVito *et al.*, Silica hybrid for corneal replacement: Optical, biomechanical, and *ex vivo* biocompatibility studies. *Invest. Ophthalmol. Vis. Sci.*, **53**(13), 8192–8192 (2012).
61. J. Torbet *et al.*, Orthogonal scaffold of magnetically aligned collagen lamellae for corneal stroma reconstruction. *Biomaterials*, **28**(29), 4268–4276 (2007).
62. N. Builles *et al.*, Use of magnetically oriented orthogonal collagen scaffolds for hemi-corneal reconstruction and regeneration. *Biomaterials*, **31**(32), 8313–8322 (2010).
63. S. Reichl, M. Borrelli, and G. Geerling, Keratin films for ocular surface reconstruction. *Biomaterials*, **32**(13), 3375–3386 (2011).
64. M. Borrelli *et al.*, *In vitro* characterization and *ex vivo* surgical evaluation of human hair keratin films in ocular surface reconstruction after sterilization processing. *J. Mater. Sci. Mater. Med.*, **24**(1), 221–230 (2013).
65. X. Gao *et al.*, Preparation and properties of a chitosan-based carrier of corneal endothelial cells. *J. Mater. Sci. Mater. Med.*, **19**(12), 3611–3619 (2008).

66. Y. Wang *et al.*, A study on the performance of hyaluronic acid immobilized chitosan film. *Biomed. Mater.*, **4**(3), 035009 (2009).
67. B. Ozcelik *et al.*, Ultrathin chitosan-poly(ethylene glycol) hydrogel films for corneal tissue engineering. *Acta Biomater.*, **9**(5), 6594–6605 (2013).
68. M. Grolik *et al.*, Hydrogel membranes based on genipin-cross-linked chitosan blends for corneal epithelium tissue engineering. *J. Mater. Sci. Mater. Med.*, **23**(8), 1991–2000 (2012).
69. D. G. Harkin *et al.*, Silk fibroin in ocular tissue reconstruction. *Biomaterials*, **32**(10), 2445–2458 (2011).
70. T. Chirila *et al.*, Bombyx mori silk fibroin membranes as potential substrata for epithelial constructs used in the management of ocular surface disorders. *Tissue Eng. Pt. A*, **14**(7), 1203–1211 (2008).
71. L. J. Bray *et al.*, Human corneal epithelial equivalents constructed on Bombyx mori silk fibroin membranes. *Biomaterials*, **32**(22), 5086–5091 (2011).
72. J. Liu *et al.*, Silk fibroin as a biomaterial substrate for corneal epithelial cell sheet generation. *Invest. Ophthalmol. Vis. Sci.*, **53**(7), 4130–4138 (2012).
73. P. W. Madden *et al.*, Human corneal endothelial cell growth on a silk fibroin membrane. *Biomaterials*, **32**(17), 4076–4084 (2011).
74. L. J. Bray *et al.*, A dual-layer silk fibroin scaffold for reconstructing the human corneal limbus. *Biomaterials*, **33**(13), 3529–3538 (2012).
75. L. Guan *et al.*, Chitosan-functionalized silk fibroin 3D scaffold for keratocyte culture. *J. Mol. Histol.*, **44**(5), 609–618 (2013).
76. B. D. Lawrence *et al.*, Silk film biomaterials for cornea tissue engineering. *Biomaterials*, **30**(7), 1299–1308 (2009).
77. E. S. Gil *et al.*, Response of human corneal fibroblasts on silk film surface patterns. *Macromol. Biosci*, **10**(6), 664–673 (2010).
78. M. Alaminos *et al.*, Construction of a complete rabbit cornea substitute using a fibrin-agarose scaffold. *Invest. Ophthalmol. Vis. Sci.*, **47**(8), 3311–3317 (2006).
79. K. Tonsomboon, and M. L. Oyen, Composite electrospun gelatin fiber-alginate gel scaffolds for mechanically robust tissue engineered cornea. *J. Mech. Behav. Biomed. Mater.*, **21**, 185–194 (2013).
80. G. G. Gallico, 3rd *et al.*, Permanent coverage of large burn wounds with autologous cultured human epithelium. *N. Engl. J. Med.*, **311**(7), 448–451 (1984).
81. H. Green, O. Kehinde, and J. Thomas, Growth of cultured human epidermal cells into multiple epithelia suitable for grafting. *Proc. Natl. Acad. Sci. USA*, **76**(11), 5665–5668 (1979).
82. S. Ahmad *et al.*, Stem cell therapies for ocular surface disease. *Drug. Discov. Today.*, **15**(7–8), 306–313 (2010).
83. G. Pellegrini *et al.*, Long-term restoration of damaged corneal surfaces with autologous cultivated corneal epithelium. *Lancet*, **349**(9057), 990–993 (1997).
84. J. Shimazaki *et al.*, Transplantation of human limbal epithelium cultivated on amniotic membrane for the treatment of severe ocular surface disorders. *Ophthalmology*, **109**(7), 1285–1290 (2002).

85. K. Nishida *et al.*, Functional bioengineered corneal epithelial sheet grafts from corneal stem cells expanded *ex vivo* on a temperature-responsive cell culture surface. *Transplantation*, **77**(3), 379–385 (2004).
86. P. Rama *et al.*, Autologous fibrin-cultured limbal stem cells permanently restore the corneal surface of patients with total limbal stem cell deficiency. *Transplantation*, **72**(9), 1478–1485 (2001).
87. A. J. Friedenstein *et al.*, Heterotopic of bone marrow. Analysis of precursor cells for osteogenic and hematopoietic tissues. *Transplantation*, **6**(2), 230–247 (1968).
88. M. J. Hoogduijn *et al.*, Human heart, spleen, and perirenal fat-derived mesenchymal stem cells have immunomodulatory capacities. *Stem Cells Dev.*, **16**(4), 597–604 (2007).
89. A. C. Zannettino *et al.*, Multipotential human adipose-derived stromal stem cells exhibit a perivascular phenotype *in vitro* and *in vivo*. *J. Cell Physiol.*, **214**(2), 413–421 (2008).
90. Y. Y. Jo *et al.*, Isolation and characterization of postnatal stem cells from human dental tissues. *Tissue Eng.*, **13**(4), 767–773 (2007).
91. Q. He, C. Wan, and G. Li, Concise review: Multipotent mesenchymal stromal cells in blood. *Stem Cells*, **25**(1), 69–77 (2007).
92. W. Oh *et al.*, Immunological properties of umbilical cord blood-derived mesenchymal stromal cells. *Cell Immunol*, **251**(2), 116–123 (2008).
93. N. Polisetty *et al.*, Mesenchymal cells from limbal stroma of human eye. *Mol. Vis.*, **14**, 431–442 (2008).
94. M. F. Pittenger *et al.*, Multilineage potential of adult human mesenchymal stem cells. *Science*, **284**(5411), 143–147 (1999).
95. H. Liu *et al.*, Bone marrow mesenchymal stem cells can differentiate and assume corneal keratocyte phenotype. *J. Cell. Mol. Med.*, **16**(5), 1114–1124 (2012).
96. T. S. Jiang *et al.*, Reconstruction of the corneal epithelium with induced marrow mesenchymal stem cells in rats. *Mol. Vis.*, **16**, 1304–1316 (2010).
97. S. Gu *et al.*, Differentiation of rabbit bone marrow mesenchymal stem cells into corneal epithelial cells *in vivo* and *ex vivo*. *Mol. Vis.*, **15**, 99–107 (2009).
98. L. Yao *et al.*, Role of mesenchymal stem cells on cornea wound healing induced by acute alkali burn. *PLoS One*, **7**(2), e30842 (2012).
99. H. Liu *et al.*, Cell therapy of congenital corneal diseases with umbilical mesenchymal stem cells: Lumican null mice. *PLoS One*, **5**(5), e10707 (2010).
100. G. I. Agorogiannis *et al.*, Topical application of autologous adipose-derived mesenchymal stem cells (MSCs) for persistent sterile corneal epithelial defect. *Graefes. Arch. Clin. Exp. Ophthalmol.*, **250**(3), 455–457 (2012).
101. Y. Ma *et al.*, Reconstruction of chemically burned rat corneal surface by bone marrow-derived human mesenchymal stem cells. *Stem Cells*, **24**(2), 315–321 (2006).
102. A. I. Teixeira *et al.*, Epithelial contact guidance on well-defined micro- and nanostructured substrates. *J. Cell Sci.*, **116**(Pt 10), 1881–1892 (2003).
103. N. W. Karuri *et al.*, Biological length scale topography enhances cell-substratum adhesion of human corneal epithelial cells. *J. Cell Sci.*, **117**(Pt 15), 3153–3164 (2004).

104. D. Myung *et al.*, Design and fabrication of an artificial cornea based on a photolithographically patterned hydrogel construct. *Biomed. Microdevices.*, **9**(6), 911–922 (2007).
105. Crabb, R. A. and A. Hubel, Influence of matrix processing on the optical and biomechanical properties of a corneal stroma equivalent. *Tissue Eng.* A, **14**(1), 173–182 (2008).
106. A. G. C. Kilic, J. C. Rodriguez-Cabello, and V. Hasirci, A collagen-based corneal stroma substitute with micro-designed architecture. *Biomater. Sci.*, **2**, 318–329 (2014).
107. C. M. Puleo *et al.*, Integration and application of vitrified collagen in multilayered microfluidic devices for corneal microtissue culture. *Lab Chip*, **9**(22), 3221–3227 (2009).
108. I. Ortega *et al.*, Combined microfabrication and electrospinning to produce 3-D architectures for corneal repair. *Acta Biomater.*, **9**(3), 5511–5520 (2013).
109. H. J. Levis *et al.*, Rapid tissue engineering of biomimetic human corneal limbal crypts with 3D niche architecture. *Biomaterials*, **34**(35), 8860–8868 (2013).
110. J.I. Ahn, L. Kuffova, K. Merrett, D. Mitra, J.V. Forrester, F. Li, and M. Griffith, Crosslinked collagen hydrogels as corneal implants: effects of sterically bulky vs. non-bulky carbodiimides as crosslinkers. *Acta Biomater.*, **9**(8), 7796–7805 (2013).
111. A.P. Lynch and M. Ahearne, Strategies for developing decellularized corneal scaffolds. Experimental *Eye Research*, **108**, 42–47 (2013).
112. M. Ahearne and A. Lynch, *Tissue Engineering Part C: Methods*, ahead of print. doi:10.1089/ten.TEC.2015.0008.

Chapter 6

Hydrogels in Bone Tissue Engineering: A Multi-Parametric Approach

Silvia M. Mihaila, Rui L. Reis, Alexandra P. Marques and Manuela E. Gomes*

3B's Research Group, Department of Polymer Engineering, University of Minho, Headquarters of the European Institute of Excellence on Tissue Engineering and Regenerative Medicine, Avepark — Parque de Ciência e Tecnologia Zona Industrial da Gandra4805-017 Barco GMR - Portugal

ICVS/3B's — PT Government Associate Laboratory, Braga/Guimarães, Portugal

**megomes@dep.uminho.pt*

Hydrogels have gained interest as templates in bone tissue engineering, essentially due to their resemblance to tissues' extracellular matrix. In fact, due to their high hydration state, they provide a suitable environment for cell encapsulation, growth and differentiation, as well as active ingredients loading and release. However, they lack mechanical stability and ability to calcify. This chapter will review the multi-parametric aspects considered in the design of hydrogels as templates for bone regeneration, by exploring current strategies aimed at improving hydrogels mineralization and consequently bone regeneration. Moreover, design tools for the introduction of micro-architectural features and generation of shape controlled tissue modules based on hydrogel matrices, will be also analyzed, with particular emphasis on targeting neovascularization, i.e., on the use of hydrogels as supportive templates for *in situ* development of vascular networks.

1. Introduction

Bone is a complex supporting tissue with functions in the movement and in the maintenance of postural stability in cooperation with muscles. Despite its hard structure, bone exists in a constant state of dynamic turnover, known as remodeling, even when growth and modeling of the skeleton have been complete.[1,2] It is estimated that approximately 6% of all human bones are remodeled in a year.[3] However, upon injury, the quality of life of patients dramatically changes. The burden of severe long-term pain and limitation in performing basic tasks, such as walking, frequently causes social and psychological distress. Even though bone has a notable regenerative ability, a considerable bone loss and a hostile microenvironment, as it happens in the case of severe trauma, tumors resection, surgical revisions or developmental deformities, can hinder this capacity.[4]

Despite decades of intensive research, the grafting of autologous bone derived from the iliac crest remains the gold standard of bone replacement treatments. Unfortunately, the amount of available bone graft is limited and the surgical harvesting involves severe donor-site pain and morbidity, which can be as high as 25%.[5] Moreover, besides being an expensive procedure, some complications, such as bone non-unions and blood loss,[6,7] may occur thus requiring additional medical procedures. Cadavers allogeneic bone grafts have been used as alternatives providing a structural host tissue to grow. Nonetheless, albeit avoiding donor-site morbidity, these grafts possess low osteoinductive properties[8] and their success is often jeopardized due to infections, transmitted diseases, delayed inosculation and immunogenicity issues.[9]

The significant clinical shortcomings of auto/allografts have been a strong impetus for the development of alternative approaches for bone replacement and regeneration. Worldwide, approximately 2.2 million bone graft procedures (autologous or banked bone) take place annually,[9] and these numbers are expected to grow due to aging population. The bone graft substitute market was evaluated at $1.9 billion in 2010 and is forecast to reach $3.3 billion in 2017. The main driving force of the market is the increase in orthopedic procedures, aging population and increased preference of emerging bone graft substitutes over auto/allografts. Demineralized bone matrix (DBM), produced by the extraction of the mineral content from allogeneic bone, contains growth factors (GFs) and proteins relevant to the bone microenvironment.[10] Whilst the osteoinductive effect of DBM has been well documented in animal studies, there are merely few reports on human clinical studies,[11] as its-heterogeneity has an impact on DBM properties and clinical performance.[12] The increased interest in providing robust solutions has nourished the expansion of the researchers' vision towards understanding the complex and multi-parametric aspects of bone regeneration (mechanics, chemistry, biological components, integration) using bio-inspired parameters for designing new bone substitutes.

The avenue of tissue engineering (TE) explores these facets of bone repair/healing by the use of cells, scaffolds and in some instances biological factors combinations towards the replacement, repair or restoration of function of a damaged tissue.[8,13,14] The selection of the most suitable material to produce the scaffolds to be used in bone TE is a determinant step towards achieving appropriate mechanical support, biocompatibility, proper integration with the host tissue, osteoinductivity, osteoconductivity, among others. Several materials, such as metals,[15] ceramics,[16] and polymeric systems,[17] have been proposed for bone replacement, but in the context TE, the poor degradability of ceramic scaffolds and the non-degradability of metallic compounds impairs new bone tissue ingrowth. Polymers, on the other hand, are known for enabling the design of their chemistry and architecture during synthesis and processing, thus allowing tailoring 3D scaffolds degradation rate to match the regeneration/repair rate of the tissue to be replaced.

Until recently, the majority of the polymeric bone substitutes were pre-fabricated constructs that require surgical implantation via invasive procedures. Clinically, there is an increased demand for materials that can be implanted under minimally invasive procedures, such as a simple injection.[18,19] Ideally, the viscosity of such materials should be low enough to allow its injection within the defect site and consequent hardening after injection taking advantage of mild crosslinking routines. This strategy would enable the incorporation of drugs, cells, and GFs in the viscous solution before administration.[20] From a TE standpoint, and considering that within natural tissues cells are embedded in a three-dimensional (3D) network combining a complex extracellular matrix (ECM), soluble bioactive factors products of homo- and hetero-typical cell–cell and cell–ECM interactions,[21] microenvironments that closely mimic those features are prone to successfully lead tissue regeneration. Various 3D templates based on hydrogels, that provide biomimetic environments depicting motifs inspired by the role of ECM in regulating bone regeneration, have been proposed.[22–24] Such matrices are conceived to integrate bone tissue and facilitate regeneration by potentiating the bone's inherent capacity to heal, thus acting as *in situ* bioreactors.

In this chapter, the bone matrix deposition and mineralization requirements will be presented in the perspective of using hydrogels as 3D templates with tunable properties that may fulfill these specifications. First, a brief overview on bone biology and dynamic role of ECM in bone remodeling/regeneration will be provided. Subsequently, this chapter will focus on strategies to engineer ECM-mimicking hydrogels by using inorganic components, matrix chemistry, spatial cells distributions and GFs as elements of multi-parametric approaches towards a successful bone TE outcome.

2. Bone Regeneration Features

2.1. A snapshot on bone biology and ECM role in regeneration

Bone is a specialized connective tissue that is exquisitely designed as the load-bearing structure of the body. Its configuration, consisting in a combination of a dense, compact structure (cortical bone) with a highly porous one (cancellous/

trabecular bone), provides the main resources to accomplish this task. Cortical bone makes up to 80% of the total bone mass in adults and comprises an extremely dense mineral structure, mainly hydroxyapatite (HA), derived from calcium phosphates (CaPs),[25] with low porosity (20%) and high mechanical compressive strength (190 MPa)[26] and an elastic modulus of around 20 GPa.[27] The functional unit of cortical bone structure is represented by the osteon that contains central haversian hollow canals that host nerves and blood vessels.[28] In contrast, cancellous bone, which accounts for the remaining 20% of the total bone mass,[29] does not have osteons, due to the high porosity (50–90%) and surface area that encourages a superior penetration of blood microvessels. Due to its spongy feature, the cancellous bone is characterized by low mechanical properties (~10MPa)[26] and an elastic modulus of around 15 GPa.[27] Cancellous bone is mainly composed of type I collagen and small amounts of non-collagenous proteins (NCPs) such as osteocalcin (OCN), osteopontin (OPN), osteonectin (ONC), bone sialoproteins (BSPs), and bone morphogenetic proteins (BMPs).[30,31] It has been demonstrated that collagen serves as an organic template or framework for bone mineralization and provides the bone tensile strength and flexibility,[32] while the anionic NCPs act as nucleators or inhibitors of mineral deposition.[31,33] Recent studies have identified a polysaccharide-enriched matrix, such as glycosaminoglycans (GAGs), at the organic — mineral interface in bone,[34] with important role in both bone formation and bone homeostasis, possibly guiding mineral size, morphology and crystallinity.[35] The carboxyl or sulfate groups of GAGs are thought to contribute to the interfacial interaction and balance between inorganic and organic phase. On the other hand, the inorganic phase contributes to bone stiffness and is responsible for its compressive strength, due to its energy-storing capacity. Although collagen may have less effect on bones biomechanics than inorganic components, it may have a profound effect on bone brittleness.[32] There is evidence that reduction of collagen content can reduce the energy required to cause bone failure (toughness) and increase the fracture risk (osteoporosis).[36]

Bone mineralization is thought to start with the synthesis of the organic matrix that then acts as a template for the deposition of insoluble, dispersed, amorphous CaPs. The formation of CaPs depots is mediated by the presence of calcium-binding moieties and the enzyme-mediated production of free phosphates.[37] The anionic proteins attached to collagen are believed to mediate the stabilization of the amorphous CaPs in the early stage of mineralization and the subsequent formation of nano-sized particles.[38] These precursors undergo several crystalline phase transitions before reaching a stable crystalline apatitic CaPs, or HA. These nanocrystals act as nucleation sites for further crystal growth until the inorganic, crystalline phase overtakes the template.[39]

Bone is constantly remodeling its mineralized matrix, which is a complex process that relies on the synchronization of bone resorption and bone deposition events coordinated by cells (osteoprogenitors and differentiated, mature cells). Within the bone structure relies the bone marrow that comprises of hematopoietic cells and supporting stroma. Until recently, stromal cells were thought to contribute solely to the hematopoietic microenvironment, however, they are now recognized as being progenitor cells of skeletal tissues.[40,41] Osteoprogenitor cells (osteoblast precursor cells) found on the internal surface of bones, can be also found on the external side of the bone, in the periosteum[42] and may also reside in the microvasculature supplying bone. They are derived from mesenchymal/stromal stem cells (MSCs) in the bone marrow where they are found in a resting phase. Upon stimulation, these cells can be activated towards secreting and modeling bone matrix. As a consequence, many bone TE strategies rely on the ability of MSCs to commit towards osteogenic lineage and restore the mineralized matrix.[43]

Osteoblasts, anchorage dependent cells, are highly responsive to the mechanical and chemical stimuli that are propagated through multiple cell–cell and cell–matrix interactions.[44] During collagen-enriched matrix deposition, osteoblasts secrete bone-related proteins and active factors that trigger the invasion of endothelial cells and the development of a vascular network within the matrix,[45,46] concomitant with the matrix mineralization.[30] Osteocytes, terminally differentiated osteoblasts, are entrapped in the mineralized ECM and have an active role in the maintenance of ECM and homeostasis.[46] They are also responsible for sensing the mechanical stress and signaling for bone remodeling.[47] While osteoblasts deposit the organic matrix, osteoclasts that are easily recognized by their multinuclearity, are responsible for bone breakdown and resorption.[48] Additionally, an unbalanced remodeling can lead to osteoporosis (bone resorption > bone formation) or osteopetrosis (bone resorption < bone formation).[49] Despite its intrinsic ability to self-regenerate and remodel, inherent osseous processes are not able to repair a large defect. It is therefore mandatory to consider the structural features of ECM along with the biological principles that master the cell–cell and cell–matrix interactions while developing novel strategies towards bone regeneration.

2.2. GFs involved in bone regeneration

Bone regeneration is a multistep process that involves the interplay between several cells such as inflammatory cells, vascular cells, mesenchymal progenitor cells and osteocytes. *In vivo,* their behavior is orchestrated by a specific set of GFs and cytokines that dictate cellular proliferation, migration and differentiation during bone repair.[50] In the first step of bone healing, inflammatory GFs and cytokines, including fibroblast growth factor (FGF), interleukin-1 (IL-1), and interleukin-6 (IL-6), macrophage colony-stimulating factor (M-CSF) and tumor necrosis factor-α (TNF-α), are involved in the recruitment of inflammatory cells. As the formation of the fracture callus occurs, high concentrations of pro-osteogenic factors, such as platelet-derived growth factor (PDGF), TGF-β, FGF-1, insulin-like growth factor (IGF), and bone morphogenic proteins (BMPs), stimulate the engagement of mesenchymal progenitor cells and their subsequent proliferation and differentiation. Concomitantly, pro-angiogenic factors, such as vascular endothelial growth factor (VEGF), FGF, BMPs and TGF-β, trigger the invasion of endothelial cells (ECs) into the newly formed soft callus re-establishing vascular network connection. Additionally, the intimate cross-talk between osteoblasts and ECs is mediated through the release of VEGF by osteoblasts that act on ECs to promote angiogenesis, and through the release of BMPs by ECs, which lead osteogenic differentiation.[51] The role of indirect cellular signaling, through the release of endogeneous GFs, is complemented by the direct cell–cell contact via gap junctions coupling, for example between ECs and osteoblasts.[52]

3. Hydrogels

Hydrogels are 3D crosslinked insoluble and hydrophilic polymeric networks that are formed by the reaction of one or more monomers through hydrogen bonds or van der Waals interactions between the constitutive chains.[24,53] The hydrogel network can be fabricated via physical or chemical crosslinking methods. While physical crosslinking is achieved by the formation of physical bonds, the chemical crosslinking consists in the formation of stable covalent bonds, mediated by crosslinking agents, between the polymer chains.[54] Hydrogels can be classified according to their origin (natural or synthetic), method of preparation (homopolymers, copolymers, interpenetrating networks), internal architecture (amorphous, semicrystalline, or hydrogen bonded structures), or electric charge (anionic, cationic, neutral, amphiphilic).[24] In addition, hydrogels exhibit a great versatility for the integration of biomacromolecules or sensitive drugs aimed at directing tissue regeneration. The mild encapsulation options enable the controlled load/release of these active ingredients without jeopardizing their activity. From a practical point of view, there is a number of hydrogels that can be implanted using minimally invasive surgical procedures, at the target (injury) site, since they that can be applied as a liquid that further hardens *in situ*.

3.1. Synthetic versus natural hydrogels

Synthetic hydrogels provide chemically programmable and reproducible platforms as their architecture can be easily tailored, allowing a better control over their properties, such as degradation, swelling, mechanical performance, bioactivity etc. Polyethylene glycol (PEG) is one of the most extensively used non-degradable polymer for hydrogels fabrication due to its hydrophobicity and facilitated chemical functionalization through available reactive end-chain moieties. Other non-degradable polymers used for bone TE applications include poly(2-hydroxyethyl methacrylate) (PHEMA)[55] and poly(methyl methacylate) (PMMA).[56] Polymers such as poly(vinyl alcohol) (PVA),[57] poly(actic-co-glycolic) acid (PLGA),[58] poly(propylene oxide) (PPO),[59] poly(propylene fumarate) (PPF)[60] or oligo(poly(ethylene glycol) fumarate) (OPF)[61] have been also proposed for applications in TE. Worth mentioning are the thermoresponsive polymers, such as poly(N-isopropylacrylamide) (PNIPAAm),[62,63] that are able to form free-standing hydrogels at physiological temperature (37°C), enabling their use as injectable delivery devices without the use of external crosslinkers, that might jeopardize the cellular viability or the activity of GFs. Even though there is a wide range of synthetic polymers, with tailorable chemical backbone, their use is often limited by biocompatibility issues and poor degradability.[64]

On the other hand, natural-origin polymers stand out as nature-inspired bio-entities with attractive biocompatibility and biodegradability features. In fact, these materials are more prone to undergo enzymatic degradation as many of the enzymes responsible for that process are found in the human body. The monomer chemistry, chain length and flexibility enable a unique control of hierarchical organization (macromolecular architecture) and functionality. This molecular print directly influences the crosslinking, gelation and processing/handling properties together with the degradation patterns and cellular response.[65] Hence, there is a wide range of natural-origin polymers (chitosan,[66] hyaluronic acid,[67] chondroitin sulfate,[68] pullulan, dextran,[69] alginate,[70] xanthan,[71] pluronic F127,[72] carboxymethyl cellulose,[73] carrageenan[74] or gellan gum[75]) that have been intensively investigated under the scope of TE. However, their use is associated with high degrees of variability, limited processing routes or lack of cellular responsive tags, respectively. For both synthetic and natural-origin polymers, the grafting of peptide sequences within their backbone, such as arginine–glycine–aspartate (RGD) or matrix metalloproteinases (MMPs) sensitive domains, allows modulating the cells attachment, as well as their migration and proliferation within the hydrogel along with its controlled degradation.[76]

For a more comprehensive review on natural and synthetic hydrogels used in TE field, please consult other existing literature.[17,65] The following section will focus on the current strategies towards the use of hydrogels as templates to support mineralization, blood vessel and bone formation.

3.2. Hydrogels as ECM mimics

Hydrogels are characterized by their ability to retain large amounts of water or biological fluids (up to 99%), due to the presence of the insoluble 3D polymeric and interconnected microporous network. The aqueous environment enables the oxygen diffusion and fluid transport, such as nutrients and cell metabolic by-products, in and out of the hydrogels.[24,77] This 3D environment facilitates the mastering of biochemical and biomechanical interactions with encapsulated cells. Moreover, by grafting functional groups, specific cellular processes such as adhesion, spreading, proliferation, differentiation and consequent matrix deposition and organization can be triggered. These features resemble those of the native ECM, therefore, hydrogels have been increasingly considered as candidate materials for TE applications. For instance, protein-based scaffolds or hydrogels, such as collagen,[78,79] gelatin[80,81] and fibrin[82] are very attractive due to their ability to mimic the ECM biodegradability and inherent biochemical properties.

Recently, hydrogels incorporating DBM have been developed, as they contain GFs, NCPs and type I collagen, the key elements of the bone ECM. In order to facilitate handling, formulation and reliable delivery in clinical applications, the DBM particles were incorporated into water-soluble gels, such as the hyaluronic acid[83] or carboxymethylcellulose.[84] Sawkins *et al.* proposed the processing of DBM as a soluble, hydrogel-like formulation, to avoid the use of liquid carriers. The DBM processing involved a mineral extraction in acid conditions, followed by lyophilization steps. Further, the DBM was dissolved in pepsin and the gelation of the DBM pre-gel solution occurred within at 37°C, upon pepsin neutralization. The DBM hydrogels exhibited sigmoidal gelation kinetics (formation of collagen aggregates at higher concentration) consistent with the nucleation and growth mechanism found in bone.[85] However, further studies are needed to assess the osteoinductivity and osteoconductivity of these hydrogels, and their *in vitro* potential can be translated in *in vivo* settings.

Meanwhile, bone TE strategies have extensively exploited the glycan components (GAGs) to address the formation and maintenance of bone tissue. The carboxyl or sulfate groups of GAGs are thought to contribute to the interfacial interaction and balance between inorganic and organic phase. The resemblance of some polysaccharides with GAGs chemistry has enabled their use as templates for bone mineralization.[66] Several GAGs-analogs were proposed for bone TE, such as chitosan,[66] hyaluronic acid,[67] chondroitin sulphate,[68] pullulan and dextran,[69] together with other biomaterials (alginate,[70] xanthan,[71] pluronic F127,[72] carboxymethyl cellulose,[73] carrageenan,[74] or gellan gum[75]).

TE approach aims at replicating the interactions that naturally occur in the bone tissue by using ECM-like structures (scaffolds), cells and/or GFs in a coordinated fashion to achieve a biomimetic construct. Moreover, these scaffolds should be finely tailored towards mimicking the intricate and organized ECM meshwork into a dynamic native microenvironment.

4. Mineralization of Hydrogels

The development of bone-like composites with improved mechanical and biomineralization features calls for a biomimetic approach inspired in natural bone. The use of hydrogels as modular scaffolds has been mainly considered for soft TE (skin, cartilage, cardiovascular applications), however, their versatility has resulted in increased interest in assessing their potential to replicate the organic phase involved in bone healing together with acting as templates for mineralization. Still, hydrogels are associated with a number of limitations, such as their lack of mineralization capacity upon implantation,[64] or the inherent poor mechanical properties that restrict their use to non-load-bearing applications.[86] However, the network biomimetic reinforcement can be achieved by the addition of other phases[64,87,88] or other approaches, such as chemically-, GFs- or cell- induced mineralization, which will be briefly discussed further.

4.1. Incorporation of inorganic components in hydrogels

The concept of combining hydrogels with inorganic components for bone TE is inspired by the composite nature of bone itself. As it happens during bone healing, inorganic elements act as seeds for HA formation, enabling and reinforcing the integration with host surrounding bone tissue.[89] The hydrated state of hydrogels permits the initiation of apatite deposition by the heterogeneous nucleation of apatite induced by the phosphate, silica and calcium ions dissolved from the ceramic components.[38,90] While hydrogels can provide an excellent 3D template to foster cell growth and ECM deposition, the presence of inorganic minerals enables their osteoconductivity (i.e., supporting the bone formation) by acting as substrates with high affinity for proteins and GFs, which directly affects the adhesion, proliferation and differentiation of progenitor cells. Moreover, the addition of inorganic particles, by reinforcing the network strength, has been shown to increase hydrogel stiffness, in turn, known to play a role in cell differentiation, including towards the osteogenic lineage.[91] Nonetheless, due to the elasticity of the hydrogel network and facilitated handling, brittle ceramic particles can be delivered in a moldable or injectable formulation. For instance, Oliveira *et al.*[73] evaluated the potential of carboxymethylcellulose, hydroxypropyl methylcellulose and alginate hydrogels as injectable HA microspheres carriers, using an orthopedic injectable device designed for minimally invasive bone surgeries.

There are many nano- and micro-structured bioactive inorganic materials, such as bioglasses or apatitic calcium phosphates (HA) that can be used to render mineralized hydrogels. These materials form a strong bond with naturally occurring HA, present at the implantation site,[92] enabling a smooth integration of the implant with the host bone tissue. Bioglasses are amorphous solids containing silica (SiO_2<60 wt.%) that are bioactive due to their high reactivity in aqueous environments. The most commonly used forms of CaPs are β-tricalcium phosphate (β-TCP), amorphous calcium phosphates and HA. These ceramic materials strongly resemble the nanostructured mineral phase of bone and are found in many normal and pathological calcified tissues in the human body.[93] In addition, carbonated apatites have a buffering effect on the acidic pH caused by the degradation by-products, thus minimizing excessive acid environment and inflammation at the implantation/injury site.[94]

In fact their combination with varied biomaterials and 3D structures aiming at bone regeneration has been following different strategies with interesting outcomes, recently and thoroughly reviewed.[95–97]

Recently, carbon nanotubes (CNTs) and graphene[98] were proposed as reinforcing agents of hydrogels, due to their outstanding mechanical properties. For instance, the rod-like shape and nanoscale dimensionality of CNTs allow the morphological mimicking of fibrillar (rod-like) proteins, such as collagen, present in bone ECM, thus, acting as guiding templates for mineral deposition.[99,100] Even more, the combination of nano-scaled single walled CNTs with biomimetic features and HA nanoparticles significantly enhanced the mechanical properties of a chitosan hydrogel and synergistically improved osteoblasts adhesion and proliferation.[101] Despite of being inert materials that cannot be mineralized, recent studies showed that graphene-based hydrogels promote cell adhesion, spreading and proliferation, and stimulate the osteogenic differentiation of stem cells, without additional inducers.[97] Graphene based hydrogels were shown to preserve the osteogenic differentiation of encapsulated bone marrow MSCs, under hypoxic conditions.[102] When subcutaneously implanted in rats, the graphene films produced minimal capsule formation and mild host tissue response, with evidence of blood vessel formation and mineralization.[97] Thus, the nano-scale biomimetic features of these systems, together with the hydrogel 3D structure, created a favorable environment to support *in vitro* osteogenic differentiation of cells and *in vivo* mineralization.

4.2. Soaking/incubation treatments

Within the bone TE context, bioactive materials stand out as preferred candidates for development of bone-mimicking constructs, due to their ability to induce and sustain apatite deposition in physiological relevant conditions. If the organic polymer is provided with apatite-forming ability, it is expected to dynamically resemble the organic phase of bone.

In literature, several *in vitro* methods have been developed to induce the spontaneous mineralization of biomaterials or to assess whether a certain material composition undergoes biomimetic mineralization. Amongst these methods, the soaking/incubation of materials in mineralization baths has gained considerable interest. With the development of simulated body fluid (SBF), a polyionic mixture, that contains polyelectrolytes in concentrations nearly equal to those found in human plasma,[103] biomimetic processes to deposit/coat apatite on organic polymers have been proposed. The biomineralization process is based on the consumption of Ca^{2+} and PO_4^{3-} ions, present in the SBF solution, thus, resulting in the spontaneous growth of bone-like apatite nuclei on the surface of biomaterials. Additionally, testing the apatite-forming abilities of biomaterials enables to predict their mineralization behavior *in vivo* after implantation. In fact, the bioactivity assessment, under SBF treatment, is a pre-requisite for screening material compositions as suitable for bone TE. Noteworthy, the outcome of a SBF treatment is highly depended on the composition of the formulation — either the chemical backbone possesses functional groups that sustain the biomimetic mineralization or the formulation contains CaPs precursors (already discussed in Sec. 4.1) that act as nucleation sites towards apatite deposition.

For instance, knowing that carboxyl groups (–COOH) have a catalytic effect for heterogeneous apatite nucleation in SBF, Ichibouji *et al.*[104] prepared hydrogels from various pectins containing –COOH groups. The authors report differences in the amount of bound calcium and mineral formation and suggest that the apatite-forming ability of the pectin gels is governed by not only the amount of carboxyl groups, but also changes in Ca^{2+} concentration and pH in surrounding solutions.

Similarly, Zainuddin *et al.* showed the formation of a 0.5μm apatitic layer on the surface of chemically defined PHEMA hydrogels upon immersion in SBF.[105] The calcification process occurred predominantly through spontaneous precipitation of CaPs. Reducing the relative number of oxygen atoms in the copolymers had a direct effect of early calcification, however, there were no differences in the extent of mineralization for prolonged soaking times. In the case of poly (ethylene imine) (PEI), the crosslinked hydrogels were shown to be efficient templates for CaPs deposition in SBF. In contrast, the non-crosslinked PEI did not lead to the phosphate nucleation and its subsequent growth in SBF.[106] The authors suggested that the un-crosslinked PEI is more protonated, and thus can release more protons, leading to the decrease of the local pH and, in consequence, a higher solubility of CaPs, hindering its precipitation.

Despite the promising results, a limitation of using soaking treatments is the deposition of the apatitic layer only at the interface between SFB and the hydrogel substrate. The incorporation of CaPs precursors in the hydrogel during its formation allows a homogeneous distribution of bioactive agents throughout the entire structure. Upon immersion in SBF, these precursors act as nucleation sites or inorganic "seeds" that lead to *in situ* formation of HA and further crystal growth. The addition of inorganic agents (detailed in Sec. 4.1) is one of the most popular method to introduce nucleation sites within inert hydrogels. As an example, Couto *et al.*[107] developed thermo-responsive hydrogels with bioactive features based on the combination of chitosan/ beta-glycerophosphate (β-GP) and bioactive glass nanoparticles. A CaPs-like layer was detected at the surface of the hybrid hydrogels, however, at a closer view, crystalline deposits were detected in the interior of the hydrogels, an indication that the mineralization occurred beyond their surface.

Alternatives to the addition of inorganic agents during hydrogel formation, involve the diffusion of Ca^{2+} and PO_4^{3-} ions that triggers the *in situ* deposition of CaPs. For instance, Azami *et al.*[108] employed a double diffusion method whereby $CaCl_2$ and Na_2PO_4 diffused in a gelatin hydrogel to form amorphous CaP in the bulk interior of the gel. Subsequent incubation into SFB lead to the maturation of the mineral phase into nanocrystalline HA. Similarly, the development of nano-structured composites in alginate hydrogels was achieved by *in situ* crystallization of HA in the presence of $CaCl_2$ and Na_2HPO_4.[109] The nano-sized crystals were evenly distributed in the hydrogel network, while the crystal size and morphology could be controlled by the material properties.

In summary, entrapping precursor phases inside a hydrogel that result in their biomimetic conversion to nano-crystalline HA, is an attractive strategy for *in vivo* applications, as the crystal maturation process with SBF can be reproduced within the body. It should be mentioned that mineralization by incubation in SBF can be significantly enhanced by either functionalizing the chemical backbone with negatively charged moieties (see Sec. 4.4) or by incorporating biomacromolecules with calcium-binding domains, such as casein, fibroin and alginate (see Sec. 4.5).

4.3. Alkaline phosphatase-mediated mineralization

The formation of *in situ* mineralization nucleation sites within hydrogels, catalyzed by enzymes that trigger its temporal and spatial development is becoming a route of great interest to generate matrices for bone regeneration .[110]

Alkaline phosphatase (ALP) is an important player in bone remodeling that acts as a catalyst in initiating calcification. More specifically, the catalytic mechanism involves the cleavage of phosphates from organic phosphomonoesters, thereby increasing the local concentration of inorganic phosphates that act as mineralization promoters, resulting into an enzyme-mediated deposition of carbonated apatite.[111,112] At the same time, the catalytic nature of ALP extends to the role of decreasing the concentration of extracellular pyrophosphates that acts as inhibitors of apatite crystal growth. Consequently, ALP loaded hydrogels constitute not only an alternative to CaPs enriched matrices, in which the CaPs tend to aggregate,[113,114] but also work as *in situ* self-promoter of mineralization.[61,75,115]

Although the harsh processing conditions of ALP-surface coated implants might compromise its activity, the entrapment of ALP into hydrogels during gelation was proved to be mild enough to preserve ALP initial catalytic activity.[61] In a similar approach, ALP was immobilized on PHEMA using a copolymerization method that did not affect the enzyme catalytic activity. As a result, after immersion in SFB for 17 days, mineralized depots could be observed.[116] Moreover, due to its molecular weight, 185 kDa,[117] and the reduced pore size of hydrogels, ALP release by diffusion[61] and loss of active component at the injury site where the hydrogel was applied is significantly hindered. After ALP incorporation, gels were incubated in a calcium glycerophosphate (Ca-GP) solution. Ca-GP diffused into the hydrogel, where under the action of ALP, phosphate was released which reacted with calcium ions to form CaPs. Incorporating ALP into gellan gum led to the formation of CaPs and increased in the stiffness of the matrix. The functionalization with polydopamine of ALP-loaded gellan gum hydrogels, not only improved the previous outcomes, but also enhanced cellular attachment and the cytocompatibility of the mineralized gels.[75]

Xie *et al.*[118] incorporated osteinductive HA particles into an alginate hydrogel matrix with physically immobilized ALP and showed that ALP-mediated mineralization provided a homogeneous mineral distribution, smaller crystal size, enabling the increase of Young's moduli. Using a functionalized peptide amphiphile hydrogel matrix that resembles the scaffolding framework of natural collage, together with ALP, it is possible to drive the controlled spatial distribution of the enzyme-mediated nucleation sites and consecutively, templating the HA deposition. Even more, due to the role of ALP to control temporal variables in bone mineralization, Spoerke *et al.* proved that both spatial and temporal elements are essential to achieve *in vitro* mineralization.[110]

Moreover, it was shown that ALP induced the mineralization with CaP of injectable thermosensitive chitosan/ β-GP hydrogels while accelerating their gelation, which is an attractive feature for clinical applications.[115] The mechanism in which ALP accelerated the gelation process can be explained by the potential of ALP to split β-GP into phosphate ions and glycerol, which can promote the ionic and hydrophobic interactions between the chitosan chains, in turn promoting gel formation, concomitantly with their reaction with Ca^{2+}, leading to *in situ* formation of CaPs.[61]

In a similar manner, mineralization of collagen type I, PEG-, or OPF-derived hydrogels was achieved by physically immobilizing ALP and, subsequently, soaking them in a Ca^{2+} and β-GP containing solutions, that acted as substrates for ALP. When OPF hydrogels were implanted in a rat subcutaneous model, only those containing immobilized ALP were able to induce mineralization.[119] Specifically, small mineral deposits were observed at the periphery of the hydrogels near the dermis/scaffold interface, as a consequence of the ALP release. However, the covalent bonding of ALP constitutes an advantage over the physical immobilization of ALP and incorporation of inorganic particles within the hydrogel that are easily diffused to the transplantation site, thus reducing the expected level of mineralization. Therefore, ALP was covalently linked to dentine-derived collagen sheets to induce their mineralization, which was confirmed upon subcutaneous implantation in rats.[120] Similarly, covalently bonded ALP-fibrin scaffolds were shown to promote a significant increase in the percentage of bone volume per total volume when compared with the empty defect *in vivo* bone formation in a mouse calvarial defect.[121]

Finally, ALP has also been used as a screening agent for assessing the mineralizability of novel hydrogel systems. Gongormus *et al.* used ALP-mediated mineralization to evaluate the mineral formation in several formulations of self-assembled peptide hydrogels.[122]

4.4. Functionalization of hydrogels

Another approach that has been proposed to induce hydrogels mineralization is the functionalization of their side chains or backbone polymers with negatively charged moieties such as carboxyl ($COOH^-$), hydroxyl (OH^-), or phosphate (PO_4^{3-}) groups. The rationale relies on mimicking the biomineralization mechanism that naturally occurs during bone healing, when non-collageneous (NCPs) calcium-binding proteins, generally negatively charged (acidic or phosphorylated),[33] act as nucleation sites for apatitic nanocrystals growth. The anionic domains attract Ca^{2+} that, subsequently, leads to an ionic saturation that triggers the CaP precipitation. Similarly, by alternating sequences of PO_4^{3-}, $COOH^-$, or OH^- moieties along the polymeric chain of the hydrogels, it is possible to decorate the hydrogel with apatitic-nucleating properties. Thus, the chemical modification of hydrogels with anionic functional groups enables the development of hydrogel with the ability to nucleate mineralization.

The modification of a PEG diacrylate (PEGDA) hydrogel with photoreactive phosphate-containing molecule, ethylene glycol methacrylate phosphate (EGMP), lead to the formation of a mineral phase throughout the hydrogel, capable of sequestring OPN which then promoted MSCs adhesion and spreading.[123] Similarly, PEG hydrogels modified with phosphoesters groups were able to drive extensive and uniform mineralization upon immersion in osteogenic medium for 3 weeks,[124] while PEG hydrogel having pendant side chains terminating in –COOH were shown to act as templates for mineralization. However, the ability of these hydrogels to undergo template mineralization was found to be strongly depended on the length of the side chain, which directly influenced the hydrophobicity of the material, a design parameter that needs to be taken into consideration for the development of bone-like materials.[125]

The introduction of hydroxyl-containing silanol (Si–OH) in injectable PNIPAAm-PEG dimethacrylate copolymer initiated apatite formation.[90] The extent of Si–OH content enabled the system to be tuned in regard to the compressive moduli that can range from 50 to 700kPa without altering the thermo-responsiveness ability of the hydrogels. These hydrogels provided Ca^{2+} bonding sites and induced subsequent mineralization when soaked in SBF. The challenge of this material system is to balance the network-forming and modulus-enhancing copolymer, while maintaining an injectable hydrogel for potential tissue regeneration.

By manipulating the structure and density of mineral-nucleating ligands presented on the hydrogels, Song *et al.* prepared libraries of PHEMA-based copolymerized with functional groups.[38] This study showed that synthetic organic matrices could be integrated with biominerals with varied affinity, morphology and cristallinity. Strong adhesion between the organic and inorganic phases was achieved for hydrogels functionalized with either carboxylate of hydroxyl ligands. The mineral-nucleating potential of hydroxyl-conjugated matrices broadened the design parameters for the development of bone-like composites and control over the material performance.

4.5. Biomacromolecules-driven mineralization

The primary amino acid sequence of the NCPs found in bone, BSP, OPN, OCN, ONC, includes a high density of aspartic and glutamic acids residues, which have a high affinity towards Ca^{2+} ions.[39] A wide range of naturally occurring biomacromolecules, such as proteoglycans, GAGs,[126] and as casein[96] or fibroin,[127] possess calcium-binding domains. Considering the hydrogel matrix as a bone ECM substitute, biomacromolecules can be immobilized within the matrix where they act as "ionic pumps" and encourage the formation of nucleation sites, thus enabling the initiation of apatitic mineral deposition and subsequent mineralization. For instance, casein, a calcium carrier phosphoprotein found in milk, lead to the mineralization of PHEMA hydrogels to which was physically immobilized.[128] Interestingly, nucleation sites of CaPs deposits in the form of nano-scaled nodules (20–30 nm) were observed exclusively in the casein-enriched domains.

The interpenetrating network (IPN) formed by embedding silk fibroin in 3D polyacrylamide (PAA) networks allowed the deposition of apatitic-like minerals, when the hydrogels were immersed in SFB. The hydrogels were uniformly covered with a mineral layer whose morphology depended on their composition. A higher content of silk fibroin within the hydrogel (30/70, 50/50) led to a more uniform coating of the surface with an apatite-like mineral phase. However, increasing the silk fibroin: PAA ratio led to a decrease in the compressive strength of the gels (from 151 kPa for pure PAA to 62 kPa for 50/50 ratio) due to the increased swelling of the hydrogel matrix.[129] Recently, Lin *et al.*[130] proposed a ABA triblock copolymeric hydrogel with both thermoresponsive and mineralization properties. PNIPAAm was chosen as the stimuli-responsive outer blocks due to its phase transition temperature, while phosphorylated poly(acrylic acid) was designed as the B block of the copolymer, due to its high affinity towards calcium ions, that benefit the biomineralization process. These hydrogels were shown to have a self-mineralization capability after being incubated in SBF for 2 weeks, thus leading to *in situ* HA deposition.

Alginates are one example of glycans with great affinity to calcium ions. A series of systems have been developed by the addition of alginate to hydrogels to promote mineralization in which alginate acts as a permeable organic matrix that supports the deposition of the minerals, while arresting the Ca^{2+} and PO_4^{3-} ions inside the polymeric "cages".[131] Gelatin, a denaturated form of collagen, was modified through intrinsic interactions with different concentrations of calcium-binding alginate, leading to the formation of porous IPN hydrogels.[131] The apatite-forming ability was favored by the increase of the alginate to gelatin ratio. In a similar manner, Cha *et al.* covalently incorporated negatively charged methacrylated alginate (MA) in poly(propylene glycol)monomethacrylate (PPGmM), so that charge density, hydrophobicity and pore size of the hydrogels could be separately controlled with mass fraction of MA and PPGmM. The authors highlighted that not only charge density (existence of Ca^{2+} binding sites), but also the hydrophobicity and pore size of the hydrogel are major variables to be considered in the integrative design of a 3D mineralized matrix that allow one to identify osteon variables, critical for bone development and repair. However, this should be preceded by studies that relate these hydrogel screening formulations to the incorporated amounts of apatite.[132]

4.6. Peptide-mediated mineralization

In a similar fashion, as presented in the section above, peptide sequences, with calcium binding affinity, can be conjugated on the hydrogel backbone, however, there are also hydrogel formulations that are based exclusively on polymerized polypeptides. The versatility of the peptide design allows the synthesis of hydrogels with specific chemistries and architectures that enable the tuning of the properties of the molecular matrix towards different behaviors,[133] such as cell adhesion, proliferation and differentiation.

The incorporation of bioactive peptide motifs such as RGD is one of the most popular route to enhance functionality, while proteolytically degradable peptide motifs (such as those recognized by cell-secreted matrix metalloproteases, MMPs, is a common approach to address biodegradability to levels that are more in tune with tissue remodeling and regeneration.[134] For instance, when MSCs were cultured with MMPs domain containing hyaluronic acid hydrogels, the osteogenic differentiation occurred accompanied by an intense mineralization of the matrix. However, switching to a less permissive matrix (without MMPs-domains), the differentiation of MSCs was shifted towards adipogenesis.[135]

The combination of RGDs with MMPs is mandatory to render the networks to be degraded, *via* cell-secreted MMPs, and easily invaded by cells. Bone regeneration was shown to depended on the proteolytic sensitivity of these matrices that assist the MMP-mediated invasion of cells, as it affects their spreading, migration and organization.[136] For example, MMP-sensitive hyaluronic acid hydrogels loaded with BMP-2 and MSCs demonstrated to promote the highest mature bone formation in a calvaria defect,[137] when compared with hydrogel without MMPs-sensitive domain, with or without BMP-2.

More recently, the incorporation of other biological components of the ECM, such as GAGs, and, in particular, heparin is yielding interesting results.[133] On the other side, there are strategies that focus on peptidic hydrogels with sequences that allow the dissolution of inorganic particles or calcium sequestration and subsequent mineralization. For example, Amosi *et al.* designed β-sheet-forming amphiphilic peptides rich in aspartic acid residues and carboxyl groups (P_{FD}-5) to improve bone regeneration by β-TCP particles.[138] This peptide hydrogel formulation was found to accelerate the dissolution of β-TCP, while acting as efficient Ca^{2+} depots, and induce the osteoblast differentiation *in vitro*. The *in vivo* studies in a rat distal femoral model showed a stronger regeneration capacity obtained in those defects treated with β-TCP-loaded hydrogels, indicating that the peptidic hydrogels and the mineral synergistically act to enhance bone regeneration, in the same range as the performance of matrices presenting cell-binding motifs or loaded with GFs.

Nonoyama *et al.* synthesized two β-sheet peptides with one ($(LE)_8$) or two glutamic acid residues ($(VEVSVKVS)_2$) per molecule, respectively, that self-assembled into nanofibers and formed a hydrogel structure in the presence of calcium ions.[139] The viscoelastic properties of both formulations increased with the addition of Ca^{2+} ions, reaching values that were compatible to their use as bone fillers, which potentiate their application to a defined injectable scenario. Then phosphate ions were added to the calcium containing $(LE)_8$, amorphous calcium phosphate was formed along the peptide fiber at neutral pH. On the other hand, $(VEVSVKVS)_2$ hydrogels enabled the deposition of HA, at basic pH. Besides providing a template for mineralization, calcium ions increased the mechanical performance of both formulations, which had a direct influence over the cristallinity of HA.

4.7. GF-mediated mineralization

In order to assess bone regeneration both *in vitro* and *in vivo*, numerous studies have elaborated hydrogel-based systems that allow the release of BMP-2 in a controlled manner, by manipulating the network properties. Solorio *et al.* encapsulated recombinant BMP-2 (rhBMP-2) within gelatin hydrogel microspheres (2–6μm diameter) with different crosslinking degrees to induce bone formation using hMSCs. These authors showed that higher degrees of crosslinking of the gelatin resulted in higher loading efficiency and slower release rate. The release of rhBMP-2 from the microspheres caused a three- to eight-fold increase in the expression of BSP sialoprotein.[140] Even more, the release kinetics of GFs can be adjusted by the hydrogel diffusion properties. For instance, Ludmila *et al.* showed that rhBMP-2-loaded hyaluronic acid hydrogels produced a more osteoinductive response when compared to a chitosan one of similar design.[141] However, it should be noted that, although more bone formation was observed in the hyaluronic acid hydrogel, the bone formed in the chitosan formulation was described as being in a more mature state with significantly higher levels of calcification. This increase in maturation within the chitosan/rhBMP-2 hydrogels was attributed to the more rapid breakdown of the hydrogel leading to a faster release of the rhBMP-2. In a further study conducted by the same group, the rhBMP-2 chitosan hydrogel was supplemented with β-tricalcium phosphate (β-TCP), which involves the osteoinduction at the cost of slower mineralization with more overall bone growth observed, but less mineralization. They attributed this effect to the slower release of rhBMP-2 from the scaffold due to the affinity of rhBMP-2 to bind to β-TCP.[142] Furthermore, since the complex process of *in vivo* bone regeneration involved the interaction of multiple GFs and cytokines, the combined delivery of multiple factors has attracted researcher's interest. The association of the osteinductive activity of BMPs with the effect of other GFs has been shown to enhance bone formation *in vivo*. Simmons *et al.* showed that the combination of BMP-2 and TGF-β3 in RGD-alginate hydrogels, results in a synergistic effect to enhance bone formation in a mice model,[143] that was not observed when GFs were delivered separately. In another study, Raiche *et al.*[144] used a two-layered gelatin hydrogel to obtain combined and sequential delivery of the BMP-2 and IGF-I. The early delivery of BMP-2 followed by increase release of BMP-2 and IGF-I after 5 days, resulted in the largest mineralized content. Even more, simultaneous delivery of both GFs did not significantly change ALP activity or matrix calcium content.

Recently, platelet-rich plasma (PRP) has been studied as a promising alternative source of bioactive agents that can be released in physiological concentrations. Upon activation, when PRP is called platelet lysates (PLs), releases a wide range of biologically relevant proteins, including GFs, such as PDGF, TGF-β, bFGF, and

VEGF among others. In fact, in combination with thrombin, PRP forms gel-like that can be exogenously applied to tissues to promote wound healing, bone growth, and tissue sealing due to the cocktail of GFs. Moreover, PLs are becoming attractive for bone regeneration, as they promote cell growth, differentiation and recruitment.[145] Santo *et al.* proposed PLs loaded chondroitin sulfate-chitosan hydrogels for the enhancement of cellular viability, proliferation and calcium deposition, suggesting an increased osteogenic differentiation of MSCs.[68] Even more, PLs hold the potential of inducing the formation of a vascular network through the release of VEGF, simultaneously with that of BMP-2.[146]

5. Bottom-up/Integrative Approaches

Bottom-up approaches comprise bio-inspired strategies to reproduce the hierarchical architecture of natural bone tissue by mimicking its microstructural features. Single units at micro- or nano-scale that serve as building blocks for further assembling have been used to construct spatially controlled structures. These modular units can be created by cell sheeting, cell-laden microfabrication or 3D direct cell-loaded hydrogel printing, and assembled in such a way to mimic the native microstructural repeating biofunctional unit of the bone tissue, which is the osteon.[147] The osteon comprises of a dense collagen network impregnated with inorganic precursors and NCPs. During bone formation, the organization of collagen into osteon-like structures enables HA deposition. These multifaceted aspects, allow the integrative spatio-temporal relations between bone organic matrix and HA mineral deposition. Osteon organization has inspired for example, a strategy in which osteoblast-like cells were seeded on dense collagen/chitosan hydrogel matrices that replicate the 3D fibrillar collagen pattern, and showed to enhance bone-like HA production.[148,149] Similarly, Xu *et al.*[150] developed a multi-functional substrate for bone formation based on a fiber-reinforced laminated hydrogel nanocomposite system aimed at mimicking the laminar structures (osteons) of bone. Layers of electrospun poly (L-lactide) fiber mesh were coated with poly(lactide-co-ethylene oxide fumarate) (PLEOF) hydrogel precursor solution, previously loaded with HA and grafted with RGD sequences to create local adhesion sites for MSCs. Subsequently, these layers were stacked, pressed together and crosslinked to produce a laminated fiber-reinforced composite with osteoinductive and cell adhesive properties. The lamination dramatically increased the Young modulus up to 570±130 MPa, within the range of wet human cancellous bone, when compared with the moduli of the hydrogel (0.5±0.07 MPa) or fiber mesh (140±3 MPa), alone. However, these values are still below the elastic moduli of the native osteon lamellae (7–22.5GPa).[151] The RGD–PLEOF hydrogel provided a bimodal degradation profile: a slow degradation rate for structural stability and a fast degradation rate of the hydrogel that was substituted by ECM deposited by the differentiating cells.

5.1. Patterning hydrogels

Micromolding and photolithography of hydrogels has been frequently used to generate 3D cell-laden microgel units. Micromolding of hydrogels uses poly(dimethyl siloxane) (PDMS) molds microfabricated with a variety of shapes and sizes. In the first step, the pre-polymer solution with encapsulated cells is casted over the PDMS mold, followed by the crosslinking by changing pH, temperature, ionic strength, or UV light application to generate a hydrogel with the exact microstructural design of the PDMS molds.[147] As these platforms can provide spatial distribution of biochemical triggers, they may enable, for example, the study of guided tissue (bone) regeneration. For instance, PEG hydrogel micropatterns (10 μm) were deposited on the surface of silicon wafers, in order to create cell anti-adhesive templates, providing a spatial control over cell interactions with the substrate. When loaded with VEGF, the cells aggregated at the boundaries of the PEG pattern and proceeded towards the osteogenic differentiation.[152]

Similarly, photolithography allows the transfer of a certain pattern into a hydrogel, by using photomasks with different patterns in specific-localized regions. Using this technology, a pre-polymer solution of a photocrosslinkable hydrogel containing a photoinitiator is placed under a mask and crosslinked under UV light exposure. The hydrogel crosslinks only in the areas under which the photomasks allow the penetration of the UV light, generating microstructural hydrogel modules with determined shape and size.[147] Recently, Jeon *et al.* engineered a novel micro-patterned alginate/PEG hydrogel system to spatially control hydrogel physical properties on the micro-scale by single or dual-crosslinking mechanisms that allowed stiffness patterning, as well. The micro-patterned checkerboard, with single and double crosslinks, allowed the examination of the effect of the size of these hydrogel patterns on the behavior of encapsulated adipose derived stem cells. By manipulating the micro-pattern size and the crosslinking mechanisms, the physical properties of alginate hydrogels were spatially tunable. Thus, as the micropattern properties dictated the behavior of encapsulated cells, in the stiffer patterns, cells formed clustered structures, larger for increased pattern size, and differentiated into the osteogenic lineage. Following this approach the local control over cell behavior was achieved ultimately, allowed the engineering of complex bone constructs with dual cell-type using a single stem cell source.[153]

By applying photolithography principles, it has been also possible to fabricate cell-laden microgels that can be assembled into 3D architectures.[154] So far, the bottom-up assembly of these units provides a powerful and highly scalable approach to form biomimetic 3D tissue constructs that replicate the microstructural organization of native tissue.[155]

Another bottom-up approach that has been explored for bone TE is 3D cell, tissue and organ printing, based on rapid prototyping technologies, such as bioplotting, ink-jet printing, laser deposition and dispensing tools.[156] The printing of cell-laden hydrogels allows the mimicking of anatomical organization of cells, matrix and bioactive molecules for obtaining functional engineered tissue. To replicate the bone tissue, osteogenic cell-laden alginate hydrogels were deposited using a 3D fiber deposition method, in 0/90 and 0/45:0/90 configurations.[156] By exchanging the printing syringe during deposition, it was possible to print hydrogels strands loaded with different cells (endothelial and MSCs) within a single scaffold, indicating that this 3D fiber deposition system can be suitable for the development of bone grafts containing multiple cell types.[157] The same authors performed an *in vivo* follow-up of printed thermosensitive gelatinous hydrogel loaded with MSCs and endothelial cells, and demonstrated the retention of spatially organized heterogeneous cell organization induced by the printing. The osteogenic differentiation of cells was localized in the parts corresponding to osteoprogenitor cells-loaded hydrogel units, while perfused blood vessels were formed in the endothelial progenitors cell-laden part of the constructs.[158]

5.2. Gradients in hydrogels

During bone remodeling and regeneration, spatial and temporal gradients regulate various cell behaviors such as proliferation, organization and differentiation. These gradients are often derived from the cell–cell and cell–ECM interactions within the surrounding environment.[159] For instance, the engineering of bone-cartilage interface, referred as osteochondral TE, poses the challenge of addressing both bone and cartilage regeneration into an integrated approach. To date, the most commonly used strategy to engineer an osteochondral construct is the fabrication of a polymer scaffold consisting of two distinct regions, with different properties (porosity, mechanical strength, diffusion and biological performance) to mimic the natural bone and cartilage ECM environments.

Recently, Galperin *et al.*[160] developed an integrated bi-layered scaffold based on degradable poly(hydroxyethyl methacrylate) hydrogels. The first layer was designed with a monodispersed pore size of 38 μm and pore surfaces decorated with HA particles to promote osteogenic differentiation of MSCs, while the second layer had 200 μm pores coated with hyaluronan to trigger the articular cartilage regeneration. The bi-layered construct supported the simultaneous matrix deposition and cell growth of two distinct cell populations. Similarly, Kim *et al.*[161] developed a bi-layered composite OPF hydrogel with modulating dual GFs release kinetics.

IGF-1 and TGF-β3 were loaded in gelatin microparticles that were then embedded in the OPF hydrogel. At 12 weeks post-implantation in a rabbit model, the results showed that the delivery of IGF-1 alone positively affected osteochondral tissue formation, when compared with the dual delivery of GFs. However, this approach does not provide a continuous and smooth transition between the two layers, as it is found in the native osteochondral space. To overcome these drawbacks, Wang *et al.*,[162] designed a gradient gel maker to control the continuous bi-directional distribution of GFs in alginate and silk hydrogels. The gradient distributions of PLGA and silk microspheres loaded with human recombinant BMP-2 (rhBMP-2) and with human recombinant IGF-1 (rhIGF-1), in the silk scaffold, enabled the osteogenic and chondrogenic differentiation of MSCs along the concentration gradients of rhBMP-2 and cross-gradient rhBMP-2/rhIGF-1. However, the MSCs differentiation did not follow the GF gradients in alginate scaffold, most likely due to the range and slope of gradients or due to the rapid diffusion of the active ingredients. These findings suggest that besides creating the gradient, the substrate properties need to be tuned in such a way to enable the concentration range and diffusion kinetics most suitable for the desired outcome.

Beside the biochemical cues, matrix mechanics directly affect the cell differentiation. Chatterjee *et al.*[163] developed a combinatorial screening method to rapidly screen the effect of hydrogels stiffness on the osteogenic differentiation of MSCs. A stiffness gradient spanning a 30-fold range in compressive modulus (10 kPa to 300 kPa) was created in a 6 cm hydrogel slab, by increasing the concentration of photocrosslinkable PEG. The gradients enabled systematic screening of osteoblast differentiation and demonstrated that hydrogels of modulus $\approx$225 kPa ($\approx$16% PEG by mass) or higher were required for inducing significant mineralization. These results demonstrate that variation of only a material property, stiffness, can be used to induce "graded" osteogenesis and generation of a mineralized tissue gradient that could be applied to integrate hard and soft tissues such as a tendon or a ligament.

6. The Vascularization Hurdle in Engineering in Bone Tissue

Both cancellous and cortical bone, although presenting different organization, contain a highly vascularized network that is essential to supply nutrients and oxygen to the cells, and remove metabolic by-products. Despite the advances made in bone TE, the classical strategies have been restricted mainly due to the lack of vascularization within the engineered bone analog, resulting in a poor graft integration and failure of the construct upon implantation. In an effort towards clinical success of these constructs, new TE concepts have emerged to incorporate vascular cues and induce vascular network formation within engineered bone constructs in an attempt to recapitulate bone physiology. The need for a vasculature network is even more demanding when engineering 3D large tissue constructs. Hydrogels have been proposed as delivery vehicles of vascular agents or as matrices that enable the invasion of the hosts own vasculature network into the construct. Among these are: (1) directing cellular responses by the release of GFs and (2) microfabrication of vascularized microscale structures using single or multiple-cell culturing approaches.

6.1. GF delivery

As mentioned above, cells respond to chemical and physical inputs by the activation of a cascade of events that are mediated by GFs signaling. A number of studies have been attempted to reproduce this signaling process *in vitro* and *in vivo* through the controlled delivery of exogenous GFs to direct cell behavior and fate. The first clinical trials aimed at assessing strategies for promoting bone formation and neovascularization, included the delivery of VEGF and BMPs at the injury site. However, no significant effects or exhibited limited efficiency[164] were observed, highlight their inability to activate the signaling machinery. The poor outcome of the direct injection was attributed to their mode of delivery, which usually was performed by simultaneous injection of both GFs, as a bolus. Such uncontrolled delivery of GFs results in rapid distribution into the body, which translates in systemic and diffused, rather than localized responses. Moreover, the relatively short half time of GFs, might render initial excess levels, followed by a shortage of GFs. Noteworthy, a relatively narrow therapeutic window of GFs dose is expected to target tissue regeneration, whereas extreme doses will most likely lead to severe side effects, as already observed.[165] Moreover, maintenance of constant GFs levels is only potentially achieved with multiple injections or continuous administration, bio-inspired alternative routes have been explored.

In an ideal situation, the localized delivery of GFs should be compatible with physiologically relevant doses and allow preserving their activity for long periods of time. Moreover, the carrier materials should simultaneously foster the communication with host cells in order to play an active role in the regeneration process. The highly swollen 3D networks of hydrogels, have proved to be powerful candidates to fulfill these requirements,[166] as they can enable diffusion of relevant GFs, with tunable release profiles, while providing the 3D support for the formation of bone and/or a vascular network. The GFs can be either freely embedded in the hydrogel or covalently bound to the polymer backbone, while the release kinetics can be tailored by alternating the material diffusion coefficient, crosslinking degree, degradation rate or the initial loading dose. For example, the release profile of PDGF from carrageenan hydrogel microbeads matched that of usually required for the development of a fully functional vascular network, that demonstrate the potential of these systems for bone TE.[167] Even more, the dual delivery of PDGF and VEGF resulted in highly dense and well-established vessel network compared with the bolus injection of either of the GFs alone.[168] Similarly, PEG diacrylate hydrogels functionalized with heparin were shown to be able to deliver VEGF in a sustained and controlled manner. When implanted in a chick chorioallantoic membrane, the heparin-PEG hydrogel showed superior angiogenic potential in stimulating new blood vessel formation compared with un-functionalized ones, which were characterized by a more rapid release of VEGF.[169]

Furthermore, the application of a VEGF/alginate hydrogel in critical-sized defects of rat calvariae resulted in increased blood vessel density, when compared with the bolus delivery. Even though there was no difference in bone regeneration at 4 weeks of implantation, there was a significant increase in bone regeneration in the VEGF/alginate treated defects. This study provides evidence that blood vessel formation is tightly coupled with bone formation.[170]

As vascularization underlies the success of guided bone regeneration, the ideal scenario would be to deliver a cascade of multiple GFs to simultaneously induce angiogenesis and mineralization, in order to produce a vascularized bone tissue substitute. The simultaneous delivery of VEGF and BMP-2 from gelatin hydrogels resulted in a synergistic effect, promoting both high osteogenic responses and blood vessel formation in a 8 mm rat cranial defect.[171] The effect of sequential delivery of GFs on supportive vascular network formation, in parallel with bone formation was also investigated by Kempen *et al.*, who proposed a sequential dual delivery system, inspired by the sequential release of GFs observed during normal bone healing.[172] BMP-2 was loaded in PLGA hydrogel microspheres that were further embedded into a PPF rod by photo-crosslinking. In parallel, a gelatin hydrogel was processed as a tubular implant and impregnated with VEGF before association with the PPF rod previously loaded with BMP-2 loaded microspheres. The system allowed an initial release of VEGF, followed by a sustained release of BMP-2 in later time points. Although VEGF alone did not induce bone formation, it did increase the formation of a supportive vascular network in ectopic implants. In combination with local sustained release of BMP-2, VEGF significantly enhanced ectopic bone formation, when compared to BMP-2 alone.[173]

These results indicate that correct GFs types and combinations, and delivery approach (simultaneous or sequential) greatly affect osteogenic differentiation of cells, new bone and blood vessel formation.

6.2. Microfabricated hydrogels for vascularized bone TE

Angiogenesis, as a prerequisite for osteogenesis, implies the intimate crosstalk between osteoblast and endothelial cells. The latter self-assemble into vascular tubes when surrounded by ECM and exposed to angiogenic factors, such as VEGF or BMPs. To stabilize the new blood vessels, endothelial cells must functionally interact with mural cells, such as pericytes or smooth muscle cells that, in turn, provide support and stability. This particularity in terms of rearrangement as functional vascular network, has been the inspiration model used in microfabrication techniques to design hydrogels structures to generate 3D-vascular-like networks. West *et al.*[174] have been decorating PEG hydrogels with RGD sequences to support ECs adhesion and alignment, while the covalent immobilization of VEGF on PEG patterned surfaces, enhanced tubule formation.[175] Nikkhah *et al.* created highly organized endothelial cord structures using micro-patterned methacrylated gelatin (GelMA). The alignment of HUVECs was significantly enhanced within 50 μm microchannels and by varying the geometrical micro-features of the GelMA patterns, it was possible to optimize the formation of stable lumen-like structures for more that 2 weeks.

The lumen configuration was reproduced by Lee *et al.*[176] by using a microfluidic device to generate hollow alginate microfibers with different diameters (50–250 μm). These microfibers were loaded with ECs and were further embedded in agar–gelatin–fibronectin hydrogels, to assess the cellular viability and organization. Similarly, Takei *et al.*[177] successfully fabricated artificial endothelized tubes with determined 3D configurations, as starting point for a self-developing capillary-like network in a collagen gel. Briefly, cell-enclosing hydrogel fibers of 250 and 500 μm diameters were embedded in a collagen gel and used as templates. The enzymatic degradation of the hydrogels fibers resulted in the development of channels with the fibers 3D configurations. The degradation of the fibers allowed the release of endothelial cells and their adhesion to the collagen gel. Culturing the cells in FGF-enriched medium trigger the migration of ECs in the collagen gel and the self-assembly into capillary-like structures.

The integration of these systems with matrices containing osteoprogenitor cells, osteoblast-like cells, osteo-related GFs or bioactive agents is expected to address the development of a functional vasculature within the bone constructs, concomitant with mineralization and new bone formation. For example, Trkov *et al.* investigated the functional interactions between hMSCs and endothelial cells, using a simple, yet controllable hydrogel system.[178] Cells were encapsulated in 3D micro-printed fibrin hydrogel channels, to enable precise control of the cellular spatial distribution distance between two cell populations (500–2000 μm). The endothelial cells-secreted factors established a distance-dependent chemotactic gradient in the hydrogel matrix that stimulated directional protrusion of hMSCs within the construct, accompanied by uniform osteogenic differentiation. However, due to the lack of consensus over the appropriate cell combinations and culturing conditions, the hydrogels matrices are merely used as model platforms for the cellular interactions to be tested, limiting their potential applications as heterotypic cells matrices for bone remodeling.

7. Conclusions and Future Directions

The complex organization of bone tissue requires the development of multi-faceted strategies that address its regeneration at all structural and functional levels. This chapter provides insights over the use of hydrogels as potential candidates for bone TE and regeneration. The highly hydrated state of the 3D network offers significant advantages over traditional ceramic and stiff polymers, in what concern biocompatibility, biodegradability, drug and cell delivery, and injectability. From a clinical standpoint, these features may prevail over hydrogels low mechanical properties and lack of bioactivity. Moreover, current TE strategies address these disadvantages rendering hydrogels with high controllable mechanical robustness and patterned mineralization features, by: (1) addition of inorganic phases, (2) *in situ* crystallization of HA inside the hydrogels, (3) enzymatically induced mineralization, (4) functionalization of hydrogel with negative moieties, and (5) calcium-binding molecules. Moreover, bottom-up approaches have been stimulating scientists as they envisage reconstruction of bone physiology using cell-laden building blocks by spatially controlling cellular and mechanical distribution. Additionally, hydrogels can be engineered in such a way that they acquire the ability to guide the restoration of functional vasculature within the bone tissue-engineered construct, simultaneously promoting functional bone formation. Ultimately, hydrogels are versatile tools that, through a multi-parametric design, allow the development of functionalized bone analogs that can be easily implanted at the defect/injury site, and further integrated within the host tissue.

Acknowledgments

Authors thank the Portuguese Foundation for Science and Technology (FCT) for the personal grants SFRH/ BD/42968/2008 through the MIT-Portugal Program (SMM). The research leading to these results has received funding from the MIT/ECE/0047/2009 project and the European Union's Seventh Framework Programme (FP7/2007-2013) under Grant agreement No. REGPOT-CT2012-316331-POLARIS.

References

1. J. Crockett, M. Rogers, F. Coxon, L. Hocking, and M. Helfrich, *J. Cell. Sci.,* **124**, 991 (2011).
2. P. Hill and M. Orth, *Brit. J. Orthodontic,* **25,** 101 (1998).
3. E. Jimi, S. Hirata, K. Osawa, M. Terashita, C. Kitamura, and H. Fukushima, *Int. J. Dentistry* **2012**, (2012).
4. Z. Gugala and S. Gogolewski, *Injury,* **33 (Suppl. 2)**, B71 (2002).

5. D. H. Kim, R. Rhim, L. Li, J. Martha, B. H. Swaim, R. J. Banco, L. G. Jenis, and S. G.Tromanhauser, *Spine J.* **9**(11), **886** (2009).
6. E. D. Arrington, W. J. Smith, H. G. Chambers, A. L. Bucknell, and N. A. Davino, *Clin. Orthop. Rela. Res.*, (**329**), 300 (1996).
7. C. H. Lohmann, D. Andreacchio, G. Koster, D. L. Carnes, D. L. Cochran, D. D. Dean, B. D. Boyan, and Z. Schwartz, *Arch. Orthop. Traum. Sur.*, **121**(10), 583 (2001).
8. C. Laurencin, Y. Khan, and S. F. El-Amin, *Expert Rev. Med. Devic.*, **3**(1), 49 (2006).
9. P. V. Giannoudis, H. Dinopoulos, and E. Tsiridis, *Injury*, **36 (Suppl. 2)**, S20 (2005).
10. A. S. Greenwald, S. D. Boden, V. M. Goldberg, Y. Khan, C. T. Laurencin, and R. N. Rosier, *J. Bone Joint Surg. Am.*, **83**-A (Suppl. 2) (Pt 2), 98 (2001).
11. W. G. De Long, T. A. Einhorn, K. Koval, M. McKee, W. Smith, R. Sanders, and T. Watson, *J. Bone Joint Surg. Am.*, **89**(3), 649 (2007).
12. E. Gruskin, B. A. Doll, F. W. Futrell, J. P. Schmitz, and J. O. Hollinger, *Adv. Drug Deliv. Rev.*, **64**(12), 1063 (2012).
13. R. Langer, and J. P. Vacanti, *Science*, **260**(5110), 920 (1993).
14. A. R. Costa-Pinto, R. L. Reis, and N. M. Neves, *Tissue Eng. B*, **17**(5), 331 (2011).
15. M. Niinomi, *J. Artifi. Org.*, **11**(3), 105 (2008).
16. S. Samavedi, A. R. Whittington, and A. S. Goldstein, *Acta Biomater.*, **9**(9), 8037 (2013).
17. D. S. Kohane and R. Langer, *Pediatr. Res.*, **63**(5), 487 (2008).
18. M. Jayabalan, K. T. Shalumon, and M. K. Mitha, *J. Mater. Sci.*, **20**(6), 1379 (2009).
19. M. B. Dreifke, N. A. Ebraheim, and A. C. Jayasuriya, *J. Biomed. Mater. Res. A*, **101**(18), 2436 (2013).
20. M. K. Nguyen and D. S. Lee, *Macromol. Biosci.*, **10**(6), 563 (2010).
21. H. Geckil, F. Xu, X. Zhang, S. Moon, and U. Demirci, *Nanomedicine*, **5**(3), 469 (2010).
22. M. P. Lutolf, F. E. Weber, H. G. Schmoekel, J. C. Schense, T. Kohler, R. Muller, and J. A. Hubbell, *Nat. Biotechnol.*, **21**(5), 513 (2003).
23. E. S. Place, N. D. Evans, and M. M. Stevens, *Nat. Mater.*, **8**(6), 457 (2009).
24. B. V. Slaughter, S. S. Khurshid, O. Z. Fisher, A. Khademhosseini, and N. A. Peppas, *Adv. Mater.*, **21**(32–33), 3307 (2009).
25. G. A. Rodan, *Bone*, **13**(Suppl. 1), S3 (1992).
26. V. I. Sikavitsas, J. S. Temenoff, and A. G. Mikos, *Biomaterials*, **22**(19), 2581 (2001)
27. J. Y. Rho, R. B. Ashman, and C. H. Turner, *J. Biomecha.*, **26**(2), 111 (1993).
28. M. L. Hillier and L. S. Bell, *J. Forensic Sci.*, **52**(2), 249 (2007).
29. A. J. Salgado, O. P. Coutinho, and R. L. Reis, *Macromol. Biosci.*, **4**(8), 743 (2004).
30. L. A. Hidalgo-Bastida and S. H. Cartmell, *Tissue Eng. B*, **16**(4), 405 (2010).
31. M. F. Young, *Calcified Tissue Int.*, **42**(1), 2 (2003).
32. S. Viguet-Carrin, P.Garnero, and P. D. Delmas, *Osteoporosis Inte.*, **11**(3), 319 (2006).
33. G. K. Hunter, P. V. Hauschka, A. R. Poole, L. C. Rosenberg, and H. A. Goldberg, *Biochem. J.*, **317**(Pt 1), 59 (1996).
34. E. R. Wise, S. Maltsev, M. E. Davies, M. J. Duer, C. Jaeger, N. Loveridge, R. C. Murray, and D. G. Reid, *Chem. Mater.*, **19**(2), 5055 (2007).
35. F. Lamoureux, M. Baud'huin, L. Duplomb, D. Heymann, and F. Redini, *BioEssays:*, **29**(8), 758 (2007).

36. D. B. Burr, *Bone,* **31**(4), 8 (2002).
37. A. M. Parfitt, *J. Clin. Endocrinol. Metab.*, **88**(10), 5043 (2003).
38. J. Song, V. Malathong, and C. R. Bertozzi, *JACS,* **127**(10), 3366 (2005).
39. L. C. Palmer, C. J. Newcomb, S. R. Kaltz, E. D. Spoerke, and S. I. Stupp, *Chem. Rev.*, **108**(11), 4754 (2008).
40. S. Shi, and C. Y. Wang, *Biotechnol. Genet. Eng. Rev.*, **21**, 133 (2004).
41. S. E. Haynesworth, J. Goshima, V. M. Goldberg and A. I. Caplan, *Bone,* **13**(1), 81 (1992).
42. H. Nakahara, V. M. Goldberg and A. I. Caplan, *J. Orthop. Res.*, **9**(4), 465 (1991).
43. M. Owen, *J. of cell Sci.*, **10**, S63 (1988).
44. J. H. Bennett, S. Moffatt and M. Horton, *Histol. Histopathol.*, **16**(2), 603 (2001).
45. S. Fuchs, A. Hofmann and C. Kirkpatrick, *Tissue Eng.*, **13**(10), 2577 (2007).
46. R. P. Pirraco, A. P. Marques and R. L. Reis, *J. Cell. Mol. Med.*, **14**(1–2), 93 (2010)
47. S. Nomura and T. Takano-Yamamoto, *Matrix Biol.*, **19**(2), 91 (2000)
48. S. L. Teitelbaum, *Science,* **289**(5484), 1504 (2000).
49. F. Lazner, M. Gowen, D. Pavasovic and I. Kola, *Human Molecular Genetics,* **8**(10), 1839 (1999).
50. A. M. Makhdom and R. C. Hamdy, *Tissue Eng. Pt. B,* **19**(5), 442 (2013).
51. K. N. Malizos and L. K. Papatheodorou, *Injury,* **36** (Suppl. 3), S13 (2005).
52. F. Villars, L. Bordenave, R. Bareille, and J. Amedee, *J. Cell. Biochem.*, **79**(4), 672 (2000).
53. A. Khademhosseini and R. Langer, *Biomaterials,* **28**(34), 5087 (2007).
54. W. E. Hennink and C. F. van Nostrum, *Adv. Drug Deliv. Rev.*, **54**(1), 13 (2002).
55. E. De Giglio, S. Cometa, M. A. Ricci, A. Zizzi, D. Cafagna, S. Manzotti, L. Sabbatini, and M. Mattioli-Belmonte, *Acta Biomater.*, **6**(1), 282 (2010).
56. C. D. Jefferiss, A. J. Lee and R. S. Ling, *J. Bone. Joint Surg. Br.*, **57**(4), 511 (1975).
57. Y. Ma, Y. Zheng, X. Huang, T. Xi, X. Lin, D. Han, and W. Song, *Biomed. Mater.*, **5**(2), 25003 (2010).
58. Z. Y. Lin, Z. X. Duan, X. D. Guo, J. F. Li, H. W. Lu, Q. X. Zheng, D. P. Quan, and S. H Yang, *J. Contr. Rel.*, **144**(2), 190 (2010).
59. A. Rey-Rico, M. Silva, J. Couceiro, A. Concheiro and C. Alvarez-Lorenzo, *Eur. Cells Mater.*, **21**, 317 (2011).
60. S. Wang, L. Lu, and M. J. Yaszemski, *Biomacromolecules,* **7**(6), 1976 (2006).
61. T. E. Douglas, P. B. Messersmith, S. Chasan, A. G. Mikos, E. L. de Mulder, G. Dickson, D. Schaubroeck, L. Balcaen, F. Vanhaecke, P. Dubruel, J. A. Jansen, and S. C. Leeuwenburgh, *Macromole. Biosci.*, **12**(8), 1077 (2012).
62. S. D. Fitzpatrick, M. A. Jafar Mazumder, F. Lasowski, L. E. Fitzpatrick, and H. Sheardown, PNIPAAm-grafted-collagen as an injectable, *in situ* gelling, bioactive cell delivery scaffold. *Biomacromolecules,* **11**(9), 2261 (2010).
63. H. T. Liao, C. T. Chen, and J. P. Chen, *Tissue Eng. C,* **17**(11), 1139 (2011).
64. J. F. Mano, R. A. Sousa, L. F. Boesel, N. M. Neves and R. L. Reis, *Compos. Sci. Technol.*, **64**(6), 789 (2004).
65. J. F. Mano, G. A. Silva, H. S. Azevedo, P. B. Malafaya, R. A. Sousa, S. S. Silva, L. F. Boesel, J. M. Oliveira, T. C. Santos, A. P. Marques, N. M. Neves, and R. L. Reis, *J. R. Soc. Interf.*, **4**(17), 999 (2007)

66. C. Zhong and C. C. Chu, *J. Mater. Chem.*, **22**(13), 6080 (2012).
67. C. Manferdini, V. Guarino, N. Zini, M. G. Raucci, A. Ferrari, F. Grassi, E. Gabusi, S.Squarzoni, A. Facchini, L. Ambrosio, and G. Lisignoli, *Biomaterials*, **31**(14), 3986 (2010).
68. V. E. Santo, M. E. Gomes, J. F. Mano, and R. L. Reis, *J. Tissue Eng. Regen. Med.*, **6** (Suppl. 3), S47 (2012).
69. J. C. Fricain, S. Schlaubitz, C. Le Visage, I. Arnault, S. M. Derkaoui, R. Siadous, S. Catros, C. Lalande, R. Bareille, M. Renard, T. Fabre, S. Cornet, M. Durand, A. Leonard, N. Sahraoui, D. Letourneur, and J. Amedee, *Biomaterials*, **34**(12), 2947 (2013).
70. A. C. Lima, P. Batista, T. A. Valente, A. S. Silva, I. J. Correia, and J. F. Mano, *Tissue Eng. A*, **19**(9–10), 1175 (2013).
71. D. Dyondi, T. J. Webster, and R. Banerjee, *Int. J. Nanomed.*, **8**, 47 (2013).
72. E. Lippens, G. Vertenten, J. Girones, H. Declercq, J. Saunders, J. Luyten, L. Duchateau, E. Schacht, L. Vlaminck, F. Gasthuys, and M. Cornelissen, *Tissue Eng. Pt. A*, **16**(2), 617 (2010).
73. S. M. Oliveira, I. F. Almeida, P. C. Costa, C. C. Barrias, M. R. Ferreira, M. F. Bahia, and M. A. Barbosa, *AAPS Pharm. Sci. Tech.*, **11**(2), 852 (2010)
74. E. G. Popa, S. G. Caridade, J. F. Mano, R. L. Reis, and M. E. Gomes, *J. Tissue Eng. Regen. Med.*, **9**, 550 (2013).
75. T. Douglas, M. Wlodarczyk, E. Pamula, H. Declercq, E. de Mulder, M. Bucko, L. Balcaen, F. Vanhaecke, R. Cornelissen, P. Dubruel, J. A. Jansen, and S. C. Leeuwenburgh, *J. Tissue Eng. Regen. Med.* **8**, 906 (2012).
76. J. Zhu and R. E. Marchant, *Expert Rev. Med. Devi.*, **8**(5), 607 (2011).
77. J. L. Drury and D. J. Mooney, *Biomaterials*, **24**(24), 4337 (2003).
78. D. Guenther, A. Oks, M. Ettinger, E. Liodakis, M. Petri, C. Krettek, M. Jagodzinski, and C. Haasper, *Int. Orthop.* **37**, 1605 (2013).
79. L. Wang and Stegemann, *Biomaterials*, **31**(14), 3976 (2010).
80. X. Liu, L. A. Smith, J. Hu, and P. X. Ma, *Biomaterials*, **30**(12), 2252 (2009).
81. D. BenDavid, T. A. Kizhner, T. Kohler, R. Muller, E. Livne, and S. Srouji, *J. Cranio. Maxillo. Surg.*, **39**(5), 364 (2011).
82. N. Lohse, J. Schulz, and H. Schliephake, *Eur. Cells Mater.*, **23**, 413 (2012).
83. Z. Schwartz, M. Goldstein, E. Raviv, A. Hirsch, D. M. Ranly, and B. D. Boyan, *Clin. Oral Implants Res.*, **18**(2), 204 (2007).
84. M. A. Reynolds, M. E. Aichelmann-Reidy, J. D. Kassolis, H. S. Prasad, and M. D. Rohner, *J. Biomed. Mater. Res. B*, **83**(2), 451 (2007).
85. M. J. Sawkins, W. Bowen, P. Dhadda, H. Markides, L. E. Sidney, A. J. Taylor, F. R. Rose, S. F. Badylak, K. M. Shakesheff, and L. J. White, *Acta Biomater.*, **9**(18), 7865 (2013).
86. J. A. Burdick, Bioengineering: Cellular control in two clicks. *Nature*, **460**(7254), 469–470 (2009).
87. Y. Khan, M. J. Yaszemski, A. G. Mikos, and C. T. Laurencin, *J. Bone Joint Surg. Am.*, **90** (Suppl. 1), S36 (2008).
88. K. Rezwan, Q. Z. Chen, B. J. Blaker, and A. R. Boccaccini, *Biomaterials*, **27**(18), 3413 (2006).
89. A. R. Boccaccini, and J. J. Blaker, *Exp. Rev. Med. Dev.*, **2**(3), 303 (2005).
90. E. Ho, A. Lowman, and M. Marcolongo, *J. Biomed. Mater. Res. A*, **83**(1), 249 (2007).

91. A. S. Rowlands, P. A. George, and J. J. Cooper-White, *Am. J. Physiol. Cell Physiol.*, **295**(4), C1037 (2008).
92. M. Kamitakahara, C. Ohtsuki, and T. Miyazaki, *J. Biomate. Appl.*, **23**(3), 197 (2008).
93. S. V. Dorozhkin and M. Epple, *Angew. Chem.*, **41**(17), 3130 (2002).
94. F. Bronner, M. C. Farach-Carson, and H. I. Roach, *Bone-Metabolic Functions and Modulators*. Springer: New York (2012).
95. K. Bleek, and A. Taubert, *Acta Biomater.*, **9**(5), 6283 (2013).
96. Z. Fan, J. Wang, Z. Wang, Z. Li, Y. Qiu, H. Wang, Y. Xu, L. Niu, P. Gong, and S. Yang, *J. Phys. Chem.*, **117**(20), 10375 (2013)
97. J. Lu, Y. S. He, C. Cheng, Y. Wang, L. Qiu, D. Li, and D. Zou, *Adv. Funct. Mater.*, **23**(28), 3494 (2013).
98. S. R. Shin, H. Bae, J. M. Cha, J. Y. Mun, Y. C. Chen, H. Tekin, H. Shin, S. Farshchi, M. R. Dokmeci, S. Tang, and A. Khademhosseini, *ACS Nano*, **6**(1), 363 (2012).
99. Y. Kuboki, M. Terada, Y. Kitagawa, S. Abe, M. Uo, and F. Watari, *Biomed. Mater. Eng.*, **6**(1), 3 (2009).
100. P. Newman, A. Minett, R. Ellis-Behnke, and H. Zreiqat, *Nanomed. Nanotechnol. Biol. Med.*, **19**(1), 1139 (2013).
101. O. Im, J. Li, M. Wang, L. G. Zhang, and M. Keidar, *Int. J. Nanomed.*, **9**(8), 2087 (2012).
102. S. Mohapatra and C., Wang, *Patent Application, US 20130230496 US 20130230496 Al*, (2013).
103. T. Kokubo and H. Takadama, *Biomaterials*, **27**(15), 2907 (2006).
104. T. Ichibouji, T. Miyazakia, E. Ishida, A. Sugino, and C. Ohtsuki, *Mater. Sci. Eng. C*, **29**(6), 1765 (2009).
105. H. Zainuddin, D. J. Hill, A. K. Whittaker, and T. V. Chirila, *J. Mater. Sci. Mater. Med.*, **17**(12), 1245 (2006).
106. A. Shkilnyy, R. Graf, B. Hiebl, A. T. Neffe, A. Friedrich, J. Hartmann, and A. Taubert, *Macromol. Biosci.*, **7**(2), 179 (2009).
107. D. S. Couto, Z. Hong, and J. F. Mano, *Acta Biomate.*, **5**(1), 115 (2009).
108. M. Azami, M. J. Moosavifar, N. Baheiraei, F. Moztarzadeh, and J. Ai, *J. Biomed. Mater. Res. Pt. A*, **100**(5), 1347 (2012).
109. M. Xie, M. O. Olderoy, J. P. Andreassen, S. M. Selbach, B. L. Strand, and P. Sikorski, *Acta Biomater.*, **1**(9), 3665 (2010).
110. E. D. Spoerke, S. G. Anthony, and S. I. Stupp, *Adv. Mater.*, **1**(4), 425 (2009).
111. E. E. Golub and K. Boesze-Battaglia, *Curr. Opin. Orthop.*, **18**, 444 (2007).
112. H. Orimo, *Journal of Nippon Medical School*, **11**(1), 4 (2010).
113. S. C. Leeuwenburgh, J. A. Jansen, and A. G. Mikos, *J. Biomate. Sci. Polym. Ed.*, **18**(12), 1547 (2007).
114. K. Lee, M. H. Oh, M. S. Lee, Y. S. Nam, T. G. Park, and J. H. Jeong, *Int. J. Pharm.*, **445**(1–2), 196 (2013).
115. T. E.Douglas, A. Skwarczynska, Z. Modrzejewska, L. Balcaen, D.Schaubroeck, S. Lycke, F. Vanhaecke, P. Vandenabeele, P. Dubruel, J. A. Jansen, and S. C. Leeuwenburgh, *Int. J. Biol. Macromol.*, **56**, 122 (2013).
116. R. Filmon, M. F. Basle, A. Barbier, and D. Chappard, *J. Biomater. Sci. Polym. Ed.*, **11**(8), 849 (2000).

117. S. P. Bruder and A. I. Caplan, *Bone,* **11**(2), 133 (1990).
118. M. Xie, Ø. M. Olderøy, Z. Zhang, J. P. Andreassen, B. L. Strand, and P. Sikorski, *RSC Adv.,* **2,** 1457 (2012).
119. M. Bongio, M. R. Nejadnik, Z. T. Birgani, P. Habibovic, L. A. Kinard, F. K. Kasper, A. G. Mikos, J. A. Jansen, S. C. Leeuwenburgh, and J. J. van den Beucken, *Macromol. Biosci.,* **13**(6), 777 (2013).
120. T. van den Bos, J. Oosting, V. Everts, and W. Beertsen, *J. Bone. Miner. Res.,* **10**(4), 616 (1995).
121. T. Osathanon, C. M. Giachelli, and M. J. Somerman, *Biomaterials,* **30**(27), 4513 (2009).
122. M. Gungormus, M. Branco, H. Fong, J. P. Schneider, C. Tamerler, and M. Sarikaya, *Biomaterials,* **31**(28), 7266 (2010).
123. C. R. Nuttelman, D. S. Benoit, M. C. Tripodi, and K. S. Anseth, *Biomaterials,* **27**(8), 1377 (2006).
124. D. A. Wang, C. G. Williams, F. Yang, N. Cher, H. Lee, and J. H. Elisseeff, *Tissue Eng.,* **11**(1–2), 201 (2005).
125. A. Phadke, C. Zhang, Y. Hwang, K. Vecchio, and S. Varghese, *Biomacromolecules,* **11**(8), 2060 (2010).
126. G. Embery, S. Rees, R. Hall, K. Rose, R. Waddington, and P. Shellis, *Euro. J. Oral Sci.,* **106** suppl 1, S267 (1998).
127. B. Marelli, C. E. Ghezzi, A. Alessandrino, J. E. Barralet, G. Freddi, and S. N. Nazhat, *Biomaterials,* **33**(1), 102 (2012).
128. D. M. Dragusin, D. E. Giol, A. Serafim, E.Vasile, and T. Zecheru, *Dig. J. Nanomater. Bios.,* **6**(4), 1909 (2011).
129. C. Zaharia, M. R. Tudora, I. C. Stancu, B. Galateanu, A. Lungu, and C. Cincu, *Mater. Sci. Eng. C,* **32**(4), 945 (2012).
130. Z. Lin, S. Cao, X. Chen, W. Wu, and J. Li, *Biomacromolecules,* **14**(7), 2206 (2013).
131. I. C. Stancu, D. M. Dragusin, E.Vasile, R.Trusca, I. Antoniac, and D. S. Vasilescu, *J. Mater. Sci. Mater. Med.,* **22**(3), 451 (2011).
132. C. Cha, E. S. Kim, I. W. Kim, and H. Kong, *Biomaterials,* **32**(11), 2695 (2011).
133. X. Liu, X. Wang, X. Wang, X. H. Ren, J. He, L. Qiao, and F. Z. Cui, *Acta Biomater.,* **9**(6), 6798 (2013).
134. C. Bokel and N. H. Brown, *Develop Cell,* **3**(3), 311 (2002).
135. S. Khetan, M. Guvendiren, W. R. Legant, D. M. Cohen, C. S. Chen, and J. A. Burdick, *Nat. Mater.,* **12**(5), 458 (2013).
136. M. P. Lutolf, J. L. Lauer-Fields, H. G. Schmoekel, A. T. Metters, F. E. Weber, G. B. Fields, and J. A. Hubbell, *PNAS,* **100**(9), 5413 (2003).
137. J. Kim, I. S. Kim, T. H. Cho, H. C. Kim, S. J. Yoon, J. Choi, Y. Park, K. Sun, and S. J. Hwang, *J. Biomed. Mater. Res. Part A,* **95**(3), 673 (2010).
138. N. Amosi, S. Zarzhitsky, E. Gilsohn, O. Salnikov, E. Monsonego-Ornan, R. Shahar, and H. Rapaport, *Acta Biomater.,* **8**(7), 2466 (2012).
139. T. Nonoyama, H. Ogasawara, M. Tanaka, M. Higuchi, and T. Kinoshita, *Soft Matter,* **8**, 11531 (2012).
140. L. Solorio, C. Zwolinski, A. W. Lund, M. J. Farrell, and J. P. Stegemann, *J. Tissue Eng. Regen. Med.,* **4**(7), 514 (2010).

141. L. Luca, A. L. Rougemont, B. H. Walpoth, R. Gurny, and O. Jordan, *J. Contr. Rel.*, **147**(1), 38 (2010).
142. K. U. Lewandrowski, D. D. Hile, B. M. Thompson, D. L. Wise, W. W. Tomford, and D. J. Trantolo, *Tissue Eng.*, **9**(1), 85 (2003).
143. C. A. Simmons, E. Alsberg, S. Hsiong, W. J. Kim, and D. J. Mooney, *Bon*, **35**(2), 562 (2004).
144. A. T. Raiche and D. A. Puleo, *Biomaterials*, **25**(4), 677 (2004).
145. L. C. Visser, S. P. Arnoczky, O. Caballero, A. Kern, A. Ratcliffe, and K. L. Gardner, *Tissue Eng. Pt. A*, **16**(3), 1021 (2010).
146. O. Wahlstrom, C. Linder, A. Kalen, and P. Magnusson, *Acta Orthop.*, **79**(3), 433 (2008).
147. J. W. Nichol and A. Khademhosseini, *Soft Matter*, **5**(7), 1312 (2009)
148. J. Silvent, N. Nassif, C. Helary, T. Azais, J. Y. Sire, and M. M. Guille, *PloS One*, **8**(2), e57344 (2013).
149. F. Chicatun, C. E. Pedraza, C. E. Ghezzi, B. Marelli, M. T. Kaartinen, M. D. McKee, and S. N. Nazhat, *Biomacromolecules*, **12**(8), 2946 (2011).
150. W. Xu, J. Ma, and E. Jabbari, *Acta Biomater.*, **6**(6), 1992 (2010).
151. S. Bensamoun, Z. Fan, I. Brice, J. Y. Rho, and M. C. Ho Ba Tho, *J. Musculoskelet. Res.*, **11**(3), 135 (2008).
152. K. Subramani, and M. A. Birch, *Biomed. Mater.*, **1**(3), 144 (2006).
153. O. Jeon and E. Alsberg, *Adv. Funct. Mater.*, **23**(38), 4765 (2013).
154. H. Qi, M. Ghodousi, Y. Du, C. Grun, H. Bae, P. Yin, and A. Khademhosseini, *Nat. Commun.*, **4**, 2275 (2013).
155. Y. Du, E. Lo, S. Ali, and A. Khademhosseini, *PNAS*, **105**(23), 9522 (2008).
156. N. E. Fedorovich, J. Alblas, W. E. Hennink, F. C. Oner, and W. J. Dhert, *Trends Biotechnol.*, **29**(12), 601 (2011).
157. N. E. Fedorovich, J. R. De Wijn, A. J. Verbout, J. Alblas, and W. J. Dhert, *Tissue Eng. A*, **14**(1), 127 (2008).
158. N. E. Fedorovich, H. M. Wijnberg, W. J. Dhert, and J. Alblas, *Tissue Eng. A*, **17**(15–16), 2113 (2011).
159. S. Sant, M. J. Hancock, J. P. Donnelly, D. Iyer, and A. Khademhosseini, *Can. J. Chem. Eng.*, **88**(6), 899 (2010).
160. A. Galperin, R. A. Oldinski, S. J. Florczyk, J. D. Bryers, M. Zhang, and B. D. Ratner, *Adv. Healthcare Mater.*, **2**(6), 872 (2013).
161. K. Kim ,J. Lam, S. Lu, P. P. Spicer, A. Lueckgen, Y. Tabata, M. E. Wong, J. A. Jansen, A. G. Mikos, and F. K. Kasper,. *J. Contr. Rel.*, **168**(2), 166 (2013).
162. X. Wang, E. Wenk, X. Zhang, L. Meinel, G. Vunjak-Novakovic, and D. L. Kaplan, *J. Contr. Rel.*, **134**(2), 81 (2009).
163. K. Chatterjee, S. Lin-Gibson, W. E. Wallace, S. H. Parekh, Y. J. Lee, M. T. Cicerone, M. F. Young, and C. G. Simon, *Biomaterials*, **31**(19), 5051 (2010).
164. V. Luginbuehl, L. Meinel, H. P. Merkle, and B. Gander, *Eur. J. Pharma. Biopharm.*, **58**(2), 197 (2004).
165. P. Carmeliet, *Nat. Med.*, **6**(4), 389 (2000).

166. A. K. Silva, C. Richard, M. Bessodes, D. Scherman, and O. W. Merten, *Biomacromolecules,* **10**(1), 9 (2009).
167. V. E. Santo, A. M. Frias, M. Carida, R. Cancedda, M. E. Gomes, J. F. Mano, and R. L. Reis, *Biomacromolecules,* **10**(6), 1392 (2009).
168. T. P. Richardson, M. C. Peters, A. B. Ennett, and D. J. Mooney, *Nat. Biotechnol.,* **19**(11), 1029 (2001).
169. O. Oliviero, M. Ventre, and P. A. Netti, *Acta Biomater.,* **8**(9), 3294 (2012).
170. D. Kaigler, E. A. Silva, and D. J. Mooney, *J. Periodontol.,* **84**(3), 230 (2013).
171. Z. S. Patel, S. Young, Y. Tabata, J. A. Jansen, M. E. Wong, and A. G. Mikos, *Bone,* **43**(5), 931 (2008).
172. T. J. Cho, L. C. Gerstenfeld, and T. A. Einhorn, *J. Bone. Miner. Res.,* **17**(3), 513 (2002).
173. D. H. Kempen, L. Lu, A. Heijink, T. E. Hefferan, L. B. Creemers, A. Maran, M. J. Yaszemski, and W. J. Dhert, *Biomaterials,* **30**(14), 2816 (2009).
174. J. J. Moon, M. S. Hahn, I. Kim, B. A. Nsiah, and J. L. West, *Tissue Eng. Pt. A,* **15**(3), 579 (2009).
175. J. E. Leslie-Barbick, C. Shen, C. Chen, and J. L. West, *Tissue Eng. Part A,* **17**(1–2), 221 (2011).
176. K. H. Lee, S. J. Shin, Y. Park, and S. H. Lee, *Small,* **5**(1), 1264 (2009).
177. T. Takei, S. Sakai, T. Yokonuma, H. Ijima, and K. Kawakami, *Biotechnol. Prog.,* **23**(1), 282 (2007).
178. S. Trkov, G. Eng, R. Di Liddo, P. P. Parnigotto, and G. Vunjak-Novakovic, *J. Tissue Eng. Regen. Med.,* **4**(3), 205 (2010).

Chapter 7

Hydrogels in Intervertebral Disk (IVD) Repair

Cem Bayram,* Murat Demirbilek* and Emir Baki Denkbaş*,†,‡

**Hacettepe University, Advanced Technologies Application and Research Center (HUNITEK), 06800 Ankara, Turkey*

†Hacettepe University, Chemistry Department, Biochemistry Div., Beytepe, 06800 Ankara, Turkey

‡denkbas@hacettepe.edu.tr

Hydrogels exhibit a versatile and effective performance in tissue engineering applications. Having a wide choice of starting material, ease of functionalizing, tuneable mechanic properties and the capability of mimicking natural extracellular matrix components make the hydrogels frequently used in regenerative medicine, especially for soft tissues. One of those soft tissues, intervertebral discs (IVDs) are avascular structures located between two vertebras, which have key roles such as shock absorber, and biomechanical stabilizer. When the IVDs are damaged or degenerated, a painful condition occurs due to chronic back pain of the patient. The great resemblance in mechanical and chemical structures between the IVD matrix and hydrogels, especially naturally based ones, has been intensively studied in the past decades. This chapter reviews the hydrogel studies for IVD repair.

1. Introduction

Intervertebral discs (IVDs) are located between two vertebras, surrounded by ligaments and muscles. There are 23 disks in the entire length of the spinal cord, each of them with approximately 8–10 mm in height and 4 cm in diameter.[1] Cervical disks are thinner than lumbar disks. For biomechanical function of spinal column, IVDs play important roles; providing stability, linking the adjacent vertebra, flexibility,

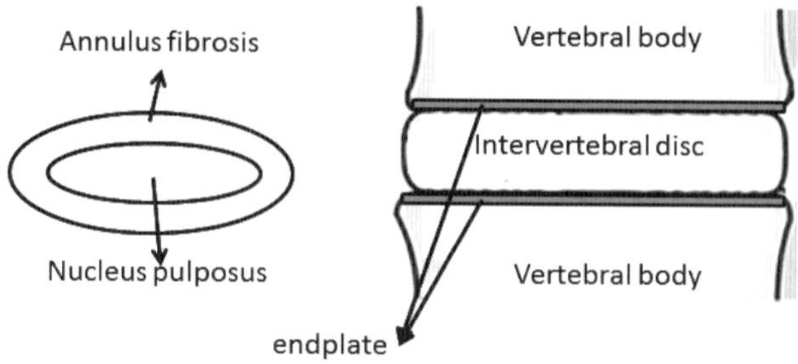

Fig. 1. Superior (left) and lateral (right) view of intervertebral disc.

extension, lateral bending and are responsible for the distribution of load throughout the spine acting as shock absorbers. Thanks to having elastic properties, they provide flexibility and ability to bend and twist. A single IVD has fibro-cartilaginous structure and can be divided into three different regions: Nucleus pulposus (NP), annulus fibrosis (AF) and two endplates (Fig. 1).[2]

NP is a semi-fluidic gel structure, surrounded by annulus fibrous and two endplates,[3] NP is located in the central region of the IVD and its viscoelastic matrix is composed of proteoglycans, collagen, elastin fibers and water. NP has chondrocyte-like cells, responsible for producting type II collagen and proteoglycans.[4] This structure is much less dense than the surrounding AF.

The superjacent tissue, AF, is denser than the nucleus since it has organized collagen type I fiber structures in water based proteoglycan gel matrix. The orientation of these organized fibers varies from the spinal axis, providing the mechanical flexibility.[5] AF has fibroblast-like cells throughout the matrix, for the producing type I collagen and proteoglycans.

IVDs are avascular and required nutrition and metabolites reaches into tissue through endplates.[6] The structure of the NP and AF lets the IVD to resist against compression and withstand complex tensile and shear forces.[7] In the AF, collagen concentration is about 70% of the dry weight with 84% of that being collagen type I and 14% collagen type II.[8] Meanwhile, in the NP, collagen concentration is about 20% of the dry weight of the NP with 80% of that being collagen type II. Proteoglycan concentration is opposite of collagen.[9]

Degenerative disc disease (DDD) is a painful condition often caused by one or more degenerated IVD. The symptoms may be chronic back pain which often radiates to the hips and pain in the buttocks and thighs while walking. IVD herniation or aging could be the cause of disruption of AF, which has limited healing capacity, resulting in poor biomechanical functioning. There is a gradual change in the NP and AF, losses of water and decreasing proteoglycan levels with increasing age.[10]

End-plates become sclerotic and exhibit less porous structure, therefore making transport of water and nutrients difficult. Metabolites become accumulated and an acidic environment occurs with decreased pH. These changes can produce a heavy back pain.[11] To treat degeneration, discectomy or fusion is the common surgical approaches. But the treatment strategy is far from the ideal treatment.[12] Suturing or gluing of degenerated discs were tested for the treatment of AF disorders. But there were no satisfied results, especially long-turn healing of AF disorder. Ahlgren and colleagues were tested direct suturing (Ahlgren, *et al.*, 2000).[13] Heuer and colleagues tested suture, fibrin glue, cyanoacrylate glue, alone and with suture.[14]

Artificial disc replacement could be an ideal strategy for treatment. Two types of replacement are possible, total disc (nucleus pulposus and anulus fibrosus) and individually nucleus pulposus or anulus fibrosus replacement.[15]

To design and produce a disc analogue has certain difficulties such as high durability and flexibility. Also surface properties are another main criteria since the mismatch between the implantable device and tissue may cause wear or corrosion problems.

As mentioned earlier, IVDs mainly consist of water, proteoglycans and collagen. The high ratio of water in the structure makes hydrogels suitable materials for IVDs. Hydrogels can be composed of naturally occurred or synthetic polymers and with an engineering approach on the chemical and structural composition it is possible to tune swelling, permeability, hydrophilicity behavior of the synthesized material.

First and foremost, the polymeric material must be biocompatible, must not trigger the immune system and have high hydrophilic character.[16–19]

Recently, injectable hydrogel systems have been focused on because the technique is minimally invasive and does not require an extensive surgical operations. In this context, once again the naturally occured and synthetic polymers have been considered as IVD replacing materials.[20,21] However, the attempts of using hydrogels based on 2-hydroxyethylmethacrylate, poly vinyl alcohol or poly vinylpyrrolidone as an IVD analogue did not generate the required mechanical behavior completely.[22,23]

"Disc tissue engineering" — a sub-discipline of tissue engineering — has originated from the recent advances and developments in material sciences and hydrogel engineering. Researchers take advantage of the extremely tunable structural properties of hydrogels by adjusting the crosslinking density, polymer concentration and swell ratio while achieving the desired mechanical performances. And for the last but not least, hydrogels can be used as carriers for the preservation of annulus and nucleus sourced cells in the presence of growth factors to provide a living tissue.

1.1. Nucleus pulposus

A traumatic rupture of annulus fibrosis upon physical force or overpressure may result a herniation or leakage of nucleus pulposus. In such situation, nucleus pulposus must be replaced with a prosthetic device prior to annulus defect repair to regain the elastic ability of spine. Hence, a tissue engineering approach for disc degeneration is to restore the loss of matrix with hydrogels and then renovate the tissue by endogenous cells with stimulants. The composition of NP is mainly consists of water and proteoglycans. Proteoglycans constitutes the 30–60% dry weight of the NP, whereas 70–90% of the wet weight is water.[24]

The feasibility of a hydrogel for NP repair should possess a number of criteria. First of all the gel matrix and degradation products should not trigger the foreign body response and since the relevant body section is sensitive to inflammation, the surgical operation must be minimally invasive. For this reason, injectable hydrogels could be a convenient choice for route of administration. The hydrogel should be capable to encapsulate cells and growth factors, which is a must for tissue engineering and support cell growth and extracellular matrix production. And finally the gels must have appropriate mechanical properties, which can be tuned by adjusting macromere or monomer concentration, crosslinking density and so swelling ratio.

Polysaccharide based 3D scaffolds such as alginate, chitosan, agarose, collagen, hyaluronic acid, and chondroitin sulfate, which have close analogy with nucleus pulposus in both mechanical and perspective, have been studied for nucleus pulposus regeneration.[23–26]

Alginate is widely used as nucleus tissue engineering since it successfully mimics the mechanical properties of nucleus pulposus and allows cell to adhere. In a study, reported by Bron and co-workers, the change in viscoelastic properties of alginate scaffolds with polymer concentration was investigated for a design of a replacement material which mimics the mechanical ability of nucleus pulposus. They showed that the desired stiffness level can be obtained by tuning the polymer and phenotypes of native IVD cells are well maintained upon culture in matrices.[27] The hydrophilicity and swelling ability of alginate were enhanced by Leone and colleagues' work, where it was modified by amidation and the researchers reported that the new form of the natural polysaccharide can be considered as a promising material for the repair of herniated NP.[23] The newly formed amidic derivative of alginate exhibited better mechanical properties in terms of storage (G') and loss (G'') moduli, respect to human lumbar NP, also performed a stimulating action for chondrocytes. Low viscosity alginate is also a proper choice for injectable applications and the mechanical integrity can be tuned by adjusting the crosslinking density as well as the polymer concentration. Methacrylate functionalized alginate was synthesized by Chou and Nicoll through esterification of hydroxyl groups and investigated as NP cell encapsulation after photo-crosslinking. The researchers reported that photo-crosslinked alginate gels have better structural integrity than those of crosslinked via divalent cations. Additionally, the encapsulated cells maintained the round morphology and secreted the characteristic proteins and proteoglycans such as collagen types I and II and aggrecan both *in vitro* and *in vivo*.[28]

Hyaluronic acid (HA) has also been focused by the researchers since it is one of the glycosaminoglycans found in natural tissue and has been used in many biomedical applications. HA is the most abundant biomolecule in NP responsible for water adsorption and able to hydrate and dehydrate under variable mechanical conditions. Although having good biocompatibility and natural availability the fast degradation of HA limits its use in tissue engineering. To overcome this limitation, researchers synthesized new forms of HA by incorporating new functionalities. HYADD3®, a dodecylamide derivative of HA and HYAFF120®, a photo-linkable ester form of HA are two types of HA and have been tested successfully in many

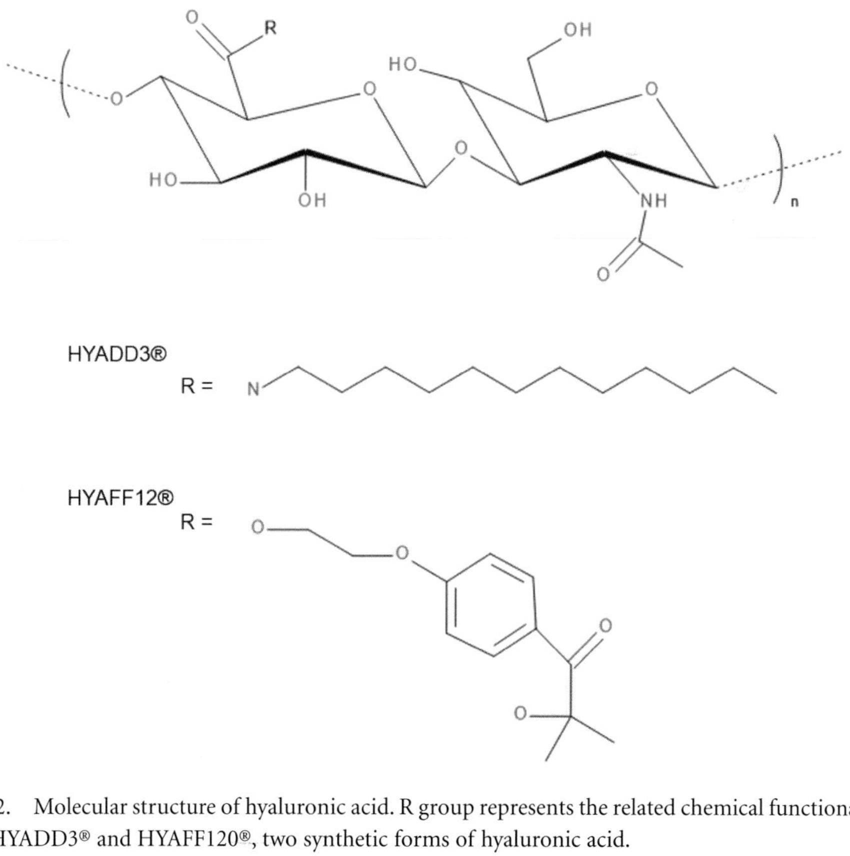

Fig. 2. Molecular structure of hyaluronic acid. R group represents the related chemical functionality for HYADD3® and HYAFF120®, two synthetic forms of hyaluronic acid.

studies both *in vitro* and *in vivo* Fig. 2. HYADD3® can be physically crosslinked itself in aqueous media,[29] while HYAFF120® requires UV light to create crosslinkage between the incorporated photoinitiator groups.[22] In a study performed by Revell and colleagues, bone marrow stem cells loaded HYADD3® and HYAFF120® were injected into the NP of the lumbar spine of nucleotomized female pigs. Six weeks later after the surgery, the injected discs had close similarity to normal NP structure containing viable chondrocytes.[30] The gel-like viscoelastic behavior of HYAFF120® is not affected by injections through catheter devices, which makes them easily used with a minimally invasive manner.

Hyaluronic acid has also been to exposed several modifications and couplings with various macromolecules such as poly ethylene glycol,[31] collagen,[32] gelatin,[33] silk-fibrin[34] and alginate to enhance mechanical stability and degradation time.[28]

Poly(*N*-isopropylacrylamide) (PNIPAAm) is a thermally responsive polymer with a lower critical solution temperature (LCST) of 32°C, which is below body temperature. The polymer dissolves in water at room temperature but when exposed to higher temperature above LCST, it becomes hydrophobic, forms a compact gel. Hence, this feature makes PNIPAAm a suitable polymer as an injectable material for minimally invasive procedures without any use of additional crosslinking agents.

PNIPAAm based injectable hydrogels have been investigated several times as NP repair devices by injection. Methacrylate functionalized chondroitin sulfate and NIPAAm were copolymerized in a recent study by Wiltsey *et al.* and characterized in terms of swelling, mechanical, adhesive, and cytotoxic properties.[35] The resulting polymer, PNIPAAM-*g*-CS, showed better water retain ability compared to homopolymer, PNIPAAm in 14-day period, and the bioadhesion strength with porcine skin was significantly increased. Hydrogels exhibited similar compressive modulus with native NP and showed no cytotoxicity to HEK-293 cells. Swelling and phase transition ability of PNIPAAm can also be enhanced by addition of hydrophilic chains such as polyethylene glycol (PEG).[36]

Another example for the use of thermoreversible PNIPAAm as a NP replacement material is the copolymerized form with hyaluronic acid (HA-PNIPAAm) via reversible addition fragmentation chain transfer polymerization and "click" chemistry. Hyaluronan based thermoreversible PNIPAAm hydrogel induced human mesenchymal stem cells differentiation toward disc cells phenotype without any growth factor supplementation *in vitro* and *ex vivo*. Higher glycosaminoglycan/DNA ratio and higher collagen type II, SOX9, and cytokeratin-19 expressions were found in synthesized copolymer matrix compared with alginate control gel.[37]

Chitosan polymer dissolves in of solution weak organic acids and forms a hydrogel via a shift in pH or temperature in presence of crosslinking agents. Recently, the studies showed that a combination of chitosan with glycerolphosphate generates a temperature sensitive structure, which is liquid at room temperature and gelates at body temperature.[38] This thermo responsive polymer has been investigated for IVD degeneration therapy as an injectable repair material[33,39] Richardson *et al.* reported that human MSCs cultured in chitosan-glycerolphosphate gels differentiated into chondrocyte like NP cells and produced cartilaginous ECM without any addition of exogenous factors.[39] Similarly, Cheng *et al.* showed that extracellular matrix related gene expression, type II collagen and aggrecan synthesis of NP cells increased when cultured in chitosan glycerolphosphate hydrogels than those in monolayer culture.[40]

Recently, a vast interest has been grown on self-assembling peptide structures to be used as three-dimensional scaffolds.[41–45] The ability to mimic extracellular matrix components, having biologically active functionalities and non-immunogenic degradability of the peptide structures offers great advantages for tissue engineering applications. The use of self-assembling peptides as NP replacement material can be exemplified by Wang and colleagues' work. Researchers synthesized a novel functionalized peptide nanofiber hydrogel by combining two self-assembling peptide structures, RADA16 and RLN, and the newly formed gel matrix showed no cytotoxicity as well as stimulate the biosynthesis of NP cells.[46]

KLD-12 (AcN-KLDLKLDLKLDL-CNH2) peptide also gelates in aqueous media forming a homogenous and structurally integral hydrogel and the structure does not show an obvious fluidity. A series of experiments were conducted by Sun and colleagues, using KLD-12 peptide hydrogel for the culture of NP cells. Researchers defined the gel matrix as an excellent injectable material for IVD tissue engineering and reported that structures showed no cytotoxicity or irritation and provided a suitable microenvironment for the cells to proliferate.[47,48]

1.2. Annulus fibrosis

Annulus fibrosis provides the required tensile strength thanks to the collagen fibrillar structures while nucleus pulposus is responsible for the compression resistance behaving like a shock absorber.[49,50] Because of having different mechanical and structural properties, annulus and nucleus tissue replacements require different polymeric materials. Alginate and chitosan based hydrogels are approved by FDA for clinical use and have been frequently tested for annulus fibrosis replacement material. Because of having close structural and functional properties, alginate and chitosan have certain similarities with the locally secreted proteoglycans.[51] Both natural polymers have biocompatible, hydrophilic and low cost properties[52,53] and have been successfully tested for cartilage tissue engineering, polyionic membrane coating system and drug encapsulation applications. Shao and Hunter designed a scaffold with alginate-chitosan polymers by wet-spinning and lyophilization. The annulus fibrosis cell seeded fibrous scaffolds showed no cytotoxicity and exhibited promising *in vitro* responses such as collagen rich ECM production with slow degradation profile.[54]

For the artificial AF engineering, complex arrangement of collagen fiber has been a major challenge, which provides the unique mechanical properties of AF. The AF has a multi-layered structure, characterized by well-organized collagen fibers embedded in a proteoglycan-water gel.[55] Therefore, an engineered AF tissue scaffold should mimic collagen architecture. Mechanical stability is another important issue for AF engineering. latridis and colleagues investigated the mechanical properties of AF tissue and found out that improvement in mechanical stability of scaffolds leads to more effective solutions for healing (latridis *et al.*, 2005). Various geometric structures were tested such as oriented honeycomb,[56] aligned electrospun nanofiber,[57] polymeric film[58] to be used for an AF replacement material. Schek *et al.* expressed that a material for AF repair should adhere on applied area, has similar modulus value to the native tissue and must support growing of cell. Researchers showed that fibrin gels crosslinked with genipin provide these requirements, especially mechanical and adhesion properties of their gels were satisfied.[59] Although, AF cells could be cultured in non-crosslinked hydrogel type scaffolds, however these types are reported as very soft and does not exhibit the required mechanical integrity.[60]

Natural polymers, such as polysaccharides and proteins have been widely studied polymers as biomaterial in hydrogels. The polymers are biocompatible, have low toxicity, susceptible to enzymatic degradation.[61] 3D alginate and agarose scaffolds were formed and primer AF cells were successfully proliferated by Gruber *et al.*[62] Again Gruber *et al.* and Colombini and co-workers successfully cultured AF cells under micromass condition with TGF-β supplementation.[63,64] S. Xinxin and J.H. Christopher formed alginate and alginate/chitosan scaffolds by wet-spinning and lyophilization techniques, then cultured canine AF cells on the scaffolds. Their results showed that the scaffold was satisfied candidate for IVD tissue engineering.[65]

AF cells are fibroblast-like cells, responsible for producing type I collagen and proteoglycans. Sato and colleagues isolated rabbit AF cells and cultured on atelocollagen scaffold which is honeycomb-shaped and on the 21st day, the cells produced a high amount of extracellular matrix, especially glycosaminoglycan and proteoglycans.[66] In another study of Sato *et al.*, an AF cell seeded scaffold (allograft) was placed into rabbits after laser vaporization and the cells proliferation on allograft and hyaline like cartilage production were reported.[66]

Synthetic polymers also have been frequently tested for materials as artificial AF studies due to their general biodegradability, good biocompatibility, together with other adjustable mechanical properties. Wan *et al.* formed poly (1,8-octanediol malate), a maleic acid-based polyester film then cultured AF cells on the film and suggested that the scaffold was an ideal candidate for AF tissue engineering.[58] Yang and co-workers investigated the effect of the surface chemistry of a polymeric polar material on AF cells by forming anionic dihydroxyl oligomer containing polyurethane nanofiber scaffolds. AF cells were cultured on the scaffold and the authors reported that increasing surface polar properties resulted in positive effect on AF cell attachment, proteoglycan, and collagen production.[67] Fibronectin is an extracellular matrix protein and enhanced cellular adhesion and proliferation. In a study carried out by Attia and colleagues, an anionic dihydroxyl oligomer containing aligned polyurethane nanofiber scaffold were formed and coated with fibronectin. They demonstrated that, fibronectin coated scaffold supported AF cells more than untreated scaffolds.[68] Sharifi and colleagues also used fibronection to modify the shape-memory poly (D,L-lactide-co-trimethylenecarbonate) scaffold for the evaluation of human AF cell activity.[69] Bone morphogenetic proteins (BMPs) are multi-functional growth factors that belong to the transforming growth factor beta (TGF-β) superfamily.[70] Some researchers have interested in especially the effect of BMP-2 on AF cells.[71,72] Li and colleagues studied effect of bone morphogenetic protein-2 (BMP-2) on extracellular matrix production of AF cells. They reported that BMP-2 incorporation increased aggrecan and collagen type II mRNA expression and decreased vesicant gene expression, up-regulated BMP-7 and TGF beta mRNA expression.[73]

Nandan *et al.* formed aligned PCL nanofiber scaffold and cultured AF cells on the scaffold. The alignment angles of their scaffold were closed to native AF. They claimed that orientation of the nanofibers lead to organized cells growth.[74] Koepsell and colleagues also formed PCL nanofiber scaffold and cultured AF cells, similar with Nandan *et al.* Koepsell and co-workers found PCL scaffold guided orientation enhanced the proliferation of AF cells and ECM distribution.[57]

Wan and colleagues prepared biphasic scaffold for annulus fibrosis replacement. The outer phase of their scaffold was a ring-shaped demineralized bone matrix gelatin and inner phase was polycaprolactone triol malate. Chondrocytes were seeded on the inner phase and the authors impressed that mechanical properties of the biphasic scaffold was close the normal rabbit AF and scaffold supported the chondrocyte cells proliferation.[75] Mizuno and colleagues used polyglycolic acid and polylactic acid for the outer phase, 2% alginate gel for the inner phase for the construction of biphasic material.[76]

Anna and Steven investigated the regulation of AF cells cultured in 3D scaffold by hydrodynamic pressure. They seeded AF cells on poly(glycolic acid)-poly(L-lactic acid) scaffolds and the samples were hydrodynamically pressurized (5 MPa, 0.5 Hz) for 4h/day from day 3 to 14 of cultivation. Researchers investigated the extracellular matrix production by ELISA and immunohistochemistry methods and found that, the cells successfully handled the pressure and have not effected for producing extracellular matrix. On the contrary, the pressure lead to more type II collagen production and extracellular matrix produced by the cells organization.[77]

2. Future Directions

Hydrogels are generally presented as ideal materials in soft tissue engineering and regenerative medicine studies thanks to their high similarities between the natural tissues. High water uptake, mimicking or having the native extracellular matrix components makes them feasible candidate materials for the repair/replacement therapies. For the last but not least, the mechanical properties of an hydrogel can be easily tuned for the desired purpose of use by adjusting the solution and/or gel parameters. Naturally occurring or synthetic hydrogels can be used alone or combined with some other hydrophilic polymers in order to get diverse responses on mechanical integration, transportation and degradation profiles. Additionally, peptide based low molecular weight gelator incorporations can enhance the bioeffectivity of hydrogels by adding extra properties into the structure.

References

1. H. Shankar, J. A. Scarlett, and S. E. Abram, Anatomy and pathophysiology of intervertebral disc disease. *Tech. Reg. Anesth. Pain. Manage*, **13**, 67–75 (2009).
2. S. C. Joana, I. C. Sandra, M. O. Joaquim, and L. R. Rui, Tissue engineering strategies applied in the regeneration of the human intervertebral disk. *Biotechnol. Adv.*, **31**, 1514–1531 (2013).
3. R. H. Rothman, and F. A. Simeone, *The Spine*, 3rd edn., Philadelphia, PA, USA: WB Saunders Company (1992).
4. J. Urban and S. Roberts, Degeneration of the intervertebral disc. *Arthritis Res. Ther.*, **5**, 120–130 (2003).
5. J. J. Cassidy, A. Hiltner, and E. Baer, Hierarchical structure of the intervertebral disc. *Connect Tissue Res*, **23**, 75–88 (1989).
6. S. Rajasekaran, J. N. Babu, R. Arun, *et al.*, ISSLS prize winner: A study of diffusion in human lumbar discs: A serial magnetic resonance imaging study documenting the influence of the endplate on diffusion in normal and degenerate discs. *Spine*, **29**, 2654–2667 (2004).
7. K. B. Broberg, On the mechanical behavior of intervertebral discs. *Spine*, **8**(2), 151–65 (1983).
8. R. E. David and M. Helen, Types I and II collagens in intervertebral disc. *Interchanging radial distributions in annulus fibrosus. Biochem.* J., **157**, 267–270 (1976).
9. D. R. Eyre, Biochemistry of the intervertebral disc. *Int. Rev. Connect. Tissue Res.*, **8**, 227–291 (1979).

10. H. F. Farfan, V. Osteria, and C. Lamy, The mechanical etiology of spondylosis and spondylolisthesis. *Clin. Orfhop. Rel. Res.*, **117**, 40–55 (1975).
11. J. A. Buckwalter, Spinal update. Aging and degeneration of the human intervertebral disc. *Spine*, **20**, 1307–1314 (1995).
12. Q. B. Bao, G. M. McCullen, P. A. Higham, J. H. Dumbleton, and H. A. Yuan, The artificial disc: Theory, design and materials. *Biomaterials*, **17**, 1157–1167 (1996).
13. B. D. Ahlgren, W. Lui, H. N. Herkowitz, M. M. Panjabi, and J.-P. Guiboux, Effect of anular repair on the healing strength of the intervertebral disc. *Spine*, **25**(17), 2165–2170 (2000).
14. F. Heuer, S. Ulrich, L. Claes, and H. J. Wilke, Biomechanical evaluation of conventional anulus fibrosus closure methods required for nucleus replacement. *J. Neurosurg. Spine*, **9**(3), 307–313 (2008).
15. B. Qi-Bin, M. M. Geoffrey, A. H. Paul, H. D. John, and A. Y. Hansen, The artificial disc: Theory, design and materials. *Biomaterids*, **17**, 1157–1167 (1996).
16. A. S. Hoffman, Hydrogels for biomedical applications. *Adv. Drug Deliv. Rev.*, **54**, 3–12 (2002).
17. A. Gloria, R. de Santis, L. Ambrosio, F. Causa, and K. E. Tanner, A multi-component fiber-reinforced PHEMA-based hydrogel/HAPEX™ device for customized intervertebral disc prosthesis. *J. Biomater. Appl.*, **25**, 795–810 (2011).
18. P. A. Netti, J. C. Shelton, P. A. Revell, C. Pirie, S. Smith, L. Ambrosio, L. Nicolais, and W. Bonfield, Hydrogels as an interface between bone and an implant. *Biomaterials*, **14**, 1098–1104 (1993).
19. N. A. Peppas, P. Bures, W. Leobandung, and H. Ichikawa, Hydrogels in pharmaceutical formulations. *Eur. J. Pharm. Biopharm.*, **50**, 27–46 (2000).
20. M. Oka, T. Noguchi, P. Kumar, K. Ikeuchi, T. Yamamuro, S. H. Hyon, and Y. Ikada, Development of an artificial articular cartilage. *Clin. Mater.*, **6**, 361–381 (1990).
21. A. Gopferich, Mechanisms of polymer degradation and erosion. *Biomaterials*, **17**, 103–114 (1996).
22. A. Gloria, A. Borzacchiello, F. Causa, and L. Ambrosio, Rheological characterisation of Hyaluronic acid derivatives as injectable materials toward nucleus pulposus regeneration. *J. Biomater. Appl.*, **26**, 745–759 (2012).
23. G. Leone, P. Torricelli, A. Chiumiento, A. Facchini, and R. Barbucci, Amidic alginate hydrogel for nucleus pulposus replacement. *J. Biomed. Mater. Res. A*, **84**, 391–401 (2008).
24. S. Ayad and J. B. Weiss, Biochemistry of the intervertebral disc. In *The Lumbar Spine and Back Pain*, Ed. M. I. V. Jayson, 3rd edn., New York, NY: Churchill-Livingstone, pp. 100–137 (1987).
25. E. C. Collin, S. Grad, D. I. Zeugolis, *et al.*, An injectable vehicle for nucleus pulposus cell-based therapy. *Biomaterials*, **32**, 2862–2870 (2011).
26. B. Huang, C. Q. Li, Y. Zhou, *et al.*, Collagen II/hyaluronan/chon-droitin-6-sulfate tri-copolymer scaffold for nucleus pulposus tissue en-gineering. *J. Biomed. Mater. Res. B, Appl. Biomater.*, **92**, 322–331 (2010).
27. J. L. Bron, L. A. Vonk, T. H. Smit, and G. H. Koenderink, Engineering alginate for intervertebral disc repair. *J. Mech. Beh. Biomed. Mater.* 1196–1205 (2011).

28. A. I. Chou, S. O. Akintoye, and S. B. Nicoll, Photo-crosslinked alginate hydrogels support enhanced matrix accumulation by nucleus pulposus cells *in vivo. Osteoarthr. Cartilage*, **17**(10), 1377–1384 (2009).
29. A. Borzacchiello, L. Mayol, A. Schiavinato, and L. Ambrosio, Effect of hyaluronic acid amide derivative on equine synovial fluid viscoelasticity. *J. Biomed. Mater. Res. A*, **92**, 1162–1170 (2010).
30. P. A. Revell, E. Damien, L. Di Silvio, N. Gurav, C. Longinotti, and A. Li, Tissue engineered intervertebral disc repair in the pig using injectable polymers, *J. Mater. Sci. Mater. Med.*, **18**(2), 303–308 (2007).
31. J. E. Frith, A. R. Cameron, D. J. Menzies, P. Ghosh, D. L. Whitehead, S. Gronthos, and J. J. Cooper-White, An injectable hydrogel incorporating mesenchymal precursor cells and pentosan polysulfate for intervertebral disc regeneration. *Biomaterials*, **34**(37), 9430–9440 (2013).
32. G. W. Omlor, A. G. Nerlich, H. Lorenz, T. Bruckner, W. Richter, M. Pfeiffer, and T. Guhring, Injection of a polymerized hyaluronic acid/collagen hydrogel matrix in an *in vivo* porcine disc degeneration model. *Eur. Spine J.*, **21**(9), 1700–1708 (2012).
33. Y. H. Cheng, S. H. Yang, C. C. Liu, A. Gefen, and F. H. Lin, Thermosensitive hydrogel made of ferulic acid-gelatin and chitosan glycerophosphate. *Carbohyd. Polym.* **92**(2), 1512–1519 (2013).
34. S. H. Park, E. S. Gil, H. Cho, B. B. Mandal, L. W. Tien, B. H. Min, and D. L. Kaplan, Intervertebral Disk Tissue Engineering Using Biphasic Silk Composite Scaffolds. *Tissue Eng. A*, **18**(5–6), 447–458 (2012).
35. C. Wiltsey, P. Kubinski, T. Christiani, K. Toomer, J. Sheehan, A. Branda, and J. Vernengo, Characterization of injectable hydrogels based on poly(N-isopropylacrylamide)-g-chondroitin sulfate with adhesive properties for nucleus pulposus tissue engineering. *J. Mater. Sci.*, **24**(4), 837–847 (2013).
36. J. D. Thomas, G. Fussell, S. Sarkar, A. M. Lowman, and M. Marcolongo, Synthesis and recovery characteristics of branched and grafted PNIPAAm–PEG hydrogels for the development of an injectable load-bearing nucleus pulposus replacement. *Acta Biomater.*, **6**, 1319–1328 (2010).
37. M. Peroglio, D. Eglin, L. M. Benneker, M. Alini, and S. Grad, Thermoreversible hyaluronan-based hydrogel supports *in vitro* and *ex vivo* disc-like differentiation of human mesenchymal stem cells. *Spine J.*, **13**(11), 1627–1639. (2013).
38. J. Cho, M. C. Heuzey, A. Begin, and P. J. Carreau, Physical gelation of chitosan in the presence of beta-glycerophosphate: The effect of temperature: *Biomacromolecules*, **6**(6), 3267–3275 (2005).
39. S. M. Richardson, N. Hughes, J. A. Hunt, A. J. Freemont, and J. A. Hoyland, Human mesenchymal stem cell differentiation to NP-like cells in chitosan–glycerophosphate hydrogels. *Biomaterials*, **29**(1), 85–93 (2008).
40. Y. H. Cheng, S. H. Yang, W. Y. Su, Y. C. Chen, K. C. Yang, W. T. K. Cheng, S. C. Wu, and F. H. Lin, Thermosensitive chitosan–gelatin–glycerol phosphate hydrogels as a cell carrier for nucleus pulposus regeneration: An *in vitro* study. *Tissue Eng. Pt. A*, **16**(2), 695–703 (2010).

41. A. Horii, X. Wang, F. Gelain, and S. Zhang, Biological designer self-assembling Peptide nanofiber scaffolds significantly enhance osteoblast proliferation, differentiation and 3-D migration. *PLoS One*, **2**, e190 (2007).
42. F. Zhang, G. S. Shi, L. F. Ren, F. Q. Hu, S. L. Li, and Z. J. Xie, Designer self-assembling peptide scaffold stimulates pre-osteoblast attachment, spreading and proliferation. *J. Mater. Sci. Mater. Med.*, **20**, 1475–1481 (2009).
43. F. Gelain, D. Bottai, A. Vescovi, and S. Zhang, Designer self-assembling peptide nanofiber scaffolds for adult mouse neural stem cell 3-dimensional cultures. *PLoS One*, **27**, e119 (2006).
44. E. Genove, C. Shen, S. Zhang, and C. E. Semino, The effect of functional-ized self-assembling peptide scaffolds on human aortic endothelial cell function. *Biomaterials*, **26**, 3341–3351 (2005).
45. H. Misawa, N. Kobayashi, A. Soto-Gutierrez, Y. Chen, A. Yoshida, J. D. Rivas-Carrillo, N. Navarro-Alvarez, K. Tanaka, A. Miki, J. Takei, T. Ueda, M. Tanaka, H. Endo, N. Tanaka, and T. Ozaki, PuraMatrix facilitates bone regeneration in bone defects of calvaria in mice. *Cell Transpl.*, **15**, 903–910 (2006).
46. B. Wang, Y. Wu, Z. Shao, S. Yang, B. Che, C. Sun, and Z. Ma, Functionalized self-assembling peptide nanofiber hydrogel as a scaffold for rabbit nucleus pulposus cells. *J. Biomed. Mater. Res. A*, **100A**, 646–653 (2012).
47. J. Sun and Q. Zheng, Experimental study on self-assembly of KLD-12 peptide hydrogel and 3-D culture of MSC encapsulated within hydrogel *in vitro*. *J. Huazhong U. Sci. Med.*, **29**(4), 512–516 (2009).
48. J. Sun, Q. Zheng, Y. Wu, Y. Liu, X. Guo, and W. Wu, Biocompatibility of KLD-12 peptide hydrogel as a scaffold in tissue engineering of intervertebral discs in rabbits. *J. Huazhong U. Sci. Med.*, **30**(2), 173–177 (2010).
49. M. Alini, P. J. Roughley, J. Antoniou, T. Stoll, and M. Aebi, A biological approach to treating disc degeneration: Not for today, but maybe for tomorrow. *Eur. Spine. J.*, **11**(Suppl 2), S215–S220 (2002).
50. M. V. Risbud, I. M. Shapiro, A. R. Vaccaro, and T. J. Albert, Stem cell regeneration of the nucleus pulposus. *Spine J.*, **4**(Suppl 6), 348S–353S (2004).
51. P. Roughley, C. Hoemann, E. DesRosiers, F. Mwale, J. Antoniou, and M. Alini, The potential of chitosan-based gels containing intervertebral disc cells for nucleus pulposus supplementation. *Biomaterials*, **27**, 388–396 (2006).
52. T. W. Chung, J. Yang, T. Akiaike, K. Y. Cho, J. W. Nah, S. I. Kim, and C. S. Cho, Preparation of alginate/galactosylated chitosan scaffold for hepatocyte attachment. *Biomaterials*, **23**, 2827–2834 (2002).
53. E. V. M. Fragonas, M. Pozzi-Mucelli, R. Toffanin, R. Rizzo, F. Silvestri, and F. Vittur, Articular cartilage repair in rabbits by using suspensions of allogenic chondrocytes in alginate. *Biomaterials*, **21**, 795–801 (2000).
54. X. Shao and C. J. Hunter, Developing an alginate/chitosan hybrid scaffold for annulus fibrosus cells. *J. Biomed. Mater. Res.*, **82A**, 701–710 (2007).
55. K. L. Markolf and J. M. Morris, The structural components of the intervertebral disc. *J. Bone Jt. Surg. Am.*, **56**, 675–687 (1974).

56. M. Sato, T. Asazuma, M. Ishihara, T. Kikuchi, K. Masuoka, S. Ichimura, M. Kikuchi, A. Kurita, and K. Fujikawa, An atelocollagen honeycomb-shaped scaffold with a membrane seal (ACHMS-scaffold) for the culture of annulus fibrosus cells from an intervertebral disc. *Inc. J. Biomed. Mater. Res.*, **64A**, 249–256 (2003).
57. L. Koepsell, L. Zhang, D. Neufeld, H. Fong, and Y. Deng, Electrospun nanofibrous polycaprolactone scaffolds for tissue engineering of annulus fibrosus. *Macromol. Biosci.*, **11**(3), 391–399 (2011).
58. Y. Wan, G. Feng, F. H. Shen, G. Balian, C. T. Laurencin, and X. Li., Novel Biodegradable Poly(1,8-octanediol malate) for annulus fibrosus regeneration. *Macromol. Biosci.*, **7**, 1217–1224 (2007).
59. R. M. Schek, A. J. Michalek, and J. C. Iatridis, Genipin-crosslinked fibrin hydrogels as a potential adhesive to augment intervertebral disc annulus repair. *Eur. Cell Mater.*, **21**, 373–383 (2011).
60. S. A. Rodrigues, K. R. Wade, A. Thambyah, and N. D. Broom, Micromechanics of annulus–end plate integration in the intervertebral disc. *Spine J.*, **12,** 143–150 (2012).
61. N. Bhattarai, J. Gunn, and M. Zhang, Chitosan-based hydrogels for controlled, localized drug delivery. *Adv. Drug Deli. Rev.* **62**, 83–99 (2010).
62. H. E. Gruber, E. C. Fisher, B. Desai, A. A. Stasky, G. Hoelscher, and E. N. Hanley, Human intervertebral disc cells from the annulus: Three-Dimensional culture in agarose or alginate and responsiveness to TGF-b1. *Exp. cell Res.*, **235**, 13–21 (1997).
63. H. E. Gruber, Y. Chow, G. L. Hoelscher, J. A. Ingram, N. Inchenko, H. J. Norton, Y. Sun, and E. N. Hanley, Micromass culture of human anulus cells morphology and extracellular matrix production. *Spine*, **35**(10), 1033–1038 (2010).
64. A. Colombini, P. Lanteri, G. Lombardi, D. Grasso, C. Recordati, A. Lovi, G. Banfi, R. Bassani, and M. Brayda-Bruno, Metabolic effects of vitamin D active metabolites in monolayer and micromass cultures of nucleus pulposus and annulus fibrosus cells isolated from human intervertebral disc. *Int. J. Biochem. Cell Biol.*, **44**, 1019–1030 (2012).
65. X. Shao and C. J. Hunter, Developing an alginate/chitosan hybrid fiber scaffold for annulus fibrosus cells. *J. Biomed. Mater. Res.*, **82A**, 701–710, (2007).
66. M. Sato, T. Asazuma, M. Ishihara, M. Ishihara, T. Kikuchi, M. Kikuchi, and K. Fujikawa, An experimental study of the regeneration of the intervertebral disc with an allograft of cultured annulus fibrosus cells using a tissue-engineering method. *Spine*, **28**(**6**), 548–553 (2003).
67. L. Yang, R. A. Kandel, G. Chang, and J. P. Santerre, Polar surface chemistry of nanofibrous polyurethane scaffold affects annulus fibrosus cell attachment and early matrix accumulation. *J. Biomed. Mater. Res.*, **91A**, 1089–1099 (2009).
68. M. Attia, J. P. Santerre, and R. A. Kandel, The response of annulus fibrosus cell to fi bronec tin-coated nano fibrous polyurethane-anio nic dihydroxyoligomer scaffolds. *Biomaterials*, **32**, 450–460 (2011).
69. S. Sharifi, T. G. van Kooten, H. J. C. Kranenburg, B. P. Meij, M. Behl, A. Lendlein, and D. W. Grijpma, An annulus fibrosus closure device based on a biodegradable shape-memory polymer network. *Biomaterials*, **34**, 8105–8113 (2013).

70. D. Chen, M. Zhao, and G. R. Mundy, Bone morphogenetic proteins. *Growth Factors.* **22**(4), 233–41 (2004).
71. S. U. Kuh, Y. Zhu, J. Li, K. J. Tsai, Q. Fei, W. C. Hutton, and T. S. Yoon, A comparison of three cell types as potential candidates for intervertebral disc therapy: Annulus fibrosus cells, chondrocytes, and bone marrow derived cells. *Joint Bone Spine,* **76,** 70–74 (2009).
72. G. Vadala, P. Mozetic, A. Rainer, M. Centola, M. Loppini, M. Trombetta, and V. Denaro. Bioactive electrospun scaffold for annulus fibrosus repair and regeneration. *Eur. Spine. J.*, **21**, S20–S26 (2012).
73. J. Li, S. T. Yoon, and W. C. Hutton, Effect of bone morphogenetic protein-2 (BMP-2) on matrix production, other BMPs, and BMP receptors in rat intervertebral disc cells. *J. Spinal Disord. Tech.*, **17**(5), 423–428 (2004).
74. N. L. Nerurkar, D. M. Elliott, and R. L. Mauck, Mechanics of oriented electrospun nanofibrous scaffolds for annulus fibrosus tissue engineering. *J. Orthop. Res.*, **25**, 1018–1028 (2007).
75. Y. Wan, G. Feng, F. H. Shen, C. T. Laurencin, and X. Li, Biphasic scaffold for annulus fibrosus tissue regeneration. *Biomaterials,* **29**, 643–652 (2008).
76. H. Mizuno, A. K. Roy, C. A. Vacanti, K. Kojima, M. Ueda, and L. J. Bonassar, Tissue-engineered composites of anulus fibrosus and nucleus pulposus for intervertebral disc replacement. *Spine,* **29**(12), 1290–1298 (2004).
77. T. R. Anna, and B. N. Steven, Hydrostatic pressure differentially regulates outer and inner annulus fibrosus cell matrix production in 3D scaffolds. *Ann. Biomed. Eng.*, online November **17**, (2007).

Chapter 8

Hydrogels in Cartilage Tissue Engineering

Antonella Motta, Mariangela Fedel and Claudio Migliaresi
Department of Industrial Engineering and BIOtech Research Center
University of Trento, Trento, Italy
claudio.migliaresi@unitn.it

Due to the absence of blood and lymphatic circulation, damages of cartilage do not spontaneously repair, and the associated pain is one of the most frequent causes of invalidity, mainly in elderly people.

This chapter examines composition, morphology and functions of articular cartilage, providing basic information of its most frequent injuries and pathologies. Surgical treatments are reported and the regenerative medicine approaches trough the use of proper materials and cells are in-depth discussed. The requirements for the scaffold materials are also considered in terms of architecture, mechanical properties, and molecular recognition capabilities, together with methods of fabrication and/or modification.

In particular, the use of hydrogels (HGs) is reviewed, both from synthetic and natural polymers, as elective materials for cartilage repair. Their mechanical properties, degradation profile as well as production and distribution of ECM can be tailored by the polymer chemistry, gelation mechanism and design parameters. HGs can offer alternative solutions to traditional "open site" surgical approaches with minimally invasive applications exploiting "*in situ*" gelation methods or smart sol-gel transitions.

1. Introduction

1.1. Articular cartilage: Structure and functions

Articular cartilage plays different crucial functions such as absorbing impact forces, transferring and distributing load to the underlying bone, providing a low-friction surface for articulation.

Cartilage can be considered a biphasic and inhomogeneous structure made of a solid phase (15–32%) and a liquid phase (68–85%), which form the extracellular matrix of the tissue. It is composed by cells (chondrocytes, 5% of the total cartilage volume) and Extracellular Matrix (ECM). Chondrocytes possess the fundamental role of preserving tissue homeostasis, degrading and remodeling the ECM to maintain tissue integrity and mechanical properties.

Cartilage is not-vascularized, a-neural and a-lymphatic. Chondrocytes get nutrishment and oxygen from the synovial liquid that diffuses through the matrix while lubricating the articular surfaces. Mature chondrocytes are rounded, unable to proliferate and located in the ECM in restricted spaces named lacunae. Some of them have cilia in the extracellular environment, which can collect mechanical stimuli to modulate tissue composition according to the load applied to the joint.[1]

Extracellular matrix composition determines the peculiar mechanical properties of hyaline cartilage (Table 1). Cartilage contains a fluid phase, namely water and electrolytes, and a solid phase mainly composed by collagen within a proteoglycan gel. Collagen, mainly type II (90–95%), forms fibrils, which provide tensile strength to the tissue and entrap other matrix macromolecules. Proteoglycans (PGs) are large macromolecules consisting of a protein core with multiple chains of glycosaminoglycans (GAGs) and oligosaccharides attached on. The predominant proteoglycan in cartilage is aggrecan, a proteoglycan with over 100 chondroitin sulfate (CS) and keratin-sulfate chains, that can interact with hyaluronic acid (HA) to form large proteoglycan aggregates. PGs can entrap large amounts of water, up to 50 times their weight. Having a high density negative charge, they can resist compression loads (up to 65 times the body weight) and contribute to the viscoelastic behavior of hyaline cartilage.[2] Their interconnection with collagen molecules prevents PG diffusion during the application of load.[3–5] Minor proteins in hyaline cartilage include glycoproteins, fibronectin and anchorin II: Their role is to stabilize the ECM and anchor chondrocytes to the surrounding matrix.

Cartilage is morphologically inhomogeneous and mechanically anisotropic, reacting to compressive, tensile and shear stresses with a non-linear response. The anisotropic behavior of cartilage is due to the orientation and distribution of the collagen fibers, through the cartilage thickness.

Table 1. Composition of native cartilage. The symbol ς represents the gel mesh size. From Ref. 6, with copyright permission.

	% by wet weight	Component	Size scale	Structure	Category
Collagens	20	Collagen II Collagen VI	400 nm 100 nm	Linear fibers Fibrillar	Large molecules
Aggrecan, proteoglycans, and their building blocks	5–7	Aggrecan aggregates Aggrecan Hyaluronian Link protein Glycosaminoglycans Decorin Biglycan	100 MDa 1–4 MDa MDa 45 kDa 5–30 kDa 40 kDa 40 kDa	Highly branched Branched Linear Globular Linear Globular Globular	Small molecules
Other	70–75	Water Salts	$<<\varsigma$ $<<\varsigma$	— —	

According to morphology, three different zones can be distinguished at increasing proteoglycans content: A superficial zone (10–20% of total thickness), an intermediate zone (40–60%), a deep zone (30–40%). The superficial zone is made by two sub-zones. It contains the highest amount of water, the lowest content of proteoglycans and the highest collagen content. In the first sub-zone there are no cells, while flattened chondrocytes populate the second sub-zone. Small collagen fibrils run parallel to the articular surface in order to better resist shear forces.

Collagen concentration decreases in the intermediate zone, fibrils have a higher diameter and a random orientation. Chondrocytes are round shaped, forming columnar structures oriented in the same direction and have a strong activity of biosynthesis. Orientation of collagen fibrils becomes perpendicular to the surface in the third zone that interfaces with a calcified zone that is the interphase between cartilage and subchondral bone.

Chondrocytes are smaller than in the other portions of the tissue and have a low metabolic activity.

1.2. Articular cartilage injuries

The pain associated with articular cartilage injuries is a major cause of disability not only in the middle-aged and older people but also in younger patients. Articular cartilage insults can take place in various pathological joint conditions such as trauma, inflammation, infection and degeneration. Defects can be primary lesions due to diseases such as osteochondritis dissecans, or may be secondary to trauma, as in the case of sport injuries. Cartilage injuries are usually classified in three categories: matrix disruption, partial thickness defects and full thickness defects.[7] In the case of matrix disruption, which mainly occurs from blunt trauma, the ECM is damaged, but generally the remaining viable chondrocytes provide to repair the damaged tissue. Partial thickness defects implicate the disruption of the cartilage tissue with no penetration into the subchondral bone. In full thickness defects, the damage involves the entire cartilage thickness and extends to the subchondral bone.

The type of injury, including its depth and size, and whether it is acute or chronic, distinctly affects the possible healing responses. Articular cartilage injuries have a limited potential for self-repair[8] mainly due to low proliferation ability of chondrocytes and absence of blood as supply of progenitor cells within the cartilage tissue.

The propensity for a specific partial thickness injury to heal is dependent on the size of the lesion, its location within the joint, and the age of the individual.[9]

More in-depth details about structure, biochemistry and metabolism of cartilage degeneration, and in particular about the pathogenesis of osteoarthritis, are described by Dijkgraaf *et al.*,[10] and by Lorenz *et al.*[11]

1.3. Articular cartilage remodeling

Considering the low cellular population, low cell migration into ECM and the absence of blood vessels, cartilage tissue has an intrinsic inability to regenerate. The tissue is organized such that repair should not be needed, and it is mainly involved in maintenance activity.[9] This process involves catabolic and anabolic activity of chondrocytes, such as breakdown of the old ECM component and synthesis of new ECM component respectively. The metabolic balance of cartilage tissue seems to be biologically controlled by mediators (polypeptides) such as transforming growth factors-β β s , promote repair after damage) β- : mitogenic agent for chondrocytes, enhance matrix synthesis insuline-like growth factors (IGF-1: enhance chondrocytes proliferation and ECM macromolecules synthesis; decrease cartilage degradation).[12] In addition, mechanical stimuli player a role in cartilage maintenance–regeneration–degradation, but the mechanisms are still unclear.

1.4. Treatment options for articular cartilage defects

In the last decades, surgeons and scientists have made various efforts to promote cartilage healing, to restore joints function and reduce pain, resulting in alternating outcomes. Different modalities for the treatment of articular cartilage defects have been developed, commonly divided into bone marrow stimulation, repair and regenerative techniques. In this context, repair refers to the healing of injured tissues or replacement of lost tissue by cells proliferation and synthesis of new extracellular matrix. Generally, articular cartilage repair fails with respect to the structure, composition, and function of normal articular cartilage. On the other side, regenerative methods aim to stimulate the formation of an entirely new articulating surface that essentially duplicates the original hyaline cartilage tissue, its structure and biomechanical properties.

The choice of the more opportune treatment technique depends on the degree of cartilage degeneration and on patient factors.[13] Bone marrow stimulation techniques include abrasion, drilling and microfracture. These methods aim at stimulating the healing response by recruiting bone marrow cells from the sub-chondral bone. The formation within the defect of a blood clot induced by sub-chondral bone drilling, in fact, allows the recruitment of progenitor stem cells, which facilitate tissue repair.[14–16] Microfracture provides effective short-term functional improvement of knee function, but medium to long-term follow up studies have revealed limited hyaline-like cartilage tissue.[17]

Repair techniques are mainly based on osteochondral autograft transplantation surgery (OATS or Mosaicplasty). OATS was first introduced in the late 1990s and consists of the transplantation of structurally intact osteochondral tissue from non-load bearing sites to the defect site. This technique seems to be able to significantly restore height and shape of articulating surface in focal osteochondral defects, with the transplantation of a composite autologous material that contains all necessary ingredients: Hyaline articular cartilage, intact tidemark and a firm bone carrier. The main problem with this reconstructive technique is the limited availability of autologous osteochondral grafts to cover the entire defect area. A similar option for the treatment of large post-traumatic injuries, osteochondritis dissecans, and osteonecrosis is represented by osteochondral allografting, which involves the removal of fresh osteochondral tissue from a donor, and its insertion into the site of injury.

The first regenerative technique, Autologous Chondrocyte Implantation (ACI), was originally described by Peterson, and introduced in Sweden in 1994 by Brittberg *et al.*[18] as a treatment for symptomatic chondral and osteochondral defects of the knee. Autologous chondrocytes are harvested from non-weight bearing regions of a joint and expanded *in vitro* up to several millions of cells. After 2–4 weeks the joint is opened, the defect area is prepared, and a periostal patch is sutured to the defect site. Expanded cells are then injected under the periostal flap (ACI-P), which is further sealed with fibrin glue. Alternatively, collagen membranes from different sources (e.g., porcine collagen) can be used instead of periosteum to confine the chondrocyte suspension at the defect site (ACI-C). Several studies showed the formation of predominantly hyaline-like cartilage tissue following ACI.[18–20] Good clinical outcomes were reported in chondral defects treated with ACI and microfracture techniques,[21,22] revealing in some cases of superior tissue regeneration for the ACI group. Technical limitations related to conventional ACI include the requirement of a two-stage procedure, long times for *in vitro* cells expansion, non-homogeneous distribution of chondrocytes in the recipient site,[15] graft hypertrophy and risk of cell leakage. Moreover, poor cell dedifferentiation is frequently obtained with first-generation ACI, with alteration of chondrocytes phenotype, gene expression and morphology.

To overcome the problems encountered with ACI, the use of scaffolds as support for chondrocytes has been introduced.

Matrix-Induced Autologous Chondrocytes Implantation (MACI) is a third generation ACI technique based on the use of a collagen scaffold. Several different scaffolds are currently used in clinical settings. The MACI™ membrane (Genzyme Biosurgery, Cambridge, MA, USA) consists of a purified and decellularized porcine type I/type III collagen bilayer. This kind of membrane has one smooth surface made of high density collagen fibers, and one rougher surface with widely-spaced collagen fibers, where chondrocytes are seeded.[23,24] Technically, this procedure facilitates surgery and reduces operative time compared to traditional ACI. Moreover, it does not require suturing because the scaffold is secured to the lesion with fibrin glue. Symptomatic improvement has been shown at short- and medium-term follow-up, but long-term outcome data and comparisons against conservative management are lacking.[25]

Other approaches for the treatment of cartilage lesions include the use of mesenchymal stem cells (MSCs), platelet-rich plasma (PRP),[26] growth factors (GFs) and bone morphogenetic proteins (BMPs),[27] as well as gene therapy.[28]

1.5. Articular cartilage tissue engineering

Although current conventional surgical techniques have provided successful outcomes in alleviating pain and improving joint function, optimal results in the regeneration of hyaline cartilage tissue have not been obtained yet, especially at the long-time. Therefore, the research has recently focused on tissue engineering (TE) approaches based on the use of scaffold materials, chondrogenic cells and signaling molecules, separately or in combination.[29] The main issues in articular cartilage TE regard the optimization of biocompatible materials serving as scaffolds for cells and matrix, as well as the combination with the more suitable cell sources and signaling molecules, such as growth factors or genes, to induce opportune cells differentiation/dedifferentiation.[30] Due to the critical characteristics of chondral tissue (e.g., absence of vascularization and low cell density) and to its limited ability to self-repair, cartilage regeneration is a significant challenge for TE.

For the development of a suitable scaffold able to promote the regeneration of functional cartilage it is important to keep in mind the features of native healthy chondral tissue and its biological function. Thus, the ideal scaffolds for articular cartilage TE should satisfy same basic requirements regarding scaffold architecture and mechanical properties, as well as materials chemistry, biocompatibility, biodegradability and bioactivity. The selected materials need to be biocompatible in the articular bio-environment, that is to say, they need to perform with an appropriate response in the specific application they are designated for.[31] In the particular case of articular cartilage, materials are expected to prevent the induction of adverse inflammatory or immune responses.

For successful tissue regeneration, scaffolds should have controllable structural and mechanical properties (Table 2). It has been proven that the chondrogenic and the osteogenic differentiation capacity of *in vitro* expanded human articular chondrocytes can be influenced by the scaffold architecture.[32] A porous and permeable 3D structure is necessary to optimize cell seeding, to allow cell attachment, retention, migration and organization through the entire scaffold, as well as to permit the diffusion of nutrients and wastes. Even if the optimal pore-size is cell-type dependent,[15,33] generally a high surface-to-volume ratio is preferable for cartilage TE scaffolds. For example, in the case of poly(urethane urea) scaffolds, small pore

Table 2. Requirements for cartilage TE scaffolds.

Property	**Role**
Biomaterials biocompatibility	To avoid adverse events such as inflammation or immune response.
3D structure	To allow cells attachment, retention, migration and organization.
Porosity	To direct cell infiltration and ECM production.
Permeability	To allow diffusion of nutrients and waste products.
Chemical composition	Presence of specific receptor ligands (e.g., RGD sequences) or functional groups to facilitate cell adhesion and differentiation.
Mechanical properties	To match native tissue (i.e., to withstand high compressive stress); can affect MSC differentiation, cell morphology, proliferation and migration.
Controllable biodegradability	To avoid the formation of toxic by-products. Ideally scaffold degradation should be balanced by new chondral tissue formation.
Bioactivity	Production of stimoli or controlled release of bioactive molecules (e.g., growth factors or gene factors) to drive and stimulate cartilage formation through cell differentiation and ECM production.
Applicability	To be clinically applicable possibly through minimally invasive procedures.

size (< 150 μm) enhanced hyaline-like extracellular matrix and thus neocartilage formation, while larger pore size promoted chondrocytes differentiation into the osteogenic pathway.[32]

The scaffold should promote cell viability, differentiation, and extracellular matrix production. Scaffolds composition and surface chemistry are critical in defining cell attachment and differentiation. The most common biomaterials used in cartilage TE include natural-derived and synthetic polymers, as well as a combination of the two (natural and synthetic materials). Natural-derived matrices include proteins (collagen and gelatin, fibrin, silk fibroin) and carbohydrate-based polymers (agarose, alginate, HA, CS, chitosan). The most used synthetic materials include poly(L-lactic acid) (PLLA), poly(glycolic acid) (PGA) and their co-polymers, poly(ε-caprolactone) (PCL), poly(ethylene glycol) (PEG), poly(urethane) (PU), and poly(vinyl alcohol) (PVA).[34] Several other experimental compositions, including composites, blends, self-assembling peptides, chemically modified or functionalized materials have been investigated for cartilage TE.

The presence of specific receptor ligands similar to those of native cartilage ECM, such as for example Arg–Gly–Asp (RGD) sequences (as in Type II and Type IV collagen), make natural polymers specifically attractive for cartilage TE, since they can induce cell adhesion, migration, proliferation and production of ECM. On the other hand, synthetic materials can be easily tailored as regards porosity, degradation and mechanical characteristics, even if in some cases degradation products may be toxic and elicit an inflammatory response.

Purpose of the scaffold is to support the expansion of new chondral tissue in the early phases and to protect the neocartilage from *in vivo* loads. Preferably, the mechanical properties of the scaffolds should be comparable to the properties of native cartilage (e.g., aggregate modulus typically in the range of 0.5–0.9 MPa and Young's modulus 0.45/0.80 MPa)[35] and match the properties of the surrounding tissues, in order to reduce shear stresses especially in critical areas, such as at the interfaces. The mechanical and viscoelastic properties of the scaffolds are fundamental in directing and enhancing mesenchimal stem cell differentiation,[36] since cells seem to "feel" the stiffness of the underlying material and to actively respond to mechanical stimuli.

Together with the presence of biomimetic hierarchical structures conferring appropriate mechanical strength, controllable biodegradability represents another critical aspect requirement for cartilage TE scaffolds. Degradation and resorption rate should be gradual and compare to new functional cartilage tissue development (approximatively in the order or days to months).

To further improve and facilitate chondral tissue remodeling, scaffolds may be also incorporated with bioactive molecules, such as growth factors or vectors carrying genes that can stimulate a specific gene expression.

Three main different scaffold architectures have been proposed for cartilage TE, including sponges, fibrous nets, and HGs. Moreover, the potential for chondral tissue genesis starting from hybrid or composite or multiphase scaffold structures has been tested.[37]

Sponges are porous 3D scaffolds presenting interesting features for cartilage TE. Porosity, pore size and interconnectivity are critical parameters that can be adjusted according to the requirements, significantly influencing cell attachment and migration, matrix production, as well as diffusion of nutrients and waste products. Different natural and synthetic polymers have been used for the fabrication of scaffolds with sponge-like architecture, and several methods have been employed for the manufacturing, including phase separation, emulsion freeze-drying, porogen-leaching[38,39] and gas foaming techniques (e.g., Supercritical carbon dioxide).[40–42]

Woven and non-woven fibrous constructs with different characteristics have been tested in articular cartilage TE. By 3D fiber deposition (rapid prototyping) or microfabrication techniques[43,44] macro- and micro-metric fiber meshes can be obtained, with controllable porosity. Non-woven nets obtained by electrospinning processes have recently gained consistent interest,[45,46] even if they present critical limitations as poor cellular infiltration due to low porosity. The development of multiscale scaffolds combining nano- and micro-scale features has been proposed as a possible solution to exploit pro-adhesive properties of nanometric fibers while improving cell migration through the scaffold.[47]

More information about sponge- and net-based scaffolds for cartilage TE, as well as on fabrication technologies, can be found in some interesting studies and literature.[42,48–50] Hydrogels for articular cartilage TE will be discussed in detail in the following sections.

2. HGs for Cartilage Repair

Why HGs are attractive scaffold materials for articular cartilage TE?

HGs are hydrophilic polymer networks, which may adsorb from 20% to thousands of times their dry weight in water assuming a final water content up to 60–90%.[51,52] They represent a very interesting type of substrate for TE, and for articular cartilage TE in particular. In fact, HGs well resemble the ECM of natural cartilage, a matrix with a high water content (about 70%) composed of a 3D network of collagen fibrils infiltrated by mainly proteoglycans and some proteins.

Polymeric HGs respond to most of the requirements for cartilage TE scaffolds (see Table 2). In fact, they present a highly hydrated, porous and permeable 3D structure, able to host and organize cells and to allow an efficient transport of oxygen, nutrients, and other water-soluble metabolites. Due to their structure and water content, HGs facilitate cell–cell and cell–matrix interactions. They typically support uniform cell distribution if the appropriate cell density suspension method is chosen. The response of HGs in terms of biocompatibility and biodegradation clearly depends on the chemical composition.

HGs can be tailored to mimic cartilage structure and functions by tuning various parameters, including chemical composition, water content, organization and biodegradability. Moreover, they can be designed to have specific biological functionalities and physical properties.

An attractive characteristic and further benefit with respect to the other types of scaffolds for cartilage TE (sponges or nets) is that HGs can be delivered through injection, thus completely filling defect sites of any size and geometry, and significantly reducing the invasiveness of the surgical intervention.

2.1. *Hydrogels* crosslinking

Structural integrity, homogeneity, and physical properties of the HGs typically depend on the crosslinking between the polymer chains.[52] Network crosslinking can be achieved through chemical bonds and/or physical interactions.

In "physical" (or reversible) HGs, the molecules are held together by entanglements and/or secondary forces including ionic, H-bonding or hydrophobic forces,[51] which determine the reversible character of these gels. Physically crosslinked gels with interest in the biomedical field can be obtained by inducing the crosslinking of the network through stimuli of different nature, such as temperature, pH, ionic strength or dehydrothermal treatment. Other gelation mechanism to prepare physical HGs can be based on the combination of polyelectrolyte chains with a multivalent ion opposite in charge (e.g., calcium alginate), on the mix of polyelectrolytes of opposite charges with proper concentrations and under suitable pH and osmolality conditions, as well as on biospecific recognition.

Typical examples of physical HGs proposed for cartilage TE are based on chitosan, collagen, alginate, HA etc.[53–55]

Especially interesting for cartilage TE are smart polymers that can undergo sol-gel transition "*in-situ*".[56] In particular, in the case of thermoresponsive gels with a lower critical solution temperature (LCST) around body temperature, the polymer solution can be easily injected in the defect area while in the liquid phase, and undergoes gelation *in situ* at about 37°C. Pluronic®, a commercially available tri-block polymer (PEO–PPO–PEO), and PNIPAAm (poly(N-isopropylacrylamide acid) are famous examples of thermoresponsive HGs with a LCST around physiological temperature. For cartilage TE applications Pluronic® has been combined to other natural and synthetic polymers (e.g., collagen,[39] gelatin,[57] chitosan,[58] alginate,[59] PEG,[60] to obtain HGs with improved mechanical properties and biodegradability with respect to the pure polymer.[56,61]

Thermoreversible gels can be formed also by some natural polymers used for cartilage scaffolds, including gelatin, agarose, amylose, amylopectin, and carrageenans. These polymers in solution at high temperature present random coil conformation. When the temperature is decreased, they tend to assume their natural conformation (triple-helix for gelatin and double-helix for polysaccharides), originating aggregates that act as physical junctions and nucleation centers for the forming gels.[62]

Due to weak stability and possible reversibility, physical HGs usually present low mechanical properties.

"Chemical" HGs are permanently crosslinked through covalent bonding. They can be formed from copolymerization (monomer + crosslinker or monomer + multifunctional macromer), crosslinking of water-soluble polymers by irradiation (e.g., UV, gamma-rays, microwaves), by chemical crosslinkers (e.g., aldehydes, carbodiimides, isocyanates) or by multifunctional reactive compounds, as well as from Schiff's base crosslinking reaction.[63]

Photoirradiation and incorporation of crosslinking agents are often causing of concern in the case of HGs for biomedical applications due to possible toxicity. Crosslinkers commonly used in the past (e.g., glutaraldehyde) have recently been replaced by natural-derived or non-toxic crosslinkers (e.g., genipin) to improve gels biocompatibility. The cytotoxicity of genipin was shown to be significantly lower (5,000 times ca.) than that of glutaraldehyde.[64] Genipin used as crosslinker for fibrin HGs intended for *in vitro* articular cartilage TE did not significantly affect cell viability, but significantly increased the dynamic compression and shear moduli of the hydrogel, inhibiting the inflammatory reaction 3 weeks after subcutaneous implantation of the fibrin gels into rats.[65] Recently, genipin has been extensively used to crosslink natural and synthetic polymer HGs.[38,66–70] The effects of genipin concentration and incubation duration on the properties of type I collagen gels have been studied by Sundararaghavan and colleagues, showing that genipin produced significant concentration- and incubation-dependent increases in the storage modulus, loss modulus, and fluorescence intensity. Minimal cytotoxicity was observed for exposure of L929 fibroblasts cultured within collagen gels to 1 mM genipin for 24 h, while 5 and 10 mM genipin caused significant cell death.[66] Dare *et al.* showed that genipin crosslinked fibrin HGs can be used as cell-compatible TE scaffolds for articular cartilage regeneration, confirming also in this case that genipin crosslinking improved the mechanical properties of the gel and slowed the rate of biodegradation and inflammation of subcutaneous implants.[65] The potential cytotoxic effects of genipin on cells were tempered by the presence of the polymeric matrix.

Summarizing, genipin is a very powerful crosslinker of natural origin; its concentration needs to be optimized to maximize all benefits for the system while minimizing cytotoxic effects.

2.2. Hydrogel matrices for cartilage TE

To be applicable for articular cartilage TE, injectable HGs need to fulfill some specific criteria. They need to be soluble in aqueous media with gelation occurring under physiological conditions (temperature, osmolality, and pH), to gel at a sufficient rapid rate to be clinically effective and at relatively mild conditions in presence of cells, proteins or other bioactive molecules, as well as to avoid the release of harmful by-products during gelation.[56]

Several types of hydrophilic polymers, either natural or synthetic in origin, have been studied for the development of HGs for cartilage TE.

2.2.1. *Naturally-derived HGs*

Natural-derived polymers used for the production of HGs for cartilage TE include proteins such as collagen, gelatin, fibrin and silk fibroin, and polysaccharides such as chitosan, agarose, alginate, HA, heparin, CS, dextran, and some types of cellulose.

Generally, natural polymers are superior in terms of bioactivity due to the presence of functional groups that can help cell adhesion, migration and differentiation through the hydrogel scaffold (e.g., RGD sequences). Their degradation products are non-cytotoxic, although the degradation kinetics is not as controllable as for synthetic polymers. Moreover, polymers derived from natural materials are economically favorable and environmentally friendly, with low manufacturing costs, high degradability, and renewability. Drawbacks of using natural polymers as hydrogel scaffolds include the potential of eliciting immune response or disease transmission, as well as rapid degradation rate and lack of adequate mechanical properties (strength and toughness) for cartilage TE.

Although tensile properties and compressive modulus of articular cartilage depend on the depth and location of the tested zone, generally aggregate equilibrium compressive modulus of human articular cartilage range from 0.1 to 2 MPa.[71] The mechanical properties of natural-derived HGs are quite far from the characteristic values of native cartilage tissues. To overcome this limit and to improve mechanical properties, natural polymers are often combined with other natural or synthetic polymers, or reinforced or combined with fibers or sponges that increase the strength of the whole scaffold.

2.2.1.1. Collagen and gelatin

As one of the main components of native chondral ECM and thanks to its high propensity to cell-binding, collagen has been widely used for the fabrication of scaffolds for cartilage TE.[72] Collagen type I and type II matrices have been proven to facilitate cartilage tissue growth both *in vitro* and *in vivo*.[73–76] Type I collagen gels from rat tail tendon and fetal calf skin have been tested, showing different degree of chondrocyte differentiation depending on cell–cell and cell–matrix interactions.[77] Properly designed collagen type I based HGs were also proven to have immunogenicity-reducing effects on engineered cartilage based on neonatal rabbit chondrocytes.[78] According to Buma and colleagues, type I collagen-based matrices are superior to guide progenitor cells from a subchondral origin into the defect, while type II collagen-based matrices present less cell migration, but direct invading cells into a chondrocyte phenotype.[76] Based on these observations the authors suggested that a composite matrix consisting of a deep layer of type I collagen and a more superficial layer of type II collagen may be the matrix of choice for cartilage regeneration.

Despite its good biocompatibility, collagen presents low biomechanical stiffness and rapid biodegradation. Collagen HGs can be obtained through UV irradiation or chemical crosslinking by means of traditional (e.g., aldehydes, isocyanates, carbodiiimides, etc.) or natural crosslinkers (genipin, transglutaminase, lysyl oxidase). Photochemical crosslinking of collagen gels by riboflavin molecules have shown to favor chondrocytes viability and improve the stability of the gel reducing its contraction following cell addition.[79]

The use of conventional chemical crosslinking agents in TE is usually discouraged due to their high cytotoxicity level. Genipin is an effective natural crosslinking agent, extracted from gardenia fruit, with a very low level of cytotoxicity compared to synthetic crosslinking agent.[80] It offers a simple alternative to improve the mechanical properties of hydrogel scaffolds obtained from natural polymers. Sundararaghavan *et al.* have shown that genipin produces significant increases in the storage modulus and loss modulus of type I collagen gels depending on concentration and incubation time of the crosslinker.[66] Moreover, collagen HGs crosslinked by genipin showed enhanced mechanical strength if compared to collagen-alginate gels.[81]

Temperature-sensitive biomimetic hydrogel scaffolds have been produced by collagen self-assembly and crosslinked with genipin. An increase of genipin concentration from 0 mM to 1 mM, induced a rise of crosslinking index and glycosaminoglycans production, resulting in increasing of modulus and decreasing of degradation ratio.[82] Among the tested conditions, collagen hydrogel scaffolds crosslinked with 0.75 mM genipin exhibited balanced optimal properties, and excellent cellular compatibility.[68]

To exploit favorable cell–collagen interactions and to provide HGs with better mechanical properties, collagen has often been crosslinked or reinforced by combination with other polymers. Moreover, collagen has been combined to HA and/or CS, obtaining matrices with promising application in cartilage TE.[67,68,75,83,84]

Gelatin, a partial derivative of collagen, is less immunogenic respect to its precursor, though retaining high density of RGD and other specific sequences which can be easily recognized by cell receptors facilitating adhesion and differentiation. Due to its scarce mechanical properties, gelatin needs to be combined with mechanically competent polymers.[85–87] Gelatin HGs coupled to methacrylic acid obtained by photo-initiated polymerization showed the propensity to support chondrocyte growth and to maintain chondrocytic phenotype *in vitro*.[86] By increasing the polymer concentration, storage modulus and loss modulus of these HGs increased, while swelling ratio and mesh size decreased consequently. Gelatin-methacrylamide HGs have been shown to support chondrocyte viability and differentiation[85] and allow modulation of hydrogel stiffness and swelling properties controlling different parameters such as on polymer concentration, UV exposure time, and thermal gelation prior to UV exposure. Gels with compressive moduli in the range 5–180 kPa have been fabricated by varying the concentration of the precursor solution from 5 to 20%.[85] It has been shown that the incorporation of small quantities of photocrosslinkable hyaluronic acid methacrylate (HA-MA), and to a lesser extent chondroitin sulfate methacrylate (CS-MA), can enhance mechanical properties of gelatin–methacrylamide HGs and promote chondrogenesis.[87]

2.2.1.2. Fibrin

Thanks to intrinsic adhesive properties, biocompatibility, and injectability, fibrin glue has been extensively used for collagen patch sealing in ACI technology. Fibrin gels can be prepared using autologous blood and polymerized under controllable conditions. Moreover, they can be rapidly dissolved in the physiological environment due to fibrinolysis. The suitability of a number of fibrin gels with different formulations for cartilage engineering has been investigated both *in vitro* and *in vivo*. The fibrin gel has been shown to be a suitable polymer gel for generating cartilage matrix from articular chondrocytes being also capable of healing and integration.[88]

In vitro analysis on a commercial product based on fibrin glue (Tissucol®) showed that chondrocytes are active in terms of cell division and matrix production when cultured in fibrin glue,[89] demonstrating that fibrin glue is a suitable polymer for the formation of injectable tissue engineered cartilage in the nude mouse model. Cartilage resurfacing attempt performed by Vinatier and colleagues demonstrated that autologous nasal chondrocytes are able to support cartilage tissue formation in subcutaneous sites, and highlighted the potential of fibrin sealants for TE and cell-based therapy of articular cartilage.[30] In the study of Eyrich and colleagues, concentration of fibrinogen and $Ca2^+$ as well as pH were modified to obtain long-term stable fibrin gels, resulting in the development of adequate cartilaginous tissue if cultured with primary bovine chondrocytes.[90] Scotti *et al.* have studied the degree of *in vitro* maturation of engineered cartilage on fibrin glue hydrogel,[91] while stable human fibrin gel systems crosslinked with genipin tested by Dare *et al.* provided a suitable environment for articular chondrogenic gene expression and elaboration of a cartilage-like ECM.[65] In a recent study, Lind *et al.* investigated the cartilage repair potential of a fibrin based hydrogel and a methoxypolyethyleneglycol (MPEG)-PLGA scaffold/chondrocyte/fibrin self-adhesive composite, both loaded with autologous chondrocytes, in a full thickness cartilage defect model in goats.[92] The study demonstrated promising early cartilage repair in femoral condyle defect goat model through the MPEG-PLGA/chondrocyte construct used in this study, while the hydrogel composed of fibrin and chondrocytes did not generate a repair tissue better than subchondral drilling.[92]

Human fibrin HGs offer a potential patient-specific autologous scaffold for cartilage regeneration, eliminating of the risk of tissue incompatibility and viral infection. Further *in vivo* studies are needed to evaluate the maturation of fibrin-based hydrogel systems and their development and integration in the physiologic environment.

2.2.1.3. Silk fibroin

Silk fibroin is a protein produced by several insects and spiders. Silkworm Fibroin that is usually isolated from the silk filaments produced by Bombyx mori silkworms can be processed to obtain films, nonwoven nets, sponges, and HGs able to sustain cell adhesion, proliferation, and ECM production both *in vivo* and *in vitro*[38] while also possessing a low inflammatory potential.[93] Silk fibroin has multifunctional features driven by the self-assembly of molecular subunits in appropriate environmental conditions.[38] Fibroin-based HGs are particularly interesting for cartilage TE, due to several attractive properties, including biocompatibility, slow biodegradation rate, good mechanical strength and thermal stability as well as versatility in processing and chemical modifications, and easy deliverability through injection.

Hydrogels can be obtained from regenerated silk fibroin by sol-gel transition in the presence of acids, ions or other additives,[94] as well as by chemical crosslinking, ultrasonication or supercritical CO_2.

Fibroin-hydrogel sponges were cultured *in vitro* with white rabbit chondrocytes, showing good cell proliferation without loss of the differentiated phenotype. Compared to collagen gels, the fibroin-hydrogel scaffolds showed an intense and more rapid production of sulfated GAGs as well as a faster increase in cell density.[95] Silk scaffolds allowed the formation of native-like articular cartilage including

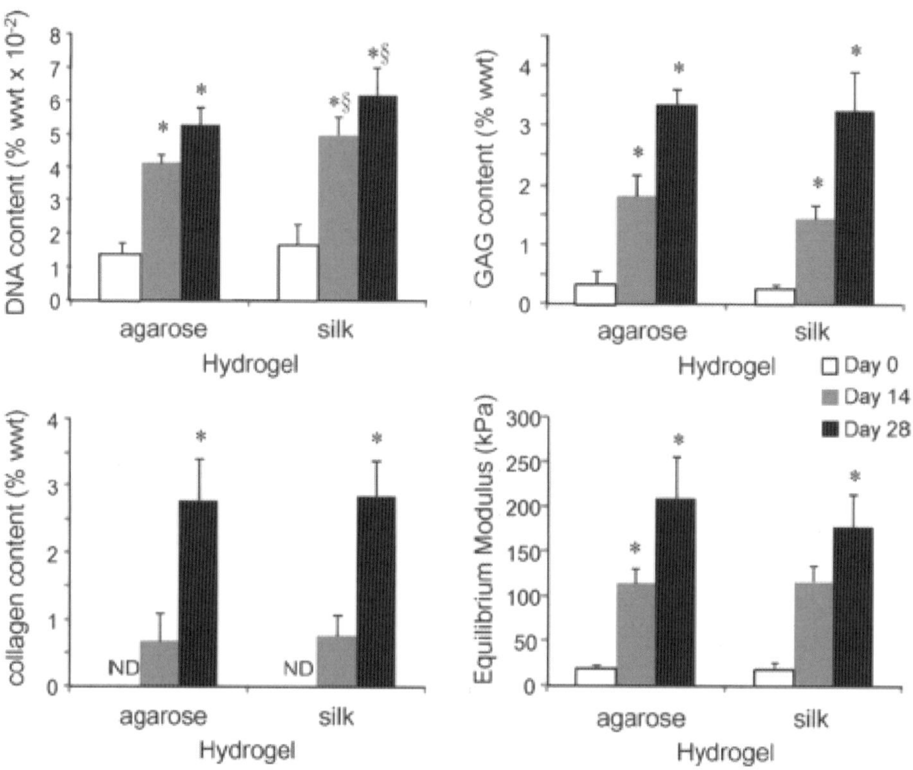

Fig. 1. DNA content, GAG content, collagen content and equilibrium modulus of cell-cultured agarose (2%) and silk (4%) HGs at different time points. (ND, not detectable; $*p < 0.05$ compared with the previous time point, $^{\S}p < 0.05$ compared with the agarose group at the same time point). Adapted with permission from Ref. 96.

spatial cell and collagen type II distribution after three weeks cultivation with MSCs.[94] Silk HGs obtained by sonication and loaded with calf chondrocytes yielded cartilaginous constructs with biochemical content and mechanical properties matching constructs based on agarose, a polymeric material considered as a gold standard for cartilage TE due to its capability to produce constructs with functional properties approaching those of native articular cartilage (Fig. 1).[96]

2.2.1.4. Alginate and agarose

Alginate and agarose are basic polysaccharides, especially interesting for cartilage TE, cell encapsulation and drug delivery, thanks to very mild gelation conditions and ability to maintain long-term chondrocyte cultures.

Alginate is a hydrophilic linear polysaccharide composed of β-D-mannuronic acid (M) and α-L-guluronic acid (G) monomers that gels in the presence of divalent ions (e.g., Ca^{2+}, Mg^{2+}, Ba^{2+}, Sr^{2+}).

Agarose is a family of polysaccharides composed of alternating sequences of 1,4-linked 3,6-anhydro-α-L-galactose and 1,3-linked β-D-galactose. Agarose gels form spontaneously by hydrogen bonding at a temperature between 17 and 40°C, depending on the degree of hydroxyethyl substitutions on the side chains.[53]

Both alginate and agarose are mostly derived from seaweeds, and are therefore readily available.

The transplantation of allogenic chondrocytes embedded in agarose allowed the formation of repair cartilage with superior properties in terms of type II collagen, proteoglycan content and cellular architecture when compared to untreated defects in rabbit knees.[97] Agarose hydrogel constructs seeded with bovine chondrocytes or MSCs showed an increase in mechanical properties with time in culture, although the increase after 56 days culture was not related to the initial seeding density or the cell type employed.[98] Nevertheless, tensile properties of the constructs were still far from the properties of native cartilage. As reported by Selmi *et al.*, ACI in combination with an agarose-alginate hydrogel scaffold (CARTIPATCH; Tissue Bank of France, Lyon, France) provided significant clinical improvement at two years follow-up showing predominance of hyaline cartilage-like repair tissue.[99]

Although alginate is an attractive material for TE due to its biocompatibility and ability to form HGs, it presents slow and uncontrollable degradation. Attempts to control alginate HGs degradability rate have been obtained using multi-functional crosslinking molecules[100] or partial oxidation and bimodal alginate molecular weight distribution.[101,102] These approaches were successful in controlling hydrogel degradation as well as in increasing hydrogel mechanical stiffness[100] and improving cartilage-like tissue formation *in vivo* as compared to bare alginate HGs.[101]

Due to the lack of specific cell-binding ligands, agarose has been modified to introduce bioactive signals to the HGs. Cell-adhesive peptides composed of specific amino acid sequences (e.g., RGD, GRGDS sequences or laminin peptides) have been conjugated to alginate, chitosan, and agarose[103–107] to try to control receptor type specific functions. This approach produced alternated results. In fact, RGD-containing peptides conjugated to sodium alginate inhibited the initial chondrogenesis of BMSCs within 3D alginate gels,[104] while the entrapment of RGD-chimeric protein in 3D alginate beads promoted the chondrogenesis of adipose-derived adult stem cells.[103]

A quite interesting approach has been proposed by Kenneth and colleagues, which evaluated the response of zonal chondrocytes encapsulated in a bilayered agarose hydrogel in an attempt to mimic the depth-dependent cellular and mechanical inhomogeneity of native articular cartilage.[108] This construct, presenting initial defined mechanical inhomogeneity, formed an engineered system showing depth-dependent cellular and compressive mechanical inhomogeneity similar to that of the native cartilage. Nevertheless, agarose presents some disadvantages, since it is not biodegradable and can be immunogenic.

2.2.1.5. Chitosan

Among polysaccharide biopolymers, chitosan has been widely used for preparing HGs for cartilage TE. Chitosan is a polycationic linear polysaccharide deriving from partial deacetylation of chitin, a substance abundantly present in the shell of crustaceans. It is a semi-crystalline polymer with degree of crystallinity dependent on the degree of deacetylation. Since it contains glucosamine and N-acetylglucosamine, it structurally resembles GAG molecules, which are abundant in natural hyaline cartilage. Among its favorable characteristics for cartilage TE we can include hydrophilicity and biocompatibility related to cell adhesion, proliferation, and differentiation.

Chitosan molecules can form reversible HGs through ionic binding and can be used as injectable carriers for TE applications when covalently crosslinked *in situ*.[109] Moreover, chitosan HGs can be obtained by chemical and thermal crosslinkings, Schiff-base reaction, free radical polymerization, and UV irradiation.[63] A chitosan-based injectable system gelling at physiological temperature with gelling time varying from 5 min to several hours has been obtained by physical crosslinking using ammonium hydrogen phosphatase as neutralizing and thermo-gelling agent.[110]

The support of cartilage-like matrix accumulation by chitosan HGs was reported both by *in vitro* and *in vivo* studies.[111,112] Chitosan hydrogel fragments tested with human chondrocytes were reported to induce reverse encapsulation, that is to say the formation of a bio-construct with dedifferentiated chondrocytes surrounded by matrix proteins typical of the natural cartilage (collagen type II and aggrecan).[54] In this case, the best results were obtained with chitosan hydrogel fragments containing 1.5% of chitosan with a deacetylation degree ranging from 30 to 45%.[54] Degree of acetylation and crystallinity can influence the kinetics of degradation of chitosan HGs through physiological depolymerization by cell-secreted lysozyme.[113,114]

The introduction of two layers of electrospun silk fibers significantly enhanced the mechanical properties of thermosensitive chitosan/glycerophosphate HGs, also showing a significant increase of Col II content in the constructs cultured with New Zealand white rabbit chondrocytes.[115]

Injectable chitosan-based HGs produced through enzymatic crosslinking with horseradish peroxidase and H_2O_2 presented enzymatic degradation rate and mechanical properties dependent on initial polymer concentration. Bovine chondrocytes cultured on these gels were metabolically active, with round morphology and uniformly distributed within the gel up to 14 day.[116] Chitosan molecules can form complexes with anionic polysaccharides such as GAGs and alginates, as well as synthetic polyanions such as poly(acrylic acid).[117] GAG-augmented chitosan HGs have shown good biocompatibility and the chondrogenic characteristics.[118] Composite HGs obtained by Schiff base reaction by mixing water-soluble chitosan and oxidized HA promoted promoted survival of bovine articular chondrocytes, retaining cells regular chondrocytic spherical morphology.[63] Gelation time, structure, equilibrium swelling, compressive modulus and degradation *in vitro* were dependent on crosslinking and relative composition of N-succinyl-chitosan (S-CS) and aldehyde hyaluronic acid (A-HA) components in the hydrogel, with a slower degradation rate for the composite matrices with a higher S-CS composition.[63] Another type of injectable chitosan composite hydrogel containing HA was recently evaluated, showing that the incorporation of HA in the crosslinked chitosan network significantly increased cell proliferation and cartilaginous ECM production by encapsulated chondrocytes.[119]

2.2.1.6. Hyaluronic acid

HA is a non-sulfated glycosaminoglycan, composed of repeating disaccharide units (250 to 10,000 or more) of N-acetyl-D-glucosamine and D-glucuronic acid. Hyaluronan is naturally present in the articular joints, contributes to the viscoelastic and lubrication properties of the synovial fluid, and concurs to absorb shocks while maintaining the structural and functional characteristics of the cartilage matrix.[120] Intra-articular injections of HA are common treatments in the case of osteoarthritis. Several studies have demonstrated that HA has anti-inflammatory properties, as well as a beneficial effect by inhibiting chondrocytic chondrolysis mediated by fibronectin fragment.

Hyaluronic acid has been shown to improve *in vitro* substrate adhesion ability and proliferative activity of human cartilage cells[121] and to have stimulatory effect on chondrocyte metabolism,[122] even though this effect seems to be concentration dependent. Moreover, it has been reported that the addition of HA at low to middle dosages in culture might increase type II collagen and aggrecan production, and improve chondrocyte redifferentation with significant enhancement for the bioengineered cartilage.[123]

Due to its excellent biocompatibility, biodegradability and gel-forming properties, HA and its derivatives have been widely explored as HGs for cartilage TE.[63]

The great potential of HA for cartilage regeneration was demonstrated by Responte *et al.*, which showed that HA treatment increased compressive stiffness and sulphated GAG content during *in vitro* cartilage formation. Improvement of biochemical and biomechanical properties was dependent on HA molecular weight, concentration, and application commencement time.[124]

A HA hydrogel with a 50% crosslinking degree was analysed with respect to its rheological behavior and *in vivo* effect in the treatment of a surgically created chondral defect in rabbit knee.[125] The results of the study showed that the HA HG injected through a needle still behaved like a gel, although it showed a reduction of the dynamic moduli. Moreover, HA hydrogel injection improved chondrocytes density and matrix appearance.

A variable and versatile group of biodegradable materials can be obtained from polymeric derivatives of hyaluronan. These materials can be produced with varying chemical structures and morphology, allowing for the modification of the hydrogel physical–chemical properties such as the rate of degradation, the degree of hydration, the cellular responses and the overall tissue responses.[126] The primary hydroxyl group of the HA backbone can be easily modified to include functional groups to facilitate covalent crosslinking, degradation and enhanced cell adhesion. HA is often bound to methacrylate (MeHA) or acrylate (AHA) functional groups to allow photo-initiated crosslinking. Methacrylated lactic acid (MeLAHA), methacrylated caprolactone (MeCLHA), and hydroxyethylmethacrylate (HeMAHA) functional groups have also been conjugated to HA molecules to form HGs with tunable degradation characteristics.[127–129]

Erickson and colleagues investigated the ability of bone marrow derived MSCs to undergo chondrogenesis in crosslinked methacrylated hyaluronic acid (MeHA) HGs with different macromer concentrations (1, 2, and 5%), showing that variations in macromer density influence mechanical properties and MSC chondrogenesis.[130] According to the authors, increasing HA macromer density promoted chondrogenesis and matrix formation and retention, but yielded functionally inferior constructs due to limited matrix distribution throughout the construct.[130]

The influence of degradation characteristics of HA HGs on *in vitro* neocartilage formation by MSCs was analyzed by Chung *et al.*[129] The authors assessed that MeHA and MeCLHA can be combined to tune temporal scaffold properties, thus controlling neocartilage production by MSCs, with optimal performances in the case of the 1:1 HGs, that showed increased mechanical properties and biochemical content over time, while retaining construct size.

Recent studies showed that MSCs and chondrocytes co-culture in a HA hydrogel significantly increased the Young's modulus, dynamic compressive modulus, GAG and collagen production, while significantly decreasing hypertrophic markers, often upregulated in the case of MSCs cultures.[131]

Commercially available products based on hyaluronan derivative polymers (HYAFF®, Fidia Farmaceutici s.p.a., Abano Terme, Italy) have been tested showing different physico–chemical properties (degree of hydration and polymer stability) depending on the type and extent of esterification of hyaluronan and considerably affecting the biological properties of these materials.[126]

Recent extents have been performed to exploit HA characteristics and biological activity also trying to improve the properties of the final hydrogel. For this scope, hybrid HGs have been studied, combining HA with several natural or synthetic polymers, including collagen, CS, chitosan, alginate, sericin, PEG, PVA, PCL, PPF, PNIPAAm etc.[68,119,132–140]

2.2.1.7. Other HGs from polysaccharides

CS is a sulfated glycosaminoglycan, important component of connective tissues and cartilage in particular. Intra-articular injection of CS (100 mg/mL) carried by α-CD-EG 4400 hydrogel was effective in postponing osteoarthritis in rabbits by improving both the biomechanical and histological properties of the knee joints.[141] As an example, collagen-CS-HA temperature-sensitive HGs with mild gelation conditions were proven to be suitable as carriers for chondrocyte transplantation or as cartilage TE scaffolds. Optimal cell compatibility *in vitro* was obtained with HGs crosslinked with genipin 0.75 mM.[68]

Recently, novel HGs prepared from natural-derived polysaccharides, including cellulose with or without functionalization,[142–144] gellan gum,[145,146] and carrageenans[147] have been developed showing highly promising results for cartilage TE.

2.2.2. *Hydrogels from synthetic polymers*

Synthetic polymers used as candidates to produce HGs for cartilage TE include PLLA, PGA and their co-polymers, PEG, poly ethylene oxide PEO, PVA, PCL, and oligo(poly(ethylene glycol) fumarate) OPF, Pluronic®, PNIPAAm, as well as co-polymers, blends, or interpenetrating networks obtained from the same polymers.

Some of the synthetic polymers involved in cartilage repair are non-degradable (e.g., PVA and PNIPAAm), while some other is degradable (e.g., PGA, PLA, etc.).

Synthetic polymers hold significant potential for cartilage TE due to ease of fabrication and chemical modification, as well as good biocompatibility. Thanks to their high versatility, they can be tailored to obtain HGs with suitable mechanical properties and controllable biodegradability.

2.2.2.1. PVA

PVA is considered to be one of the first synthetic polymers tested for cartilage regeneration in the early '70s.[148,149] It is a hydrophilic non-degradable polymer, highly soluble in water, derived from polyvinyl acetate through partial or full hydrolysis. PVA shows variable physical characteristics, chemical and mechanical properties according to the amount of hydroxylation. PVA HGs with tunable properties, swelling and structural stability can obtained by physical or chemical crosslinking. The degree of crosslinking dictates the amount of fluid uptake, and thus the physical, chemical, and diffusional properties of the polymer, and ultimately its biological properties.[150] PVA HGs can be sterilized and molded into desired shapes, but their mechanical properties are scarce if compared to natural cartilage. To overcome this problem, different processes have been developed to produce PVA HGs with enhanced mechanical characteristics.[149]

An interesting review of polyvinyl alcohol and its uses in cartilage and orthopedic applications has been recently proposed by Baker *et al.*[150]

Due to their high water content, elastic and compressive mechanical properties PVA HGs have been investigated as scaffolds for the regeneration of damaged cartilage. PVA cryogels used in cartilage resurfacing prepared from high concentrations of high-molecular weight polymers (generally 30% PVA or higher) have water contents similar to the surrounding healthy cartilage. The HGs can be manufactured to have tensile strength in the range of 1–17 MPa, almost approaching the values of natural cartilage, and compressive modulus varying from 0.0012 and 0.85 MPa depending on the polymer concentration and number of cycles tested.[151]

Composition and thickness of the PVA layer are also critical parameters that need to be optimized to improve the wear resistance of the HGs.

PVA hydrogel was also implanted in white rabbits for up to 52 weeks as an artificial articular cartilage replacement resulting in low inflammatory responses and high *in vivo* biocompatibility.[152] PVA gels with 80–90% water content by weight were implanted subcutaneously or intramuscularly into rabbits, and no adverse effects were noticed in the surrounding tissue leading to a confirmation of the biocompatibility of the material.[153,154]

To overcome PVA scarce propensity to integrate with surrounding cartilage tissues, highly porous constructs have been obtained incorporating PLGA microparticles into PVA HGs by double-emulsion or salt-leaching techniques.[113,155] The hybrid HG scaffolds produced by double emulsion presented compressive moduli similar to that of healthy cartilage. They showed successful outcomes in facilitating the integration of neocartilage with surrounding tissue, improving proteoglycan production, cell adhesiveness and cartilage tissue growth within the pores of the HGs that initially contained PLGA respect to bare PVA HGs.[113]

Other tentative to improve PVA cell-adhesion as well as mechanical resistance and durability have been performed by developing a composite hydrogel with hydroxyapatite as bioactive and biocompatible filler.[156,157] The addition of hydroxyapatite modified the physical and chemical features of the hydrogel promoting material crosslinking and stability and improving the mechanical properties of the composite HG.[156]

2.2.2.2. Poly(lactic acid) (PLA) and PGA

PLA in both L and D forms, PGA, and their copolymer poly (lactic-co-glycolic acid) (PLGA) are among the most popular synthetic polymers used for cartilage TE scaffolds. It has been shown that PGA improved proteoglycan synthesis when compared to collagen scaffolds.[158] Studies demonstrated that nonwoven PLGA scaffolds are suitable for the chondrogenesis of human adipose-derived stem cells. PLLA has a slower degradation rate than PGA, being thus more suitable for applications requiring longer duration of scaffold structural support. PLGA has high biocompatibility, ability to degrade into harmless monomer units, a useful range of mechanical properties, and controllable degradation time depending on the copolymer ratio.[159]

Nevertheless, these polymers are predominantly used for the fabrication of porous scaffolds (sponges or nets) rather than for HGs.[160,161]

2.2.2.3. Poly(ethylene glycol) (PEG) and poly(ethylene oxide) (PEO)

PEG and PEO are hydrophilic polymers presenting several interesting properties for biomedical applications, such as good biocompatibility, low immunogenicity, ability to bind large amounts of water and resistance to protein adsorption. PEG and PEO are approved by FDA for several medical applications and are currently widely used to prepare HGs for TE. They can be photocrosslinked by modifying each end of the polymer with either acrylates or methacrylates. Hydrogels are then formed when the modified PEO or PEG is mixed with the appropriate photoinitiator and crosslinked via UV exposure.[52]

PEG HGs hold several advantages for cartilage TE. In fact, they can gel under mild conditions allowing *in situ* gelation and cell-encapsulation, they are versatile, easy functionalizable or modifiable with bioactive groups.

Photopolymerizable PEG-based HGs have been studied to successfully design synthetic hydrogel for cartilage regeneration.[162,163]

Nguyen *et al.* characterized the mechanical properties of PEG HGs over a broad range, with reference to articular cartilage, assessing the effects of PEG precursor molecular weight (508 Da, 3.4 kDa, 6 kDa, and 10 kDa) and concentration (10–40%).[164] According to the authors, as PEG MW decreased and concentration increased, HGs exhibited a decrease in swelling ratio, an increase in equilibrium confined compressive modulus H(A0) and equilibrium tensile modulus E(ten), an increase in dynamic compressive stiffness, and a decrease in hydraulic permeability. H(A0) and E(ten) were positively correlated with the final PEG concentration, accounting for swelling. From these results we can argue that PEG HGs can be prepared to mimic many of the static and dynamic mechanical properties of articular cartilage.

Analyzing chondrocyte-laden PEG HGs, Nicodemus and colleagues highlighted that gel structure has a significant impact on pericellular and extracellular matrix deposition and on how chondrocytes respond to the environment, particularly with respect to catabolic expression.[165] Studies conducted on degradable and non-degradable PEG-based HGs cultured with photo-encapsulated calf chondrocytes showed that changes in the HG crosslinking density can affect water content and diffusion of molecules and newly synthetized ECM.[166] By proper design of partially degrading HGs, the initial gel mechanics can be maximized without compromising the final distribution of proteoglycans.[167] The same authors investigated the potential for using photopolymerized PEO-based HGs in regenerating cartilaginous tissue for partial and full thickness defects, showing that chondrocytes encapsulated in a 8 mm thick photocrosslinked HG produced cartilaginous tissue throughout the construct after 6 weeks *in vitro* culture.[168]

PEG-derived copolymer HGs has been developed with the aim to tune the final properties of the hydrogel constructs for specific TE applications. Biodegradable PEG HGs can be obtained via copolymerization with degradable synthetic polymers such as PLA, PGA and PPF or in combination with many naturally occurring biopolymers, such as HA, fibrinogen, chitosan, and heparin.[63] Various strategies to produce PEG copolymers with various block compositions and specific bulk properties, including low melting points and improved surface hydrophilicity or enhanced cell adhesion, have been described by Tessmar and Göpferich.[60]

PEG HGs and amphiphilic interpenetrating polymer networks (IPNs) of PEG with poly(methyl methacrylate) (PMMA) were prepared and characterized for their mechanical and swelling properties, showing that lower PEG MW, higher crosslink densities and higher PMMA fraction, all lead to higher modulus and lower water content.[169] Introduction of IPN greatly improved the hydrogels' strength, while some of the HGs tested were found to fit cartilage modulus, water content, and poroelastic properties.

2.2.2.4. Poly(propylene fumarate) (PPF) and OPF

A relatively new class of synthetic polymers for cartilage TE is represented by PPF. PPF is a hydrophobic, unsaturated linear polyester, which undergoes degradation by hydrolysis of the ester linkage. It has been previously utilized as biomaterial for orthopedic applications such as bone cements and recently it has been employed for cell encapsulation. It can form HGs when synthesized as a block copolymer with hydrophilic PEG and crosslinked either chemically[170] or via UV exposure.[171]

Crosslinking performed with a vinyl monomer produces a mechanically strong polymer networks. The study of He and colleagues demonstrated the ability to develop biodegradable, injectable HGs having adjustable parameters by varying the ratio of PEG-dimethacrylate to PPF.[171]

OPF has been evaluated for cellular activitiy *in vitro* and for soft and hard tissue response *in vivo*[172,173] as well as for use as thermoreversible hydrogel for articular cartilage engineering.[174]

OPF macromolecular structure, polymer concentration and crosslinking can be tailored to control physical–chemical properties of HGs, including degradation,[172,175] providing useful features for cartilage TE and cell delivery applications. Photoencapsulated chondrocytes within 2-mm thick OPF hydrogel disks remained viable throughout the 3-week culture period, expressed mRNA of type II collagen and produced cartilaginous matrix within the hydrogel constructs.[175]

OPF-based hydrogel scaffolds were proven to undergo biocompatible degradation and support healthy tissue growth in rabbit osteochondral defects.[176] The quality of the osteochondral tissue filled with OPF HG was improved over time with mostly hyaline cartilage filling the chondral region at 14 weeks. Moreover, TGF-b1 loading in the top layer of scaffolds appeared to exert some effect on cartilage quality in the defect area. Results with this type of hydrogel are encouraging, although further studies are necessary to optimize these scaffolds for osteochondral regeneration.

2.2.3. *Other matrices and novel approaches for cartilage HGs*

New strategies based on the development of double network (DN) HGs have been proposed, yielding high-strength constructs with high water content (90 wt% ca.) and significantly improved mechanical strength.[177] Typical DN HGs are prepared according to a two-steps reaction: (1) a crosslinked negatively charged hydrogel is prepared as first network, which serves as a template; (2) a large amount of neutral monomers swells into the first network, and loosely crosslinked or not crosslinked neutral polymer chains are prepared in the template as second network. The network structure of the DN hydrogel is similar to articular cartilage. The function of the first network of the DN HG, i.e., rigid and high crosslinked network, is similar to the collagen fibers of articular cartilage, and the function of the second network of the DN hydrogel, i.e., loosely crosslinked or not crosslinked flexible polymer chains, is similar to the PG aggregates of articular cartilage. The mechanical strength of the DN HGs can be modulated achieving in some cases mechanic properties matching the properties of native soft tissues (1–10 MPa tensile strength and 20–60 MPa compressive strength), and low friction coefficient.[177]

One of the most common polymer combinations of DN hydrogel consists of negatively charged poly2-acryl amido-2-methylpropane sulfonic acid (PAMPS) as the first network and neutral polyacrylamide (PAAm) as the second network.

PAMPS/PDMAAm DN hydrogels have been tested *in vivo*, showing the ability to help induction cartilage regeneration.[177] Another *in vivo* study using the mature rabbit femoral trochlea osteochondral defect model investigated the induction ability for cartilage regeneration of PAMPS and PDMAAm single network gels in comparison to PAMPS/PDMAAm DN gel, showing that the chondrogenic potential of a negatively charged PAMPS gel component plays an important role in the cartilage regeneration ability of the PAMPS/PDMAAm DN gels *in vivo*.[178]

DN HGs have also been created by photocrosslinking using two modified biomacromolecules, such as gellan gum methacrylate (GGMA) for the first network and gelatin methacrylamide (GelMA) for the second network.[179] These DN HGs exhibited a compressive failure stress approaching the values of native cartilage, also showing good cell viability of NIH-3T3 fibroblasts following cell encapsulation.

Further studies on the biocompatibility of this promising class of HGs still need to be carried out to evaluate their potential of inducing hyaline cartilage formation.

Besides several interesting properties such as versatility, flexibility in design, fabrication and chemical modification, HGs based on synthetic polymers also hold some disadvantages for cartilage TE, such as the lack of optimal biocompatibility, or the possible release of toxic by-products.

To exploit the respective advantages of natural and synthetic polymers hybrid HGs have been developed and tested as scaffolds for cartilage TE.

2.2.4. *Hybrid HGs*

Chitosan has been combined to various natural and synthetic polymers including HA,[119,134,138] glycerophosphate and hydroxyethyl cellulase[112] Pluronic®,[58] and polyol salt[180] to produce thermosensitive HGs with favorable characteristics and biocompatibility for cartilage TE. As an example, in fact, chondrocytes have been efficiently encapsulated inside the chitosan-HA gel and cultured for 1 month retaining characteristic spherical morphology with sufficient production of ECM and expression of type II collagen and aggrecan.[134] The incorporation of HA in methacrylated glycol chitosan/HA HGs was shown to increase the proliferation and deposition of cartilaginous ECM by encapsulated chondrocytes.[119] A preliminary analysis of fibrin-polyurethane composites for articular cartilage TE has been performed in 2005 by Lee *et al.*, showing that fibrin hydrogel improved the potential of a polyurethane scaffold-based system allowing for a significant increase in the percentage of newly synthesized GAG in the constructs, and in type II collagen and aggrecan gene expression after 14 days.[181] Bioactive PVA/HA HGs were prepared by *in situ* sol-gel synthesis and tested *in vivo*. These hybrid HGs showed good biomechanical properties, combined tightly with ambient tissues, and induced natural and firm binding with the subchondral bone.[182]

Other hybrid HGs tested for cartilage TE applications were obtained for example combining PEG with agarose,[183] type I collagen and HA,[184] or with a degradable macromer, poly(lactic acid)-b-poly(ethylene glycol)-b-poly(lactic acid) endcapped with acrylate groups (PEG-LA-DA).[166]

More complex constructs have been recently developed to simulate articular cartilage physiological structure and to mimic its original zonal organization, with the aim to enhance the final properties of the whole scaffold and to resemble the native structure of hyaline cartilage.[185] Different approaches have been used for this aim. Some authors used different cell types (e.g., zonal chondrocytes or chondrocytes and precursor cells) applied to different specific regions of the construct.[186] For example, Kim *et al.* designed multi-layered constructs by encapsulating chondrocytes from different zones of articular cartilage in 2D and 3D cultures.[187]

Another approach consists of using HGs with different mechanical properties in separate layers of the scaffold. Modular hydrogel systems have been designed to systematically control hydrogel properties, and to control construct organization through advanced fabrication techniques. For example, zonal chondrocytes were seeded in a layered agarose hydrogel with initially arranged depth-varying compressive properties, leading to a biomimetic cartilage tissue construct with depth-dependent cellular and compressive mechanical inhomogeneity similar to that of the native tissue.[108]

Combined approaches, for example varying both the cell source and mechanical properties, have also recently been investigated. Nevertheless, development of zonal construct properties is dependent on both the hydrogel material and mechanical properties, and is not as simple as prescribing the hydrogel stiffness in different layers.[185]

Other strategies have been developed to combine the properties of soft, hydrophilic HGs with polymer structures presenting higher mechanical properties (e.g., sponges or electrospun nets) into multicomponent scaffolds.

2.2.5. *Comparative studies*

A comparative study to investigate the differences in the mechanical properties of a natural-derived (agarose) and a synthetic polymer (PEG) HGs has been recently performed by Roberts and colleagues.[188] The two polymers exhibited very different mechanical behaviors. Physically crosslinked agarose HGs displayed typical viscoelastic properties, while covalently crosslinked PEG HGs had largely elastic behavior with minimal stress relaxation and frequency dependence.

Few studies exist that directly compare cartilage neoformation in different hydrogel scaffolds. An interesting study by Mouw *et al.* tested contemporaneously different materials, including agarose, alginate, collagen I, fibrin and polyglycolic acid (PGA). The constructs were cultured *in vitro* with bovine articular chondrocytes for 20 or 40 days.[189] Significant differences were found in cell proliferation, quantity of ECM accumulation and fine structure of newly synthetized GAGs among the five scaffold groups. According to the authors, cell proliferation was slower in PGA with respect to the other constructs. Over the first 20 days PGA constructs had the lowest increase in cell content but the greatest accumulation of sGAG per cell and the greatest rate of sGAG release to the media. Chondrocytes in agarose synthesized a high quantity of sGAG, showing a higher and more extensive presence of pericellular matrix with respect to other gels. Moreover, agarose presented an internal D-disaccharide profile most closely resembling that of articular cartilage, as opposite to alginate constructs. Collagen constructs contained uniformly distributed cells and diffuse proteoglycan matrix, while fibrin displayed the presence of sGAG surrounding single cells or aggregates. On the whole, the results of the study suggest that the scaffold material directly or indirectly influences chondrocyte proteoglycan metabolism and the quality of tissue engineered cartilage.[189]

The study of Awad *et al.* assessed the biological, biochemical, and biomechanical properties of different scaffolds for cartilage TE, including alginate HG, agarose HG, and gelatin (Surgifoam®, denatured porcine collagen type I) porous sponge.[190] Chondrogenic growth and differentiation of adipose-derived adult stem cells in agarose, alginate, and gelatin scaffolds was evidently influenced by the biomaterial scaffold selected. While neither of the biomaterials studied approached native cartilage mechanical properties they demonstrated significant composition-function relationships that could provide important indications for engineering functional tissues.

3. HGs in Combination with Biological Components

Matrices able to release those biologically active polypeptides, can offer promising therapeutic treatment in the cartilage regeneration after damages due to traumatic events or degenerative diseases (such as osteoarthritis). However, the effects of such GFs can be changed and modulated by the interaction with the chemistry-structure of the matrix it self and cells eventually loaded before implantation. HGs based on both natural and synthetic polymers are of interest for encapsulation of cells, as drug carrier, becoming especially attractive in the field of TE as matrices for repairing and regenerating cartilage.[51]

Different strategies were proposed for 3D scaffolds with a role as carriers to controlled delivery of GFs, to improve the bioactivity of the system in terms of chondrogenic differentiation and ECM synthesis stimuli.[173] In particular, hybrid constructs are designed to induce integration between patient cartilage and the new tissue of the graft, and the eventual incongruities of the contact surfaces (cartilage–cartilage and cartilage–transition zone–bone). HGs were also successfully used as delivery systems of encapsulate cells, with or without growth factors or other bioactive molecules, allowing at the same time nutrients and metabolites diffusion.[56] In addition, the hydrated environment of HGs preserves for longer time the biomolecules activity as well as cell viability.[191–193]

3.1. HGs biologically modified

3.1.1. *Introduction*

Homeostasis and development of articular cartilage is regulated by numerous growth factors working in concert.[194] In particular, BMP, Transforming Growth Factors beta superfamily (TGF-b), Fibroblast Growth Factors (FGF), Platelets Derived Growth Factors (PDGF), and Insulin Growth Factor-1 (IGF-1) in combination with hydrogel-based matrices, are mainly proposed to up-regulate cell proliferation and ECM synthesis in cartilage TE applications.[195] As carrier system, several polymeric (injectable) HGs were proposed as matrix or for cell–drug encapsulation, by starting from natural-derived polymers (collagen, gelatin, HA, fibrin, heparin-dextran, etc.) as well as synthetic ones (PEG-based, PLGA, poly(ethylene oxide)–poly(propylene oxide)–poly(ethylene oxide) (PEO–PPO–PEO), etc.).[173,196]

Active TGF-α up regulates the proteoglycan synthesis, prevents the cartilage ECM degradation by inhibiting the MMP activity, and plasminogen activator inhibitor I. TGF-α TGF-α and TGF-α are present in the cartilage ECM, affecting the chondrocytes phenotype, playing an important role in the late stage of chondrocytes differentiation,[173] while BMPs are mainly active in the early stages of chondrogenesis stimulating cell–cell interactions.[197] Cell proliferation and cell death inhibition is controlled by IGF-1, a polypeptide expressed in developing cartilage, mature cartilage, and synovial fluids, and by some of FGF family molecules (FGF-1, -2, and -18). In addition, IGF-1 as well as PDGF are involved in chondrocytes differentiation and proliferation in MSC, and in the synthesis of proteoglycan and collagen type II.

3.1.2. *HGs as drug releasing systems*

The chondrogenic factor TGFα-1 was widely proposed to improve the bioactivity of various HG materials,[194,198–200] to overcome the cartilage poor healing capacity.

A chemically modified gelatin with methacrylic acid hydrogel obtained by photoinitiating polymerization, was loaded with TGFα-1 to promote cartilage regeneration increasing interaction between cells. Biological performance, in terms of cell viability, cell proliferation and GAG synthesis, was assessed *in vitro*, culturing rabbit ears chondrocytes up to 12 days. GAG synthesis was normalized to viable cells, showing for all samples (w/wo TGFα-1) an increase of GAGs production along the culture time but was higher in the hydrogel with TGFα-1.[86] To better control and delayed the releasing of the bioactive molecule during the therapeutic treatment, TGFα-1 was encapsulated in poly(D,L-lactide-co-glycolide) (PLGA) microspheres, and then included in a PEG HGs stabilized by using different ratio of a natural-derived crosslinking molecule, genipin.[200] The GFs releasing was controlled by a double step, from microspheres to PEG hydrogel matrix, and then from the matrix to the external environment by diffusion. The main role in releasing kinetic was played by the crosslinking degree of the hydrogel, in fact was seen more controlled and prolonged in the low degree of crosslinking samples. In addition, the hydrogel lets the containment of microspheres in the chondral defect. A biodegradable thermo reversible pluronic F127 PEO–PPO–PEO HG was combined with crosslinked HA to improve the final mechanical stability and loaded with TGF-α1 (physically conjugated), a typical chondrogenic factor. The release profile related to the *in vitro* study, showed a minimal initial burst of the GF, and a continuous realizing out of the construct up to 20 days. The specific bioactivity of the construct was evaluated encapsulating human adipose-derived stem cells in the final hydrogel and the chondrogenic effect studied in *in vivo* models (nude mouse, subcutaneous injection and rabbit, implantation in articular cartilage). Histological analysis showed the presence of cartilage specific ECM and Glycosamino-glycans (GAGs), evidence of the favorable effect of the construct on cell viability, and differentiation into a chondrogenic lineage.[201]

Heparin, one of GAGs plays an important role in the interaction with active proteins, associated with cell adhesion, cell proliferation, and differentiation, with various influences on the biological activities, such as angiogenesis, adipogenesis, and osteogenesis. The effects of heparin on chondrogenesis was considered and evaluated by Jin *et al.*[202] and an injectable hydrogel was prepared by mixing a heparin–tyramine conjugate and a dextran–tyramine conjugate in different proportions. The so obtained matrices were characterized in terms of swelling capability, gelation time, and rheological behavior. *In vitro* tests were performed by entrapping in the HGs bovine primary chondrocytes and culturing them up to 14 days. The biological performance of the constructs war assessed by Live–dead assay, DNA quantification, and gene expression of collagen type I and type II, aggrecan, Sox9. The addition of heparin was seen to be able to affect hydrogel properties (improve swelling capability, speed up the gelation, and increase the storage modulus). From the biological side, heparin influence in content-depend manner, the up-regulation of aggrecan (maximised when Dex/TA/Hep-TA weight ratio was 50/50) and collagen type II expression (maximised when Dex/TA/Hep-TA weight ratio was 50/50 and 25/75). Finally, the deposition of proteoglycans was underlined by staining with toluidine blue HGs after 21 days of culture, and the chondrogenic effect of heparin was confirmed.

Matrices with a controlled double factor release were also designed, combining nano- or micro-particles in a hydrogel matrix. Alginate gel containing BMP-7 (faster release) was mixed with Pluronic F-68/heparin/TGF-α2/chitosan nanoparticles (slower release) to induce chondrogenesis in Mesenchymal stem cells (MSCs) and cartilage regeneration.[203] The sequential releasing of GFs has also been proposed to avoid dedifferentiation of MSCs. The delivery of GFs from the composite matrix was due to diffusion throughout alginate HG and in the nanoparticles regulated by degradation kinetic. Heparin has been included in the system for the binding of growth factors because of their controlled release.

Non-degradable PVA HGs has been combined with growth factor-loaded PLGA degradable microsphere in a single step. In particular, as growth factor was selected the insulin-like growth factor-1 (IGF-1), polypeptide expressed in developing cartilage, mature cartilage, and synovial fluids. The effect of the local and controlled release on IGF-1 on cartilage regeneration, were assessed by implanting the construct pre-loaded with chondrocytes isolated from cartilage explants taken from the knee joints of a piglet. The cell loaded-construct was cultured for 1 week before subcutaneous implantation in nude mice. The *in vitro* delivery and activity of IGF-1 was detected up to 46 days. *In vivo* results showed the formation of a thick cartilaginous layer around the implant but also a chondrocytes migration into the hydrogel, probably driven by IGF-1, a chemo attractant for this specific phenotype. However, some sign of calcification in the new-formed cartilage was observed, probably due to a non-optimal GF concentration and releasing, type of cells used.[204]

As a source of growth factors involved in the cartilage healing process, Platelets Reach Plasma (PRP) was even considered and combined with hydrogel scaffolds loaded with cells. PRP is an autologous source of several growth factors such as PDGF, TGF-α1, IGF-1, VEGF, and can be easily available. In addition, PRP contains all components of plasma, such as hormones, electrolytes, and other proteins, such as Fibrinogen.[205] Beneficial effects of PRP treatment have been reported for chondral and osteo-chondral lesions, as consequence of trauma or osteoarthritis, in humans and animal models.[206] PRP was conjugated with HGs with and without embedded chondrocytes, and in both cases it was clearly demonstrated by *in vitro* and *in vivo* studies, the role in enhancing the chondrogenesis of non-articular cartilage lesion, improving the ECM-specific cartilage ECM, and the production of GAGs by synoviocytes.[207]

Lee *et al.*[208] published a quite extensive comparative evaluation of the regenerative efficacy of injectable HGs with different composition, loaded with chondrocytes (isolated from rabbit articular cartilage) and combined with PRP. HGs were synthetized by conjugating amine reactive poly(ethylene glycol)-tyramine (PEG-TA) with gelatin (type A from porcine skin) back bone, so obtaining two graft copolymer with different storage moduli. Two hybrid HGs were prepared by mixing hydrogel/cells/PRP. *In vivo* evaluations were performed by implantation in articular cartilage defect in rabbit groove femur. Positive effects in terms of cartilage regeneration were observed for both composite-PRP loaded, even in *in vitro* studies, in particular the up-regulation of the expression of collagen type II, aggrecan, SOX9, CB1, and CB2 was measured by RT-PCR. The over expression of CB1 and CB2 (target genes for anti-inflammatory and chondrocyte metabolism) induced by hydrogel composites suggests a possible preventive and direct role of PRP on cartilage remodeling. The up-regulation of ChM-1 (antiangiogenic factor) induced by PRP should be considered an additional element leading to an enhancement of cartilage regeneration.[208] PRP was also loaded as active factor in gelatin microspheres and the biological performance was evaluated by microspheres administration by injections in equine cartilage defect after subchondral drilling as bone marrow stimulation step.[209] Gelatine microspheres-PRP loaded, supported by bone marrow MSC infiltration, induces some favorable effects on cartilage regeneration in horses, in terms of accelerated cartilage regeneration. However the system is still under investigation to better understand and evaluate possible negative effect.

3.1.3. *HGs as cells carrier*

3.1.3.1. Cell type and stem cells sources

Different hydrogel-based scaffolds, mainly of natural origin, in combination with cells, isolated from different sources, have been explored. As starting polymers, biopolymers such as HA, chitosan, agarose, alginate, collagen, fibrin, gelatin[210] are becoming very popular and promising. However, injectable HGs for cell encapsulation were fabricated even by using synthetic polymers such as Pluronic, PEG, alone or in combination with biological moieties or biopolymers,[58] or synthetic peptides.[211] The selection of cells for cartilage regeneration as well as their *in vitro* manipulation is still a crucial step, particularly if differentiated chondrocytes are used. It has been demonstrated that during a bi-dimensional culture, they undergo a down regulation of chondrogenic genes, a decrease in their ability to synthesize ECM-cartilage specific, and a modification of membrane markers. The combination hydrogel-chondrocytes and specific GFs, in particular TGF-α, can help to overcome the cell de-differentiation.[212,213] Stem cells have shown promising results to overcome some of these constraints and represent one of the most studied cell source for osteochondral regeneration. In particular, bone marrow mesenchymal stem cells (BMSCs) and adipose-derived stem cells (ASCs) have demonstrated great potential in chondrogenic and osteogenic differentiation.[214] Their main advantages include a high availability, the ease of *in vitro* expansion and intrinsic repair capacity[212]; on the other hand, stem cells have a potential instability and careful and specific stimulations are required to induce the expression of a chondral phenotype.[213] However, results have been encouraging and autologous BMSCs transplantation was employed for the repair of full-thickness articular cartilage defects in patellae of two human patients.[215] Other resources for stem cells include synovial tissue and periosteum: Cells derived from synovial membrane can be harvested with minimum complications for the donor and have a high regenerative capacity,[146] while in the periosteum progenitor cells are present with a high proliferation rate and differentiation potential.[216]

3.1.3.2. Cell-based strategies

HGs based on Collagen type I, one of the main ECM structural proteins, and alginate, a polysaccharides isolated from brown algae, were compared as chondrocytes carrier, in terms of phenotype stability and induction for ECM-specific molecules synthesis, mainly proteoglycans.[73] Interestingly, different biological outcomes were observed during *in vitro* evaluations, in particular chondrocytes proliferation was sensibly up regulated by collagen hydrogel, leading a high cell density in the system if compared with cell embedded in alginate matrix. However, the lower cell density in alginate better contribute in reducing cell–cell contacts, the maintenance of cell shape (round) and function. As consequence, a better collagen type II and GAG expression and accumulation was seen in the alginate constructs. A similar chondrocytes behavior was observed even on chondrocytes encapsulate in fibrin glue.[91] *In vitro* chondrocytes cultured up to 5 weeks, showed a well controlled proliferation as in alginate, and synthesis of collagen type II as well as GAG. In addition, fibrin glue showed promising capability in the phenotype stability in long-term cell culture, needed to obtain a mature cartilage biological substitute. In addition, fibrin-based matrix degradation can be tuned by playing with fibrinogen and thrombin concentration mainly, in order to provide a stable shape and volume, suitable for chondrogenesis. Limitations were reported in terms of nutrient diffusion and waste removal from the construct, in particular culturing cells in static condition. A multicomponent matrix based on alginate based HG combined with Gelatin type I microspheres loaded with chondrocytes, was developed as new cell delivery methodology.[217] Gelatin microsphere were produced in water-in-oil (soya oil) emulsion involving cell–gelatin suspension. Gelatin microspheres, when at 37°C, dissolve completely in 2 days, so producing a double effect, chondrocytes releasing in the alginate-based matrix and producing micro-cavity (acting as porogen) in the bulk. The potential of the composite system in hyaline cartilage regeneration was investigated by *in vitro* test, and the expression of cartilage-specific markers expression as well as GAGs and collagen production were considered.

Chondocytes encapsulated in a heparin-poly(ethyleneglycol) (PEG-DA) (with different molecular ratio) HGs were evaluated *in vitro* system as well as *in vivo* (subcutaneous implantation in nude mouse and injection in partial-thickness cartilage defects in rabbit) by Kim *et al.*[218] Biological evaluations showed results supporting the good expression and secretion of ECM-specific molecules (e.g., GAGs, collagen type II, aggrecan). However, the hydrogel with 10% (by wt) heparin showed the best balance between hydrogel mechanical properties and biological performance, providing a suitable chondrogenic environment. *In vivo* results were consistent with the previous evaluations, in particular the 10% heparin base matrix, but interestingly, all samples performance were much more efficient than *in vitro.* Final results (e.g., chondrogenesis) were discussed in terms of heparin content, degradation kinetic, and on the mechanical properties (mechanical strength). In particular, heparin 10% hydrogel showed an initial modulus value of 2.3 kPa, describing a mechanical strength lower then the native articular cartilage (500–1000 kPa range).

However, it seems that weaker HGs, with a lower crosslinking degree, can better influence and guide chondrocytes functions in the early stages of activity. In particular because crosslinking degree strongly affect hydrogel architecture/degradation/mechanical properties, with a direct impact on cells spatial organization and new tissue composition and organization.[219] The influence of hydrogel crosslinking density on biological performance was also considered and discussed by Nicodemus *et al.*[165] (see Fig. 2). As matrices model were fabricated PEG HGs with three crosslinking densities (10, 15, and 20 wt.%), with compressive moduli ranged from 60 to 590 kPa, but in all samples the water content was above 85%. Chondrocytes encapsulated in the PEG matrices and cultured *in vitro*, and the

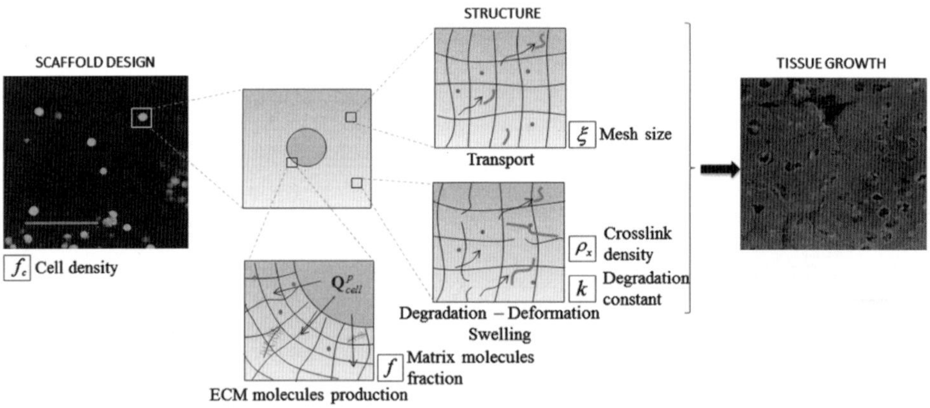

Fig. 2. Multiscale approach to modelling tissue production by cells encapsulated in HGs. Left: chondrocytes encapsulated within a PEG hydrogel, shown at day 3 after encapsulation. Cytosol of live cells fluoresce green showing the chondrocytic round morphology. Nuclei of dead cells fluoresce red. Scale bar indicates 100 microns. Right: CS (red) by chondrocytes encapsulated within a degradable PEG hydrogel after 28 days *in vitro.* Cell nuclei are stained blue. Scale bar indicates 50 microns. From Ref. 6 with copyright permission.

crosslinking degree seemed to strongly affect in particular the GAGs content, data consistent with histological analysis after Safranin-O staining. At the same time, an increasing crosslinking density seems to up-regulate the catabolic enzymes production (MMP-1 and MMP-13) so triggering faster matrix degradation (see Fig. 3).

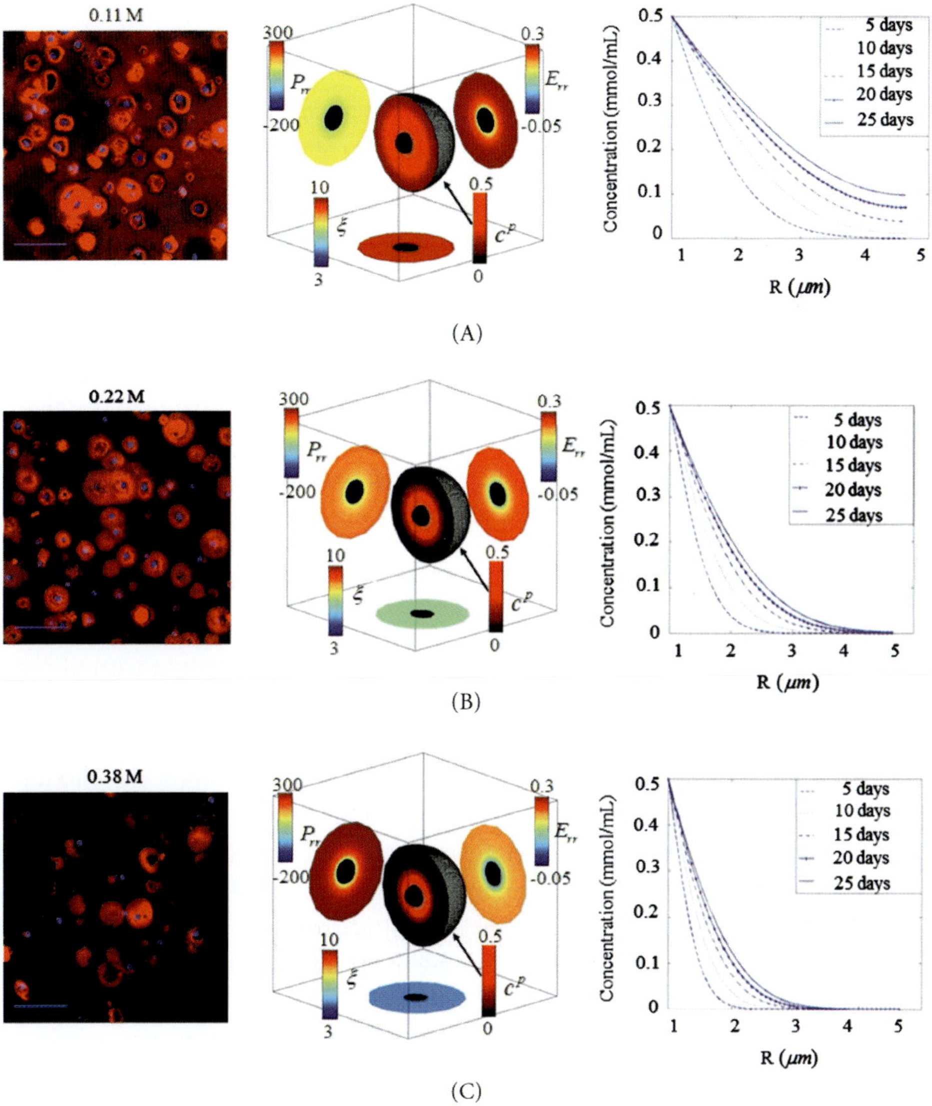

Fig. 3. (A–C) An experimental results and the model results for different crosslink densities of a stable hydrogel. Regarding the experimental pictures, CS elaboration (red) by chondrocytes encapsulated within PEGDM HGs and cultured for 25 days *in vitro* with varied crosslinking density. Cell nuclei are stained in blue. Scale bars indicate 50 microns. In 3D plots, the stress P_{rr} (KPa), strain E_{rr}, mesh size ς(nm) and concentration c^{p} (mmol/mL) can be observed. From Ref. 6, with copyright permission.

MSCs were also considered as cellular component for hybrid HGs design. Li *et al.*[221] have been developed collagen (rat-tail collagen type I) microencapsulation techniques to entrap Mesenchymal Stem Cells (MSCs) in high concentration (1×10^6 cells/mL^{-1}), to significantly drive the cell fate Versus chondrogenic differentiation. In addition, microencapsulation of MSCs in a condensed collagen meshwork was evaluated on *in vivo* model (rabbit cartilage focal defect) and was seen to be able to improve cell survival, good microspheres localization at the damaged implantation site, and provided a mechanically stable matrix with reduced elastic modulus (about 9 kPa) if compared with collagen gel. Different natural-derived polysaccharides, HA-tyramine conjugates (Ha–Tyr), were proposed to fabricate matrices to encapsulate MSCs for a cell delivery system. Mechanical and degradation properties of (Ha–Tyr)-HGs were tuned by varying the crosslinking degree (different concentration of H_2O_2, oxidant in the oxidative coupling reaction, from 500 to 1000 micro-M), see Fig. 4.[220] Encapsulate MSCs condensation (contraction) and fate was evaluated *in vitro*, and correlated with the crosslinking density. Interestingly, decreases in crosslinking degree determine an increase in contraction of the system and mass lost, and decreasing in cell number, compared to the higher crosslinked HGs. The data reported are just apparently in contradiction, but this increased cell number density was induce by hydrogel contraction (affected by the low density of crosslinking) with the reduction in matrix volume, reflecting cellular condensation.

Synovium-derived mesenchymal stem cells (SMSC), is even a promise stem cell source for TE applications, because both cartilage and synovium have been originate from a common group of progenitors. Injectable gellan (a natural polysaccharide and food additive FDA approved) HGs were proposed by different authors, as promising matrix for cell delivery, even in the cartilage regeneration field.[222] Gellan injectable hydrogel was combined with SMSCs and the chondrogenic potential evaluated *in vitro* in chondrogenic medium, by detecting the gene expression of the main cartilage-specific markers, such as collagen type II, aggrecan, biglycan, SOX9.[146] To produce an injectable hydrogel, gellan gum was properly modified to decreasing the gelling point from 42°C to about 37.5°C. Encapsulated SMSCs in the gellan matrix and cultured *in vitro* in chondrogenic conditions, were seen to be able to proliferate, maintaining their chondrogenic morphology, up to 42 days. It was also well evident, the formation of a cartilaginous tissue into the construct, confirming the potentiality of the proposed technology. As potential alternative source of stem cells, the umbilical cord derived MSCs were also isolated and embedded in a collagen type I hydrogel, to design a hybrid scaffolds for cartilage applications.[223] Preliminary *in vitro* data, showed the potential of that specific source of stem cells when combined with a suitable matrix as therapeutic treatment of damaged cartilage.

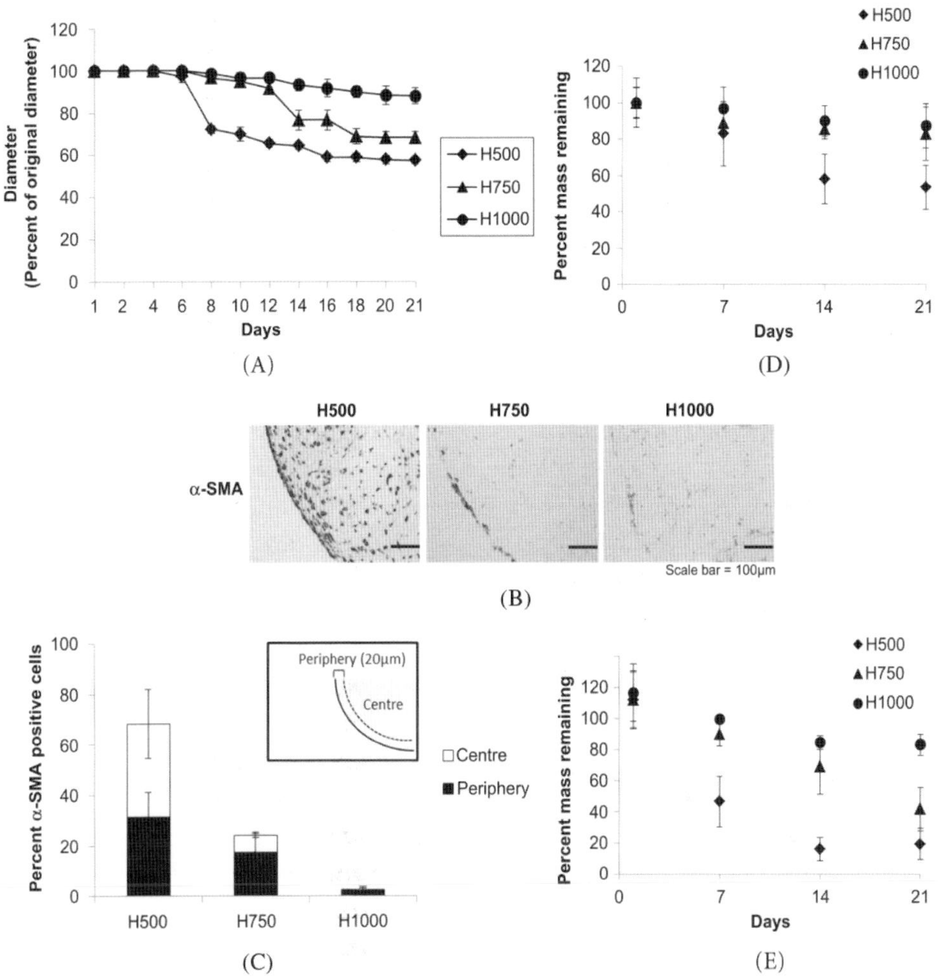

Fig. 4. Cell-mediated contraction and degradation of HA–Tyr HGs (H500, H750, and H1000). (A) Contraction of MSC-laden HA–Tyr HGs with time in culture, as a percentage of the original diameter (mean ± SD; n = 8/group/time point). Two-factor ANOVA revealed significant effects of time, gel crosslinking and their interactions on cell-mediated contraction ($p < 0.0001$). (B) Immunohistochemistry for α-SMA (red chromogen). Representative images for α-SMA expression in MSC-laden constructs (n = 5/group; scale bar: 100 μm). (C) Histomorphometric analysis of the percentage of α-SMA-expressing MSCs in the center and periphery of the constructs at 3 weeks. Inset shows the schematic model to illustrate the center and periphery of the construct. Degradation of HA-Tyr HGs (H500, H750 and H1000) as a function of crosslinking degree and culture time for (D) Cell-free and (E) MSC-laden constructs (mean ± SD; n = 4–8/group/time point). From Ref. 220, with copyright permission.

A more complex hydrogel, designed to recapitulate the highly organized zonal architecture of articular cartilage, different PEG-based HGs was prepared including CS and/or matrix metalloproteinase-sensitive peptides (MMP-pep), and/or HA, so obtaining HGs with different chemistry, mechanical performance and degradation kinetic. *In vitro* studies showed the ability of each hydrogel to direct the differentiation fate of MSCs (isolated from bone marrow), and to induce ECM organization comparable to the natural cartilage layers.[224]

As advanced concept of TE and the use of HG matrices for tissue engineered cartilage, it should be reported one examples on scaffold free methodology. The strategy is based on the use of a specific hydrogel as scaffold with an intermediate role. The HG was used to growth *in vitro* a cartilage substitute and then removed before transplantation. Su *et al.*,[225] developed a methodology to set up a micro scaled 3D living hyaline cartilage graft, supported by a biomaterial-based interim scaffold system (see Fig. 5).[225] The base of the strategy was to recapitulate and mimic the pre-cartilage condensation during primary development of cartilage. Alginate was used to encapsulate chondrocytes in micro beads in order to growth micro tissue "nodules" and then, thanks to the proliferative activity of cells, a 3D macro network was obtained. The alginate scaffold is then gently removed by simple citric leaching treatment. As final results, a biological graft is obtained, ready to be implanted.

3.1.3.3. Advanced HGs as drug releasing system and cell carrier

For cell encapsulation, many authors have been proposed the combination of encapsulated cells with signaling molecules or GF, to better drive the chondrogenic effect of the therapeutic system. Na *et al.* have designed an injectable thermo-reversible hydrogel based on p(NiPAAm-co-AAc) blended with HA, and loaded with TGFα-3 and embedded rabbit chondrocytes. The synergic effect between GFs and cells embedded in the hydrogel was *in vivo* evaluated in nude mice (up to 56 days) in terms of cartilage specific ECM synthesis (GAGs and collagen). The chondrogenic effects of TGFα-3 coupled with rabbit mesenchymal stem cells (rMSCs) in the injectable thermo-reversible hydrogel of p(NiPAAm-co-AAc), was evaluated by Park and Na[226] in nude mouse implantation (back sub-cutis injection). To determine neo-cartilage formation, GAG content, DNA content, and total collagen content were assessed as well as collagen type II, aggrecan, and Sox9 gene expression by RT-PCR. Construct loaded with FGFs was used as reference system. rMSCs embedded in presence of TGFα-3, showed a better chondrogenic differentiation, and higher proliferation level at all experimental time points considered (up to 8 weeks), indicating the role of the bioactive molecule in triggering the proper molecular and cellular events involved in chondrogenesis. Natural-derived polymers were also taken into consideration to fabricate multi-functional HGs, coupling growth factors, stem cells or

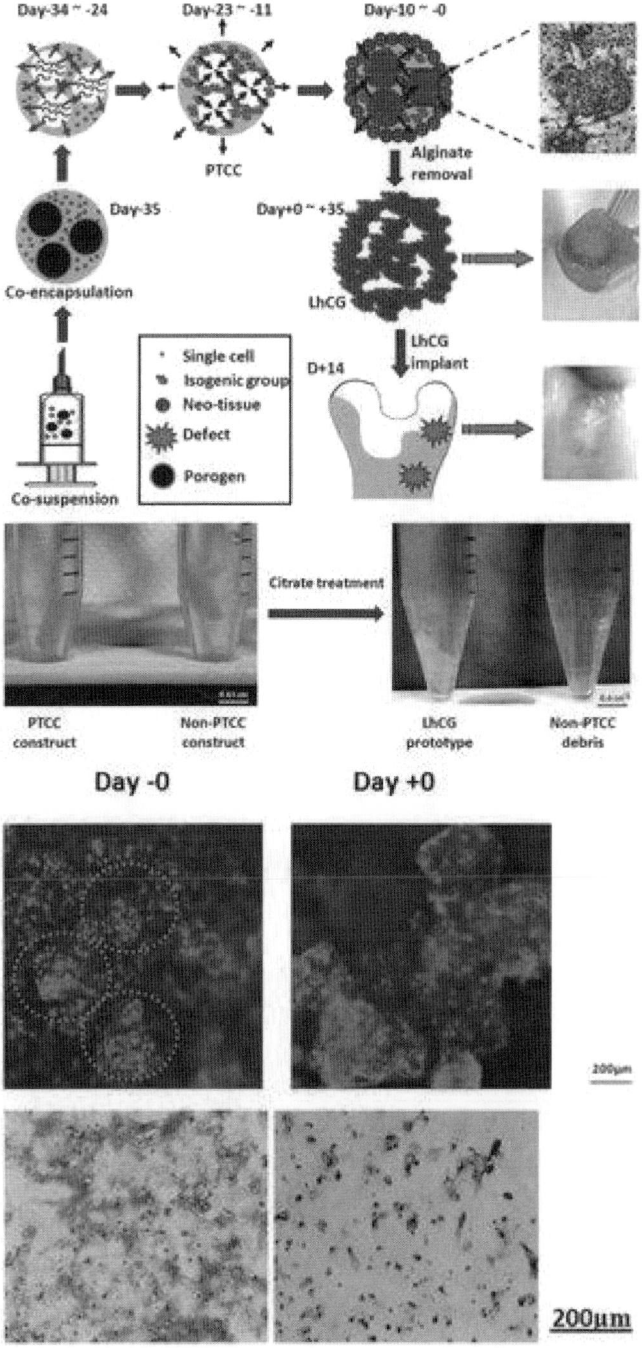

Fig. 5. Preparation of LhCG. (A) Schematic illustration of LhCG fabrication process. Clockwise from left bottom: Porcine chondrocytes are co-encapsulated with gelatin microspheres (porogen) in alginate hydrogel. After gelation of alginate, the dissolving of gelatin microspheres creates cavities within the gel bulk. PTCC16, 18 take place giving rise to micro tissue nodules

Fig. 5. (*Continued*) within alginate hydrogel scaffold. The microtissues interact and secrete ECM extensively, forming an intricate interpenetrating network of ECM and hydrogel. With removal of alginate scaffold, the structural integrity of the construct remains intact and a pure, scaffold-free LhCG is formed. (B) Gross overview of PTCC and non-PTCC construct before and after alginate removal via sodium citrate treatment on day 0. The structural integrity in the PTCC construct, which therefore formed LhCG construct was maintained after alginate removal, but non-PTCC collapsed into debris. (C) Live/dead staining indicated the formation of neotissue nodules in PTCC sample before gel removal at day –0 and LhCG sample right after gel removal at day +0. (D) Histochemical staining of PTCC sample after extended culture without alginate removal. From left to right: Safranin O, Massons Trichrome staining. PTCC sample (day +70, totally 105 day cultivation without gel removal) showed weak and discontinuous stainings for GAG and collagen suggesting alginate removal is mandatory for further tissue development as that in LhCG. From Ref. 225, with copyright permission.

progenitor cells.[94,227–229] Fibrin, agarose and gellan gum HGs biological performances were compared, and progenitor cells (isolated intra-patellar fat pad) were encapsulated in microspheres loaded with TGF-α3. The final performance was determined by the complex interactions between matrix and cells, and by the GFs releasing mediated by the hydrogel itself.[230] After *in vitro* evaluations, Fibrin-based matrix showed the weaker results in terms of induced chondrogenesis, instead gellan gum and agarose HGs appeared to support more robust cartilage formation in terms of GAGs synthesis. However, gellan gum HGs showed even a large accumulation of collagen type II and type I, suggesting a de-differentiation of the phenotype into fibro chondrocyte lineage. More promising results, both *in vitro* and *in vivo* (sub-cutis of nude mouse), were obtained by Park *et al.*,[231] preparing a composite system with fibrin hydrogel pre-loaded with marrow-derived stromal cells in high concentration, combined with nanoparticles containing TGF-α3, as inducer of chondrogenesis. *In vivo* preliminary results, particularly histological observations, revealed the presence of round cells aggregates in the hydrogel matrix, and ECM-specific hyaline cartilage molecules production just in the complex system. Without the presence of growth factor, the chondrogenic effect sensibly decreases.

4. Conclusions

As discussed in the previous paragraphs, HG scaffolds from polymers of either natural or synthetic origin possess many attractive characteristics for cartilage TE. Their mechanical properties, degradation profile as well as production and distribution of ECM are significantly influenced by polymer chemistry, gelation mechanism and design parameters, such as macromere molecular weight, polymer concentration and crosslinking density.[130,165,232,233]

Design considerations for the initial mechanical properties of the HGs should be balanced with parameters that control ECM synthesis and cell distribution within the gel. In fact, it should be considered that HGs with relatively low initial mechanical properties could promote the formation of constructs with high compressive and dynamic modulus compared to HGs whose mechanical strength is initially higher.[130,232]

By selecting or combining specific materials, design parameters can be tuned to develop HG-based scaffolds with a response more and more suitable for cartilage TE. Despite the various efforts and steps forward in the field of cartilage TE, the ideal hydrogel formulation presenting optimal mechanical properties, biodegradability, and biocompatibility still needs to be developed. Novel highly performing solutions for the development of hydrogel matrices are able to regenerate functional hyaline cartilage are still under evaluation.

At present, most of the experimentation with hydrogel scaffolds for cartilage TE is based on *in vitro* studies, although several materials are being tested *in vivo* on animals and, in some cases on human patients.

References

1. G. Vunjak-Novakovic, I. Martin, B. Obradovic, S. Treppo, A. J. Grodzinsky, R. Langer, and L. E. Freed, Bioreactor cultivation conditions modulate the composition and mechanical properties of tissue-engineered cartilage. *J. Orthopaed. Res.*, **17**, 130–138 (1999).
2. K. Noyori, T. Takagi, and H. E. Jasin, Characterization of the macromolecular components of the articular cartilage surface. *Rheumatol. Int.*, **18**(2), 71–77 (1998).
3. J. S. Temenoff and A. G. Mikos, Review: Tissue engineering for regeneration of articular cartilage. *Biomaterials*, **21**(5), 431–440 (2000).
4. P. J. Roughley, The structure and function of cartilage proteoglycans. *Eur. Cells Mat.*, **12**, 92–101 (2006).
5. T. M. Quinn and V. Morel, Microstructural modeling of collagen network mechanics and interactions with the proteoglycan gel in articular cartilage. *Biomech. Model. Mechanobiol.*, **6**, 73–82 (2007).
6. V. Dhote, S. Skaalure, U. Akalp, J. Roberts, S. J. Bryant, and F. J. Vernerey, On the role of hydrogel structure and degradation in controlling the transport of cell-secreted matrix molecules for engineered cartilage. *J. Mech. Behav. Biomed. Mater.*, **19**, 61–74 (2013).
7. A. E. Beris, M. G. Lykissas, C. D. Papageorgiou, and A. D. Georgoulis, Advances in articular cartilage repair. *Injury*, **36**(Suppl. 4), S14–S23 (2005).
8. L. Engebretsen and R. Bahr, Surgical treatment of cartilage injuries. *Eur. Surg.*, **36**(1), 13–19 (2004).

9. K. Muldrew, Osteoarthritis as an inevitable consequence of the structure of articular cartilage. *Med. Hypotheses*, **59**(4), 389–397 (2002).
10. L. C. Dijkgraaf, L. G. M. De Bont, G. Boering, and R. S. B. Liem, The structure, biochemistry, and metabolism of osteoarthritic cartilage: A review of the literature. *J. Oral Maxil. Surg.* (1995).
11. H. Lorenz and W. Richter, Osteoarthritis: Cellular and molecular changes in degenerating cartilage. *Progr. Histochem. Cytochem.*, **40**, 135–163 (2006).
12. J. Suh and S. Scherping, Basic science of articular cartilage injury and repair. *Operative Techniques in ...*, **3**(2), 78–86 (1995).
13. A. J. Detterline, S. Goldberg, B. R. Bach, and B. J. Cole, Treatment options for articular cartilage defects of the knee. *Orthopaedic Nursing/National Association of Orthopaedic Nurses*, **24**, 361–366 (2005).
14. K. H. Pridie, A method of resurfacing osteoarthritic knee joints. *J. Bone Joint Surg.*, **41**, 618–619 (1959).
15. A. Matsiko, T. Levingstone, and F. O'Brien, Advanced strategies for articular cartilage defect repair. *Materials*, **6**(2), 637–668 (2013).
16. J. R. Steadman, W. G. Rodkey, and J. J. Rodrigo, Microfracture: Surgical technique and rehabilitation to treat chondral defects. *Clin. Orthopaed. Rel. Res.*, S362–S369 (2001).
17. K. Mithoefer, T. McAdams, R. J. Williams, P. C. Kreuz, and B. R. Mandelbaum, Clinical efficacy of the microfracture technique for articular cartilage repair in the knee: An evidence-based systematic analysis. *Am. J. Sport Med.*, **37**(10), 2053–2063 (2009).
18. M. Brittberg, A. Lindahl, A. Nilsson, C. Ohlsson, O. Isaksson, and L. Peterson, Treatment of deep cartilage defects in the knee with autologous chondrocyte transplantation. *New Eng. J. Med.*, **331**, 889–895 (1994).
19. L. Peterson, H. S. Vasiliadis, M. Brittberg, and A. Lindahl, Autologous chondrocyte implantation: A long-term follow-up. *Am. J. Sport Med.*, **38**, 1117–1124 (2010).
20. A. M. Bhosale and J. B. Richardson, Articular cartilage: Structure, injuries and review of management. *Brit. Med. Bull.*, **87**, 77–95 (2008).
21. G. Knutsen, J. O. Drogset, L. Engebretsen, T. Grøntvedt, V. Isaksen, *et al.*, A randomized trial comparing autologous chondrocyte implantation with microfracture. Findings at five years. *J. Bone Joint Surg. Am.*, **89**, 2105–2112 (2007).
22. D. B. F. Saris, J. Vanlauwe, J. Victor, M. Haspl, M. Bohnsack, Y. Fortems, *et al.*, Characterized chondrocyte implantation results in better structural repair when treating symptomatic cartilage defects of the knee in a randomized controlled trial versus microfracture. *Am. J. Sport Med.*, **36**, 235–246 (2008).
23. P. Cherubino, F. A. Grassi, P. Bulgheroni, and M. Ronga, Autologous chondrocyte implantation using a bilayer collagen membrane: A preliminary report. *J. Orthopaed. Surg.*, **11**, 10–15 (2003).
24. A. J. Gibson, S. M. McDonnell, and A. J. Price, Matrix-induced autologous chondrocyte implantation. *Oper. Tech. Orthopaed.*, **16**(4), 262–265 (2006).
25. M. Jacobi, V. Villa, R. A. Magnussen, and P. Neyret, MACI — a new era? *Sports Med., Arthrosc. SMARTT*, **3**(1), 10 (2011).

26. G. Filardo, E. Kon, R. Buda, A. Timoncini, A. Di Martino, *et al.*, Platelet-rich plasma intra-articular knee injections for the treatment of degenerative cartilage lesions and osteoarthritis. *Knee Surg. Sport Tr. A.*, **19**(4), 528–535 (2011).
27. A. Gobbi and L. Bathan, Biological approaches for cartilage repair. *The J. Knee Surg.*, **22**(1), 36–44 (2009).
28. J. S. Temenoff, K. A. Athanasiou, R. G. LeBaron, and A. G. Mikos, Effect of poly(ethylene glycol) molecular weight on tensile and swelling properties of oligo(poly(ethylene glycol) fumarate) hydrogels for cartilage tissue engineering. *J. Biomed. Mater. Res.*, **59**(3), 429–437 (2002).
29. G. D. Nicodemus and S. J. Bryant, Cell encapsulation in biodegradable hydrogels for tissue engineering applications. *Tissue Eng. Part B, Rev.*, **14**, 149–165 (2008).
30. C. Vinatier, D. Mrugala, C. Jorgensen, J. Guicheux, and D. Noël, Cartilage engineering: A crucial combination of cells, biomaterials and biofactors. *Trends in Biotechnology*, **27**(5), 307–314 (2009).
31. D. F. Williams, On the mechanisms of biocompatibility. *Biomaterials*, **29**, 2941–2953 (2008).
32. H. Stenhamre, U. Nannmark, A. Lindahl, P. Gatenholm, and M. Brittberg, Influence of pore size on the redifferentiation potential of human articular chondrocytes in poly(urethane urea) scaffolds. *J. Tissue Eng. Regen. Med.*, **5**(7), 578–588 (2011).
33. J. Zeltinger, J. K. Sherwood, D. A. Graham, R. Müeller, and L. G. Griffith, Effect of pore size and void fraction on cellular adhesion, proliferation, and matrix deposition. *Tissue Eng.*, **7**(5), 557–572 (2001).
34. D. F. Duarte Campos, W. Drescher, B. Rath, M. Tingart, and H. Fischer, Supporting biomaterials for articular cartilage repair. *Cartilage*, **3**(3), 205–221 (2012).
35. K. A. Athanasiou, M. P. Rosenwasser, J. A. Buckwalter, T. I. Malinin, and V. C. Mow, Interspecies comparisons of *in situ* intrinsic mechanical properties of distal femoral cartilage. *J. Orthopaed. Res.*, **9**(3), 330–340 (1991).
36. A. J. Engler, S. Sen, H. L. Sweeney, and D. E. Discher, Matrix elasticity directs stem cell lineage specification. *Cell*, **126**, 677–689 (2006).
37. M. Kim, B. Hong, J. Lee, S. E. Kim, S. S. Kang, Y. H. Kim, and G. Tae, Composite system of PLCL scaffold and heparin-based hydrogel for regeneration of partial-thickness cartilage defects. *Biomacromolecules*, **13**(8), 2287–2298 (2012).
38. C. Foss, C. Migliaresi, and A. Motta, The optimization of a scaffold for cartilage regeneration. *Organogenesis*, **9**(1), 19–21 (2013).
39. Y. Wang, E. Bella, C. S. D. Lee, C. Migliaresi, L. Pelcastre, *et al.*, The synergistic effects of 3-D porous silk fibroin matrix scaffold properties and hydrodynamic environment in cartilage tissue regeneration. *Biomaterials*, **31**(17), 4672–4681 (2010).
40. H. Tai, M. L. Mather, D. Howard, W. Wang, L. J. White, *et al.*, Control of pore size and structure of tissue engineering scaffolds produced by supercritical fluid processing. *Eur. Cells Mater.*, **14**, 64–77 (2007).
41. J. J. A. Barry, H. S. Gidda, C. A. Scotchford, and S. M. Howdle, Porous methacrylate scaffolds: Supercritical fluid fabrication and *in vitro* chondrocyte responses. *Biomaterials*, **25**(17), 3559–3568 (2004).

42. T. Tateishi, G. Chen, and T. Ushida, Biodegradable porous scaffolds for tissue engineering. *J. Artif. Org.*, **5**(2), 77–83 (2002).
43. T. B. F. Woodfield, J. Malda, J. de Wijn, F. Péters, J. Riesle, and C. A. van Blitterswijk, Design of porous scaffolds for cartilage tissue engineering using a three-dimensional fiber-deposition technique. *Biomaterials*, **25**(18), 4149–4161 (2004).
44. L. Moroni, J. A. A. Hendriks, R. Schotel, J. R. De Wijn, and C. A. van Blitterswijk, Design of biphasic polymeric 3-dimensional fiber deposited scaffolds for cartilage tissue engineering applications. *Tissue Eng.*, **13**(2), 361–371 (2007).
45. M. L. Alves da Silva, A. Martins, A. R. Costa-Pinto, P. Costa, S. Faria, *et al.*, Cartilage tissue engineering using electrospun PCL nanofiber meshes and MSCs. *Biomacromolecules*, **11**(12), 3228–3236 (2010).
46. J. M. Coburn, M. Gibson, S. Monagle, Z. Patterson, and J. H. Elisseeff, Bioinspired nanofibers support chondrogenesis for articular cartilage repair. *Proc. Nat. Acad. Sci. U. S. A.*, **109**(25), 10012–10017 (2012).
47. A. Thorvaldsson, H. Stenhamre, P. Gatenholm, and P. Walkenström, Electrospinning of highly porous scaffolds for cartilage regeneration. *Biomacromolecules*, **9**(3), 1044–1049 (2008).
48. R. Cancedda, B. Dozin, P. Giannoni, and R. Quarto, Tissue engineering and cell therapy of cartilage and bone. *Matrix Biol.*, **22**(1), 81–91 (2003).
49. D. W. Hutmacher, Scaffolds in tissue engineering bone and cartilage. *Biomaterials*, **21**(24), 2529–2543 (2000).
50. T. Sato, G. Chen, T. Ushida, T. Ishii, N. Ochiai, T. Tateishi, and J. Tanaka, Evaluation of PLLA–collagen hybrid sponge as a scaffold for cartilage tissue engineering. *Mater. Sci. Eng. C*, **24**(3), 365–372 (2004).
51. A. S. Hoffman, Hydrogels for biomedical applications. *Ann. N. Y. Acad. Sci.*, **944**, 62–73 (2001).
52. J. L. Drury and D. J. Mooney, Hydrogels for tissue engineering: Scaffold design variables and applications. *Biomaterials*, **24**(24), 4337–4351 (2003).
53. W. Zhao, X. Jin, Y. Cong, Y. Liu, and J. Fu, Degradable natural polymer hydrogels for articular cartilage tissue engineering. *J. Chem. Technol. Biotechnol.*, **88**(3), 327–339 (2013).
54. A. Montembault, K. Tahiri, C. Korwin-Zmijowska, X. Chevalier, M. T. Corvol, and A. Domard, A material decoy of biological media based on chitosan physical hydrogels: Application to cartilage tissue engineering. *Biochimie*, **88**(5), 551–564 (2006).
55. H. J. Kwon, K. Yasuda, J. P. Gong, and Y. Ohmiya, Polyelectrolyte hydrogels for replacement and regeneration of biological tissues. *Macromol. Res.* (2014).
56. A. A. Amini and L. S. Nair, Injectable hydrogels for bone and cartilage repair. *Biomed. Mater.*, **7**(2), 024105 (2012).
57. D. H. Kim, S.-J. Heo, J.-W. Shin, C. W. Mun, K. M. Park, K. D. Park, and K. S. Jee, Preparation of thermosensitive gelatin–pluronic copolymer for cartilage tissue engineering. *Macromol. Res.*, **18**(4), 387–391 (2010).
58. K. M. Park, S. Y. Lee, Y. K. Joung, J. S. Na, M. C. Lee, and K. D. Park, Thermosensitive chitosan-Pluronic hydrogel as an injectable cell delivery carrier for cartilage regeneration. *Acta Biomater.*, **5**(6), 1956–1965 (2009).

59. J. Sun and H. Tan, Alginate-based biomaterials for regenerative medicine applications. *Materials*, **6**(4), 1285–1309 (2013).
60. J. K. Tessmar and A. M. Göpferich, Customized PEG-derived copolymers for tissue-engineering applications. *Macromol. Biosci.*, **7**(1), 23–39 (2007).
61. M. Jasionowski, K. Krzyminski, W. Chrisler, L. M. Markille, J. Morris, and A. Gutowska, Thermally-reversible gel for 3-D cell culture of chondrocytes. *J. Mater. Sci.*, **15**(5), 575–582 (2004).
62. A. Gutowska, B. Jeong, and M. Jasionowski, Injectable gels for tissue, **349**, 342–349 (2001).
63. H. Tan and K. G. Marra, Injectable, biodegradable hydorgels for tissue engineering applications. *Materials*, **3**(3), 1746–1767 (2010).
64. H.-W, Sung, R.-N. Huang, L. L. H. Huang, and C.-C. Tsai, *In vitro* evaluation of cytotoxicity of a naturally occurring cross-linking reagent for biological tissue fixation. *J. Biomat. Sci. Polym. Ed.*, **10**(1), 63–78 (1999).
65. E. V. Dare, M. Griffith, P. Poitras, J. A. Kaupp, S. D. Waldman, D. J. Carlsson, *et al.*, Genipin cross-linked fibrin hydrogels for *in vitro* human articular cartilage tissue-engineered regeneration. *Cells Tissues Organs*, **190**(6), 313–325 (2009).
66. H. G. Sundararaghavan, G. A. Monteiro, N. A. Lapin, Y. J. Chabal, J. R. Miksan, and D. I. Shreiber, Genipin-induced changes in collagen gels: Correlation of mechanical properties to fluorescence. *J. Biomed. Mater. Res. Pt. A*, **87**(2), 308–320 (2008).
67. C.-S. Ko, J.-P. Huang, C.-W. Huang, and I.-M. Chu, Type II collagen-chondroitin sulfate-hyaluronan scaffold cross-linked by genipin for cartilage tissue engineering. *J. Biosci. Bioeng.*, **107**(2), 177–182 (2009).
68. L. Zhang, K. Li, W. Xiao, L. Zheng, Y. Xiao, H. Fan, and X. Zhang, Preparation of collagen–chondroitin sulfate–hyaluronic acid hybrid hydrogel scaffolds and cell compatibility *in vitro*. *Carbohyd. Polym.*, **84**(1), 118–125 (2011).
69. M. Ferretti, K. G. Marra, K. Kobayashi, A. J. Defail, and C. R. Chu, Controlled *in vivo* degradation of genipin crosslinked polyethylene glycol hydrogels within osteochondral defects. *Tissue Eng.*, **12**(9), 2657–2663 (2006).
70. S.-M. Lien, W.-T. Li, and T.-J. Huang, Genipin-crosslinked gelatin scaffolds for articular cartilage tissue engineering with a novel crosslinking method. *Mater. Sci. Eng. C*, **28**(1), 36–43 (2008).
71. L. Zhang, J. Hu, and K. A. Athanasiou, The role of tissue engineering in articular cartilage repair and regeneration. *Crit. Rev. Biomed. Eng.*, **37**, 1–57 (2009).
72. D. Puppi, F. Chiellini, A. M. Piras, and E. Chiellini, Polymeric materials for bone and cartilage repair. *Progr. Polym. Sci.*, (2010).
73. H. Yamaoka, H. Asato, T. Ogasawara, S. Nishizawa, T. Takahashi, T. Nakatsuka, *et al.*, Cartilage tissue engineering using human auricular chondrocytes embedded in different hydrogel materials, 1–11 (2006).
74. S. Ibusuki, A. Papadopoulos, M. Ranka, G. Halbesma, M. Randolph, *et al.*, Engineering cartilage in a photochemically crosslinked collagen gel. *J. Knee Surg.*, **22**(01), 72–81 (2010).

75. J. S. Pieper, P. M. van der Kraan, T. Hafmans, J. Kamp, P. Buma, J. L. C. van Susante, *et al.*, Crosslinked type II collagen matrices: Preparation, characterization, and potential for cartilage engineering. *Biomaterials*, **23**(15), 3183–3192 (2002).
76. P. Buma, J. S. Pieper, T. van Tienen, J. L. C. van Susante, P. M. van der Kraan, J. H. Veerkamp, *et al.*, Cross-linked type I and type II collagenous matrices for the repair of full-thickness articular cartilage defects — A study in rabbits. *Biomaterials*, **24**(19), 3255–3263 (2003).
77. L. Schuman, P. Buma, D. Versleyen, B. De Man, P. M. Van Der Kraan, W. B. Van Den Berg, and G. N. Homminga, Chondrocyte behavior within different types of collagen gel *in vitro*. *Biomaterials*, **16**, 809–814 (1995).
78. T. Yuan, L. Zhang, K. Li, H. Fan, Y. Fan, J. Liang, and X. Zhang, Collagen hydrogel as an immunomodulatory scaffold in cartilage tissue engineering. *J. Biomed. Mater. Res. Part B*, **102**(2), 337–344 (2014).
79. S. Ibusuki, G. J. Halbesma, M. A. Randolph, R. W. Redmond, I. E. Kochevar, and T. J. Gill, Photochemically cross-linked collagen gels as three-dimensional scaffolds for tissue engineering. *Tissue Eng.*, **13**(8), 1995–2001 (2007).
80. S. Yunoki, Y. Ohyabu, and H. Hatayama, Temperature-responsive gelation of type I collagen solutions involving fibril formation and genipin crosslinking as a potential injectable hydrogel. *Int. J. Biomater.*, 620765 (2013).
81. L. Zeng, Y. Yao, D.-A. Wang, and X. Chen, Effect of microcavitary alginate hydrogel with different pore sizes on chondrocyte culture for cartilage tissue engineering. *Mater. Sci. Eng. C*, **34**, 168–175 (2014).
82. C. Zhang, N. Sangaj, Y. Hwang, A. Phadke, C.-W. Chang, and S. Varghese, Oligo(trimethylene carbonate)-poly(ethylene glycol)-oligo(trimethylene carbonate) triblock-based hydrogels for cartilage tissue engineering. *Acta Biomater.*, **7**(9), 3362–3369 (2011).
83. Y. Guo, T. Yuan, Z. Xiao, P. Tang, Y. Xiao, Y. Fan, and X. Zhang, Hydrogels of collagen/chondroitin sulfate/hyaluronan interpenetrating polymer network for cartilage tissue engineering. *J. Mater. Sci.*, **23**(9), 2267–2279 (2012).
84. J. S. Pieper, A. Oosterhof, P. J. Dijkstra, J. H. Veerkamp, and T. H. van Kuppevelt, Preparation and characterization of porous crosslinked collagenous matrices containing bioavailable chondroitin sulphate. *Biomaterials*, **20**(9), 847–858 (1999).
85. W. Schuurman, P. A. Levett, M. W. Pot, P. R. van Weeren, W. J. A. Dhert, D. W. Hutmacher, *et al.*, Gelatin-methacrylamide hydrogels as potential biomaterials for fabrication of tissue-engineered cartilage constructs. *Macromol. Biosci.*, **13**(5), 551–561 (2013).
86. X. Hu, L. Ma, C. Wang, and C. Gao, Gelatin hydrogel prepared by photo-initiated polymerization and loaded with TGF-beta1 for cartilage tissue engineering. *Macromol. Biosci.*, **9**(12), 1194–1201 (2009).
87. P. A. Levett, F. P. W. Melchels, K. Schrobback, D. W. Hutmacher, J. Malda, and T. J. Klein, A biomimetic extracellular matrix for cartilage tissue engineering centered on photocurable gelatin, hyaluronic acid and chondroitin sulfate. *Acta Biomater.*, **10**(1), 214–223 (2014).

88. G. M. Peretti, J.-W. Xu, L. J. Bonassar, C. H. Kirchhoff, M. J. Yaremchuk, and M. A. Randolph, Review of injectable cartilage engineering using fibrin gel in mice and swine models. *Tissue Eng.*, **12**(5), 1151–1168 (2006).
89. G. N. Homminga, P. Buma, H. W. Koot, P. M. van der Kraan, and W. B. van den Berg, Chondrocyte behavior in fibrin glue *in vitro*. *Acta Orthopaed. Scand.*, **64**(4), 441–445 (1993).
90. D. Eyrich, F. Brandl, B. Appel, H. Wiese, G. Maier, M. Wenzel, *et al.*, Long-term stable fibrin gels for cartilage engineering. *Biomaterials*, **28**(1), 55–65 (2007).
91. C. Scotti, L. Mangiavini, F. Boschetti, F. Vitari, C. Domeneghini, G. Fraschini, and G. M. Peretti, Effect of *in vitro* culture on a chondrocyte-fibrin glue hydrogel for cartilage repair. *Knee Surg., Sport Traumatol. Arthrosc.*, **18**(10), 1400–1406 (2010).
92. M. Lind, A. Larsen, C. Clausen, K. Osther, and H. Everland, Cartilage repair with chondrocytes in fibrin hydrogel and MPEG polylactide scaffold: An *in vivo* study in goats. *Knee Surg. Sport Traumatol. Arthros.*, **16**(7), 690–698 (2008).
93. M. Santin, A. Motta, G. Freddi, and M. Cannas, *In vitro* evaluation of the inflammatory potential of the silk fibroin. *J. Biomed. Mater. Res.*, **46**, 382–389 (1999).
94. Y. Wang, U. J. Kim, D. J. Blasioli, H. J. Kim, and D. L. Kaplan, *In vitro* cartilage tissue engineering with 3D porous aqueous-derived silk scaffolds and mesenchymal stem cells. *Biomaterials*, **26**, 7082–7094 (2005).
95. H. Aoki, N. Tomita, Y. Morita, K. Hattori, Y. Harada, M. Sonobe *et al.*, Culture of chondrocytes in fibroin-hydrogel sponge. *Bio. Med. Mater. Eng.*, **13**(4), 309–316 (2003).
96. P.-H. G. Chao, S. Yodmuang, X. Wang, L. Sun, D. L. Kaplan, and G. Vunjak-Novakovic, Silk hydrogel for cartilage tissue engineering. *J. Biomed. Mater. Res. Pt. B*, **95**(1), 84–90 (2010).
97. B. Rahfoth, J. Weisser, F. Sternkopf, T. Aigner, K. von der Mark, and R. Bräuer, Transplantation of allograft chondrocytes embedded in agarose gel into cartilage defects of rabbits. *Osteoarthr. Cartilage*, **6**(1), 50–65 (1998).
98. A. H. Huang, M. Yeger-McKeever, A. Stein, and R. L. Mauck, Tensile properties of engineered cartilage formed from chondrocyte- and MSC-laden hydrogels. *Osteoarthr. Cartilage*, **16**(9), 1074–1082 (2008).
99. T. A. S. Selmi, P. Verdonk, P. Chambat, F. Dubrana, J.-F. Potel, L. Barnouin, and P. Neyret, Autologous chondrocyte implantation in a novel alginate-agarose hydrogel: Outcome at two years. *J. Bone Joint Surg. Br.*, **90**(5), 597–604 (2008).
100. S.-H. Lee, B.-S. Kim, S. H. Kim, S. W. Kang, and Y. H. Kim, Thermally produced biodegradable scaffolds for cartilage tissue engineering. *Macromol. Biosci.*, **4**(8), 802–810 (2004).
101. K. H. Bouhadir, K. Y. Lee, E. Alsberg, K. L. Damm, K. W. Anderson, and D. J. Mooney, Degradation of partially oxidized alginate and its potential application for tissue engineering. *Biotechnol. Progr.*, **17**(5), 945–950 (2001).
102. T. Boontheekul, H.-J. Kong, and D. J. Mooney, Controlling alginate gel degradation utilizing partial oxidation and bimodal molecular weight distribution. *Biomaterials*, **26**(15), 2455–2465 (2005).

103. J.-C. Chang, S.-H. Hsu, and D. C. Chen, The promotion of chondrogenesis in adipose-derived adult stem cells by an RGD-chimeric protein in 3D alginate culture. *Biomaterials*, **30**(31), 6265–6275 (2009).
104. J. T. Connelly, A. J. García, and M. E. Levenston, Inhibition of *in vitro* chondrogenesis in RGD-modified three-dimensional alginate gels. *Biomaterials*, **28**(6), 1071–1083 (2007).
105. O. Jeon, K. H. Bouhadir, J. M. Mansour, and E. Alsberg, Photocrosslinked alginate hydrogels with tunable biodegradation rates and mechanical properties. *Biomaterials*, **30**, 2724–2734 (2009).
106. Y. Yamada, K. Hozumi, F. Katagiri, Y. Kikkawa, and M. Nomizu, Biological activity of laminin peptide-conjugated alginate and chitosan matrices. *Biopolymers*, **94**(6), 711–720 (2010).
107. Y. Yamada, K. Hozumi, A. Aso, A. Hotta, K. Toma, F. Katagiri, *et al.*, Laminin active peptide/agarose matrices as multifunctional biomaterials for tissue engineering. *Biomaterials*, **33**(16), 4118–4125 (2012).
108. K. W. Ng, G. A. Ateshian, and C. T. Hung, Zonal chondrocytes seeded in a layered agarose hydrogel create engineered cartilage with depth-dependent cellular and mechanical inhomogeneity. *Tissue Eng. Part A*, **15**(9), 2315–2324 (2009).
109. C. Migliaresi, A. Motta, and A. Di Benedetto, Injectable scaffolds for bone and cartilage regeneration. In Eds. F. Bronner, M. C. Farach-Carson, and A. G. Mikos, *Engineering of functional Skeletal Tissues*. Berlin: Springer, pp. 95–109 (2007).
110. L. S. Nair and C. T. Laurencin, Biodegradable polymers as biomaterials. *Progr. Polym. Sci.*, **32**(8–9), 762–798 (2007).
111. A. Lahiji, A. Sohrabi, D. S. Hungerford, and C. G. Frondoza, Chitosan supports the expression of extracellular matrix proteins in human osteoblasts and chondrocytes. *J. Biomed. Mater. Res.*, **51**(4), 586–595 (2000).
112. T. Hao, N. Wen, J.-K. Cao, H.-B. Wang, S.-H. Lü, T. Liu, *et al.*, The support of matrix accumulation and the promotion of sheep articular cartilage defects repair *in vivo* by chitosan hydrogels. *Osteoarthr. Cartilage*, **18**(2), 257–265 (2010).
113. K. L. Spiller, J. L. Holloway, M. E. Gribb, and A. M. Lowman, Design of semi-degradable hydrogels based on poly(vinyl alcohol) and poly(lactic-co-glycolic acid) for cartilage tissue engineering. *J. Tissue Eng. Regen. Med.*, **5**(8), 636–647 (2011).
114. F. Ganji, M. J. Abdekhodaie, and S. A. Ramazani, Gelation time and degradation rate of chitosan-based injectable hydrogel. *J. Sol-Gel Sci. Tech.*, **42**(1), 47–53 (2007).
115. F. Mirahmadi, M. Tafazzoli-Shadpour, M. A. Shokrgozar, and S. Bonakdar, Enhanced mechanical properties of thermosensitive chitosan hydrogel by silk fibers for cartilage tissue engineering. *Mater. Sci. Eng. C*, **33**(8), 4786–4794 (2013).
116. R. Jin, L. S. Moreira Teixeira, P. J. Dijkstra, M. Karperien, C. A. van Blitterswijk, Z. Y. Zhong, and J. Feijen, Injectable chitosan-based hydrogels for cartilage tissue engineering. *Biomaterials*, **30**(13), 2544–2551 (2009).
117. J. K. Suh and H. W. Matthew, Application of chitosan-based polysaccharide biomaterials in cartilage tissue engineering: A review. *Biomaterials*, **21**(24), 2589–2598 (2000).

118. V. F. Sechriest, Y. J. Miao, C. Niyibizi, A. Westerhausen-Larson, H. W. Matthew, C. H. Evans, *et al.*, GAG-augmented polysaccharide hydrogel: A novel biocompatible and biodegradable material to support chondrogenesis. *J. Biomed. Mater. Res.*, **49**(4), 534–541 (2000).
119. H. Park, B. Choi, J. Hu, and M. Lee, Injectable chitosan hyaluronic acid hydrogels for cartilage tissue engineering. *Acta Biomater.*, **9**(1), 4779–4786 (2013).
120. R. A. A. Muzzarelli, F. Greco, A. Busilacchi, V. Sollazzo, and A. Gigante, Chitosan, hyaluronan and chondroitin sulfate in tissue engineering for cartilage regeneration: A review. *Carbohyd. Polym.*, **89**(3), 723–739 (2012).
121. A. M. Patti, A. Gabriele, A. Vulcano, M. T. Ramieri, and C. Della Rocca, Effect of hyaluronic acid on human chondrocyte cell lines from articular cartilage. *Tissue Cell*, **33**(3), 294–300 (2001).
122. M. Akmal, A. Singh, A. Anand, A. Kesani, N. Aslam, A. Goodship, and G. Bentley, The effects of hyaluronic acid on articular chondrocytes. *J. Bone Joint Surg. Br.*, **87**(8), 1143–1149 (2005).
123. E. M. Ehlers, P. Behrens, L. Wünsch, W. Kühnel, and M. Russlies, Effects of hyaluronic acid on the morphology and proliferation of human chondrocytes in primary cell culture. *Ann. Anat.*, **183**(1), 13–17 (2001).
124. D. J. Responte, R. M. Natoli, and K. A. Athanasiou, Identification of potential biophysical and molecular signalling mechanisms underlying hyaluronic acid enhancement of cartilage formation. *J. R. Soc. Interf.*, **9**(77), 3564–3573 (2012).
125. R. Barbucci, S. Lamponi, A. Borzacchiello, L. Ambrosio, M. Fini, P. Torricelli, and R. Giardino, Hyaluronic acid hydrogel in the treatment of osteoarthritis. *Biomaterials*, **23**(23), 4503–4513 (2002).
126. D. Campoccia, P. Doherty, M. Radice, P. Brun, G. Abatangelo, and D. F. Williams, Semisynthetic resorbable materials from hyaluronan esterification. *Biomaterials*, **19**(23), 2101–2127 (1998).
127. I. L. Kim, R. L. Mauck, and J. A. Burdick, Hydrogel design for cartilage tissue engineering: A case study with hyaluronic acid. *Biomaterials*, **32**(34), 8771–8782 (2011).
128. S. Sahoo, C. Chung, S. Khetan, and J. A. Burdick, Hydrolytically degradable hyaluronic acid hydrogels with controlled temporal structures. *Biomacromolecules*, **9**(4), 1088–1092 (2008).
129. C. Chung, M. Beecham, R. L. Mauck, and J. A. Burdick, The influence of degradation characteristics of hyaluronic acid hydrogels on *in vitro* neocartilage formation by mesenchymal stem cells. *Biomaterials*, **30**(26), 4287–4296 (2009).
130. I. E. Erickson, A. H. Huang, S. Sengupta, S. Kestle, J. A. Burdick, and R. L. Mauck, Macromer density influences mesenchymal stem cell chondrogenesis and maturation in photocrosslinked hyaluronic acid hydrogels. *Osteoarthr. Cartilage*, **17**(12), 1639–1648 (2009).
131. L. Bian, D. Y. Zhai, R. L. Mauck, and J. A. Burdick, Coculture of human mesenchymal stem cells and articular chondrocytes reduces hypertrophy and enhances functional properties of engineered cartilage. *Tissue Eng. Pt. A*, **17**(7–8), 1137–1145 (2011).

132. L. Zhao, H.-J. Gwon, Y.-M. Lim, Y.-C. Nho, and S. Y. Kim, Hyaluronic acid/chondroitin sulfate-based hydrogel prepared by gamma irradiation technique. *Carbohyd. Polym.*, **102**, 598–605 (2014).
133. L.-S. Kontturi, E. Järvinen, V. Muhonen, E. C. Collin, A. S. Pandit, I. Kiviranta, *et al.*, An injectable, *in situ* forming type II collagen/hyaluronic acid hydrogel vehicle for chondrocyte delivery in cartilage tissue engineering. *Drug Deliv. Transl. Res.*, (2014).
134. N. S. Remya and P. D. Nair, Engineering cartilage tissue interfaces using a natural glycosaminoglycan hydrogel matrix — An *in vitro* study. *Mater. Sci. Eng. C*, **33**(2), 575–582 (2013).
135. H. Park and K. Y. Lee, Facile control of RGD-alginate/hyaluronate hydrogel formation for cartilage regeneration. *Carbohyd. Polym.*, **86**(3), 1107–1112 (2011).
136. S. Dinescu, B. Gălăţeanu, M. Albu, A. Lungu, E. Radu, A. Hermenean, and M. Costache, Biocompatibility assessment of novel collagen-sericin scaffolds improved with hyaluronic Acid and chondroitin sulfate for cartilage regeneration. *BioMed. Res. Int.*, 598056 (2013).
137. B. R. Mintz and J. A. Cooper, Hybrid hyaluronic acid hydrogel/poly(ε-caprolactone) scaffold provides mechanically favorable platform for cartilage tissue engineering studies. *J. Biomed. Mater. Res. Pt. A*, 1–9 (2013).
138. H. Tan, C. R. Chu, K. A. Payne, and K. G. Marra, Injectable *in situ* forming biodegradable chitosan-hyaluronic acid based hydrogels for cartilage tissue engineering. *Biomaterials*, **30**(13), 2499–2506 (2009).
139. E. Liao, M. Yaszemski, P. Krebsbach, and S. Hollister, Tissue-engineered cartilage constructs using composite hyaluronic acid/collagen I hydrogels and designed poly(propylene fumarate) scaffolds. *Tissue Eng.*, **13**(3), 537–550 (2007).
140. K. Muramatsu, M. Ide, and F. Miyawaki, Biological evaluation of tissue-engineered cartilage using thermoresponsive poly (N-isopropylacrylamide)-grafted hyaluronan, 1–9 (2012).
141. J. H. Hui, S.-W. Chan, J. Li, J. C. H. Goh, L. Li, X. F. Ren, and E. H. Lee, Intra-articular delivery of chondroitin sulfate for the treatment of joint defects in rabbit model. *J. Mol. Histol.*, **38**(5), 483–489 (2007).
142. C. Vinatier, O. Gauthier, A. Fatimi, C. Merceron, M. Masson, *et al.*, An injectable cellulose-based hydrogel for the transfer of autologous nasal chondrocytes in articular cartilage defects. *Biotechnol. Bioeng.*, **102**(4), 1259–1267 (2009).
143. C. Vinatier, D. Magne, P. Weiss, C. Trojani, N. Rochet, G. F. Carle, *et al.*, A silanized hydroxypropyl methylcellulose hydrogel for the three-dimensional culture of chondrocytes. *Biomaterials*, **26**(33), 6643–6651 (2005).
144. G. Leone, R. Barbucci, A. Borzacchiello, L. Ambrosio, P. A. Netti, and C. Migliaresi, Preparation and physico-chemical characterisation of microporous polysaccharidic hydrogels. *J. Mater. Sci.*, **15**, 463–467 (2004).
145. J. T. Oliveira, L. Martins, R. Picciochi, P. B. Malafaya, R. A. Sousa, N. M. Neves, *et al.*, Gellan gum: A new biomaterial for cartilage tissue engineering applications. *J. Biomed. Mater. Res. Pt. A*, **93**(3), 852–863 (2010).

146. J. Fan, Y. Gong, L. Ren, R. R. Varshney, D. Cai, and D.-A. Wang, *In vitro* engineered cartilage using synovium-derived mesenchymal stem cells with injectable gellan hydrogels. *Acta Biomater.*, **6**(3), 1178–1185 (2010).
147. E. G. Popa, P. P. Carvalho, A. F. Dias, T. C. Santos, V. E. Santo, A. P. Marques, *et al.*, Evaluation of the *in vitro* and *in vivo* biocompatibility of carrageenan-based hydrogels. *J. Biomed. Mater. Res. Pt. A*, **2009**, 1–11 (2014).
148. J. C. Bray and E. W. Merrill, Poly(vinyl alcohol) hydrogels for synthetic articular cartilage material. *J. Biomed. Mater. Res.*, **7**(5), 431–443 (1973).
149. B.R. Rawal, R. Ribeiro, M. Chouksey, *et al.*, Biomaterials for cartilage repair: A review. *J. Med. Sci.*, **13**(8), 615–620 (2013).
150. M. I. Baker, S. P. Walsh, Z. Schwartz, and B. D. Boyan, A review of polyvinyl alcohol and its uses in cartilage and orthopedic applications. *J. Biomed. Mater. Res. Part B*, **100**(5), 1451–1457 (2012).
151. J. L. Holloway, K. L. Spiller, A. M. Lowman, and G. R. Palmese, Analysis of the *in vitro* swelling behavior of poly(vinyl alcohol) hydrogels in osmotic pressure solution for soft tissue replacement. *Acta Biomater.*, **7**(6), 2477–2482 (2011).
152. J. Beyerlein and A. B. Imhoff, SaluCartilage??ein neuer künstlicher Knorpelersatz für die arthroskopische Behandlung der fokalen Osteonekrose. *Arthroskopie*, **16**(1), 34–39 (2003).
153. K. Tamura, O. Ike, S. Hitomi, J. Isobe, Y. Shimizu, and M. Nambu, A new hydrogel and its medical application. *ASAIO Trans.*, **32**(1), 605–608 (1986).
154. M. Kobayashi, Y.-S. Chang, and M. Oka, A two year *in vivo* study of polyvinyl alcohol-hydrogel (PVA-H) artificial meniscus. *Biomaterials*, **26**(16), 3243–3248 (2005).
155. Y. Cao, D. Xiong, Y. Niu, Y. Mei, Z. Yin, and J. Gui, Compressive properties and creep resistance of a novel, porous, semidegradable poly(vinyl alcohol)/poly(lactic-co-glycolic acid) scaffold for articular cartilage repair. *J. Appli. Polym. Sci.*, **131**(11), n/a–n/a (2014).
156. J. S. Gonzalez and V. A. Alvarez, Mechanical properties of polyvinylalcohol/hydroxyapatite cryogel as potential artificial cartilage. *J. Mech. Behav. Biomed. Mater.*, **34C**, 47–56 (2014).
157. A. S. Maiolo, M. N. Amado, J. S. Gonzalez, and V. A. Alvarez, Development and characterization of Poly (vinyl alcohol) based hydrogels for potential use as an articular cartilage replacement. *Mater. Sci. Eng. C*, **32**(6), 1490–1495 (2012).
158. D. A. Grande, C. Halberstadt, G. Naughton, R. Schwartz, and R. Manji, Evaluation of matrix scaffolds for tissue engineering of articular cartilage grafts. *J. Biomed. Mater. Res.*, **34**(2), 211–220 (1997).
159. Y. Zhang, F. Yang, K. Liu, H. Shen, Y. Zhu, W. Zhang, *et al.*, The impact of PLGA scaffold orientation on *in vitro* cartilage regeneration. *Biomaterials*, **33**, 2926–2935 (2012).
160. K. Y. Lee, K. H. Bouhadir, and D. J. Mooney, Controlled degradation of hydrogels using multi-functional cross-linking molecules. *Biomaterials*, **25**(13), 2461–2466 (2004).
161. L. E. Freed, J. C. Marquis, A. Nohria, J. Emmanual, A. G. Mikos, and R. Langer, Neocartilage formation *in vitro* and *in vivo* using cells cultured on synthetic biodegradable polymers. *J. Biomed. Mater. Res.*, **27**(1), 11–23 (1993).

162. J. Elisseeff, News and views, **7**(April), 271–273 (2008). Retrieved from file:///C:/Users/mariangela.fedel/AppData/Local/Mendeley Ltd./Mendeley Desktop/Downloaded/Elisseeff-2008-News & views.pdf.
163. Y. Park, M. P. Lutolf, J. A. Hubbell, E. B. Hunziker, and M. Wong, Bovine primary chondrocyte culture in synthetic matrix metalloproteinase-sensitive poly(ethylene glycol)-based hydrogels as a scaffold for cartilage repair. *Tissue Eng.*, **10**(3–4), 515–522 (2004).
164. Q. T. Nguyen, Y. Hwang, A. C. Chen, S. Varghese, and R. L. Sah, Cartilage-like mechanical properties of poly (ethylene glycol)-diacrylate hydrogels. *Biomaterials*, **33**(28), 6682–6690 (2012).
165. G. D. Nicodemus, S. C. Skaalure, and S. J. Bryant, Gel structure has an impact on pericellular and extracellular matrix deposition, which subsequently alters metabolic activities in chondrocyte-laden PEG hydrogels. *Acta Biomater.*, **7**(2), 492–504 (2011).
166. S. J. Bryant and K. S. Anseth, Controlling the spatial distribution of ECM components in degradable PEG hydrogels for tissue engineering cartilage. *J. Biomed. Mater. Res. Pt. A*, **64**(1), 70–79 (2003).
167. S. J. Bryant and K. S. Anseth, Hydrogel properties influence ECM production by chondrocytes photoencapsulated in poly(ethylene glycol) hydrogels. *J. Biomed. Mater. Res.*, **59**(1), 63–72 (2002).
168. S. J. Bryant and K. S. Anseth, The effects of scaffold thickness on tissue engineered cartilage in photocrosslinked poly (ethylene oxide) hydrogels. *Biomaterials*, **22**, 619–626 (2001).
169. A. Rakovsky, D. Marbach, N. Lotan, and Y. Lanir, Poly (ethylene glycol)-Based hydrogels as cartilage substitutes: Synthesis and mechanical characteristics. *J. Appl. Pol. Sci.*, (2008).
170. L. J. Suggs, M. S. Shive, C. A. Garcia, J. M. Anderson, and A. G. Mikos, *In vitro* cytotoxicity and *in vivo* biocompatibility of poly(propylene fumarate-co-ethylene glycol) hydrogels. *J. Biomed. Mater. Res.*, **46**(1), 22–32 (1999).
171. S. He, M. J. Yaszemski, A. W. Yasko, P. S. Engel, and A. G. Mikos, Injectable biodegradable polymer composites based on poly(propylene fumarate) crosslinked with poly(ethylene glycol)-dimethacrylate. *Biomaterials*, **21**(23), 2389–2394 (2000).
172. H. Shin, P. Quinten Ruhé, A. G. Mikos, and J. A. Jansen, *In vivo* bone and soft tissue response to injectable, biodegradable oligo(poly(ethylene glycol) fumarate) hydrogels. *Biomaterials*, **24**(19), 3201–3211 (2003).
173. S.-H. Lee and H. Shin, Matrices and scaffolds for delivery of bioactive molecules in bone and cartilage tissue engineering. *Adv. Drug Deliv. Rev.*, **59**(4–5), 339–359 (2007).
174. J. P. Fisher, S. Jo, A. G. Mikos, and A. H. Reddi, Thermoreversible hydrogel scaffolds for articular cartilage engineering. *J. Biomed. Mater. Res. Pt. A*, **71**(2), 268–274 (2004).
175. M. Dadsetan, J. P. Szatkowski, M. J. Yaszemski, and L. Lu, Characterization of photo-cross-linked oligo[poly(ethylene glycol) fumarate] hydrogels for cartilage tissue engineering. *Biomacromolecules*, **8**(5), 1702–1709 (2007).
176. T. A. Holland, Y. Tabata, and A. G. Mikos, Dual growth factor delivery from degradable oligo(poly(ethylene glycol) fumarate) hydrogel scaffolds for cartilage tissue engineering. *J. Contr. Rel.*, **101**(1–3), 111–125 (2005).

177. Y. Chen, K. Dong, Z. Liu, and F. Xu, Double network hydrogel with high mechanical strength: Performance, progress and future perspective. *Sci. China Technol. Sci.*, **55**(8), 2241–2254 (2012).
178. M. Ogawa, N. Kitamura, T. Kurokawa, K. Arakaki, Y. Tanaka, J. P. Gong, and K. Yasuda, Poly(2-acrylamido-2-methylpropanesulfonic acid) gel induces articular cartilage regeneration *in vivo*: Comparisons of the induction ability between single- and double-network gels. *J. Biomed. Mater. Res. Part A*, **100**(9), 2244–2251 (2012).
179. H. Shin, B. D. Olsen, and A. Khademhosseini, The mechanical properties and cytotoxicity of cell-laden double-network hydrogels based on photocrosslinkable gelatin and gellan gum biomacromolecules. *Biomaterials*, **33**(11), 3143–3152 (2012).
180. A. Chenite, C. Chaput, D. Wang, C. Combes, M. D. Buschmann, C. D. Hoemann *et al.*, Novel injectable neutral solutions of chitosan form biodegradable gels *in situ*. *Biomaterials*, **21**(21), 2155–2161 (2000).
181. C. R. Lee, S. Grad, K. Gorna, S. Gogolewski, A. Goessl, and M. Alini, Fibrin-polyurethane composites for articular cartilage tissue engineering: A preliminary analysis. *Tissue Eng.*, **11**(9–10), 1562–1573 (2005).
182. Y. Zheng, H. Lv, Y. Wang, H. Lu, L. Qing, and T. Xi, Performance of novel bioactive hybrid hydrogels *in vitro* and *in vivo* used for artificial cartilage. *Biomed. Mater.*, **4**(1), 15015 (2009).
183. D. A. Rennerfeldt, A. N. Renth, Z. Talata, S. H. Gehrke, and M. S. Detamore, Tuning mechanical performance of poly(ethylene glycol) and agarose interpenetrating network hydrogels for cartilage tissue engineering. *Biomaterials*, **34**(33), 8241–8257 (2013).
184. L. A. S. Callahan, A. M. Ganios, D. L. McBurney, M. F. Dilisio, S. D. Weiner, W. E. Horton, and M. L. Becker, ECM production of primary human and bovine chondrocytes in hybrid PEG hydrogels containing type I collagen and hyaluronic acid. *Biomacromolecules*, **13**(5), 1625–1631 (2012).
185. T. J. Klein, S. C. Rizzi, J. C. Reichert, N. Georgi, J. Malda, W. Schuurman, *et al.*, Strategies for zonal cartilage repair using hydrogels. *Macromol. Biosci.*, **9**(11), 1049–1058 (2009).
186. T. J. Klein, B. L. Schumacher, T. A. Schmidt, K. W. Li, M. S. Voegtline, *et al.*, Tissue engineering of stratified articular cartilage from chondrocyte subpopulations. *Osteoarthr. Cartilage*, **11**(8), 595–602 (2003).
187. T.-K. Kim, B. Sharma, C. G. Williams, M. Ruffner, *et al.*, Experimental model for cartilage tissue engineering to regenerate the zonal organization of articular cartilage. *Osteoarthr. Cartilage*, **11**(9), 653–664 (2003).
188. J. J. Roberts, A. Earnshaw, V. L. Ferguson, and S. J. Bryant, Comparative study of the viscoelastic mechanical behavior of agarose and poly(ethylene glycol) hydrogels. *J. Biomed. Mater. Res. Part B*, **99**(1), 158–169 (2011).
189. J. K. Mouw, N. D. Case, R. E. Guldberg, A. H. K. Plaas, and M. E. Levenston, Variations in matrix composition and GAG fine structure among scaffolds for cartilage tissue engineering. *Osteoarthr. Cartilage*, **13**(9), 828–836 (2005).
190. H. A. Awad, M. Q. Wickham, H. A. Leddy, J. M. Gimble, and F. Guilak, Chondrogenic differentiation of adipose-derived adult stem cells in agarose, alginate, and gelatin scaffolds. *Biomaterials*, **25**(16), 3211–3222 (2004).

191. M. Sittinger, D. W. Hutmacher, and M. V. Risbud, Current strategies for cell delivery in cartilage and bone regeneration. *Curr. Opin. Biotechnol.*, (2004).
192. J. Raghunath, H. J. Salacinski, K. M. Sales, P. E. Butler, and A. M. Seifalian, Advancing cartilage tissue engineering: The application of stem cell technology. *Curr. Opin. Biotechnol.*, (2005).
193. J. Kopeček, Hydrogel biomaterials: A smart future? *Biomaterials*, **28**, 5185–5192 (2007).
194. L. A. Fortier, J. U. Barker, E. J. Strauss, T. M. McCarrel, and B. J. Cole, The role of growth factors in cartilage repair. *Clin. Orthopaed. Rel. Res.*, **469**(10), 2706–2715 (2011).
195. H.-R. Lee, K. M. Park, Y. K. Joung, K. D. Park, and S. H. Do, Platelet-rich plasma loaded *in situ*-formed hydrogel enhances hyaline cartilage regeneration by CB1 upregulation. *J. Biomed. Mater. Res. Part A*, **100**(11), 3099–3107 (2012).
196. J. J. Roberts, G. D. Nicodemus, S. Giunta, and S. J. Bryant, Incorporation of biomimetic matrix molecules in PEG hydrogels enhances matrix deposition and reduces load-induced loss of chondrocyte-secreted matrix. *J. Biomed. Mater. Res. Pt. A*, **97**(3), 281–291 (2011).
197. A. R. Haas and R. S. Tuan, Chondrogenic differentiation of murine C3H10T1/2 multipotential mesenchymal cells: II. Stimulation by bone morphogenetic protein-2 requires modulation of N-cadherin expression and function. *Differentiation*, **64**, 77–89 (1999).
198. I. C. Bonzani, J. H. George, and M. M. Stevens, Novel materials for bone and cartilage regeneration. *Curr. Opin. Chem. Biol.* (2006).
199. P. M. Rocha, V. E. Santo, M. E. Gomes, R. L. Reis, and J. F. Mano, Encapsulation of adipose-derived stem cells and transforming growth factor-β1 in carrageenan-based hydrogels for cartilage tissue engineering. *J. Bioact. Compat. Polym.*, (2011).
200. A. J. DeFail, C. R. Chu, N. Izzo, and K. G. Marra, Controlled release of bioactive TGF-beta 1 from microspheres embedded within biodegradable hydrogels. *Biomaterials*, **27**(8), 1579–1585 (2006).
201. H. H. Jung, K. Park, and D. K. Han, Preparation of TGF-β1-conjugated biodegradable pluronic F127 hydrogel and its application with adipose-derived stem cells. *J. Contr. Rel.*, **147**(1), 84–91 (2010).
202. R. Jin, L. S. Moreira Teixeira, P. J. Dijkstra, C. A. van Blitterswijk, M. Karperien, and J. Feijen, Chondrogenesis in injectable enzymatically crosslinked heparin/dextran hydrogels. *J. Contr. Rel.*, **152**(1), 186–195 (2011).
203. S. M. Lim, S. H. Oh, H. H. Lee, S. H. Yuk, G. Il. Im, and J. H. Lee, Dual growth factor-releasing nanoparticle/hydrogel system for cartilage tissue engineering. *J. Mater. Sci.*, **21**(9), 2593–2600 (2010).
204. K. L. Spiller, Y. Liu, J. L. Holloway, S. A. Maher, Y. Cao, W. Liu, *et al.*, A novel method for the direct fabrication of growth factor-loaded microspheres within porous nondegradable hydrogels: Controlled release for cartilage tissue engineering. *J. Contro. Rel.*, **157**(1), 39–45 (2012).
205. L. A. Fortier, C. H. Hackett, and B. J. Cole, The effects of platelet-rich plasma on cartilage: Basic science and clinical application. *Oper. Tech. Sport Med.*, **19**(3), 154–159 (2011).

206. J. A. Textor, N. H. Willits, and F. Tablin, Synovial fluid growth factor and cytokine concentrations after intra-articular injection of a platelet-rich product in horses. *Vet. J.*, **198**(1), 217–223 (2013).
207. K. Akeda, H. S. An, M. Okuma, M. Attawia, K. Miyamoto, E. J.-M. Thonar, *et al.*, Platelet-rich plasma stimulates porcine articular chondrocyte proliferation and matrix biosynthesis. *Osteoarthr. Cartilage*, **14**(12), 1272–1280 (2006).
208. H.-R. Lee, K. M. Park, Y. K. Joung, K. D. Park, and S. H. Do, Platelet-rich plasma loaded hydrogel scaffold enhances chondrogenic differentiation and maturation with up-regulation of CB1 and CB2. *J. Contr. Rel.*, **159**(3), 332–337 (2012).
209. N. Tsuzuki, N. Oshita, J. Seo, K. Yamada, S. Haneda, H. Furuoka, *et al.*, Effect of platelet-rich plasma-incorporated gelatin hydrogel microspheres and subchondral drilling on equine cartilage defects. *J. Equine Vet. Sci.*, (2014).
210. N. C. Hunt and L. M. Grover, Cell encapsulation using biopolymer gels for regenerative medicine. *Biotechnol. Lett.*, **32**(6), 733–742 (2010).
211. K. Chawla, T. Bin Yu, L. Stutts, M. Yen, and Z. Guan, Modulation of chondrocyte behavior through tailoring functional synthetic saccharide-peptide hydrogels. *Biomaterials*, **33**, 6052–6060 (2012).
212. D. Nesic, R. Whiteside, M. Brittberg, D. Wendt, I. Martin, and P. Mainil-Varlet, Cartilage tissue engineering for degenerative joint disease. *Adv. Drug Deliv. Rev.*, (2006).
213. I. Martin, S. Miot, A. Barbero, M. Jakob, and D. Wendt, Osteochondral tissue engineering. *J. Biomech.* (2007).
214. E. C. Rodriguez-Merchan, Regeneration of articular cartilage of the knee. *Rheumatol. Int.*, **33**(4), 837–845 (2013).
215. S. Wakitani, T. Mitsuoka, N. Nakamura, Y. Toritsuka, Y. Nakamura, and S. Horibe, Autologous bone marrow stromal cell transplantation for repair of full-thickness articular cartilage defects in human patellae: Two case reports. *Cell Transplant.*, **13**, 595–600 (2004).
216. J. Ringe, I. Leinhase, S. Stich, A. Loch, K. Neumann, and A. Haisch, Human mastoid periosteum-derived stem cells: Promising candidates for skeletal tissue engineering, 136–146 (2008).
217. W. Leong, T. T. Lau, and D.-A. Wang, A temperature-cured dissolvable gelatin microsphere-based cell carrier for chondrocyte delivery in a hydrogel scaffolding system. *Acta Biomater.*, **9**(5), 6459–6467 (2013).
218. M. Kim, S. E. Kim, S. S. Kang, Y. H. Kim, and G. Tae, The use of de-differentiated chondrocytes delivered by a heparin-based hydrogel to regenerate cartilage in partial-thickness defects. *Biomaterials*, **32**(31), 7883–7896 (2011).
219. J. J. Roberts, G. D. Nicodemus, E. C. Greenwald, and S. J. Bryant, Degradation improves tissue formation in (un)loaded chondrocyte-laden hydrogels. *Clin. Orthopaed. Rel. Res.*, **469**(10), 2725–2734 (2011).
220. W. S. Toh, T. C. Lim, M. Kurisawa, and M. Spector, Modulation of mesenchymal stem cell chondrogenesis in a tunable hyaluronic acid hydrogel microenvironment. *Biomaterials*, **33**(15), 3835–3845 (2012).

221. Y. Y. Li, H. W. Cheng, K. M. C. Cheung, D. Chan, and B. P. Chan, Mesenchymal stem cell-collagen microspheres for articular cartilage repair: Cell density and differentiation status. *Acta Biomater.*, 1–11 (2014).
222. J. T. Oliveira, T. C. Santos, L. Martins, R. Picciochi, A. P. Marques, A. G. Castro, *et al.*, Gellan gum injectable hydrogels for cartilage tissue engineering applications: *In vitro* studies and preliminary *in vivo* evaluation. *Tissue Eng. Part A*, **16**(1), 343–353 (2010).
223. X. Chen, F. Zhang, X. He, Y. Xu, Z. Yang, L. Chen, *et al.*, Chondrogenic differentiation of umbilical cord-derived mesenchymal stem cells in type I collagen-hydrogel for cartilage engineering. *Injury*, **44**(4), 540–549 (2013).
224. L. H. Nguyen, A. K. Kudva, N. L. Guckert, K. D. Linse, and K. Roy, Unique biomaterial compositions direct bone marrow stem cells into specific chondrocytic phenotypes corresponding to the various zones of articular cartilage. *Biomaterials*, **32**(5), 1327–1338 (2011).
225. K. Su, T. T. Lau, W. Leong, Y. Gong, and D.-A. Wang, Creating a living hyaline cartilage graft free from non-cartilaginous constituents: An intermediate role of a biomaterial scaffold. *Adv. Funct. Mater.*, **22**(5), 972–978 (2012).
226. K. H. Park and K. Na, Effect of growth factors on chondrogenic differentiation of rabbit mesenchymal cells embedded in injectable hydrogels. *J. Biosci. Bioeng.*, **106**, 74–79 (2008).
227. P. B. Malafaya, G. A. Silva, and R. L. Reis, Natural-origin polymers as carriers and scaffolds for biomolecules and cell delivery in tissue engineering applications. *Adv. Drug Deliv. Rev.*, **59**, 207–233 (2007).
228. J. M. Dang and K. W. Leong, Natural polymers for gene delivery and tissue engineering. *Adv. Drug Deliv. Rev.*, **58(4)**, 487–499 (2006).
229. J. S. Choi, B. S. Kim, J. D. Kim, Y. C. Choi, H. Y. Lee, and Y. W. Cho, *In vitro* cartilage tissue engineering using adipose-derived extracellular matrix scaffolds seeded with adipose-derived stem cells. *Tissue Eng. Pt. A*, **18**, 80–92 (2012).
230. M. Ahearne and D. J. Kelly, A comparison of fibrin, agarose and gellan gum hydrogels as carriers of stem cells and growth factor delivery microspheres for cartilage regeneration. *Biomed. Mater* **8**(3), 035004 (2013).
231. J. S. Park, H. N. Yang, D. G. Woo, H.-M. Chung, and K.-H. Park, *In vitro* and *in vivo* chondrogenesis of rabbit bone marrow-derived stromal cells in fibrin matrix mixed with growth factor loaded in nanoparticles. *Tissue Eng. Pt. A*, **15**(8), 2163–2175 (2009).
232. Z. Izadifar, X. Chen, and W. Kulyk, Strategic design and fabrication of engineered scaffolds for articular cartilage repair. *J. Funct. Biomater.*, **3**(4), 799–838 (2012).
233. G. D. Nicodemus and S. J. Bryant, The role of hydrogel structure and dynamic loading on chondrocyte gene expression and matrix formation. *J. Biomechanics*, **41**(7), 1528–1536 (2008).

Chapter 9

Application of Hydrogels for Tendon and Ligament Repair and Tissue Engineering

Matteo Stoppato,* Friedrich von Flotow* and Catherine K. Kuo*,†,‡
**Department of Biomedical Engineering, Tufts University*
4 Colby Street, Medford, MA 02155, USA
†Cell, Molecular and Developmental Biology, Sackler School of Graduate Biomedical Sciences, Tufts University School of Medicine
136 Harrison Ave., Boston, MA 02111, USA
‡CatherineK.Kuo@tufts.edu

Tendons and ligaments are musculoskeletal tissues necessary for transferring forces experienced by the surrounding tissues. They are critical for locomotion and joint stabilization, and injury to these tissues is both common and debilitating. The current gold standard of repair is surgical grafting followed by physiotherapy. However, tendon and ligament healing generates scar tissue, which is mechanically inferior to the normal tissue. Tendon and ligament repair therefore has great need for tissue engineering solutions. An engineered tissue would restore normal tissue properties and function as a tissue replacement, or deliver therapeutic agents to enhance natural healing. To this end, a variety of scaffold materials that support or deliver various cells and/or biomolecules have been investigated, with hydrogel scaffolds offering a promising option. Hydrogels have a range of material and biological properties that can be tuned to mimic native tissue characteristics, giving them a unique advantage for tendon/ligament tissue engineering applications. This chapter will focus on important tendon and ligament characteristics and the use of hydrogels in tendon/ligament tissue engineering.

1. Introduction

Tendons and ligaments (T/L) are important tissues with unique mechanical properties necessary for transferring and modulating forces, allowing for skeletal movement and providing joint stability. Injuries to T/L tissue are common and debilitating, with approximately 32 million musculoskeletal injuries occurring annually in the USA. Of these, around 14.4 million correspond to injuries of the tendon, ligament and joint capsules.[1] T/L injuries disrupt the tissue structure and usually occur through either a chronic, degenerative pathway or as an acute rupture event, often with the former preceding the latter. With few exceptions, fully ruptured T/L is unable to heal, and requires surgical intervention to restore attachment. Adult T/L healing of partially ruptured or surgically repaired T/L tissue is a long process, and results in scar tissue that fails to approximate the microstructure, composition or mechanical properties of normal tissue. Current repair strategies rely on graft replacement, followed by periods of immobilization and rehabilitation. Autologous grafts result in a second injury site and extended healing time for the patient, while allogeneic cadaveric grafts carry the risks of immune rejection and disease transmission. Finally, in the best-case scenario the outcome is an inadequate scar tissue with reduced tensile strength. Because of these and other significant disadvantages to graft repair, tissue engineering (TE) holds great therapeutic potential.

A TE approach to T/L therapy may follow two general strategies, both aimed at stimulating generation of either tendon (tenogenesis) or ligament (ligamentogenesis). One strategy involves the replacement of the injured T/L with an engineered tissue that has been cultured *in vitro* to provide function immediately upon implantation, while the other strategy aims to deliver cells or biological factors to enhance healing *in vivo*. Both approaches can utilize scaffold materials to provide a suitable local environment that elicits appropriate tenogenic or ligamentogenic biological responses from either the cells seeded in the scaffold *in vitro* or those that migrate in from the surrounding tissue *in vivo*. Scaffold selection must consider mechanical, structural, chemical, and biological properties for the target tissue. To date, many TE approaches use fibrous scaffolds in an attempt to approximate the mechanical strength and/or the extracellular matrix (ECM) structure of T/L tissue. For instance, nanofibrous scaffolds are biomimetic with respect to the structure of the T/L ECM, but have not succeeded in regenerating T/L. Such scaffolds are often challenged by difficulties with cell infiltration and are limited in their ability to deliver therapeutics that may promote desirable biological responses. Hydrogel scaffolds, in contrast, are attractive for their ability to encapsulate cells and biologic factors. Additionally, hydrogels have high biocompatibility because of their high water content, tunable material properties and ability to be injected into molds or other scaffolds *in vitro*, or directly into wound sites *in vivo*.

This chapter will discuss the utility of hydrogel scaffolds in various approaches to T/L TE. For instance, composite structures composed of one or more types of scaffolds (where at least one is a hydrogel) have resulted in engineered constructs with improved properties with respect to T/L when compared to either scaffold alone. Composite hydrogel materials have also been used in the field of interface tissue engineering (ITE), in which scaffolds with gradient features attempt to approximate the mismatch of properties at the interface of soft and hard tissues. Hydrogels as therapeutic delivery vehicles have been explored with positive results including induction of tenogenic and ligamentogenic differentiation and an increase in deposition of key T/L proteins. Hydrogels have also been implemented as tendon sheaths for the prevention of post-operative adhesions, which arise during tissue healing but restrict movement and can result in subsequent tearing *in vivo*. The objective of this chapter is to highlight recent reports that utilize hydrogels in T/L TE. The chapter begins with an overview of T/L tissue characteristics that can be targeted in an engineered therapeutic solution, and background on T/L injury, healing and current repair strategies. The chapter will then focus on state-of-the-art of T/L TE strategies involving hydrogels, and close with remarks regarding their potential for future applications in this field.

2. Tendon and Ligament

TE approaches to T/L repair are promising, but fully functional engineered T/L has yet to be achieved. An engineered solution should be comparable in composition, structure, mechanical properties and function of normal T/L tissue to act as a complete replacement that is superior to surgically repaired tissue. With these design aspects in mind, the following is an overview of normal T/L tissue properties, mechanisms of injury and healing and current medical treatments.

2.1. Function

T/Ls are both fibrous connective tissues that possess viscoelastic mechanical properties, allowing for adaptation to high rates of strain. Tendons transfer mechanical forces from muscles to bones, while ligaments connect bone-to-bone to stabilize joint structures. T/L are capable of withstanding high tensile strain, a characteristic that is paramount to their function. T/L store energy during locomotion to increase movement efficiency and act as a force buffer to prevent injury to other tissues.[2–7]

These important functions of T/L rely on the structural and compositional features, described in the next section.

2.2. Structure and composition

Likely due to their similar functions, T/L possess relatively comparable structure and composition. T/L are composed of collagen molecules arranged in nanoscale parallel structures known as fibrils. In a dimensionally hierarchical manner (Fig. 1), parallel fibrils bundle together to comprise individual fibers, which pack together into fascicles. In turn, fascicles are enclosed within the endotenon and bundle together to form the overall tendon or ligament, which itself is sheathed in the epitenon. Some tendons, notably the Achilles, have a paratenon, or false tendon sheath, that surround them. This is in contrast to digital flexor tendons, which have a true sheath to facilitate gliding of the enclosed tendon.[8] Both tissues experience predominately uniaxial loading, but may experience additional loading in other directions. Specific orientation and fibril diameter correlate with T/L function and thus vary by anatomical location, age and species.[9–13]

Water makes up the majority of human T/L total wet weight (60–80%), while the dry weight is composed principally of collagen type I (~60%) with smaller amounts of types III (~8%), IV, V, and VI. T/L also contains proteoglycans, glycoproteins, glycosaminoglycans and a variety of small leucine-rich proteoglycans (SLRPs) such as decorin and biglycan.[6,15–17] Biochemical composition varies in both tissues as a function of anatomical location. For instance, in human the dry weight percentage of elastin is 2% of the Achilles tendon,[18] 5% of the anterior cruciate ligament (ACL),[19] 7.3% of the posterior longitudinal ligament[20] and 47% of the ligamentum flavum.[21] The main constituents of the T/L connective sheath, including the endotenon, epitenon, and paratenon, are collagen types I and III and

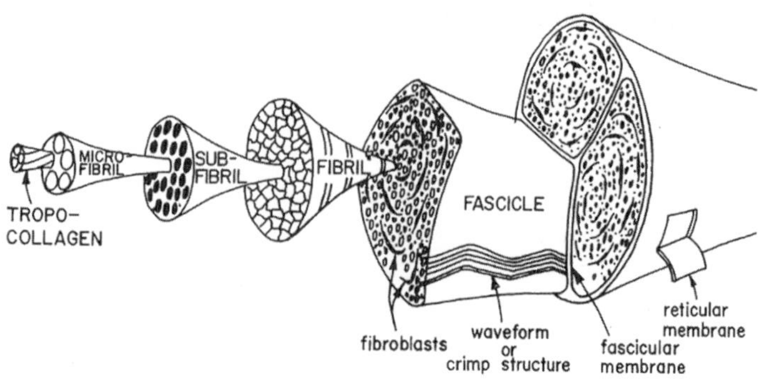

Fig. 1. The hierarchal structure of T/Ls (Kastelic *et al.*[14]).

elastin organized in longitudinal, oblique, and transverse orientations.[22] Proteoglycans facilitate hydration and, in turn, fiber gliding.[23–25]

The T/L endotenon and epitenon contain the vascularity, lymphatics, and nerves.[26,27] Overall T/L vascularity is different between the two tissues, though both have low vascularity when compared to muscle and other surrounding tissue. Tendons may receive their blood supply from adjacent vessels, from myotendinous junctions, or through diffusion from surrounding capillaries.[8,28–31] Conversely, ligaments seem to be vascularized principally at the insertion sites.[32] Innervation studies have shown that most nerve fibers do not enter the main T/L body but terminate as nerve endings on the surface.[33] Nerve fibers seem predominantly associated with blood vessels.[34]

Adult tendon is relatively acellular. Cells in T/L include macrophages, plasma cells, fat cells, myofibroblasts, mesenchymal stem cells (MSCs) and others, but the most commonly present are fibroblasts (e.g., tenocytes in tendon), which produce collagen and other ECM components. The cellularity of T/L tissue during development changes drastically from a high amount in the embryo and fetus to a rapid drop-off in the adult.[35–38]

2.3. Mechanical properties

T/L tissues are viscoelastic and mechanically adaptive, and respond to applied load through alteration of metabolic activity, collagen synthesis, cross-sectional area and tensile strength.[13,39] The uniquely high tensile strength of T/L has been attributed to ECM biochemical content, crosslinking between collagens and other ECM molecules,[16,38,40,41] and fiber morphology and alignment.[42] Aligned collagen fibrils of mixed diameters packed tightly together along the T/L axis grant high tensile strength to the tissues with minimal strain. Large diameter fibrils resist rupture while smaller diameter fibrils resist creep, in which individual fibers slide past one another.[43] The loading of T/L tissue has been described by a stress–strain curve that contains three major regions: the toe region, the linear region and the yield and failure region.[44,45] Generally, collagen fibers demonstrate a crimped morphology at rest[46–48] that begins to straighten during loading, reflected by the toe region of a stress–strain curve that coincides with relatively low tissue stiffness. Tissue stiffness increases as fibers uncrimp and straighten, correlating with the linear region of the curve. Microscopic failure occurs at strains over ~4% and is characterized by slipping of adjacent molecules, while macroscopic failure occurs at strains over ~8% and results in T/L rupture.[49,50] This is the failure region, in which individual fibrils within the ligament or tendon begin to fail. Due to the accumulation of damage, stiffness is reduced and the T/L begins to fail. T/L overuse and applied stress superior to the ultimate load that the structure can withstand eventually lead to injury.[51–53]

2.4. Injury and healing

Injuries to T/L can be characterized as acute ruptures or chronic degeneration. Acute T/L injury occurs as a result of a single overload event that results in partial or complete rupture. Degenerative T/L disorders, such as tendinopathy, often precede tendon rupture, indicating abnormal tissue was present before the rupture occurred.[54] The most frequent tendon injuries to humans involve the shoulder rotator cuff, hand flexor tendon and the Achilles tendon,[55] while the medial collateral ligament (MCL) and ACL are the most commonly injured ligaments.[56]

While acute ruptures require surgical repair (see next section), partial ruptures in which a complete gap does not form may heal by forming a scar. The healing response of postnatal T/L is a lengthy process that rarely restores complete and normal function to the T/L tissue. Healing takes place intrinsically and extrinsically simultaneously. The intrinsic process is marked by cellular activity of the T/L fibroblasts within the T/L tissue while extrinsic healing relies on recruitment of cells not within the T/L tissue.[57–60] Healing includes three main phases, which are the inflammatory phase, the proliferative phase and the remodeling phase. The 48–72 h inflammatory phase is characterized by infiltration of leukocytes, red blood cells, and platelets. A fibrin clot forms and many growth factors are released, including insulin-like growth factor-I (IGF-I), transforming growth factor beta (TGFβ), vascular endothelial growth factor (VEGF), platelet-derived growth factor (PDGF) and basic fibroblast growth factor (bFGF). These growth factors attract leukocytes and stimulate angiogenesis and cell proliferation and the second phase of healing.[6,53,61,62] T/L healing in the proliferative phase occurs days to months post-injury and is marked by intrinsic and extrinsic fibroblast recruitment to the injury site, fibroblast proliferation, and ECM deposition with an altered composition and structure from its healthy, normal state. Unlike healthy adult T/L tissue, the healing tissue is highly vascularized and highly cellular, with a relatively large amount of collagen type III.[63–65]

Also in the proliferative phase, many growth factors are highly expressed, stimulating cell proliferation (IGF-I, bFGF), angiogenesis (VEGF) and collagen synthesis (TGF-β). Likewise, high expression of ECM degradation molecules called matrix metalloproteinases (MMPs) and their inhibitors (TIMPs) can be seen in this stage of T/L healing.[66–68] The final remodeling phase of T/L healing takes place over months to years and overlaps with the proliferative phase. Remodeling is dominated by the production of collagen type I as the tissue matrix is replenished in an attempt to restore normal tissue. Collagen fibers begin to align, cell density decreases, and scar tissue forms.

Scar tissue differs from normal T/L tissue in composition and structure, including collagen and proteoglycan content, innervation, and vascularity.[64]

Though scar can return some function to the T/L over time, the resulting tissue does not approximate the mechanical properties of original tissues. For instance, one study found up to a 50% decrease in healed canine T/L tensile strength in the absence of post-operative activity.[69] Another group demonstrated the maximum load of healed rabbit patellar tendons to be 25% of normal tissue, even after 6 months.[70] Additionally, the MCL of rabbits was found to have scar tissue with inferior creep behavior and viscoelastic properties when compared to normal tissue.[64,71] Current treatments of T/L injuries involve surgical intervention followed by a rehabilitation period of rest and controlled movement.

2.5. Current medical treatments

Clinical intervention for the repair of T/L tissue is usually required due to their poor healing capacity and inability to close gaps that result from significant ruptures. Surgical repair with autologous or allogeneic grafts is typical, followed by immobilization and physical therapy with controlled movement.[69,72–75] T/L grafts attempt to restore the morphology, length, and strength of a tendon or ligament. The initial stiffness and tension of the graft are important considerations and may affect implant stability.[76] Autologous tissue remains the most commonly used graft for T/L reconstruction, due to excellent compatibility with the host tissues and the propensity of autografts to revascularise and remodel when transplanted.[77–79] However, this method presents the risk of morbidity at donor sites and requires multiple invasive procedures.[80–82] Allografts are increasingly utilized, but these tissues are expensive and carry a risk of immunological response by the host and disease transmission from the donor.[83] The relative scarcity of human tissue available for transplantation has amplified interest in xenografts as alternatives. However, the risk of an immune system response that rejects the xenograft and the potential for infectious diseases to be passed between species remain significant drawbacks to their use.[84–90] These clinical treatments are associated with a high failure rate, slow graft incorporation time, lack of graft-to-bone integration and, in the best case, a healed tissue with inferior mechanical properties, thereby resulting in compromised tissue functionality compared to healthy, normal tissue.[91,92] These challenges have prompted efforts to develop TE solutions to T/L repair.

2.6. Current state of tendon and ligament TE

Broadly speaking, TE aims to regenerate new tissue from engineered constructs, which are usually scaffolds carrying cells and/or biofactors. The ultimate goal of TE is to restore normal properties and function to an injured or diseased tissue.

One strategy is to culture an engineered tissue *in vitro* until it can be functional immediately upon implantation, while a second strategy is to implant a construct to deliver therapeutics *in vivo* to promote an improved healing response at the injury site. In both cases the scaffold plays a pivotal role. It is therefore important to consider the scaffold's physical, chemical and biological properties. A bioactive scaffold will guide cell adherence, morphology, migration, differentiation and ECM production. Furthermore, the scaffold will rapidly integrate for immediate load-bearing post-implantation and exhibit a degradation rate comparable to the rate of new tissue formation.

Various synthetic and natural polymer materials and processing techniques have been investigated for T/L TE scaffolds. Amongst synthetic polymers, the most common are aliphatic polyester polymers such as poly (lactic acid) (PLA),[93,94] poly (glycolic acid) (PGA),[95] poly (lactic-co-glycolide) (PLGA)[96] and poly(ε-caprolactone) (PCL).[97–100] Polyhydroxyesters are inexpensive polymers that are degradable by hydrolysis and able to be tailored with a range of specific material properties. Unfortunately, they may release acidic or unnatural byproducts, eliciting unwanted biological responses. Moreover, they lack functional chemical groups for cell adhesion. For these reasons there has been increased interest in natural materials such as collagen[101,102] and silk,[103–105] which have inherent biocompatibility and biodegradability.[106–109] However, many natural polymers can potentially activate immunogenic reactions. They may also have significant batch-to-batch variations, which can make material properties of the fabricated scaffolds unpredictable and difficult to control. Scaffold form and microstructure, controlled via both material selection and processing, contribute greatly to its properties and function. Scaffolds of varying forms have been explored in T/L TE, including nano- and microfiber matrices, braided fibers, foams, sponges and composites, but none have resulted in a functional engineered T/L. To date, there is not a gold standard scaffold for T/L TE.

There has been interest in developing three-dimensional scaffolds that mimic structural aspects of the ECM. For this reason, nanofibers have been broadly studied for their biomimicry with respect to the fibrillar structure of the collagenous matrix of native T/L.[110–113] Electrospinning is a commonly used technique to produce polymer fibers with diameters in the nanometer range.[114] It has been demonstrated that *in vitro* aligned nanofibers elicit more desirable T/L-like cell responses than randomly oriented nanofibers. For instance, more collagen was synthesized by human ligament fibroblasts seeded on aligned nanofiber scaffolds compared to randomly oriented nanofiber scaffolds.[110] In other studies, human and rat tendon fibroblast-like cell morphology, alignment and distribution were enhanced with increasing nanofiber alignment.[112,113] The effect of fiber diameter on human tendon fibroblast response has also been explored.[115] It was found that nanofiber scaffolds promoted higher cell number, collagen content, and proteoglycan synthesis, while microfiber scaffolds were associated with the upregulation of several tendon marker genes including collagen types I, III, V and tenomodulin. Based on these results, the researchers hypothesized that microfiber scaffolds mimic healthy tissue (inducing a maintenance response) and nanofiber scaffolds mimic injured tissue (inducing a reparative response). Because of their biomimicry to T/L tissue, composite scaffolds that combine nano- and microfibers have been produced. They are further justified for their use in T/L TE in that microfibers add mechanical strength and degradation resistance while nanofibers provide high surface area for cell attachment and tissue integration.[114,116–118] For example, the addition of PLGA nanofibers to a PLGA microfiber mesh resulted in faster proliferation and upregulation of collagen type I, decorin and biglycan genes of seeded porcine bone marrow stromal cells when compared to cells on the PLGA mesh alone.[119] To improve poor scaffold tensile strength, braided structures have been developed. A braided silk fibroin scaffold for T/L TE introduced by Altman *et al.*[103] was shown to support collagen gene expression and matrix deposition by human progenitor bone marrow stromal cells. Other studies have used various polymers such as PGA, PLGA, and PLA to fabricate three-dimensional braided fiber scaffolds that approximate the mechanical properties of native ACL tissue.[120,121] Barber *et al.* electrospun bundles of aligned PLLA, braided them together and seeded the scaffold with hMSCs. The hMSCs adhered to these scaffolds in an aligned orientation, parallel to the length of the nanofibers, and upregulated scleraxis expression.[122] Despite these studies and additional work that has optimized the braiding angles of these scaffolds to better mimic T/L stress–strain behavior,[123] desirable mechanical properties have not been achieved. Furthermore, nanofiber scaffolds and braided structures face limited nutrient diffusion, cell migration and tissue infiltration.[124]

To improve T/L TE outcomes, other scaffold forms have also been combined. Foam and sponge scaffolds are advantageous because of controllable porosity and cell infiltration. Liu *et al.* engineered a silk fibroin hybrid scaffold with geometries at two levels: A silk knitted scaffold and an interspersed microporous silk sponge.[124] This scaffold promoted proliferation and gene expression of ligament-related ECM markers by seeded hMSCs. Similarly, collagen microsponges were fabricated in the empty space of a PLGA-knitted scaffold resulting in mechanically strong constructs with high canine ACL cell seeding efficiency and homogenous matrix deposition.[125] Another composite scaffold combined a knitted silk fibroin base matrix with a lyophilized collagen type I microsponge.[126] Rabbit MSCs seeded in this scaffold upregulated ligament matrix genes compared to in the silk scaffold alone. In addition, the composite scaffold implanted in a rabbit MCL defect model resulted in more native microstructure with larger diameter collagen fibrils and greater integration between neoligament and native ligament.

An appropriate T/L scaffold for TE applications should reproduce the inherent mechanical and biological properties of the T/L ECM and simultaneously induce tissue regeneration while degrading. Thus far, efforts have not addressed all of these concerns together. Integration of hydrogels into T/L TE scaffold systems may be beneficial. Hydrogels may act as 3D scaffolds that are capable of supporting cells and maintaining the activity of bioactive agents (Fig. 2). Furthermore, their hydrophilicity mimics the aqueous physiologic aspect of natural tissue. The next section details different aspects of hydrogels used in T/L TE. In particular, we will review studies focused on hydrogels used as scaffolds and as delivery vehicles. We will then detail some specific applications of hydrogels in T/L repair.

3. Hydrogels in Tendon and Ligament TE

Most of the T/L TE scaffolds described up to now fail to simultaneously approximate T/L tissue mechanical properties and allow cell migration and tissue ingrowth. To circumvent these potential drawbacks there are efforts to develop new strategies that integrate hydrogels. Hydrogels are crosslinked polymeric networks that contain hydrophilic groups with the ability to retain high water content within their structure.[127,128] Due to this property and to their range of material and biological properties that can mimic native tissue characteristics, hydrogels possess a unique biocompatibility that can be exploited in TE applications. The porous structure of hydrogels can be tuned by changing the molecular weight and concentration of the polymer and the density of crosslinks.[129,130] The tunable porosity of hydrogels allows for the physical entrapment of biomolecules within the hydrogel network and their release at a controllable rate depending on the diffusion coefficient of the biomolecules through the hydrogels.[131] Active agents may also be covalently conjugated to the hydrogel matrix through a linker that is slowly cleaved to release the biomolecules.[132] Furthermore, hydrogels generally exhibit high permeability for oxygen and other water-soluble metabolites, making them attractive scaffolds for use in cell encapsulation.[133–136] This is achieved by directly mixing the cells with the polymer solution prior to gelation, which facilitates their homogeneous distribution throughout the hydrogel and precludes a secondary cell-seeding step. Finally, most hydrogels are injectable *in situ* into custom shaped molds and even into other scaffolds to form composite structures.

While hydrogels are heavily used in the field of regenerative medicine, exploration of their utility in T/L TE has been more limited, likely due to their insufficient mechanical properties as standalone scaffolds. Instead, they could be an excellent tool to control tissue regeneration by either acting as a cell support for the development of an engineered tissue *in vitro* or as a delivery vehicle of therapeutic agents

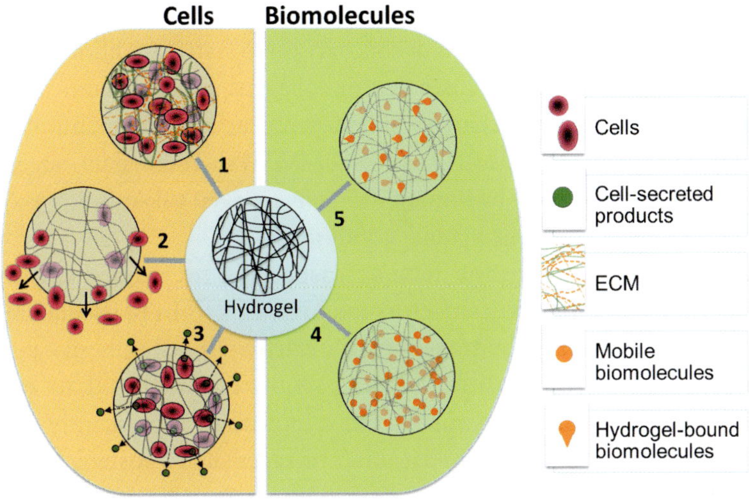

Fig. 2. Versatile use of hydrogels in T/L TE. (A) 1 — Temporary ECM to induce cells to synthesize new tissue. (B) 2 — Delivery of cells *in situ*. (C) 3 — Maintenance of cell viability for *in situ* production and release of therapeutic factors. (D) 4 — Encapsulation of active agents to be released *in situ* via diffusion or scaffold degradation. (E) 5 — Covalent tethering of active agent to be released *in situ* by linker cleavage.

in vivo, which may include cells or biomolecules (Fig. 2). In the former case the hydrogel provides a temporary support, acting as a synthetic ECM and directing cells to synthesize new tissue (Fig. 2(A)) as described in Sec. 3.1. In Sec. 3.2, we will present the latter case, in which hydrogels deliver cells (Fig. 2(B)) or biomolecules (Figs. 2(D) and 2(E)), or maintain cell viability for subsequent *in situ* production of biomolecules (Fig. 2(C)). Additionally, we will discuss the use of hydrogels in the engineering of interfacial tissue between soft and bone tissues due to their tunable structure (Sec. 3.3). Finally, we will review some studies in which hydrogels were specifically engineered to prevent adhesions formation during T/L healing which limits the motion (Sec. 3.4).

3.1. Hydrogels as TE scaffolds

Hydrogels offer a promising approach to restoring injured T/L tissue by acting as a temporary synthetic ECM for TE applications. For instance, T/L constructs engineered from cell-laden hydrogels have been cultured in bioreactors under mechanical conditioning in an effort to stimulate tenogenesis or ligamentogenesis and produce tissues *in vitro* prior to implantation. In a recent study, we cultured human MSCs in three-dimensional collagen type I gels under static and dynamic tension and assessed the expression of genes implicated in embryonic tendon development.[137] We found that cyclic stretching was required to maintain expression of scleraxis over time and that it enhanced collagen deposition in comparison to static conditions. Surprisingly, the dramatically higher staining intensity for collagens under dynamic loading conditions did not correlate with changes in collagen gene expression, but did coincide with significant changes in mRNA levels of MMPs. These data suggested that increased collagen deposition by human MSCs in collagen gel scaffolds may be regulated by MMPs under dynamic loading conditions.

In another *in vitro* study, the Temenoff laboratory developed a hydrogel as a carrier of MSCs and biochemical factors for T/L TE.[138,139] In particular, a synthetic poly(ethylene glycol)-based hydrogel incorporated with cell-adhesive peptide and matrix metalloproteinase-sensitive peptide sequences was used to encapsulate MSCs. Cyclically straining the construct in a bioreactor significantly upregulated T/L genes examined, but not genes for other lineages such as osteogenic, chondrogenic and adipogenic. Noth *et al.* also applied cyclic stretching to engineer an ACL tissue using a collagen type I gel scaffold containing human MSCs.[140] This resulted in upregulation of ligament-specific matrix gene expression and protein synthesis and elongated fibroblast-like cells embedded in a wavy orientated collagenous tissue. In another study, autologous MSCs were seeded in a collagen gel, which was combined with a collagen sponge, mechanically stimulated and subsequently implanted *in vivo* into bilateral full-thickness, full-length defects created in the central third of rabbit patellar tendons (PTs).[141] The findings showed that mechanical stimulation of cell-hydrogel-sponge constructs enhanced several properties of central PT repairs when compared to non-stimulated constructs, including linear stiffness and maximum force, stress and modulus.

In other efforts, combining hydrogels with other materials have produced composite scaffolds that incorporate characteristics of their components, making them particularly suitable in T/L TE. The combinatorial effects enable these composite scaffolds to deliver viable cells and bioactive growth factors while also improving mechanical properties. One group combined collagen gels with poly(3-hydroxybutyrate-co-3-hydroxyhexanoate) (PHBHHx) tube because of the polymer's tissue compatibility and delayed biodegradability.[142] This tenocyte-seeded scaffold was implanted in a rat Achilles tendon repair model and found to accelerate return to function and desirable cell migration and organization when compared to no cell or no polymer controls. A similar idea using a collagen gel as a delivery vehicle for rabbit MSCs was exploited.[143] Autologous MSC–hydrogel–sponge constructs were placed in rabbit full-length defects in the central third of patellar tendons, significantly improving repair biomechanics compared to unseeded controls. Thomopoulos *et al.* incorporated platelet-derived growth factor BB (PDGF-BB) along with adipose-derived mesenchymal stem cells (ASCs) into a heparin/fibrin-based delivery system (HBDS).[144] This hydrogel construct was then layered with PLGA electrospun nanofibers. *In vitro* studies verified that the cells remained viable and that sustained growth factor release was achieved. *In vivo* studies in a large animal clinically relevant canine tendon model demonstrated that the cells remained viable in the tendon repair environment. Young *et al.* tested the hypothesis that implanting autologous mesenchymal stem cell-seeded constructs in a rabbit tendon gap resulted in significantly improved mechanical properties.[145] The results showed that applying a MSC-loaded collagen type I scaffold to a sutured defect site significantly improves tendon repair biomechanics compared with the sutures alone.

3.2. Hydrogels as delivery vehicles

Delivering therapeutic agents such as cells and biomolecules to injured T/L via hydrogels may enhance tissue healing or regeneration. This strategy can influence functions of the cells present at the site of interest. For instance, Murray *et al.* placed a collagen-platelet composite around suture repairs of ACL canine and porcine injuries models.[146–149] The scaffold was designed to deliver platelets and plasma proteins, key components required for healing of many tissues. The collagen-platelet composite improved ACL function and biomechanical properties after suture repair. Another group explored the effects of delivering growth factors bFGF and GDF-5 (growth and differentiation factor 5) together and individually via a commercially available peptide hydrogel (PuraMatrix, BD) to injured rabbit MCL.[150] Their findings suggested that combined use of bFGF and GDF-5 in a hydrogel may enhance the healing of MCL injury by increasing the proliferation and migration of ligament fibroblasts and by regulating collagen synthesis and fiber alignment.

Other groups have focused on the use of collagen gels for gene delivery *in situ* by loading them with vector-encoding cDNA prior to the insertion into defects.[53,151,152] In particular, Steinert *et al.* loaded a collagen gel with IGF-I cDNA and placed it between the cut ends of two human ACL fascicle pieces cultured *ex vivo.*[151] The gene was successfully delivered to the ACL fibroblasts that migrated into the hydrogel from the severed tissue. This led to an increase in the cellularity of the hydrogels and synthesis and deposition of collagen types I and III, elastin, tenascin, and vimentin. This study demonstrated an efficient method to deliver genes to ACL cells and offers a possible approach to ACL repair. Similarly, another study fabricated collagen type I gels to deliver adenoviral vectors encoding the human cDNA for bone morphogenetic protein (BMP)-12 and -13 to human MSCs and ACL fibroblasts *in vitro.*[152] Transgene expression for at least 14 days was confirmed and transductions, even at high viral doses, were not detrimental to cell viability and proliferation. The authors, analyzing immunohistochemical data and some key T/L genes (decorin, scleraxis, tenascin-C, and tenomodulin) asserted that

BMP-12 and -13 gene-transfer into MSC and ACL fibroblasts induced ligamentogenic differentiation.

Many studies suggest that implantation of autologous cells are beneficial to tissue healing. Unfortunately, in TE applications, implanted cells commonly suffer from low survival over the long term, which has led to the use of hydrogels to maintain cellular viability. Temenoff and her group studied multiple cell delivery methods in comparison to optimize and control them.[153] They synthesized hydrogel scaffolds based on oligo(poly(ethylene glycol) fumarate) and acrylated poly(ethylene glycol)-dithiothreitol with varying degradation rates and control over MSC delivery to tendon overuse injuries. Cell delivery was found to be dependent on changes in the ratio of the two polymers, with more cells found in the tissue with the faster degrading hydrogels. They suggested that this type of hydrogel provides a versatile biomaterial platform to deliver MSCs to treat tendinopathy.

The ability of hydrogels to deliver therapeutics relies on degradation of the hydrogel or diffusion of the therapeutic and the possibility to be employed in combination with other TE scaffolds. Hydrogels as delivery vehicles will likely continue to be incorporated into T/L TE strategies because of their promising applications.

3.3. Hydrogel in T/L-bone interface TE

Soft tissue-to-bone interfaces are characterized by gradual changes in composition, structural organization and material properties from one tissue to the other.[154] An engineered interface between T/L and bone that mimics the natural interfacial tissue would transfer loads across joints, minimizing stress concentrations and preventing implant failure. Driven by these motivations, ITE is rapidly developing with the ultimate goal of obtaining a gradient in composition and stiffness from bone to soft tissue.

Hydrogels are considered in T/L ITE because of their versatility of structure and ability to be modified in a spatially controlled way. In one study, a continuous material was developed to avoid abrupt interfaces in connective tissues between bone and soft tissue, such as tendon or ligament.[155] The work reported a new strategy to create a hydrogel material that can be functionalized using different combinations of carboxybetaine methacrylate and carboxybetaine dimethacrylate crosslinker. Fabricated hydrogels exhibited high strength and high compressive moduli, with a spatially controlled gradient of physical and chemical properties. A separate study by Paxton *et al.* investigated the use of a poly(ethylene glycol) diacrylate hydrogel containing the calcium phosphate hydroxyapatite and the peptide RGD to create an *in vitro* tissue interface for ligaments.[156] The result was an increase in mechanical strength and capacity to bind primary rat tendon fibroblasts. This work suggested that further investigation of calcium phosphate ratios is needed to obtain a functional interface between engineered ligament and bone. To improve recovery after ACL reconstruction in rabbit models, Oka *et al.* conjugated simvastatin, a drug that has been shown to promote osteoblast differentiation, to a gelatin hydrogel.[157] The native ACL was completely transected and the bone tunnels were created. An ipsilateral semitendinosus tendon was harvested as a graft and gelatin hydrogel was wrapped around and advanced though the tunnel. This study demonstrated low-dose simvastatin-conjugated gelatin hydrogels enhanced angiogenesis and osteogenesis without affecting biomechanical properties of reconstructed ACL in the long term.

Scaffolds that combine different materials, including hydrogels, and that exhibit complex gradient features to mimic the native T/L microenvironment have also been investigated in ITE. For example, Lee *et al.* produced a single porous cylindrical poly(L-lactide-co-ε-caprolactone) scaffold, which was divided in three regions to contain fibrocartilage regions at the two ends and a ligament region in the middle.[158] The ligament region was seeded with fibroblasts (from rabbit patella ligament) while a heparin-based hydrogel was injected in combination with fibrochondrocytes (isolated from rabbit meniscus) and a small amount of bone morphogenetic protein-2 (BMP-2) in the fibrocartilage regions. Their results showed that the local delivery of cells and growth factors using a heparin-based hydrogel was beneficial in mouse subcutaneous implantation models, accelerating the differentiation and maturation of fibrochondrocytes into osteoblasts. It was also possible to recreate the unique properties of the interface between bone and ligament tissue in a single scaffold.

Taken together, ITE may be achievable via scaffolds, including hydrogels, combined with other approaches such as gene and cell therapy.

3.4. Hydrogels in tendon adhesion prevention during healing

Development of post-operative adhesions between tendon and its surrounding sheath is common, especially in the digital flexor tendons of the hand. Despite being part of the normal T/L healing process, adhesion leads to functional disability by restricting motion and inhibiting gliding. Surgical and rehabilitative options that seek to reduce adhesion formation include different suturing techniques, post-operative physiotherapy, ultrasound therapy and drug treatments that target the adhesion formation process. However, these strategies have had mixed success and they have not been demonstrated in humans.[159]

A variety of hydrogels have found use in the prevention of adhesions by sheathing the tendon, acting as barriers to the adhesions. The ideal hydrogel used for peritendinous adhesion prevention is biocompatible, appropriately biodegradable and allows for tendon movement and healing while restricting adhesion formation. One study demonstrated the use of the commercially available two-component synthetic hydrogel surgical sealant FocalSeal-L as an inhibitor of adhesion formation in divided and repaired deep flexor tendons of chickens.[160] Peritendinous adhesion formation was decreased while breaking strength stayed the same when compared to no sealant controls. In another study, Zhao *et al.* treated the surface of canine flexor tendon autografts with a carbodiimide-derivatized hyaluronic acid-gelatin hydrogel and found lower gliding resistance after six weeks compared to non-treated graft repairs.[161] Similarly, hyaluronan-based hydrogels to treat flexor tendon injuries in rabbits were evaluated by Liu *et al.*[162] In this study, a crosslinked hydrogel was found to promote adhesion-free healing, due to its longer residence time and inhibition of fibroblast attachment, spreading, and proliferation.

Another group worked to establish an anti-adhesion material with a mixture of the hydrogels polyvinyl alcohol (PVA) and poly(2-methacryloyloxyethyl phosphorylcholine-co-n-butyl methacrylate-co-p-vinylphenylboronic acid) (PMBV).[163] A 2-methacryloyloxyethyl phosphorylcholine (MPC) polymer hydrogel was injected around repaired tendon in a rabbit achilles tendon injury model and in a chicken flexor digitorum profundus injury tendon model. The resulting hydrogel had nanometer-sized pores, allowing penetration of signaling molecules but inhibiting cellular infiltration. The MPC hydrogel prevented tissue adhesion to tendon without adverse effects on the healing process of injured tendon.

These studies demonstrate that hydrogels have found yet another practical application to T/L TE, emphasizing their potential in the field. Hydrogels may reduce T/L adhesions by sheathing the tissues, preventing cellular attachment and proliferation while maintaining the gliding movement and overall function of the tissue.

4. Conclusions and Outlook

Toward addressing the inadequate return of injured T/L tissue to normal function, the field of T/L TE has made rapid advances. As we have discussed, strategies may include the *in vitro* production of a clinically functional tissue for later implantation at the site of injury, or the initiation of improved *in vivo* healing by implanting an inductive construct which delivers therapeutics. In either case, a scaffold plays a role in support of the tissue structure and/or as a delivery vehicle of therapeutic cells and biomolecules. Many scaffold materials have been studied, with hydrogels offering a versatile solution. They have controllable structure and material properties and are well-suited for supporting and guiding cells and tissue while delivering therapeutics. Their tunable structure also imparts hydrogels with the ability to approximate gradient features, which may be particularly useful in engineering interfaces between soft

T/L tissue and hard bone tissue. Finally, we have described several applications of hydrogels for the prevention of adhesions, which arise due to normal healing but inhibit T/L movement and function.

Hydrogels and their combination with other scaffolds will be a likely theme in future investigations. In fact, several extremely stretchable and tough hydrogels have recently been developed, exploiting the idea of mixing two different types of hydrogels to combine desirable properties.[164–166] To date there are no results of the use of combination hydrogels in T/L *in vivo* injury models, though they show exciting promise. In addition to future *in vivo* applications, hydrogels in T/L TE will likely continue to be utilized in the study of *in vitro* culture conditions that are tenogenic or ligamentogenic for stem cells. These and additional future efforts that advance the field of T/L TE will utilize hydrogels because of the demonstrated versatility of their material and biological properties and their promising potential for clinical application.

Acknowledgments

We thank Nathan R. Schiele PhD for critically reading this chapter. We gratefully acknowledge funding support from the Tissue Engineering Resource Center (TERC)–NIH P41 EB002520 from the National Institute of Biomedical Imaging and Biomedical Engineering (C.K.K. is collaborator).

References

1. D. L. Butler, N. Juncosa, and M. R. Dressler, *Ann. Rev. Biomed. Eng.*, **6**, 303 (2004).
2. D. L. Butler, E. S. Grood, F. R. Noyes, and R. F. Zernicke, *Exercise Sport Sci. Rev.*, **6**, 125 (1978).
3. S. L. Woo, M. A. Gomez, Y. K. Woo, and W. H. Akeson, *Biorheology*, **19**, 397 (1982).
4. T. A. Wren, G. S. Beaupre, and D. R. Carter, *J. Rehab. Res. Dev.*, **37**, 217 (2000).
5. A. Hoffmann and G. Gross, *Int. Orthop.*, **31**, 791 (2007).
6. T. W. Lin, L. Cardenas, and L. J. Soslowsky, *J. Biomech.*, **37**, 865 (2004).
7. M. L. Killian, L. Cavinatto, L. M. Galatz, and S. Thomopoulos, *J. Shoulder Elb. Surg.*, **21**, 228 (2012).
8. M. Benjamin, E. Kaiser, and S. Milz, *J. Anat.*, **212**, 211 (2008).
9. D. A. Parry, M. H. Flint, G. C. Gillard, and A. S. Craig, *FEBS Lett.*, **149**, 1 (1982).
10. D. Amiel, J. B. Kleiner, R. D. Roux, F. L. Harwood, and W. H. Akeson, *J. Orthop. Res.*, **4**, 162 (1986).
11. M. J. Moore and A. De Beaux, *J. Anat.*, **153**, 163 (1987).
12. F. H. Silver, J. W. Freeman, and G. P. Seehra, *J. Biomech.*, **36**, 1529 (2003).
13. J. Bojsen-Moller, K. K. Kalliokoski, M. Seppanen, M. Kjaer, and S. P. Magnusson, *J. Appl. Physiol.*, **101**, 196 (2006).

14. J. Kastelic, A. Galeski, and E. Baer, *Connect. Tissue Res.*, **6**, 11 (1978).
15. M. Kjaer, *Physiol. Rev.*, **84**, 649 (2004).
16. P. S. Robinson, T. F. Huang, E. Kazam, R. V. Iozzo, D. E. Birk, and L. J. Soslowsky, *J. Biomech. Eng.*, **127**, 181 (2005).
17. L. M. Dourte, L. Pathmanathan, A. F. Jawad, R. V. Iozzo, M. J. Mienaltowski, D. E. Birk, and L. J. Soslowsky, *J. Biomech. Eng.*, **134**, 031005 (2012).
18. O. H. Lowry, D. R. Gilligan, and E. M. Katersky, *J. Biol. Chem.*, **139**, 795 (1941).
19. J. A. Dodds and S. P. Arnoczky, *Arthroscopy*, **10**, 132 (1994).
20. H. Nakagawa, Y. Mikawa, and R. Watanabe, *Spine*, **19**, 2164 (1994).
21. Y. Mikawa, H. Hamagami, J. Shikata, and T. Yamamuro, *Arch. Orthop. Trauma Surg.*, **105**, 343 (1986).
22. M. Kvist, L. Jozsa, M. Jarvinen, and H. Kvist, *Pathol. Res. Prac.*, **180**, 416 (1985).
23. R. W. Rowe, *Connect. Tissue Res.*, **14**, 9 (1985).
24. K. A. Derwin, L. J. Soslowsky, J. H. Kimura, and A. H. Plaas, *J. Orthop. Res.*, **19**, 269 (2001).
25. P. E. Matuszewski, Y. L. Chen, S. E. Szczesny, S. P. Lake, D. M. Elliott, L. J. Soslowsky, and G. R. Dodge, *Connect. Tissue Res.*, **53**, 343 (2012).
26. J. Kastelic, A. Galeski, and E. Baer, *Connect. Tissue Res.*, **6**, 11 (1978).
27. N. Ochiai, T. Matsui, N. Miyaji, R. J. Merklin, and J. M. Hunter, *J. Hand Surg.*, **4**, 321 (1979).
28. D. A. W. Edwards, *J. Anat.*, **80**, 147 (1946).
29. R. Scapinelli, *Acta Anat.*, **70**, 305 (1968).
30. J. Y. Nordin, M. Benoist, G. Pouget, and J. Cauchoix, *Rev. Rhum.*, **43**, 291 (1976).
31. J. H. Ginsburg, L. A. Whiteside, and T. L. Piper, *Am. J. Sport Med.*, **8**, 15 (1980).
32. R. C. Bray, A. W. Fisher, and C. B. Frank, *J. Anat.*, **172**, 69 (1990).
33. P. Sharma and N. Maffulli, *J. Bone. Joint Surg. Am.*, **87**, 187 (2005).
34. P. W. Ackermann, M. Spetea, I. Nylander, K. Ploj, M. Ahmed, and A. Kreicbergs, *J. Histochem. Cytochem.*, **49**, 1387 (2001).
35. C. K. Kuo, B. C. Petersen, and R. S. Tuan, *Dev. Dyn.*, **237**, 1477 (2008).
36. A. Oryan and A. H. Shoushtari, *Anatomia, Histologia, Embryologia*, **37**, 134 (2008).
37. J. P. Brown, R. M. Lind, A. F. Burzesi, and C. K. Kuo, *PLoS One*, **7**, e38475 (2012).
38. J. E. Marturano, J. D. Arena, Z. A. Schiller, I. Georgakoudi, and C. K. Kuo, *Proc.Natl. Acad. Sci. USA*, **110**, 6370 (2013).
39. S. L. Woo, M. A. Ritter, D. Amiel, T. M. Sanders, M. A. Gomez, S. C. Kuei, S. R. Garfin, and W. H. Akeson, *Connect. Tissue Res.*, **7**, 177 (1980).
40. C. Couppe, P. Hansen, M. Kongsgaard, V. Kovanen, C. Suetta, P. Aagaard, M. Kjaer, and S. P. Magnusson, *J. Appl. Physiol.*, **107**, 880 (2009).
41. J. E. Marturano, J. F. Xylas, G. V. Sridharan, I. Georgakoudi, and C. K. Kuo, *Acta Biomater.*, **10**, 1370 (2014).
42. S. P. Lake, K. S. Miller, D. M. Elliott, and L. J. Soslowsky, *J. Biomech.*, **43**, 727 (2010).
43. V. Ottani, M. Raspanti, and A. Ruggeri, *Micron.*, **32**, 251 (2001).
44. S. L. Woo, J. M. Hollis, D. J. Adams, R. M. Lyon, and S. Takai, *Am. J. Sports Med.*, **19**, 217 (1991).

45. S. L. Woo, R. E. Debski, J. D. Withrow, and M. A. Janaushek, *Am. J. Sports Med.*, **27**, 533 (1999).
46. J. Diamant, A. Keller, E. Baer, M. Litt, and R. G. Arridge, *Proc. R. Soc. Lond. B. Biol. Sci.*, **180**, 293 (1972).
47. J. Kastelic, I. Palley, and E. Baer, *J. Biomech.*, **13**, 887 (1980).
48. M. Franchi, A. Trire, M. Quaranta, E. Orsini, and V. Ottani, *Scientific World Journal*, **7**, 404 (2007).
49. K. A. Hansen, J. A. Weiss, and J. K. Barton, *J. Biomech. Eng.*, **124**, 72 (2002).
50. J. H. Wang, *J. Biomech.*, **39**, 1563 (2006).
51. T. A. Wren, G. S. Beaupre, and D. R. Carter, *J. Rehab. Res. Dev.*, **37**, 217 (2000).
52. M. L. Killian, L. Cavinatto, L. M. Galatz, and S. Thomopoulos, *J. Shoulder Elb. Surg.*, **21**, 228 (2012).
53. R. Hauser, E. Dolan, H. Phillips, A. Newlin, R. Moore, and B. Woldin, *Open Rehabilitation J.*, **6**, 1 (2013).
54. P. Kannus and L. Jozsa, *J. Bone. Joint. Surg. Am.*, **73**, 1507 (1991).
55. G. Yang, B. B. Rothrauff, and R. S. Tuan, *Birth. Defects Res. C Embryo. Today*, **99**, 203 (2013).
56. H. J. Jung, M. B. Fisher, and S. L. Woo, *Sports. Med. Arthrosc. Rehabil. Ther. Technol.*, **1**, 9 (2009).
57. G. Lundborg and F. Rank, *J. Hand Surg.*, **3**, 21 (1978).
58. C. Frank, S. L. Woo, D. Amiel, F. Harwood, M. Gomez, and W. Akeson, *Am. J. Sports Med.*, **11**, 379 (1983).
59. G. Lundborg, F. Rank, and B. Heinau, *Scand. J. Plast. Recons. Surg.*, **19**, 113 (1985).
60. A. Khanna, N. Gougoulias, and N. Maffulli, *Acta. Orthop. Belgica*, **75**, 433 (2009).
61. T. Molloy, Y. Wang, and G. Murrell, *Sports Med.*, **33**, 381 (2003).
62. G. Schulze-Tanzil, O. Al-Sadi, E. Wiegand, W. Ertel, C. Busch, B. Kohl, and T. Pufe, *Scand. J. Med. Sci. Sports*, **21**, 337 (2011).
63. N. Shrive, D. Chimich, L. Marchuk, J. Wilson, R. Brant, and C. Frank, *J. Orthop. Res.*, **13**, 923 (1995).
64. C. B. Frank, *J. Musculoskelet. Neuronal. Interact*, **4**, 199 (2004).
65. H. L. Ansorge, J. E. Hsu, L. Edelstein, S. Adams, D. E. Birk, and L. J. Soslowsky, *J. Orthop. Res.*, **30**, 448 (2012).
66. R. James, G. Kesturu, G. Balian, and A. B. Chhabra, *J. Hand Surg.*, **33**, 102 (2008).
67. W. Petersen, F. Unterhauser, T. Pufe, T. Zantop, N. P. Sudkamp, and A. Weiler, *Arch. Orthop. Traum. Su.*, **123**, 168 (2003).
68. R. Garofalo, E. Cesari, E. Vinci, and A. Castagna, *Sports Med. Arthrosc. Rev.*, **19**, 207 (2011).
69. R. D. Montgomery, *Semin. Vet. Med. Surg.*, **4**, 304 (1989).
70. H. Miyashita, M. Ochi, and Y. Ikuta, *Arch. Orthop. Trauma. Surg.*, **116**, 454 (1997).
71. G. M. Thornton, G. P. Leask, N. G. Shrive, and C. B. Frank, *J. Orthop. Res.*, **18**, 238 (2000).
72. D. L. Hamner, C. H. Brown, Jr., M. E. Steiner, A. T. Hecker, and W. C. Hayes, *J. Bone. Joint. Surg. Am.*, **81**, 549 (1999).

73. L. S. Crossett, R. K. Sinha, V. F. Sechriest, and H. E. Rubash, *J. Bone. Joint. Surg. Am.*, **84-A**, 1354 (2002).
74. M. Kettler, J. Lunger, V. Kuhn, W. Mutschler, and M. J. Tingart, *Am. J. Sports Medi.*, **35**, 1544 (2007).
75. C. Zhao, S. L. Moran, S. S. Cha, A. Kai Nan, and P. C. Amadio, *J. Hand Surg.*, **32**, 518 (2007).
76. S. L. Woo, R. E. Debski, J. Zeminski, S. D. Abramowitch, S. S. Saw, and J. A. Fenwick, *Ann. Rev. Biomed. Eng.*, **2**, 83 (2000).
77. R. P. A. Janssen, J. van der Wijk, A. Fiedler, T. Schmidt, H. A. G. M. Sala, and S. U. Scheffler, *Knee Surg. Sports Tr. Ar.*, **19**, 1299.
78. S. J. Kew, J. H. Gwynne, D. Enea, M. Abu-Rub, A. Pandit, D. Zeugolis, R. A. Brooks, N. Rushton, S. M. Best, and R. E. Cameron, *Acta. Biomater.*, **7**, 3237.
79. M. M. Murray, *Journal/The ACL Handbook.* NewYork: Springer, (2013).
80. R. W. Nutton, I. McLean, and E. Melville, *J. Roy. Coll. Surg. Edin.*, **44**, 236 (1999).
81. S. M. Strickland, J. D. MacGillivray, and R. F. Warren, *Orthop. Clin. N. Am.*, **34**, 41 (2003).
82. A. Robertson, R. W. Nutton, and J. F. Keating, *J. Bone Joint Surg. Br.*, **88**, 988 (2006).
83. J. L. Carey, W. R. Dunn, D. L. Dahm, S. L. Zeger, and K. P. Spindler, *J. Bone Joint Surg. Am.*, **91**, 2242 (2009).
84. L. M. Dejardin, S. P. Arnoczky, B. J. Ewers, R. C. Haut, and R. B. Clarke, *Am. J. Sports Med.*, **29**, 175 (2001).
85. S. F. Badylak, *Semin. Cell Dev. Biol.*, **13**, 377 (2002).
86. M. H. Zheng, J. Chen, Y. Kirilak, C. Willers, J. Xu, and D. Wood, *J. Biomed. Mater. Res. B*, **73**, 61 (2005).
87. D. A. Coons and F. Alan Barber, *Sports Med. Arthrosc. Rev.*, **14**, 185 (2006).
88. K. A. Derwin, A. R. Baker, R. K. Spragg, D. R. Leigh, and J. P. Iannotti, *J. Bone Joint Surg. Am.*, **88**, 2665 (2006).
89. T. F. Schlegel, R. J. Hawkins, C. W. Lewis, T. Motta, and A. S. Turner, *Am. J. Sports Med.*, **34**, 275 (2006).
90. C. K. Kuo, J. E. Marturano, and R. S. Tuan, *Sports Med. Arthrosc. Rehabil. Ther. Technol.*, **2**, 20 (2010).
91. A. J. Krych, J. D. Jackson, T. L. Hoskin, and D. L. Dahm, *Arthroscopy*, **24**, 292 (2008).
92. N. Mehran, J. G. Skendzel, B. P. Lesniak, and A. Bedi, *Oper. Techn. Sport Med.*, **21**, 10 (2013).
93. O. Laitinen, P. Tormala, R. Taurio, K. Skutnabb, K. Saarelainen, T. Iivonen, and S. Vainionpaa, *Biomaterials*, **13**, 1012 (1992).
94. J. A. Cooper, Jr., J. S. Sahota, W. J. Gorum 2nd, J. Carter, S. B. Doty, and C. T. Laurencin, *Proc. Natl. Acad. Sci. USA*, **104**, 3049 (2007).
95. B. Wang, W. Liu, Y. Zhang, Y. Jiang, W. J. Zhang, G. Zhou, L. Cui, and Y. Cao, *Biomaterials*, **29**, 2954 (2008).
96. H. W. Ouyang, J. C. Goh, A. Thambyah, S. H. Teoh, and E. H. Lee, *Tissue Eng.*, **9**, 431 (2003).
97. B. L. Seal, T. C. Otero, and A. Panitch, *Mat. Sci. Eng. R*, **34**, 147 (2001).

98. V. Guarino, F. Causa, and L. Ambrosio, *Expert Rev. Med. Dev.*, **4**, 405 (2007).
99. F. A. Petrigliano, C. S. English, D. Barba, S. Esmende, B. M. Wu, and D. R. McAllister, *Tissue Eng.*, **13**, 2721 (2007).
100. J. Lee, V. Guarino, A. Gloria, L. Ambrosio, G. Tae, Y. H. Kim, Y. Jung, S. H. Kim, and S. H. Kim, *J. Biomater. Sci. Polym. Edit.*, **21**, 1173 (2010).
101. M. Chvapil, D. P. Speer, H. Holubec, T. A. Chvapil, and D. H. King, *J. Biomed. Mater. Res.*, **27**, 313 (1993).
102. M. G. Dunn, J. B. Liesch, M. L. Tiku, and J. P. Zawadsky, *J. Biomed. Mater. Res.*, **29**, 1363 (1995).
103. G. H. Altman, R. L. Horan, H. H. Lu, J. Moreau, I. Martin, J. C. Richmond, and D. L. Kaplan, *Biomaterials*, **23**, 4131 (2002).
104. H. Liu, H. Fan, Y. Wang, S. L. Toh, and J. C. Goh, *Biomaterials*, **29**, 662 (2008).
105. R. L. Horan, I. Toponarski, H. E. Boepple, P. P. Weitzel, J. C. Richmond, and G. H. Altman, *J. Knee Surg.*, **22**, 82 (2009).
106. K. A. Derwin, A. R. Baker, R. K. Spragg, D. R. Leigh, and J. P. Iannotti, *J. Bone Joint Surg. Am.*, **88**, 2665 (2006).
107. T. W. Gilbert, A. M. Stewart-Akers, A. Simmons-Byrd, and S. F. Badylak, *J. Bone Joint Surg. Am.*, **89**, 621 (2007).
108. B. C. Fleming, K. P. Spindler, M. P. Palmer, E. M. Magarian, and M. M. Murray, *Am. J. Sports Med.*, **37**, 1554 (2009).
109. K. R. Kinneberg, V. S. Nirmalanandhan, N. Juncosa-Melvin, H. M. Powell, S. T. Boyce, J. T. Shearn, and D. L. Butler, *J. Orthop. Res.*, **28**, 1092 (2010).
110. C. H. Lee, H. J. Shin, I. H. Cho, Y. M. Kang, I. A. Kim, K. D. Park, and J. W. Shin, *Biomaterials*, **26**, 1261 (2005).
111. J. P. Spalazzi, M. C. Vyner, M. T. Jacobs, K. L. Moffat, and H. H. Lu, *Clin. Orthop. Relat. Res.*, **466**, 1938 (2008).
112. K. L. Moffat, A. S. Kwei, J. P. Spalazzi, S. B. Doty, W. N. Levine, and H. H. Lu, *Tissue Eng. A*, **15**, 115 (2009).
113. J. Xie, X. Li, J. Lipner, C. N. Manning, A. G. Schwartz, S. Thomopoulos, and Y. Xia, *Nanoscale*, **2**, 923 (2010).
114. Q. P. Pham, U. Sharma, and A. G. Mikos, *Tissue Eng.*, **12**, 1197 (2006).
115. C. Erisken, X. Zhang, K. L. Moffat, W. N. Levine, and H. H. Lu, *Tissue Eng. A*, **19**, 519 (2013).
116. Z. W. Ma, M. Kotaki, R. Inai, and S. Ramakrishna, *Tissue Eng.*, **11**, 101 (2005).
117. R. Murugan and S. Ramakrishna, *Tissue Eng.*, **13**, 1845 (2007).
118. W. J. Li, R. L. Mauck, J. A. Cooper, X. Yuan, and R. S. Tuan, *J. Biomech.*, **40**, 1686 (2007).
119. S. Sahoo, H. Ouyang, J. C. Goh, T. E. Tay, and S. L. Toh, *Tissue Eng.*, **12**, 91 (2006).
120. H. H. Lu, J. A. Cooper, Jr., S. Manuel, J. W. Freeman, M. A. Attawia, F. K. Ko, and C. T. Laurencin, *Biomaterials*, **26**, 4805 (2005).
121. J. A. Cooper, H. H. Lu, F. K. Ko, J. W. Freeman, and C. T. Laurencin, *Biomaterials*, **26**, 1523 (2005).
122. J. G. Barber, A. M. Handorf, T. J. Allee, and W. J. Li, *Tissue Eng. A*, (2011).

123. J. W. Freeman, M. D. Woods, and C. T. Laurencin, *J. Biomech.*, **40**, 2029 (2007).
124. H. F. Liu, H. B. Fan, Y. Wang, S. L. Toh, and J. C. H. Goh, *Biomaterials*, **29**, 662 (2008).
125. G. P. Chen, T. Sato, M. Sakane, H. Ohgushi, T. Ushida, J. Tanaka, and T. Tateishi, *Mater. Sci. Eng. C-Bio. S.*, **24**, 861 (2004).
126. X. Chen, Y. Y. Qi, L. L. Wang, Z. Yin, G. L. Yin, X. H. Zou, and H. W. Ouyang, *Biomaterials*, **29**, 3683 (2008).
127. N. A. Peppas and A. G. Mikos, *Hydrogels in Medicine and Pharmacy*, **1**, 1 (1986).
128. N. A. Peppas and B. D. Barr-Howell, *Hydrogels in Medicine and Pharmacy*, **1**, 27 (1986).
129. C. K. Kuo and P. X. Ma, *Biomaterials*, **22**, 511 (2001).
130. C. K. Kuo and P. X. Ma, *J. Biomed. Mater. Res. A*, **84**, 899 (2008).
131. T. R. Hoare and D. S. Kohane, *Polymer*, **49**, 1993 (2008).
132. D. Filpula and H. Zhao, *Adv. Drug. Deliv. Rev.*, **60**, 29 (2008).
133. F. Brandl, F. Sommer, and A. Goepferich, *Biomaterials*, **28**, 134 (2007).
134. C. R. Nuttelman, M. A. Rice, A. E. Rydholm, C. N. Salinas, D. N. Shah, and K. S. Anseth, *Prog. Polym. Sci.*, **33**, 167 (2008).
135. G. D. Nicodemus and S. J. Bryant, *Tissue Eng. B Rev.*, **14**, 149 (2008).
136. B. V. Slaughter, S. S. Khurshid, O. Z. Fisher, A. Khademhosseini, and N. A. Peppas, *Adv. Mater.*, **21**, 3307 (2009).
137. C. K. Kuo and R. S. Tuan, *Tissue Eng. A*, **14**, 1615 (2008).
138. D. M. Doroski, M. E. Levenston, and J. S. Temenoff, *Tissue Eng. A*, **16**, 3457.
139. P. J. Yang, M. E. Levenston, and J. S. Temenoff, *Tissue Eng. A*, **18**, 2365.
140. U. Noth, K. Schupp, A. Heymer, S. Kall, F. Jakob, N. Schutze, B. Baumann, T. Barthel, J. Eulert, and C. Hendrich, *Cytotherapy*, **7**, 447 (2005).
141. J. T. Shearn, N. Juncosa-Melvin, G. P. Boivin, M. T. Galloway, W. Goodwin, C. Gooch, M. G. Dunn, and D. L. Butler, *J. Biomech. Eng.*, **129**, 848 (2007).
142. W. R. Webb, T. P. Dale, A. J. Lomas, G. D. Zeng, I. Wimpenny, A. J. El Haj, N. R. Forsyth, and G. Q. Chen, *Biomaterials*, **34**, 6683 (2013).
143. N. Juncosa-Melvin, G. P. Boivin, C. Gooch, M. T. Galloway, J. R. West, M. G. Dunn, and D. L. Butler, *Tissue Eng.*, **12**, 369 (2006).
144. C. N. Manning, A. G. Schwartz, W. Liu, J. Xie, N. Havlioglu, S. E. Sakiyama-Elbert, M. J. Silva, Y. Xia, R. H. Gelberman, and S. Thomopoulos, *Acta. Biomater.*, **9**, 6905 (2013).
145. R. G. Young, D. L. Butler, W. Weber, A. I. Caplan, S. L. Gordon, and D. J. Fink, *J. Orthop. Res.*, **16**, 406 (1998).
146. M. M. Murray, K. P. Spindler, C. Devin, B. S. Snyder, J. Muller, M. Takahashi, P. Ballard, L. B. Nanney, and D. Zurakowski, *J. Orthop. Res.*, **24**, 820 (2006).
147. M. M. Murray, K. P. Spindler, E. Abreu, J. A. Muller, A. Nedder, M. Kelly, J. Frino, D. Zurakowski, M. Valenza, B. D. Snyder, and S. A. Connolly, *J. Orthop. Res.*, **25**, 81 (2007).
148. S. M. Joshi, A. N. Mastrangelo, E. M. Magarian, B. C. Fleming, and M. M. Murray, *Am. J. Sports Med.*, **37**, 2401 (2009).

149. M. M. Murray, E. Magarian, D. Zurakowski, and B. C. Fleming, *Arthroscopy*, **26**, S49 (2010).
150. K. Saiga, T. Furumatsu, A. Yoshida, S. Masuda, S. Takihira, N. Abe, and T. Ozaki, *Biochem. Bioph. Res. Co.*, **402**, 329 (2010).
151. A. F. Steinert, M. Weber, M. Kunz, G. D. Palmer, U. Noth, C. H. Evans, and M. M. Murray, *Biomaterials*, **29**, 904 (2008).
152. M. Haddad-Weber, P. Prager, M. Kunz, L. Seefried, F. Jakob, M. M. Murray, C. H. Evans, U. Nöth, and A. F. Steinert, *Cytotherapy*, **12**, 505 (2010).
153. Y. Qiu, J. J. Lim, L. Scott, Jr., R. C. Adams, H. T. Bui, and J. S. Temenoff, *Acta Biomater.*, **7**, 959.
154. A. Seidi, M. Ramalingam, I. Elloumi-Hannachi, S. Ostrovidov, and A. Khademhosseini, *Acta Biomater.*, **7**, 1441 (2011).
155. L. R. Carr, J. E. Krause, J.-R. Ella-Menye, and S. Jiang, *Biomaterials*, **32**, 8456.
156. J. Z. Paxton, K. Donnelly, R. P. Keatch, and K. Baar, *Tissue Eng. Part A*, **15**, 1201 (2009).
157. S. Oka, T. Matsumoto, S. Kubo, T. Matsushita, H. Sasaki, Y. Nishizawa, T. Matsuzaki, T. Saito, K. Nishida, Y. Tabata, M. Kurosaka, and R. Kuroda, *Tissue Eng. Pt. A*, **19**, 1233 (2013).
158. J. Lee, W. Il Choi, G. Tae, Y. H. Kim, S. S. Kang, S. E. Kim, S. H. Kim, Y. Jung and S. H. Kim, *Acta Biomater.*, **7**, 244 (2011).
159. A. Khanna, M. Friel, N. Gougoulias, U. G. Longo, and N. Maffulli, *Brit. Med. Bull.*, **90**, 85 (2009).
160. R. E. H. Ferguson and B. Rinker, *Ann. Plas. Surg.*, **56**, 54 (2006).
161. C. Zhao, Y.-L. Sun, P. C. Amadio, T. Tanaka, A. M. Ettema, and K.-N. An, *J. Bone Joint Surg. Am.*, **88**, 2181 (2006).
162. Y. C. Liu, A. Skardal, X. Z. Shu, and G. D. Prestwich, *J. Orthop. Res.*, **26**, 562 (2008).
163. N. Ishiyama, T. Moro, K. Ishihara, T. Ohe, T. Miura, T. Konno, T. Ohyama, M. Kimura, M. Kyomoto, K. Nakamura, and H. Kawaguchi, *Biomaterials*, **31**, 4009 (2010).
164. M. A. Haque, T. Kurokawa, and J. P. Gong, *Polymer*, **53**, 1805 (2012).
165. J. Y. Sun, X. H. Zhao, W. R. K. Illeperuma, O. Chaudhuri, K. H. Oh, D. J. Mooney, J. J. Vlassak, and Z. G. Suo, *Nature*, **489**, 133 (2012).
166. M. C. Darnell, J. Y. Sun, M. Mehta, C. Johnson, P. R. Arany, Z. G. Suo, and D. J. Mooney, *Biomaterials*, **34**, 8042 (2013).

Chapter 10

Hydrogels in Bone Tissue Engineering

Minh Khanh Nguyen,[*,a] **Julia E. Samorezov**[*,a] **and Eben Alsberg**[*,†,‡,§]

**Department of Biomedical Engineering, Case Western Reserve University*
10900 Euclid Ave., Cleveland, Ohio 44016 (USA)
[†]Department of Orthopaedic Surgery, Case Western Reserve University
10900 Euclid Ave., Cleveland, Ohio 44016 (USA)
[‡]Division of General Medical Sciences, National Center for Regenerative Medicine, Case Western Reserve University
10900 Euclid Ave., Cleveland, Ohio 44016 (USA)
[§]eben.alsberg@case.edu

1. Introduction

Bone injuries can result in patient morbidity and even mortality in the absence of effective healing, creating a critical need for engineered bone tissue. Bone grafts are currently used in the treatment of non-union fractures or critical-size bone defects caused by orthopedic trauma, tumor resections, or congenital malformations; 2.2 million bone grafts are performed annually worldwide, creating a $2.5 billion industry.[1,2] Autografts, the clinical gold standard, are limited by donor site morbidity and potential lack of available tissue. Alternatives such as allografts carry the risk of immune rejection or disease transmission. Similarly, distraction osteogenesis, a lengthy and painful process, also carries the risk of pin infections. For these reasons, a great deal of effort has been dedicated to develop bone tissue engineering strategies. Hydrogels, water swollen hydrophilic polymer networks, are uniquely suited biomaterials for these approaches because of their modularity — different modifications are possible to tailor properties such as their porosity and pore size, cell adhesivity, mechanics, degradation rate, and capacity to control the delivery of bioactive factors, sometimes independently. In addition, their high water content and similarities to

[a] M. K. Nguyen and J. E. Samorezov contributed equally to this work.

native extracellular matrix (ECM) make hydrogels appealing scaffolds for engineered bone. This chapter will present research pursuits to develop hydrogel systems for bone tissue engineering, emphasizing recent work in the literature.

2. Bone Biology

2.1. Bone structure

The human skeleton is composed of 206 bones with diverse structure, including long bones, short bones, flat bones, and irregular bones. All of these bones are made of one of two types of tissue: Overall, 80% is cortical bone (also called compact bone) and 20% is cancellous bone (also called trabecular bone).[3] Cortical bone surrounds cancellous bone, and also makes up the shafts of long bones.[4] It is very strong, especially in compression, and dense, with <5% porosity.[4,5] In contrast, while cancellous bone has similar composition (50–70% mineral content, 20–40% organic matrix, 5–10% water, and <3% lipids),[4] its structure is made up of a highly porous (50–90%) network of thin rods called trabeculae.[6] This increased porosity leads to increased surface area allowing for higher metabolic activity.

The primary mineral component in bone is hydroxyapatite, in the form of small, thin plate-shaped crystals approximately 300 × 200 Å, with thickness 10–40 Å, or rods of similar length that are only 8–15 Å thick.[7,9] Hydroxyapatite is strong but brittle, and requires its interaction with the organic matrix to provide toughness to help prevent bone fractures.[10] This mineral phase can also serve as a reservoir for calcium and phosphate ions, which can be stored or released depending on the body's needs.[11] The organic matrix component is primarily the protein type I collagen,[4] and glycoproteins and proteoglycans that can bind to hydroxyapatite, serve regulatory functions such as modulating cell attachment, proliferation and differentiation, or help to sequester growth factors.[11]

The two main cells that form and remodel this matrix are osteoblasts and osteoclasts. Osteoblasts are derived from mesenchymal stem cells (MSCs) in the marrow or periosteum. These cells are responsible for secreting osteoid, the organic matrix of bone which is ultimately mineralized. After laying down bone matrix, osteoblasts can become osteocytes, terminally differentiated osteoblasts embedded within the mineralized bone, or bone lining cells, flattened cells that cover the bone surface and extend processes to communicate with osteocytes. The osteocytes serve important functions such as maintaining ion balance, sensing mechanical loading and signaling to initiate bone remodeling.[3] The cells responsible for bone resorption are the hematopoietic stem cell-derived osteoclasts. These multinucleated cells release proteolytic enzymes and create an acidic local environment to break down the organic and mineral components of bone matrix, respectively.[4]

Except at articulating surfaces and some ligament and tendon attachment points, the outer bone surfaces are lined with periosteum, a very metabolically active connective tissue layer.[12] The periosteum is comprised of two layers: the outer fibrous layer is dense and made up primarily of fibroblasts in collagen and elastin, permitting for ligament and/or tendon attachments, and the inner layer contains MSCs and osteogenic progenitor cells, osteoblasts, and fibroblasts.[13] It is also highly vascularized and innervated. Because of the cell populations it contains, the periosteum is very important for bone repair and regeneration. The inner bone surfaces are lined with endosteum, a thin layer of connective tissue which interfaces with the bone, marrow, and blood vessels. This tissue also has high metabolic activity; bone resorption, described below, more often occurs from the endosteal surface of the bone.[4] The medullary cavity in the bone interior is filled with marrow, a hematopoietic tissue responsible for blood cell formation, composed of fat, MSCs, hematopoietic stem cells and many blood vessels.[14] Successful bone tissue engineering aims to regenerate these structures to create a tissue that can resist the mechanical loads bone is subject to and integrate with the surrounding tissue.

2.2. Bone development

Efforts to regenerate bone tissue often seek to recapitulate the body's natural bone development processes. There are two main pathways for bone development in the human skeleton during growth: Most of the flat craniofacial bones and some short bones develop via intramembranous ossification and the remaining bones develop primarily via endochondral ossification.[15,16]

In intramembranous ossification, MSCs differentiate directly into osteoblasts. This happens when a group of MSCs rapidly divide to form a cluster or nodule.[17] These cells then stop dividing and differentiate, laying down osteoid and creating a small region of newly calcified bone matrix. This continues around a network of capillaries, creating woven bone. Vascularized mesenchyme condenses on the surface of the newly formed bone to become the periosteum.[17] This randomly oriented bone can then be remodeled to form cortical or cancellous bone.[15]

In contrast, during endochondral ossification, bone is laid down following development of a cartilage template.[18,19] Endochondral ossification occurs during embryonic development and continues after birth; it is the method by which long bones grow in length.[19] This process begins with the formation of cartilage, derived from MSCs that differentiated into chondrocytes. In long bones, a ring of perichondrium around middle of the cartilage becomes periosteum, which then follows the intramembranous ossification pathway to form a bony collar.[19] At the same time, the chondrocytes on the inside hypertrophy, the cartilage ECM becomes calcified, and the cells secrete specific matrix metalloproteases and growth factors including VEGF.[19,20] Blood vessels permeate the calcified cartilage interior, bringing progenitor cells that can become osteoclasts to break down the cartilage matrix and osteoblasts to begin secreting osteoid, mineralizing it and forming new woven bone.[19,21] The woven bone is then remodeled into mature cortical or cancellous bone. This process continues during growth of long bones — areas of cartilage in what are called growth plates remain and when developing bones are growing longer, the growth plate chondrocytes proliferate, hypertrophy, and ultimately are replaced by bone tissue formed by endochondral ossification. In adults, the growth plates are ossified, therefore preventing further bone growth.[18] Conversion of the avascular cartilage to highly vascular and mineralized bone is an intricate process, but starting with a tissue template with low metabolic requirements that develops into mature bone is a promising strategy for tissue engineering.

2.3. Bone maintenance and healing

Bone tissue can be highly metabolically active, with turnover rate dependent on mechanical loading, hormones, growth factors, and body ion balance.[22] In adults, bone is renewed regularly through remodeling. Osteoclasts and osteoblasts work in sequence to resorb old bone and replace it with new bone. This process starts with recruitment and activation of osteoclast precursors, which mature into osteoclasts that then bind to the bone matrix and locally break down the bone. Preosteoblasts are then recruited and differentiate into osteoblasts which synthesize new organic matrix and concentrate calcium and phosphate to mineralize it.[23] The result is new cortical or cancellous bone that has no microdamage and is well suited for the body's mechanical loading needs.

Bones maintain a capacity for self-repair throughout adult life, healing defects below a critical size. Understanding this healing process is useful in bone tissue engineering. Immediately after a bone fracture, a hematoma, or localized collection of blood, develops around the site to bring in inflammatory cells and fibroblasts. These cells secrete various signals that initiate the healing response, and the fibroblasts form granulation tissue while necrotic tissue is removed by osteoclasts.[24] Defects that are stabilized or rigidly fixed after fracture heal primarily by the mechanism of intramembranous ossification as described earlier — MSCs brought to the fracture site differentiate directly into osteoblasts.[25] In contrast, fractures with limited motion allowed at the injury site, such as those in a cast, heal mainly by endochondral ossification: after 3–4 weeks, a cartilaginous fracture callus forms which ultimately calcifies, and is surrounded by a cuff of woven bone formed by intramembranous ossification.[26] The calcified cartilage is replaced by mineralized bone tissue that bridges the two bone fragments. Over time, the new woven bone is remodeled, with mature cortical or cancellous bone taking its place.

3. Cell Sourcing

Hydrogels for bone tissue engineering should fill the defect site and provide a scaffolding for cells. These cells may be the patient's own cells recruited to the defect for repair, or transplanted cells delivered directly to the defect site. Cell-free hydrogels use physical or biochemical cues to induce migration of host cells to the defect site. The hydrogels are often described as osteoconductive if they allow bone formation by the cells they contact, and osteoinductive if the material system alone is enough to stimulate osteogenesis. Host cell recruitment by an osteoinductive scaffold can take advantage of the fracture repair process by bringing MSCs and osteoblasts to the defect site, and is most successful in the case of an acute injury.[27] Cell transplantation may be desirable for larger defects, where there may not be adequate numbers of cells present to repair the site and it may take those cells present longer to migrate into the defect interior. In this case, cell-laden hydrogels can be cultured in a defined *in vitro* environment for a period of time to form bone-like tissue prior to implantation. Alternately, constructs can be injected or implanted immediately, when valuable in treating defects is resulting from trauma.

3.1. Differentiated cells and cell lines

Early cell-based work in bone tissue engineering involved predominantly differentiated osteoblasts encapsulated in biomaterial hydrogels. Examples of cell sources for early hydrogel *in vitro* and animal work include the MC3T3 mouse preosteoblast line[16,28] and osteoblasts derived directly from rat calvaria.[16,29,30] These studies have shown that a number of hydrogel properties, including those described in Sec. 5 of this chapter, are critical to bone tissue formation. While these cells are quite useful for studying osteoblast behavior and how it is affected by hydrogel properties, they are limited in their clinical applications. In the case of autologous cells, the invasive harvesting procedures employed may lead to donor site morbidity, and elderly or diseased patients may lack healthy bone tissue from which to obtain differentiated cells.[31] Allogeneic osteoblasts can be obtained from cadaver tissue, but may trigger an immune response in the donor, and carry the potential for disease transmission that, while small, is a risk with serious consequences.[32]

3.2. Adult stem cells

The discovery of multipotent cells in the bone marrow would become very significant for tissue engineering. As early as 1966 Friedenstein and colleagues were able to stimulate cells from the bone marrow of mice to undergo osteogenesis.[33] Ultimately named MSCs, these cells have proven valuable in a number of systems for osteogenesis, including hydrogels. They can be isolated from human bone marrow in a minimally invasive procedure, most often from the iliac crest, expanded by *in vitro* culture, and differentiated into osteoblasts *in vitro* and *in vivo*.[34–36] Most commonly, culture supplements including dexamethasone, ascorbic acid, and β-glycerophosphate are used for osteogenic induction of human MSCs (hMSCs). Many hydrogel modifications have been studied to enhance or drive hMSC osteogenesis, including varying material properties such as adhesion and stiffness as well as delivering growth factors including bone morphogenetic proteins (BMPs), which are members of the transforming growth factor β (TGF-β) superfamily.[37–39]

More recently, adipose tissue has been shown to be one of several additional sources to contain a similar population of cells with mesenchymal lineage differentiation potential.[40] Most often called human adipose-derived stem cells (hASCs), these cells are obtained from liposuction aspirates and also show strong *in vitro* and *in vivo* osteogenic differentiation in response to the same supplements used for hMSC osteogenesis.[41,42] hASCs have recently been encapsulated in biomaterial hydrogels and differentiated into osteoblasts.[43] Because of the relative ease of obtaining these cells and their strong osteogenic potential, they have the capacity for use in bone tissue engineering. Other tissue sources of MSCs in humans include dental pulp, skeletal muscle, synovial tissue, heart, placenta, and cord blood.[44–49] However, these cells are generally less well characterized than bone marrow-derived hMSCs, and their potential in bone regeneration is less well explored.

3.3. Embryonic stem cells

Cells derived from the inner cell mass of a preimplantation human embryo have been shown to have the capacity to differentiate *in vitro* into almost all cell types with therapeutic potential, including osteoblasts. This was demonstrated as early as 2001 with mouse embryonic stem cells[50] and in 2003 with human embryonic stem cells (hESCs),[51] relying on embryoid body formation followed by culture in media identical to a formulation used to differentiate hMSCs into osteoblasts. Little work has been explored with hydrogels as scaffolds for hESC osteogenesis. In one recent example, alginate hydrogel microbeads were used to encapsulate hESCs and then mixed with a paste of calcium phosphate cement particles in chitosan. The resulting system exhibited osteogenic gene expression and mineralization.[52] The drug delivery capacity of hydrogels can be exploited to enhance bone formation *in vivo* by hESCs. For example, preliminary work has shown that bone morphogenetic protein-2 (BMP-2) delivery from a poly(N-isopropylacrylamide-*co*-acrylic acid) hydrogel improved *in vivo* subcutaneous bone formation by encapsulated hESCs.[53] While embryonic stem cells do show some potential, future work will have to address possible tumor formation, an important risk to consider when using these cells.[54,55]

3.4. Induced pluripotent stem cells

The recent development of induced pluripotent stem cells (iPSCs), where adult somatic cells can be reprogrammed *in vitro* to a state of pluripotency, gives the promise of treatments using autologous stem cells for all tissue types.[56] Once fibroblasts are reverted to a state of pluripotency, they can be differentiated into osteoblasts following similar protocols to those used for hMSCs.[57] Demonstrating the exciting potential of this cell population, iPSCs were recently delivered in silk scaffolds for treating periodontal bone defects.[58] Because the use of iPSCs is so new, little work has been done to study iPSC osteogenesis in hydrogels. However, as they did for other cell types, hydrogels will likely play a role in the model systems designed to understand and optimize iPSC osteogenesis as well as deliver these cells to bone defects for tissue engineering purposes.

4. Choice of Material

Hydrogels are an important class of biomaterials for bone tissue engineering, serving as bioactive factor delivery depots and providing temporary structural support and signaling for cell growth and new bone tissue formation. One reason for the rapid advancement in this field is the better understanding of biological, chemical and physical signals that play a role in osteogenesis. The wide range of the biomaterial properties known to affect bone tissue formation includes cell adhesivity,[59] hydrogel stiffness,[60] degradation rate,[61] pore size and porosity,[62] and controlled presentation of osteogenic bioactive factors.[63] The potential modularity of hydrogels is especially useful, as one base material can be modified to change one or more of these properties, often independently. Hydrogel materials employed for tissue engineering can be synthetic or from naturally-derived sources.

4.1. Naturally-derived hydrogels

Hydrogels based on naturally-derived polymers, such as agarose, alginate, chitosan, fibrin, gelatin, collagen, and hyaluronic acid, are of interest for bone tissue engineering because they often exhibit inherent biological activity, biocompatibility and/or biodegradability.[64] For example, gelatin hydrogels containing BMP-2 were used to repair a skull defect in monkeys.[65] Similarly, fibrin hydrogels were also harnessed to release BMP-2 to enhance bone formation *in vitro* and in rats.[66] Alginate hydrogels modified with cell adhesion ligands have been shown to support bone formation in both *in vitro* and *in vivo*.[63] Natural polymers do have drawbacks, such as batch-to-batch variation, a limited range of physical properties and potential pathogen transmission.[67]

4.2. Synthetic hydrogels

Synthetic polymer hydrogels provide a versatile platform for bone tissue engineering because of the control they offer: their chemical structures can be easily tuned to achieve a wide range of physical properties such as stiffness and degradation rate,[68] they possess no risk of biological pathogens and do not exhibit batch-to-batch variation in product.[64,68] In spite of these advantages, they exhibit no inherent biological activity and some synthetic polymers can produce acidic byproducts.[64,67] Poly (ethylene glycol) (PEG) is one of the well-studied synthetic polymers for hydrogels applied to bone regeneration.[69,70] For example, PEG hydrogels could support the osteogenic differentiation of encapsulated hMSCs cultured in osteogenic media.[69] In addition, numerous other synthetic polymer hydrogels, such as oligo(poly(ethylene glycol) fumarate) (OPF)[71] and sulfamethazine-poly(ε-caprolactone-*co*-lactic acid)-PEG-poly(ε-caprolactone-*co*-lactic acid)-sulfamethazine,[72] have also been utilized in bone tissue engineering. To combine some of the benefits of synthetic hydrogels with the biologic activity of naturally-derived hydrogel materials, hybrid materials are being engineered. For example, PEG hydrogels were covalently functionalized with a cell adhesion peptide containing the Arg–Gly–Asp (RGD) sequence, which is present in several ECM molecules, to facilitate cell attachment to the crosslinked network.[69]

5. Hydrogel Physical and Biochemical Properties

5.1. Cell adhesivity

In tissue regeneration including bone engineering, the interactions of cells with hydrogels can dramatically affect cellular response. Presentation of adhesion ligands, peptide sequences found in ECM proteins that can bind to cell surface integrin receptors, can promote cell attachment to hydrogels, which is critical for cell migration,[73] proliferation[74] and differentiation.[74–76] Adhesion ligands (e.g., RGD, Ile–Lys–Val–Ala–Val (IKVAV) and Tyr–Ile–Gly–Ser–Arg (YIGSR))[77–79] and ECM proteins containing these sequences (e.g., fibronectin and laminin)[80–82] can be covalently tethered to hydrogel macromers for the attachment of cells through their receptors. In addition, hydrogels derived from some natural polymers such as chondroitin, hyaluronic acid, collagen, gelatin, and fibrin can support cell adhesion without the need of additional cell adhesion ligands or proteins.[83,84] Cell adhesion ligand type and density have been shown to control cell behavior *in vitro*, and their presentation can enhance bone tissue formation *in vivo*. Rat osteoblasts in alginate hydrogels covalently modified with peptides containing the RGD peptide sequence form significantly more bone subcutaneously in a mouse model than those in hydrogels without the peptide.[16]

5.2. Stiffness and strength

Cells in or on hydrogels are sensitive to mechanical signals, including stiffness, which is an important cue in bone tissue engineering. Hydrogels typically have elastic moduli many orders of magnitude lower than bone: Cancellous bone and cortical bone have Young's moduli around 1–15 GPa and 17 GPa, respectively.[85,86] Most hydrogels for bone regeneration are used *in vivo* either in non-load bearing areas or with some sort of fixation, so their decreased stiffness compared to native bone does not impede their use in this application.[87] However, the range of moduli achieved by hydrogels, from <0.1 kPa to hundreds of kPa, can significantly affect the behavior of osteoblasts and their progenitors.[88] Without changing the base material, crosslinking density, crosslinker molecular weight, and macromer molecular weight and concentration can all affect the mechanical properties of hydrogels.[89–92] For example, the compressive moduli of alginate hydrogels increased when a higher amount of Ca^{2+} was used to crosslink them.[93] Tensile and compressive moduli of photo crosslinked PEG hydrogels increased with greater PEG concentrations.[90] The resulting changes in stiffness have been shown to affect osteoblast, preosteoblast, or hMSC proliferation, osteogenic differentiation, and mineralization.[28,39,94]

5.3. Degradation

Hydrogel degradability is required for tissue engineering applications to create free space that is necessary for cell proliferation, the production of new bone tissue and enhanced diffusion of nutrients to and metabolic waste from developing tissue.[95] Degradable hydrogels can serve as a temporary matrix for cell multiplication and ECM production, and are replaced with newly formed tissue. Ideally, their degradation rate should be optimized to match the growth rate of developing tissue. Many of the factors that affect hydrogel stiffness also play a role in determining the degradation rate: This property can be tuned by, for example, varying crosslinker density and molecular weight,[92,96,97] macromer concentration[96] or macromer molecular weight.[61,98] Hydrogels often degrade through hydrolytic,[96,97] enzymatic[99] or photolytic cleavage[100] of the macromer backbone or crosslinkers and the resulting degraded molecules can leach out of the implant site. If these degradation products have molecular weights lower than approximately 50 kDa, they can be cleared from the body by the kidneys.[101] Hydrolytically degradable hydrogels are constructed using macromers containing water-labile bonds, such as esters, orthoesters, acetals or carbonates.[102,103] The hydrolytic degradation rate of hydrogels in aqueous media is strongly dependent on pH and salt concentration. Hydrogels derived from components of natural tissues (e.g., collagen, hyaluronic acid, etc.) or crosslinked with an enzyme-sensitive molecule (such as a matrix metalloproteases (MMPs)-sensitive peptide) can degrade in the presence of specific enzymes (such as MMPs, hyaluronidase or collagenases).[99,104] This approach permits cells in the hydrogels and/or at an *in vivo* defect site to locally produce enzymes to regulate their degradation profiles. Hydrogels containing photo-cleavage molecules such as *ortho*-nitrobenzyl esters can degrade upon exposure to a light source, and their degradation properties can be controlled by light intensity and exposure time.[100] While this adds a level of complexity to systems used *in vivo*, it does allow for a key benefit: Highly precise spatiotemporal control over hydrogel degradation kinetics.

5.4. Porosity and pore size

Biomaterial scaffolds require void spaces to give cells room to grow and secrete new ECM, and to allow for diffusion of nutrients and metabolites.[95,105] These void spaces are usually characterized by porosity and pore size. Although both porosity and pore size increase during hydrogel swelling and degradation, hydrogels often exhibit nanoscale pore size (nanoporous hydrogels).[95] Previous work with solid materials such as ceramics showed that scaffolds with pore sizes higher than 100 μm are favorable for enhancing new bone formation and permitting vascularization.[106,107] This work supports the development of hydrogels with porosity on this scale. Macroporous hydrogels engineered to have larger pores on the order of hundreds of microns have been shown to facilitate cell proliferation and differentiation.[62,108] Techniques developed to fabricate macroporous hydrogels include gas foaming using $NaHCO_3$[108] or a salt leaching[62] method. In addition, macroporous hydrogels have also been created by the encapsulation and subsequent dissolution of polymer microspheres.[109,110] In alginate hydrogels with pores created by dissolved gelatin microspheres, the resulting interconnected macropores to increased spreading and proliferation of mouse preosteoblast cells.[111] Additionally, it has been shown that a hydrogel with pores ranging from 100–600 μm could promote hMSC growth and differentiation into osteoblasts.[108] Other work has demonstrated that 100 μm pore size hydrogels have potential for bone regeneration in orbital floor fractures.[62] Recently, osteogenic differentiation of hMSCs in macroporous hydrogels with similar porosity but different micro architectural structure was examined, with spongy hydrogels promoting greater differentiation than columnar hydrogels.[112]

6. Delivery of Bioactive Factors

In bone tissue engineering, bioactive factors such as growth factors, genes, and osteotrophic drugs are valuable since they can mediate cell proliferation, migration, and differentiation, bone formation at the site of implant (osteogenesis) and/or vascularization.

Bioactive molecules often exhibit a short half-life and poor tissue penetration.[113] As a result, their frequent injection at high doses can be necessary to maintain therapeutic efficacy,[114] which may result in toxic and/or off-target effects. For example, high doses of BMP-2 can actually encourage bone resorption,[115] or lead to the development of structurally abnormal bone and tissue inflammation.[116] Therefore, scaffolds such as hydrogels that can locally deliver the desired bioactive molecules to target tissues in a controlled manner over a desired period of time may reduce the potential negative effects of high doses while maintaining positive effects on neotissue formation.

6.1. Growth factor delivery

BMPs, the most well studied growth factors for bone tissue regeneration, can bind to osteoprogenitor cells and osteoblasts and induce their migration, proliferation, differentiation, and formation of new bone tissue.[117–119] Recombinant human BMP-2 (rhBMP-2) (Infuse, Medtronic Sofamor Danek Inc., Minneapolis, MN) and rhBMP-7 (osteogenic protein-1, OP-1, Stryker Biotech, Hopkinton, MA) in collagen sponges were approved by US Food and Drug Administration (FDA) and used widely for some clinical applications, such as the treatment of femur and tibia fracture non-unions[120–122] and spinal fusions.[123–125] While they have demonstrated success in their approved applications, these sponges provide little control over the release of the BMPs, which release much higher BMP concentrations to bone defects than present during natural healing.[126,127] As a result, these systems can lead to vertebral body bone resorption,[128] or swelling that causes significant side effects when used in the anterior cervical spine.[129] To address these problems, functionalized hydrogels have been engineered to locally and gradually release lower BMP-2 doses to tissues to, for example, induce ectopic bone formation in rat[130] and mouse (Fig. 1)[63] models. Future work will likely focus on further enhancing control of BMP release kinetics.

Delivery of therapeutic growth factors to induce angiogenesis or vascularization is also a promising approach for the treatment of bone defects because of the close relationship between fracture healing and formation of blood vessels. Capillary networks facilitate the transport of nutrients and oxygen to tissues. Vascular endothelial growth factor (VEGF), basic fibroblast growth factor (bFGF) and platelet-derived growth factor (PDGF) are important factors for enhancing tissue vascularization. VEGF and bFGF increase endothelial cell migration and proliferation, and ultimately promote vessel formation.[131] PDGF does not directly increase angiogenesis, but it stimulates production of VEGF[132] and promotes stabilization of newly formed vasculature.[133] In addition, bFGF and PDGF have mitogenic effects on osteoprogenitor cells and promote osteogenesis.[132,134] VEGF releasing hydrogels stimulated early blood vessel formation and bone regeneration at a later time point in a rat critically-sized calvarial bone defect model.[135] Similarly,

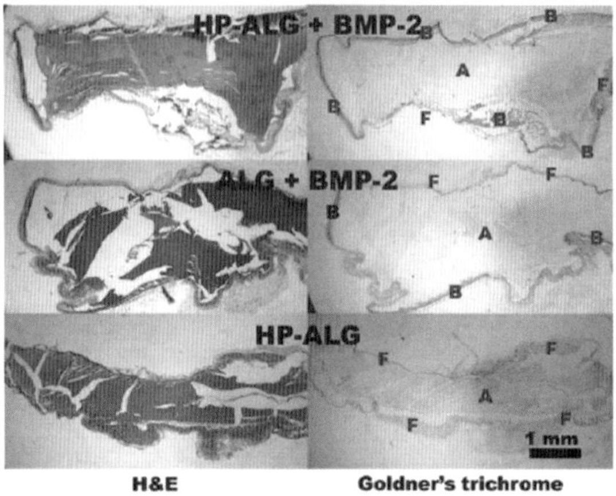

(A)

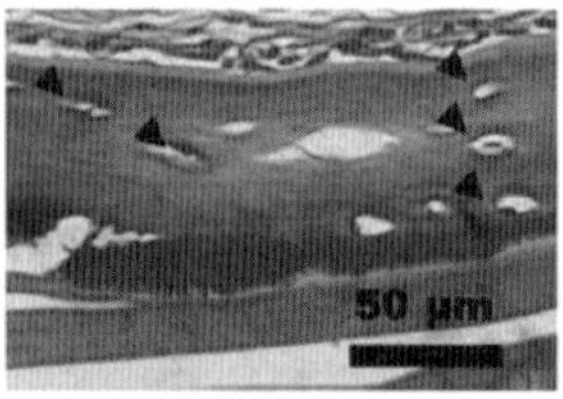

(B)

Fig. 1. (A) Photomicrographs of hematoxylin and eosin (H&E) and Goldner's trichrome-stained sections of photocrosslinked alginate hydrogels implanted subcutaneously in the backs of mice and harvested at 8 weeks. The experimental conditions were BMP-2-loaded heparin-alginate (HP-ALG), BMP-2-loaded unmodified alginate, and empty HP-ALG. More bone was formed in the HP-ALG + BMP-2 condition. A: residual alginate hydrogel; B: bone; F: fibrous tissue. (B) A higher magnification photomicrograph of an H&E stained BMP-2-loaded HP-ALG implant section after 8 weeks showing newly formed bone tissue surrounding the construct. Closed triangles show osteocytes within lacunae. Reproduced with permission from Ref. 63. Copyright © 2011 Elsevier.

hydrogels containing bFGF promoted micro-vascularization and bone regeneration in mice maxillae.[136] When an angiogenic molecule (i.e., VEGF) was delivered in combination with an osteogenic molecule such as BMP-2, synergistic effects of the co-delivery substantially increased bone regeneration as a result of the formation of vascular network.[137] Such co-delivery systems show great promise for enhancing vascularization and bone formation.

6.2. Gene delivery

Gene therapy has gained great interest for driving bone tissue regeneration; delivering nucleic acids can lead to changes at the transcriptional or translational level.[138,139]

It may be desirable to deliver DNA because the cells taking up the material can be programmed to produce osteogenic or vasculogenic growth factors providing for their sustained presentation to cells at a defect site.[140,141] This would negate the need to deliver the proteins, which may be expensive and/or lose bioactivity over time. The DNA could also encode transcription factors promoting, for example, VEGF production for angiogenesis.[9,11] The delivery of DNA to cells is performed using viral or non-viral vectors.[138] Viral vectors such as adenoviral and retroviral vectors exhibit highly efficient transduction capability, resulting in enhanced gene expression, but they can cause a host immune response and integration of the DNA into the host genome.[138] The delivery of non-viral material, such as plasmid DNA (pDNA), can avoid some of the negative effects of viral transfection, but it is not efficient because the macromolecular nature and negative charge of pDNA limit its ability to penetrate the negatively charged cell membrane.[142] One strategy to overcome this issue is compacting the negatively charged pDNA with transfection agents (e.g., cationic polymers, cationic liposomes, etc.) into nanoparticles via electrostatic interactions prior to delivery. This packing technology can deliver DNA to cells better than the administration of naked pDNA alone. Despite the lower levels of transfection efficiency, the non-viral delivery approach offers the benefit of decreased immunogenicity, and improved safety due to its transient effects on gene expression.[138] Scaffolds releasing pDNA encoding for a therapeutic gene are well known as gene activated matrices. Delivery of DNA encoding BMP-2 from alginate hydrogels could promote early osteogenic differentiation of multipotent stromal cells both *in vitro* and in mice,[143] and induce bone formation at ectopic and orthotopic locations in goats.[144] Sustained delivery of calcium-phosphate nanoparticles containing DNA encoding for BMP-2 from alginate hydrogels could also enhance bone formation of transplanted MC3T3 cells in mice.[145] In addition to pDNA, short interfering RNA (siRNA), a genetic molecule that can silence cell gene expression post-transcriptionally via a mechanism called RNA interference (RNAi), released from a biodegradable hydrogel could suppress gene expression of noggin (a BMP inhibitor), resulting in ectopic bone formation in mice.[146] Recently, sustained delivery of siRNA condensed with cationic polyethyleneimine into nanoparticles from *in situ* forming PEG hydrogels could enhance the osteogenic differentiation of encapsulated hMSCs *in vitro*.[147]

6.3. Drug and small molecule delivery

In addition to delivery of growth factors and genetic material, osteotrophic molecules such as fluvastatin, simvastatin, and parathyroid hormone peptide can also be delivered from hydrogels for bone tissue engineering. Fluvastatin and simvastatin, family members of statins, have been found to induce bone formation.[148–150] They have been released from hydrogels to promote osteogenic differentiation of hMSCs[151] and MC3T3 cells,[152] and bone regeneration of critically-sized nasal defects in rabbits.[153] Parathyroid hormone produced by the parathyroid gland can also stimulate bone formation,[154] and local delivery of parathyroid hormone peptide (PTH1-34) from PEG hydrogels enhanced bone repair in the dog mandible.[155]

7. Recent Developments

7.1. Use of hydrogels as high throughput screening systems

To gain a better understanding of the ideal microenvironments for bone tissue engineering, a number of high throughput screening systems have been developed.[156–158] These systems are advantageous because they can screen a large number of physical and biochemical signaling factors to determine their effects on osteogenesis. These experiments often use very small volumes of material (as low as nanoliter-scale),[159] and can be used to efficiently study interactions between different material properties, growth factors, and cell types. Hydrogels are especially relevant for these systems because small droplets of macromer solution can be deposited onto glass slides or into microwells using techniques including inkjet printing[160] or robotic array spotting,[161] or stamped with tight spatial control using microcontact printing,[162] and then crosslinked.[160] Cells can be then seeded on their surface or incorporated into the hydrogel solutions prior to crosslinking.

High throughput systems have been used to study effects of substrate stiffness, mechanical loading, and different osteogenic agents on important behaviors such as stem cell differentiation. For example, Gobaa *et al.* used a PEG hydrogel-based system to characterize hMSC osteogenic differentiation on hydrogels of varying stiffness, and observed increased osteogenesis on hydrogels with a modulus of 50 kPa compared to that on 1 kPa hydrogels independent of the hydrogels' cell adhesivity.[156] Notably, their system allows the study of single cells in individual microwells. High throughput systems are also extremely useful for studying factors in a combinatorial manner, because they permit evaluation of many more samples than otherwise possible with macroscale constructs; they require substantially fewer cells, smaller material and reagent volumes, and less space in incubators. For example, 240 different conditions could be examined in a study of mouse ESCs on acrylamide gels coated with collagen or laminin in the presence of different growth factors together or separately using robotic spotting which allowed 1,200 different samples to be tested on a 1 × 3 inch slide. This work confirmed some of the known effects of individual growth factors, and gave insights into their combinatorial effects.[157] In addition, high throughput systems that allow for active mechanical stimulation of microscale arrays of cells in three-dimensional (3D) hydrogels are advantageous because they can be used to study combined effects of mechanical loading with other signals, including those presented in this chapter, that are important for bone tissue engineering. For example, hydrostatic pressure can be controlled to mechanically stimulate microfabricated arrays of PEG hydrogels to compress them in a controlled and repeatable fashion.[158] High throughput systems will be key in characterizing the role of hydrogel physical and biochemical properties, growth factor presentation, and mechanical loading, separately and in combination, on osteogenesis, assisting in the development of osteoinductive hydrogel microenvironments for cell-based bone tissue regeneration.

7.2. Hydrogel micropatterning

Hydrogels lend themselves well to a number of micropatterning techniques that may be used to try to control or enhance cell function. Once desirable signals are identified, their presentation can be controlled spatially to direct cell behavior. As in the previous section, these signals can be biochemical, such as growth factors, or physical, such as stiffness or degradation rate. For example, immobilized BMP-2 patterned onto the surface of fibrin hydrogels via inkjet printing and cultured in myogenic media led to spatially targeted osteogenic lineage progression by muscle-derived stem cells seeded on the hydrogels. The remaining cells differentiated towards the myogenic lineage. Differentiation followed the BMP-2 pattern suggesting that this method can be used to attain precise control over BMP-2 position and the resulting cellular response(s).[163] Such an approach could be useful for engineering a bone interface, such as with cartilage or with ligaments and tendons.

Materials that respond to visible or UV light allow for spatial control of their properties by controlling where the light stimulates chemical reactions. For example, a photoresponsive agarose hydrogel was stimulated by a focused laser to create reactive groups in specific 3D regions, and then these reactive moieties were used to tether RGD-containing peptide sequences to the material in channels; this platform was shown to guide neurite outgrowth.[164] Localized control of immobilized growth factors within hydrogels can be achieved by two-photon laser scanning photolithography; this has been used to create agarose hydrogels with spatial control over tethered stem cell differentiation factors,[165] and PEG hydrogels with patterned presentation of cell adhesion ligands.[166] The latter resulted in guided 3D migration of fibroblasts. In the case of physical signals, recent work using an alginate/PEG system with encapsulated hASCs and a checkerboard pattern of equal regions of high and low hydrogel crosslink density showed that varying the pattern size affected hASC osteogenic differentiation (Fig. 2).[29] These approaches utilized a variety of tissue types, and the success of these platforms in regulating cell behaviors suggests that controlled spatial presentation of specific physical and biochemical signals will likely be valuable for the future of bone tissue engineering.

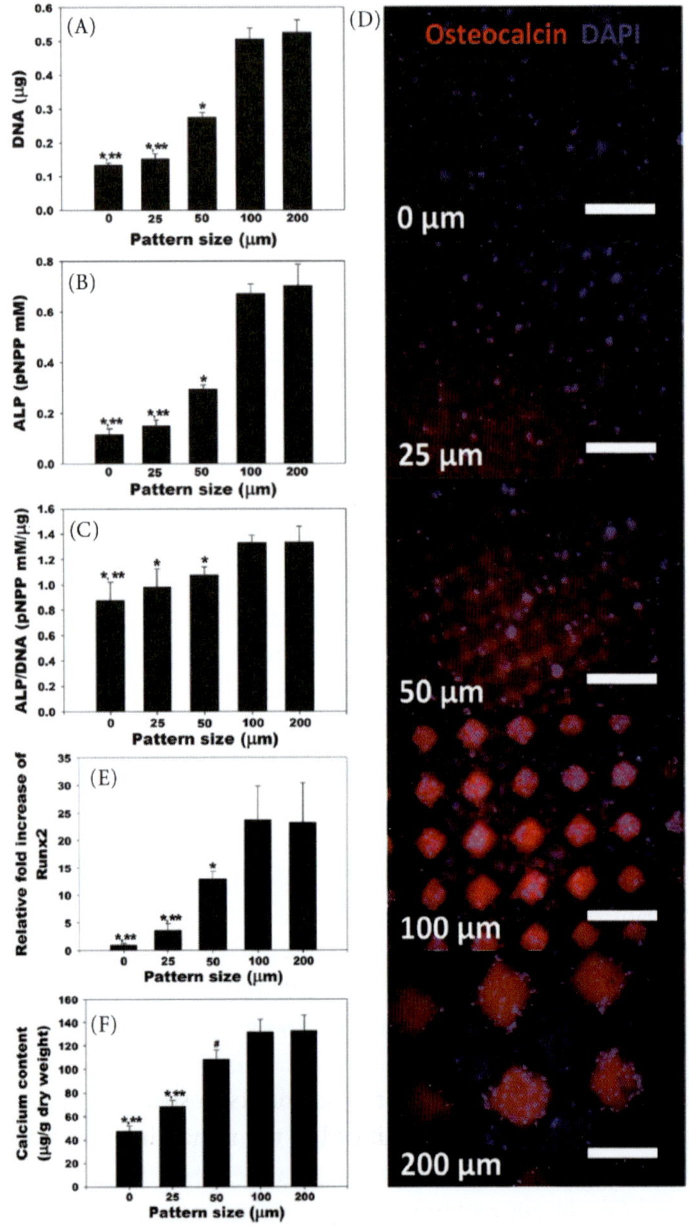

Fig. 2. Effect of crosslinking pattern size on the osteogenic differentiation of hASCs encapsulated within micropatterned oxidized alginate methacrylate/poly(ethylene glycol) hydrogels after culture in osteogenic differentiation media for 4 weeks. (A) DNA content, (B) alkaline phosphatase (ALP) activity, and (C) ALP/DNA. (D) Fluorescence images of encapsulated hASCs stained for osteocalcin (red) and nuclei (DAPI, blue). All scale bars represent 200 μm. (E) Relative runt-related transcription factor-2 (Run × 2) gene expression and (F) calcium content in the constructs. The relative gene expression levels were normalized to the control unpatterned (0 μm) group. $*p < 0.05$ compared with 100 and 200 μm, $**p < 0.05$ compared with 50 μm, and $^{\#}p < 0.05$ compared with 200 μm. Reproduced with permission from Ref. 182. Copyright 2013 John Wiley and Sons.

Another approach to pattern hydrogels is bottom-up assembly using microgels. These modular microgels that can be seeded with cells provide some advantages for osteogenesis: Compared to top–down approaches where cells are seeded onto preformed scaffolds, they allow more uniform cell distribution throughout larger scaffolds, and compared to uniform hydrogels, they can create localized microenviroments with distinct properties in a single construct. A number of exciting studies have demonstrated the ability to create osteogenic constructs assembled from individual microgels containing encapsulated cells[167] or cells seeded on the surface of these hydrogel microcarriers.[168] Using patterns of microgels with different types of signals and/or encapsulated cells, such as osteoprogenitors and endothelial cells, offers the possibility of forming bone tissue with an integrated vasculature. One bottom–up approach is directed assembly of lock-and-key-shaped microgels: One cell type can be encapsulated in cross-shaped PEG microgels, and another in rod-shaped microgels, which are induced to fit together at the air–liquid interface of a high density, hydrophobic solution in a controlled manner.[169] Alternately, the layout of these microgels can be spatially controlled using vacuum molding, which has been used for highly controlled multi-component modular assembly of collagen-chitosan composite microbeads into predefined patterns.[170] Microgels can also be used to obtain spatial control over signal presentation, such as growth factor delivery, to cells in high cell density systems. Hydrogel microspheres loaded with TGF-β1 were incorporated within aggregates with hMSCs and mineral particles; The growth factor-loaded microspheres enhanced the formation of a cartilage template for endochondral ossification.[171] Future work will likely see combinations of these bottom-up approaches to pattern both cells and signals in assembled biomaterial systems.

Spatially controlled co-culture systems for bone tissue engineering can also be created by using micromolded silicone channels filled with a cell-containing alginate solution and a light-sensitive caged calcium complex that, upon exposure to UV light releases the calcium, which crosslinks the alginate solution. First the researchers crosslinked a solution of pre-osteoblasts in specific spaces, and then a solution of endothelial cells was crosslinked in the remaining free space, allowing spatial control over the location of bone cells and cells that might form a supporting vasculature.[172] A simple way to study interactions between different cell types is to use 3D molds of polydimethylsiloxane to stamp controlled patterns into a cell-laden hydrogel that is not fully crosslinked; as the material gels, it will take on the pattern created by the stamp.[173] These techniques allow precise spatial control over the position of different biomaterials, cell types, and/or bioactive factors, which is useful both for exploring biological responses to complex but controlled 3D microenvironmental signals and discovering new design criteria for hydrogel systems to drive bone formation.

7.3. Hydrogels to control cell-cell interactions

Bone is made up of the variety of cell types as described in Sec. 2.1, and these cells can communicate through cell–cell contact between adjacent cells or secreted soluble and insoluble factors. The simplest co-culture systems use cells either mixed together in monolayer, or separated by a transwell membrane to allow for the exchange of soluble factors in the media. These systems have led to a better understanding of two-dimensional (2D) cell behaviors relevant to bone repair. For example, such systems were used to show that physical contact is not necessary for chondrocytes to modulate osteoblast phenotype, or the other way around, highlighting the role of soluble factors.[174] Additionally, monolayer co-culture of chondrocytes with cells from a mouse MSC line improved their osteogenic differentiation, since chondrocytes secrete some soluble signaling factors that promote the differentiation process.[175] Monolayer co-culture systems have also elucidated bone-vascular interactions, as the addition of HUVECs to bone marrow-derived osteoblast progenitors enhanced their osteogenic differentiation. Blocking VEGF receptors inhibited this response, demonstrating the role of this growth factor.[176]

Hydrogels permit studying these cell–cell interactions while also presenting a more physiological 3D environment to the cells of interest. Encapsulating one cell type in 3D biomaterial hydrogels and culturing it on a transwell membrane above a monolayer of another cell type can allow the culture of the embedded cells in a more biologically relevant environment while also allowing for study of the biological crosstalk. For instance, hMSCs encapsulated OPF/PEGDA hydrogels could be cultured in transwells above osteoblasts in monolayer. Increased osteogenic differentiation of hMSCs was observed as a result of paracrine signaling agents produced by osteoblasts.[177]

Hydrogels allow for another simple 3D co-culture system: Mixing two cell types uniformly through a material prior to crosslinking, allowing for both physical and biochemical interactions between more than one cell population. For example, mixing HUVECs and hMSCs in fibrin hydrogels allowed researchers to investigate the time course of engineered bone vascularization: Vascular development occurred before late stage osteogenesis.[115] The 3D co-encapsulation approach can also result in a therapeutic advantage for new bone tissue formation *in vivo* since hydrogels can localize at a defect site both cell populations and the factors they secrete, which may influence neighboring cell survival and proliferation, and promote angiogenesis and osteogenesis. In the case of human osteoprogenitors and HUVECs, their co-encapsulation in alginate microspheres led to enhanced bone healing in a mouse femoral defect compared to microspheres containing osteoprogenitors alone.[178]

More sophisticated co-cultures can be accomplished using micropatterning techniques, including those described in Sec 7.2. For example, a photomask was used to control UV crosslinking of PEGDA mixed with mouse ESCs to create regions of cell-laden PEG. When an alginate solution containing a hepatocyte cell line was added, it filled in the spaces between the ES cells in PEG, and allowed for localized cell–cell interactions and improved ESC growth.[179] Using the stamping micropatterning technique described earlier, populations of stem cells from different sources (i.e., bone marrow, umbilical cord vein or umbilical cord artery) and human umbilical vein endothelial cells (HUVECs) were seeded into different fibrin hydrogel channels separated by 500, 1,000, and 2,000 μm.[173] The stem cells derived from bone marrow, but not those from umbilical cord vein and umbilical cord artery, tended to migrate toward the HUVECs and then stabilized vascular networks formed by HUVECs,[173] indicating that the cellular interactions were strongly dependent on the cell populations utilized (Fig. 3). Systems like these may

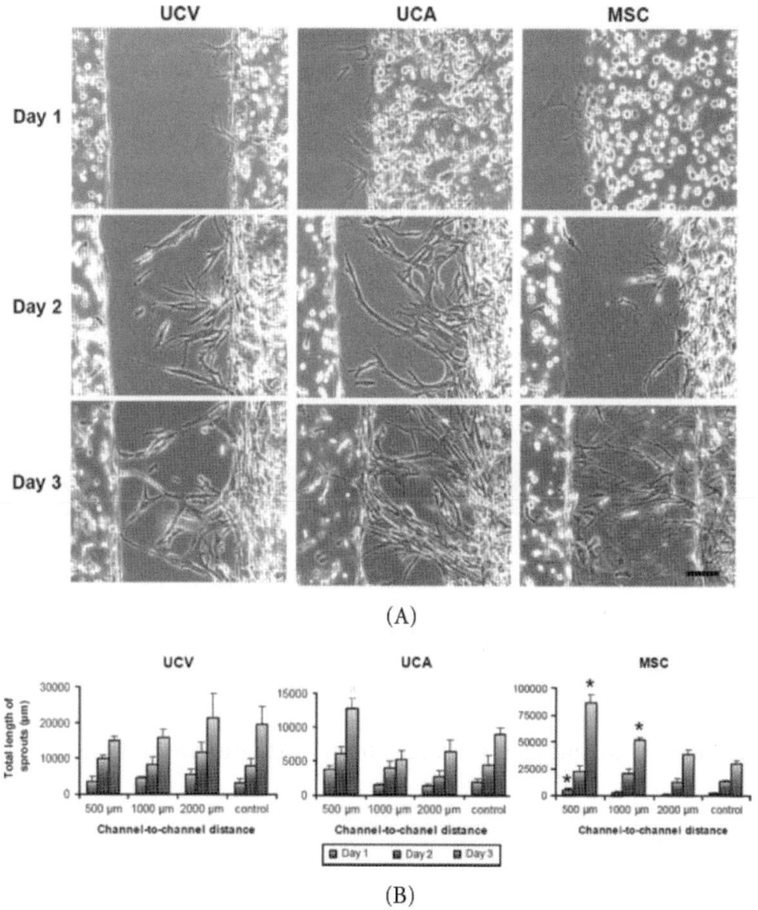

Fig. 3. Influences of cell source, channel spacing and culture time on the guided migration of stem cells. (A) Photomicrographs of micropatterned fibrin channels seeded with HUVECs (left channel) and stem cells from umbilical cord vein (UCV) cells, umbilical cord artery (UCA) cells and bone marrow (MSC) (right channel) 500 μm apart, demonstrating migration of stem cells for the first 3 days after cell seeding. Scale bar indicates 100 μm. (B) The total lengths of sprouts for stem cells from the different sources with different channel spacing over time. $*p < 0.005$ indicates statistically significant differences between groups. Reproduced with permission from Ref. 173. Copyright 2010 John Wiley and Sons.

be useful to understand how one cell population communicates with another during bone regeneration, and these insights will be useful when engineering constructs to regrow bone.

8. Clinical Translation

While the clinical translation process starts with *in vitro* and small animal *in vivo* models described previously, a number of cell-laden hydrogel therapies have been used to try to repair bone defects in humans. As explained earlier, delivering BMPs from collagen sponges is a clinically established and FDA approved bone repair procedure, with wide acceptance by surgeons.[122,125] Fewer cell-based therapies have been used in humans, and most of these therapies are hMSC based, because of the relative ease of obtaining these cells, potential for use of autologous cells, and their well-known osteogenic potential. While many of these therapies rely on ceramic scaffolds, bone chips, or demineralized bone matrix, there has been some work using hydrogels to maintain hMSCs at the defect site. When these procedures use biomaterial carriers, they typically rely on materials approved by regulatory agencies for other applications, such as dressing wounds. For example, a type I collagen sponge has been used to deliver hMSCs to treat human bone deficiencies up to 70 cm^3 throughout the body, including the femur, ilium, fibula, and humerus. This treatment resulted in pain reduction and new bone formation in 7 out of 10 patients.[180] Another example of use in humans is hMSCs mixed with lyophilized bone chips, platelet rich plasma and a fibrin hydrogel. In 12 patients, this mixture was more successful than bone chips alone in healing tibial osteotomies.[181] Thus far, hydrogels have been used primarily as passive depots for cells or growth factors, but as regulatory bodies approve more functionalized hydrogels for use in humans, their role in bone regeneration is likely to expand.

9. Conclusion

Hydrogels are a promising biomaterial category for bone tissue regeneration. Their diversity and capacity for modular modifications allow them to be valuable for presenting biological, chemical, architectural, and physical signals to surrounding or encapsulated cells. Their biocompatibility and potential for delivery via minimally invasive injection support their clinical translation. Building on bone's relatively high capacity for self-repair, much work has focused on tailoring hydrogels to improve their interactions with cells, mechanical properties and controlled presentation of bioactive factors to induce osteogenesis. Future work will likely entail patterning these properties in time and space, optimizing their presentation to create highly controlled instructive materials. In addition, currently available treatments such as BMP delivery can be improved by providing more control over therapeutic release kinetics. Progress in material chemistry and fabrication is likely to continue to advance the use of hydrogels as scaffolds for bone tissue engineering.

Acknowledgments

We apologize that space limitations prevent us from including all of the excellent work that has been done in this rapidly growing area. The authors gratefully acknowledge funding from the National Institutes of Health (R01AR063194, R56DE022376, R21AR061265, R01AR053733), the Department of Defense Congressionally Directed Medical Research Programs (OR110196), the AO Foundation, and a New Scholar in Aging grant from the Ellison Medical Foundation to EA, and a NSF Graduate Research Fellowship to JES.

References

1. P. V. Giannoudis, H. Dinopoulos, and E. Tsiridis, *Injury*, **36**, S20–S27 (2005).
2. B. M. Desai, *Am. J. Orthop.*, **36**, 8–11 (2007).
3. J. A. Buckwalter, M. J. Glimcher, R. R. Cooper, and R. Recker, Bone biology. I: Structure, blood supply, cells, matrix, and mineralization, Instructional course lectures, **45**, 371–386 (1996).
4. B. Clarke, *J. Am. Soc. Nephrol.*, **3**, S131–S139 (2008).
5. P. Augat and S. Schorlemmer, *Age Ageing*, **35**, ii27–ii31 (2006).
6. I. Singh, *J. Anat.*, **127**, 305–310 (1978).
7. S. Weiner, W. Traub, *FEBS Lett.*, **206**, 262–266 (1986).
8. E. L. Smith, R. L. Hill, I. R. Lehman, R. J. Lefkowitz, and P. Handler, *Mamm. Biochem.*, (1983).
9. D. Trentin, H. Hall, S. Wechsler, J. A. Hubbell, *Proc. Natl. Acad. Sci. USA*, **103**, 2506–2511 (2006).
10. S. Viguet-Carrin, P. Garnero, and P. Delmas, *Osteoporosis Int.*, **17**, 319–336 (2006).
11. E. J. Rebar, Y. Huang, R. Hickey, A. K. Nath, D. Meoli, S. Nath, B. L. Chen, L. Xu, Y. X. Liang, A. C. Jamieson, L. Zhang, S. K. Spratt, C. C. Case, A. Wolffe, and F. J. Giordano, *Nat. Med.*, **8**, 1427–1432 (2002).
12. E. Seeman, Osteoporosis international: A journal established as result of cooperation between the European Foundation for Osteoporosis and the National Osteoporosis Foundation of the USA, **18**, 123–128 (2007).
13. M. R. Allen, J. M. Hock, and D. B. Burr, *Bone*, **35**, 1003–1012 (2004).
14. L. Weiss, *J. morphol.*, **117**, 467–537 (1965).
15. H. C. Blair, M. Zaidi, and P. H. Schlesinger, *Biochem. J.*, **364**, 329–341 (2002).

16. E. Alsberg, K. W. Anderson, A. Albeiruti, R. T. Franceschi, and D. J. Mooney, *J. Dent. Res.*, **80**, 2025–2029 (2001).
17. A. L. Arsenault and F. P. Ottensmeyer, *J. Cell Biol.*, **98**, 911–921 (1984).
18. N. Ortega, D. J. Behonick, and Z. Werb, *Trends Cell Biol.*, **14**, 86–93 (2004).
19. E. J. Mackie, Y. A. Ahmed, L. Tatarczuch, K. S. Chen, and M. Mirams, *Int. J. Biochem. Cell B*, **40**, 46–62 (2008).
20. J. Dai and A. B. Rabie, *J. Dent. Res.*, **86**, 937–950 (2007).
21. G. Karsenty and E. F. Wagner, *Dev. Cell.*, **2**, 389–406 (2002).
22. H. M. Frost, *Calcified Tissue Res.*, **3**, 211–237 (1969).
23. L. G. Raisz, *Clin. Chem.*, **45**, 1885–1885 (1999).
24. A. M. Phillips, *Injury*, **36**, S5–S7 (2005).
25. Z. Thompson, T. Miclau, D. Hu, and J. A. Helms, *J. Orthop. Res.*, **20**, 1091–1098 (2002).
26. R. A. Schneider and J. A. Helms, *Orthop.*, **9**, 20–24 (1998).
27. R. C. Rennert, M. Sorkin, R. K. Garg, and G. C. Gurtner, *Regen. Med.*, **7**, 833–850 (2012).
28. V. Pandit, J. M. Zuidema, K. N. Venuto, J. Macione, G. Dai, R. J. Gilbert, and S. P. Kotha, *Tissue Eng. A*, (2013).
29. J. A. Burdick and K. S. Anseth, *Biomaterials*, **23**, 4315–4323 (2002).
30. D. S. Benoit and K. S. Anseth, *Biomaterials*, **26**, 5209–5220 (2005).
31. A. S. Greenwald, S. D. Boden, V. M. Goldberg, Y. Khan, C. T. Laurencin, and R. N. Rosier, *J. Bone Joint Surg. Am.*, **83-A**, 98–103 (2001).
32. C. G. Finkemeier, *J. Bone Joint Surg. Am.*, **84-A**, 454–464 (2002).
33. A. J. Friedenstein, S. Piatetzky, II, and K. V. Petrakova, *J. Embryol. Exp. Morph.*, **16**, 381–390 (1966).
34. A. J. Friedenstein, R. K. Chailakhyan, and U. V. Gerasimov, *Cell Tissue Kinet.*, **20**, 263–272 (1987).
35. H. M. Lazarus, S. E. Haynesworth, S. L. Gerson, N. S. Rosenthal, and A. I. Caplan, *Bone Marrow Transpl.*, **16**, 557–564 (1995).
36. A. I. Caplan, *J Orthop Res.*, **9**, 641–650 (1991).
37. H. Shin, S. Jo, and A. G. Mikos, *J. Biomed. Mater. Res.*, **61**, 169–179 (2002).
38. C. R. Nuttelman, M. C. Tripodi, and K. S. Anseth, *J. Biomed. Mater. Res. A*, **68**, 773–782 (2004).
39. L. S. Wang, C. Du, J. E. Chung, and M. Kurisawa, *Acta Biomater.*, **8**, 1826–1837 (2012).
40. P. A. Zuk, M. Zhu, P. Ashjian, D. A. De Ugarte, J. I. Huang, H. Mizuno, Z. C. Alfonso, J. K. Fraser, P. Benhaim, and M. H. Hedrick, *Mol. Biol. Cell.*, **13**, 4279–4295 (2002).
41. B. Levi, A. W. James, E. R. Nelson, D. Vistnes, B. Wu, M. Lee, A. Gupta, and M. T. Longaker, *PLoS ONE*, **5**, e111–e177 (2010).
42. O. Jeon, J. W. Rhie, I. K. Kwon, J. H. Kim, B. S. Kim, and S. H. Lee, *Tissue Eng.*, **14**, 1285–1294 (2008).
43. J. S. Gabbay, J. B. Heller, S. A. Mitchell, P. A. Zuk, D. B. Spoon, K. L. Wasson, R. Jarrahy, P. Benhaim, and J. P. Bradley, *Ann. Plas. Surg.*, **57**, 89–93 (2006).
44. S. Shi and S. Gronthos, *J Bone Miner. Res.*, **18**, 696–704 (2003).
45. J. T. Williams, S. S. Southerland, J. Souza, A. F. Calcutt, and R. G. Cartledge, *Am. Surg.*, **65**, 22–26 (1999).

46. C. De Bari, F. Dell'Accio, P. Tylzanowski, and F. P. Luyten, *Arthritis Rheum.*, **44**, 1928–1942 (2001).
47. D. J. Warejcka, R. Harvey, B. J. Taylor, H. E. Young, and P. A. Lucas, *J. Surg. Res.*, **62**, 233–242 (1996).
48. Y. Fukuchi, H. Nakajima, D. Sugiyama, I. Hirose, T. Kitamura, and K. Tsuji, *Stem Cells*, **22**, 649–658 (2004).
49. K. Mareschi, E. Biasin, W. Piacibello, M. Aglietta, E. Madon, and F. Fagioli, *Haematologica*, **86**, 1099–1100 (2001).
50. L. D. Buttery, S. Bourne, J. D. Xynos, H. Wood, F. J. Hughes, S. P. Hughes, V. Episkopou, and J. M. Polak, *Tissue Eng.*, **7**, 89–99 (2001).
51. V. Sottile, A. Thomson, and J. McWhir, *Cloning and stem cells*, **5**, 149–155 (2003).
52. M. Tang, W. Chen, M. D. Weir, W. Thein-Han, and H. H. Xu, *Acta Biomater.*, **8**, 3436–3445 (2012).
53. M. J. Kim, J. S. Park, S. Kim, S. H. Moon, H. N. Yang, K. H. Park, H. M. Chung, *Stem Cells Dev.*, **20**, 1349–1358 (2011).
54. B. Blum and N. Benvenisty, *Adv. Cancer Res.*, **100**, 133–158 (2008).
55. D. E. Baker, N. J. Harrison, E. Maltby, K. Smith, H. D. Moore, P. J. Shaw, P. R. Heath, H. Holden, and P. W. Andrews, *Nat. Biotechnol.*, **25**, 207–215 (2007).
56. K. Takahashi and S. Yamanaka, *Cell*, **126**, 663–676 (2006).
57. K. Tashiro, M. Inamura, K. Kawabata, F. Sakurai, K. Yamanishi, T. Hayakawa, and H. Mizuguchi, *Stem Cells*, **27**, 1802–1811 (2009).
58. X. Duan, Q. Tu, J. Zhang, J. Ye, C. Sommer, G. Mostoslavsky, D. Kaplan, P. Yang, and J. Chen, *J. Cell Physiol.*, **226**, 150–157 (2011).
59. E. Alsberg, K. W. Anderson, A. Albeiruti, R. T. Franceschi, and D. J. Mooney, *J. Dent. Res.*, **80**, 2025–2029 (2001).
60. A. J. Engler, S. Sen, H. L. Sweeney, and D. E. Discher, *Cell*, **126**, 677–689 (2006).
61. E. Alsberg, H. J. Kong, Y. Hirano, M. K. Smith, A. Albeiruti, and D. J. Mooney, *J Dent Res.*, **82**, 903–908 (2003).
62. M. W. Betz, A. B. Yeatts, W. J. Richbourg, J. F. Caccamese, D. P. Coletti, E. E. Falco, and J. P. Fisher, *Biomacromolecules*, **11**, 1160–1168 (2010).
63. O. Jeon, C. Powell, L. D. Solorio, M. D. Krebs, and E. Alsberg, *J. Control Release.*, **154**, 258–266 (2011).
64. H. Tan and K. G. Marra, *Materials*, **3**, 1746–1767 (2010).
65. Y. Takahashi, M. Yamamoto, K. Yamada, O. Kawakami, and Y. Tabata, *Tissue Eng.*, **13**, 293–300 (2007).
66. H. S. Yang, W.-G. La, S. H. Bhang, J.-Y. Jeon, J. H. Lee, and B.-S. Kim, *Tissue Eng. A*, **16**, 1225–1233 (2009).
67. Y. Li, J. Rodrigues, and H. Tomas, *Chem Soc Rev.*, **41**, 2193–2221 (2012).
68. E. S. Place, J. H. George, C. K. Williams, and M. M. Stevens, *Chem. Soc. Rev.*, **38**, 1139–1151 (2009).
69. D. S. W. Benoit, A. R. Durney, and K. S. Anseth, *Biomaterials*, **28**, 66–77 (2007).
70. C. R. Nuttelman, M. C. Tripodi, and K. S. Anseth, *J. Biomed. Mater. Res.*, **68A**, 773–782 (2004).

71. J. S. Temenoff, H. Park, E. Jabbari, D. E. Conway, T. L. Sheffield, C. G. Ambrose, and A. G. Mikos, *Biomacromolecules*, **5**, 5–10 (2003).
72. H. K. Kim, W. S. Shim, S. E. Kim, K.-H. Lee, E. Kang, J.-H. Kim, K. Kim, I. C. Kwon, and D. S. Lee, *Tissue Eng.*, **15**, 923–933 (2008).
73. S.-H. Lee, J. J. Moon, and J. L. West, *Biomaterials*, **29**, 2962–2968 (2008).
74. O. Jeon and E. Alsberg, *Adv. Func. Mater.*, (2013).
75. S. Levenberg, N. F. Huang, E. Lavik, A. B. Rogers, and J. Itskovitz-Eldor, R. Langer, Differentiation of human embryonic stem cells on three-dimensional polymer scaffolds, *Proc. Nat. Acad. Sci.*, **100**, 12741–12746 (2003).
76. K. Narine, O. D. Wever, D. V. Valckenborgh, K. Francois, M. Bracke, S. Desmet, M. Mareel, and G. V. Nooten, *Tissue Eng.*, **12**, 2707–2716 (2006).
77. W. J. King, L. Jongpaiboonkit, and W. L. Murphy, *J. Biomed. Mater. Res.*, **93A**, 1110–1123 (2010).
78. M. H. Fittkau, P. Zilla, D. Bezuidenhout, M. P. Lutolf, P. Human, J. A. Hubbell, and N. Davies, *Biomaterials*, **26**, 167–174 (2005).
79. H. Shin, S. Jo, and A. G. Mikos, *Biomaterials*, **24**, 4353–4364 (2003).
80. C. R. Nuttelman, D. J. Mortisen, S. M. Henry, and K. S. Anseth, *J Biomed Mater Res.*, **57**, 217–223 (2001).
81. S. E. Stabenfeldt, A. J. García, and M. C. LaPlaca, *J. Biomed. Mater. Res. A*, **77A**, 718–725 (2006).
82. S. K. Seidlits, C. T. Drinnan, R. R. Petersen, J. B. Shear, L. J. Suggs, and C. E. Schmidt, *Acta Biomater.*, **7**, 2401–2409 (2011).
83. P. B. Malafaya, G. A. Silva, and R. L. Reis, *Adv. Drug Deliv. Rev.*, **59**, 207–233 (2007).
84. A. M. Ferreira, P. Gentile, V. Chiono, and G. Ciardelli, *Acta Biomater.*, **8**, 3191–3200 (2012).
85. E. M. Bueno and J. Glowacki, *Rheumatology*, **5**, 685–697 (2009).
86. E. Lucchinetti, D. Thomann, and G. Danuser, *J. Mater. Sci..*, **35**, 6057–6065 (2000).
87. P. Garcia, T. Histing, J. H. Holstein, M. Klein, M. W. Laschke, R. Matthys, A. Ignatius, B. Wildemann, J. Lienau, A. Peters, B. Willie, G. Duda, L. Claes, T. Pohlemann, and M. D. Menger, *Eur. Cell Mater.*, **26**, 1–12; discussion 12–14 (2013).
88. S. Nemir and J. L. West, *Ann. Biomed. Eng.*, **38**, 2–20 (2010).
89. K. Y. Lee and D. J. Mooney, *Prog. Polym. Sci.*, **37**, 106–126 (2012).
90. Q. T. Nguyen, Y. Hwang, A. C. Chen, S. Varghese, and R. L. Sah, *Biomaterials*, **33**, 6682–6690 (2012).
91. P. Eiselt, K. Y. Lee, and D. J. Mooney, *Macromolecules*, **32**, 5561–5566 (1999).
92. H. J. Kong, E. Alsberg, D. Kaigler, K. Y. Lee, and D. J. Mooney, *Adv. Mater.*, **16**, 1917–1921 (2004).
93. K. Y. Lee, J. A. Rowley, P. Eiselt, E. M. Moy, K. H. Bouhadir, and D. J. Mooney, *Macromolecules*, **33**, 4291–4294 (2000).
94. K. Chatterjee, S. Lin-Gibson, W. E. Wallace, S. H. Parekh, Y. J. Lee, M. T. Cicerone, M. F. Young, and C.G. Simon Jr., *Biomaterials*, **31**, 5051–5062 (2010).
95. J. L. Drury and D. J. Mooney, *Biomaterials*, **24**, 4337–4351 (2003).
96. K. Nguyen, P. N. Dang, and E. Alsberg, *Acta Biomater.*, **9**, 4487–4495 (2013).

97. O. Jeon, K. H. Bouhadir, J. M. Mansour, and E. Alsberg, *Biomaterials*, **30**, 2724–2734 (2009).
98. H. J. Kong, D. Kaigler, K. Kim, and D. J. Mooney, *Biomacromolecules*, **5**, 1720–1727 (2004).
99. Y. Lei, Q. K. T. Ng, and T. Segura, *Microsc. Res. Tech.*, **73**, 910–917 (2010).
100. D. R. Griffin and A. M. Kasko, *J. Am. Chem. Soc.*, **134**, 13103–13107 (2012).
101. K. Ulbrich and V. Šubr, *Adv. Drug Deliv. Rev.*, **62**, 150–166 (2010).
102. L. S. Nair and C.T. Laurencin, *Prog. Polym. Sci.*, **32**, 762–798 (2007).
103. V. K. Garripelli, J. K. Kim, R. Namgung, W. J. Kim, M. A. Repka, and S. Jo, *Acta Biomater.*, **6**, 477–485 (2010).
104. T. Segura, B. C. Anderson, P. H. Chung, R. E. Webber, K. R. Shull, and L. D. Shea, *Biomaterials*, **26**, 359–371 (2005).
105. M. Lovett, K. Lee, A. Edwards, and D. L. Kaplan, *Tissue Eng. B.*, **15**, 353–370 (2009).
106. S. F. Hulbert, S. J. Morrison, and J. J. Klawitter, *J. Biomed. Mater. Res.*, **6**, 347–374 (1972).
107. V. Karageorgiou and D. Kaplan, *Biomaterials*, **26**, 5474–5491 (2005).
108. V. Keskar, N. W. Marion, J. J. Mao, and R. A. Gemeinhart, *Tissue Eng. A*, **15**, 1695–1707 (2009).
109. S. G. Lévesque, R. M. Lim, and M. S. Shoichet, *Biomaterials*, **26**, 7436–7446 (2005).
110. L.-H. Han, J. H. Lai, S. Yu, and F. Yang, *Biomaterials*, **34**, 4251–4258 (2013).
111. O. Jeon, P. N. Dang, and E. Alsberg, Macroporous cell-laden hydrogels fabricated with photocrosslinkable alginate for tissue engineering, (Submitted).
112. A. Phadke, Y. Hwang, S. H. Kim, S. H. Kim, T. Yamaguchi, K. Masuda, and S. Varghese, *Eur. Cell Mater.*, **25**, 114–128; discussion, 128–119 (2013).
113. J. Babensee, L. McIntire, and A. Mikos, *Pharm. Res.*, **17**, 497–504 (2000).
114. H. M. van Beuningen, H. L. Glansbeek, P. M. van der Kraan, and W. B. van den Berg, *Osteoarthr. Cartilage.*, **8**, 25–33 (2000).
115. C. Correia, W. L. Grayson, M. Park, D. Hutton, B. Zhou, X. E. Guo, L. Niklason, R. A. Sousa, R. L. Reis, and G. Vunjak-Novakovic, *Plos One*, **6**, (2011).
116. J. N. Zara, R. K. Siu, X. Zhang, J. Shen, R. Ngo, M. Lee, W. Li, M. Chiang, J. Chung, J. Kwak, B. M. Wu, K. Ting, and C. Soo, *Tissue Eng. A*, **17**, 1389–1399 (2011).
117. P. V. Giannoudis and T. A. Einhorn, *Injury*, **40**, S1–S3 (2009).
118. A. Nauth, B. Ristevski, R. Li, and E. H. Schemitsch, *Injury*, **42**, 574–579 (2011).
119. N. Sykaras, L. A. Opperman, Bone morphogenetic proteins (BMPs): How do they function and what can they offer the clinician?
120. N. K. Kanakaris, G. M. Calori, R. Verdonk, P. Burssens, P. D. Biase, R. Capanna, L. B. Vangosa, P. Cherubino, F. Baldo, J. Ristiniemi, G. Kontakis, and P. V. Giannoudis, *Injury*, **39**, S83–S90 (2008).
121. G. Zimmermann, C. Wagner, K. Schmeckenbecher, A. Wentzensen, and A. Moghaddam, *Injury*, **40**, S50–S53 (2009).
122. N. K. Kanakaris, N. Lasanianos, G. M. Calori, R. Verdonk, T. J. Blokhuis, P. Cherubino, P. De Biase, and P. V. Giannoudis, *Injury*, **40**, S54–S61 (2009).
123. S. D. Cook, J. E. Dalton, E. H. Tan, T. S. Whitecloud, and D. C. Rueger, *Spine*, **19**, 1655–1663 (1994).

124. M. A. Rodgers, J. V. Brown, M. K. Heirs, J. P. Higgins, R. J. Mannion, M. C. Simmonds, and L. A. Stewart, *Bmj.*, **346**, f3981 (2013).
125. B. McKay and H. S. Sandhu, *Spine*, **27**, S66–S85 (2002).
126. G. N. King, *Pharm. Biotechnol.*, **2**, 131–142 (2001).
127. J. D. Smucker, J. M. Rhee, K. Singh, S. T. Yoon, and J. G. Heller, *Spine*, **31**, 2813–2819 (2006).
128. J. W. McClellan, D. S. Mulconrey, R. J. Forbes, and N. Fullmer, *J. Spinal. Disord. Tech.*, **19**, 483–486 (2006).
129. L. B. Shields, G. H. Raque, S. D. Glassman, M. Campbell, T. Vitaz, J. Harpring, and C. B. Shields, *Spine*, **31**, 542–547 (2006).
130. G. Bhakta, B. Rai, Z. X. H. Lim, J. H. Hui, G. S. Stein, A. J. van Wijnen, V. Nurcombe, G. D. Prestwich, and S. M. Cool, *Biomaterials*, **33**, 6113–6122 (2012).
131. M.J. Cross and L. Claesson-Welsh, *Trends Pharmacol. Sci.*, **22**, 201–207 (2001).
132. J. O. Hollinger, C. E. Hart, S. N. Hirsch, S. Lynch, and G. E. Friedlaender, *J. Bone Joint Surg. Am.*, **90A**, 48–54 (2008).
133. A. I. Caplan and D. Correa, *J. Orthop. Res.*, **29**, 1795–1803 (2011).
134. E. Canalis, M. Centrella, and T. McCarthy, *J. Clin. Invest.*, **81**, 1572–1577 (1988).
135. D. Kaigler, E. A. Silva, and D. J. Mooney, *J. Periodontol.*, **84**, 230–238 (2012).
136. N. Kodama, M. Nagata, Y. Tabata, M. Ozeki, T. Ninomiya, and R. Takagi, *Bone*, **44**, 699–707 (2009).
137. D. H. R. Kempen, L. Lu, A. Heijink, T. E. Hefferan, L. B. Creemers, A. Maran, M. J. Yaszemski, and W. J. A. Dhert, *Biomaterials*, **30**, 2816–2825 (2009).
138. N. Kimelman Bleich, I. Kallai, J. R. Lieberman, E. M. Schwarz, G. Pelled, and D. Gazit, *Adv. Drug. Deliv. Rev.*, **64**, 1320–1330 (2012).
139. C. Evans, *Injury*, **42**, 599–604 (2011).
140. J. Fang, Y. Y. Zhu, E. Smiley, J. Bonadio, J. P. Rouleau, S. A. Goldstein, L.K. McCauley, B.L. Davidson, and B.J. Roessler, Stimulation of new bone formation by direct transfer of osteogenic plasmid genes, *Proc. Natl. Acad. Sci.*, **93**, 5753–5758 (1996).
141. J. H. Jang, C. B. Rives, and L. D. Shea, *Mol. Ther.*, **12**, 475–483 (2005).
142. M. Al-Dosari and X. Gao, *AAPS J.*, **11**, 671–681 (2009).
143. F. Wegman, A. Bijenhof, L. Schuijff, F. C. Oner, W. J .A. Dhert, and J. Alblas, *Eur. Cells Mater.*, **21**, 230–242 (2011).
144. F. Wegman, R. E. Geuze, Y. J. van der Helm, F. Cumhur Öner, W. J. A. Dhert, and J. Alblas, *J. Tissue Eng. Regen.*, (2012).
145. M. D. Krebs, E. Salter, E. Chen, K. A. Sutter, and E. Alsberg, *J. Biomed. Mater. Res-A*, **92A**, 1131–1138 (2010).
146. T. Manaka, A. Suzuki, K. Takayama, Y. Imai, H. Nakamura, and K. Takaoka, *Biomaterials*, **32**, 9642–9648 (2011).
147. M. K. Nguyen, O. Jeon, M. D. Krebs, S. Daniel and E. Alsberg, *Biomaterials*, **35**, 6278–6286 (2014).
148. T. Maeda, A. Matsunuma, T. Kawane, and N. Horiuchi, *Biochem. Biophys. Res. Commun.*, **280**, 874–877 (2001).

149. G. Mundy, R. Garrett, S. Harris, J. Chan, D. Chen, G. Rossini, B. Boyce, M. Zhao, and G. Gutierrez, *Science*, **286**, 1946–1949 (1999).
150. I. R. Garrett and G. Mundy, *Arthritis Res.*, **4**, 237–240 (2002).
151. D. S. W. Benoit, C. R. Nuttelman, S. D. Collins, and K. S. Anseth, *Biomaterials*, **27**, 6102–6110 (2006).
152. Y. Park, A. David, K. Park, C.-Y. Lin, K. Than, K. Lee, J. Park, I . Jo, K. Park, and V. Yang, *AAPS J.*, **15**, 367–376 (2013).
153. A. Mukozawa, K. Ueki, K. Marukawa, K. Okabe, A. Moroi, and K. Nakagawa, *Clin. Oral Implan. Res.*, **22**, 1327–1335 (2011).
154. P. T. Cheng, C. Chan, and K. Müller, *J. Bone Miner. Res.*, **10**, 119–126 (1995).
155. P. Valderrama, R. E. Jung, D. S. Thoma, A. A. Jones, and D. L. Cochran, *J. Periodontol.*, **81**, 737–747 (2010).
156. S. Gobaa, S. Hoehnel, M. Roccio, A. Negro, S. Kobel, and M. P. Lutolf, *Nat. Methods.*, **8**, 949–955 (2011).
157. C. J. Flaim, D. Teng, S. Chien, and S. N. Bhatia, *Stem Cells Dev.*, **17**, 29–39 (2008).
158. C. Moraes, G. Wang, Y. Sun, and C.A. Simmons, *Biomaterials*, **31,** 577–584 (2010).
159. D. G. Anderson, S. Levenberg, and R. Langer, *Nat. Biotechnol.*, **22**, 863–866 (2004).
160. E. A. Roth, T. Xu, M. Das, C. Gregory, J. J. Hickman, and T. Boland, *Biomaterials*, **25**, 3707–3715 (2004).
161. V. I. Dyukova, E. I. Dementieva, D. A. Zubtsov, O. E. Galanina, and N. V. Bovin, *Anal Biochem.*, **347**, 94–105 (2005).
162. D. Biswal, H. D. Chirra, and J. Z. Hilt, *Biomed. Microdevices*, **10**, 213–219 (2008).
163. J. A. Phillippi, E. Miller, L. Weiss, J. Huard, A. Waggoner, and P. Campbell, *Stem Cells*, **26**, 127–134 (2008).
164. Y. Luo and M. S. Shoichet, *Nat Mater.*, **3**, 249–253 (2004).
165. R. G. Wylie, S. Ahsan, Y. Aizawa, K. L. Maxwell, C. M. Morshead, and M.S. Shoichet, *Nat. Mater.*, **10**, 799–806 (2011).
166. S. H. Lee, J. J. Moon, and J. L. West, *Biomaterials*, **29**, 2962–2968 (2008).
167. H. A. Declercq, T. De Caluwe, O. Krysko, C. Bachert, and M. J. Cornelissen, *Biomaterials*, **34**, 1004–1017 (2013).
168. M. Chen, X. Wang, Z. Ye, Y. Zhang, Y. Zhou, and W.-S. Tan, *Biomaterials*, **32**, 7532–7542 (2011).
169. Y. Du, E. Lo, S. Ali, and A. Khademhosseini, Directed assembly of cell-laden microgels for fabrication of 3D tissue constructs, *Proc. Natl. Acad. Sci. USA*, **105**, 9522–9527 (2008).
170. D. J. Caldwell, R. R. Rao, and J. P. Stegemann, *Adv. Healthc. Mater.*, **2**, 673–677 (2013).
171. P. N. Dang, X. Yu, N. Dwivedi, L. Phillips, W. L. Murphy, E. Alsberg, Controlled delivery of chondrogenic and osteogenic factors for enhanced bone formation via endochondral ossification, (In Preparation).
172. B. H. Chueh, Y. Zheng, Y. S. Torisawa, A. Y. Hsiao, C. Ge, S. Hsiong, N. Huebsch, R. Franceschi, D. J. Mooney, and S. Takayama, *Biomed. Microdevices.*, **12**, 145–151 (2010).

173. S. Trkov, G. Eng, R. Di Liddo, P. P. Parnigotto, and G. Vunjak-Novakovic, *J. Tissue Eng. Regen. M.*, **4**, 205–215 (2010).
174. R. Nakaoka, S. X. Hsiong, and D. J. Mooney, *Tissue Eng.*, **12**, 2425–2433 (2006).
175. L. C. Gerstenfeld, J. Cruceta, C. M. Shea, K. Sampath, G. L. Barnes, and T. A. Einhorn, *J. Bone Miner. Res.*, **17**, 221–230 (2002).
176. M. Grellier, N. Ferreira-Tojais, C. Bourget, R. Bareille, F. Guillemot, and J. Amédée, *J. Cell Biochem.*, **106**, 390–398 (2009).
177. S. Seto, M. Casas, and J. Temenoff, *Cell Tissue Res.*, **347**, 589–601 (2012).
178. M. Grellier, P. L. Granja, J.-C. Fricain, S. J. Bidarra, M. Renard, R. Bareille, C. Bourget, J. Amédée, and M.A. Barbosa, *Biomaterials*, **30**, 3271–3278 (2009).
179. S. Sugiura, J. M. Cha, F. Yanagawa, P. Zorlutuna, and H. Bae, *J. Tissue. Eng. Regen. M.*, (2013).
180. M. Jager, E. M. Jelinek, K. M. Wess, A. Scharfstadt, M. Jacobson, S. V. Kevy, and R. Krauspe, *Stem Cell Res. Ther.*, **4**, 34–43 (2009).
181. D. Dallari, L. Savarino, C. Stagni, E. Cenni, A. Cenacchi, P. M. Fornasari, U. Albisinni, E. Rimondi, N. Baldini, and A. Giunti, *J. Bone Joint Surg. Am.*, **89**, 2413–2420 (2007).
182. O. Jeon and E. Alsberg, *Adv. Func. Mater.*, **23**, 4765–4775 (2013).

Chapter 11

Hydrogels in Cardiac Tissue Engineering

Shauna M. Dorsey and Jason A. Burdick*
Department of Bioengineering, University of Pennsylvania
210 S. 33rd Street, Philadelphia, PA 19104, USA
**burdick2@seas.upenn.edu*

Injectable hydrogels are a promising therapy to attenuate ventricular remodeling after myocardial infarction (MI). Both natural and synthetic materials have been explored as acellular bulking agents to mechanically stabilize the myocardium and as delivery vehicles for cells and/or therapeutic molecules to modulate the maladaptive biological processes after infarction. Various materials, cells, and therapeutic molecules have demonstrated positive outcomes in animal models and, in doing so, provide insight for future material development and optimization. However, many questions remain before injectable hydrogel systems can be safely translated to patients. Future work should focus on elucidating the mechanism behind the positive outcomes of hydrogel delivery *in vivo* and the application of these systems to clinically relevant scenarios, including minimally-invasive, percutaneous delivery. Further development of injectable hydrogels for cardiac repair will have considerable clinical impact with respect to broadening our understanding of the mechanisms of remodeling after infarction and improving therapies to prevent progression to heart failure. This chapter focuses on the development of both acellular and cellular injectable hydrogel systems applied as cardiac repair therapies *in vivo.*

1. Introduction

Heart failure is a major public health issue that affects almost 23 million individuals worldwide and 5.8 million individuals in the US alone.[1] Of these cases, nearly 70% are due to coronary artery disease (CAD), which causes myocardial infarction (MI).[2,3]

MI initiates a dynamic process of left ventricular (LV) remodeling, which includes both maladaptive biological and mechanical changes that may lead to altered contractile properties and heart failure.

MI occurs following coronary artery occlusion, resulting in depletion of nutrients and oxygen to the myocardium and subsequent cardiomyocyte death.[4] This cardiomyocyte necrosis leads to an influx of inflammatory cells to remove the necrotic debris and activation of bioactive molecules, such as matrix metalloproteinases (MMPs).[5,6] Although initially essential to provide the necessary substructure for subsequent scar formation, MMP upregulation is not met by a concomitant upregulation of their endogenous inhibitors (TIMPs), leading to a substantial increase in the breakdown of cardiomyocyte extracellular matrix (ECM) and collagen.[7] Since cardiomyocytes are tethered together with collagen, this leads to continuous stretch and slippage of cardiomyocytes from one another, which weakens the myocardial wall and makes it susceptible to global geometric changes, including thinning and dilatation.[8,9] The mechanical phenomenon that both initiates and sustains the adverse LV remodeling process is initial infarct expansion.[10–13] Infarct expansion is a progressive pathologic process that causes abnormal stress distributions in the borderzone (BZ) regions surrounding the infarct. As part of this self-perpetuating process, increased regional BZ stresses lead to maladaptive cellular and enzymatic processes, such as additional cardiomyocyte apoptosis and MMP activation, which ultimately lead to altered contractile properties and heart failure.[11–13]

Thus, limiting initial infarct expansion has been identified as a potential therapy to reduce the morbidity associated with LV remodeling that leads to heart failure. Previous strategies to limit infarct expansion include surgical reconstruction of the dilated LV and physical restraint of the ventricle or infarct region using polymeric meshed materials to prevent dilation.[14–17] Injectable biomaterials have been explored to circumvent the surgical placement of LV restraining devices or cardiac patches. Injectable biomaterials offer the potential for minimally invasive delivery, which decreases damage to the target and surrounding tissues and enables more rapid clinical translation. Recently, injectable hydrogels have been used to mechanically stabilize the myocardial wall and modulate maladaptive biological processes post-MI through delivery of therapies, such as cells and growth factors (Fig. 1).[18,19] Hydrogels are water-swollen networks of polymer chains with a high degree of flexibility, similar to natural tissue[20] and can be formed through numerous mechanisms, including *in situ* formation.[21] Due to their high tissue-like water content, structural similarity to native macromolecular components, and mechanical properties resembling those of soft tissues, they have been used for numerous applications in tissue engineering and drug delivery.[22] This chapter will specifically focus on acellular and cellular injectable hydrogels utilized for MI treatment *in vivo*.

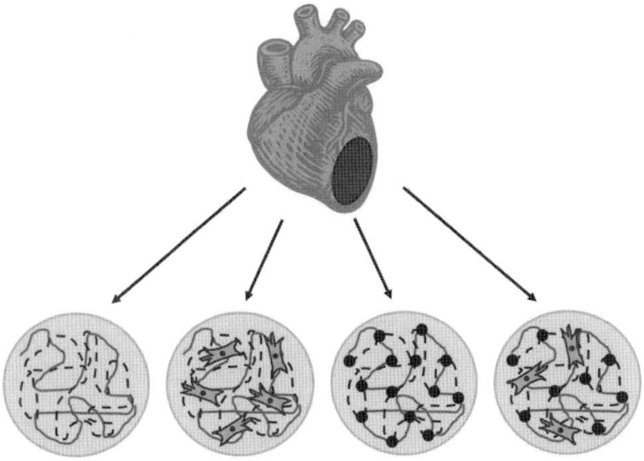

Fig. 1. Schematic of injectable hydrogel strategies for MI repair. Hydrogels can be used as acellular bulking agents (A) or as a delivery vehicle for cells (B), therapeutic molecules (C), or a combination of cells and molecules (D).

2. Acellular Approaches

2.1. LV restraints

Over the past decade, it has become evident that the regional mechanical changes in the myocardium after MI must be considered when designing biomaterial-based approaches for cardiac repair.[5,18,23] After infarction, the myocardial wall thins in the infarct region and the LV dilates as the heart shifts to a more spherical shape; this leads to an increase in BZ stress. As described by the Law of Laplace (Eq. 1), stress (T) is directly proportional to pressure (P) and the radius of curvature (R) and inversely proportional to the myocardial thickness (h). Therefore, an increase in the ventricular radius (R) and decrease in wall thickness (h) leads to an increase in myocardial stress.

$$T = \frac{P * R}{h}. \tag{1}$$

Previous strategies to attenuate LV expansion include the use of polymeric meshes wrapped around the heart or sutured to the infarct surface, referred to as LV restraints. The objective for the use of ventricular restraints is to decrease the ventricular radius (R) by forcing the heart to maintain its elliptical shape, and in turn, decrease wall stress (T) to attenuate LV remodeling.[13,15,24,25]

Several studies examined the use of biomaterial supports to restrain the LV to prevent dilatation and adverse remodeling.[26] Kelley *et al.* demonstrated preserved

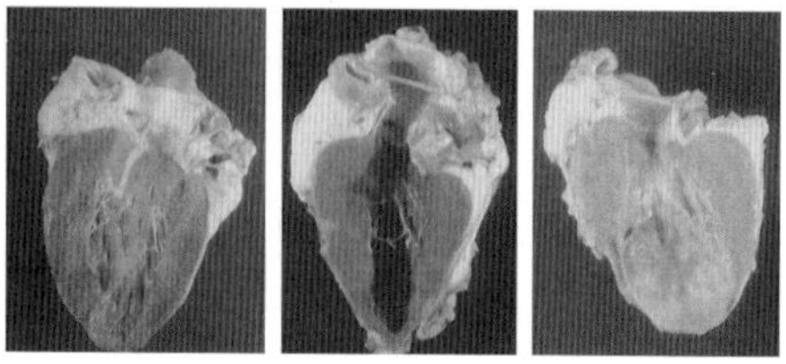

Fig. 2. Use of LV restraints to restore ventricular geometry post-MI. Longitudinally cut LV of normal (A), restraint-treated (B), and infarcted (C) sheep hearts. Figure adapted from (15).

LV geometry and cardiac function in a sheep MI model with suturing of a poly(propylene) (Marlex) mesh to the infarct (Fig. 2).[15] Subsequently, Bowen *et al.* demonstrated that the poly(propylene) restraint has biological significance, leading to increased collagen and decreased protease (MMP-1, MMP-2) activity in the BZ myocardium.[27] Since these initial studies, the use of heart-wrapping techniques has shown that infarct expansion limitation has a pronounced effect in ameliorating post-MI remodeling,[13,15,24,25,28] motivating translation to clinical trials. Acorn Cardiovascular, Inc. (St. Paul, Minnesota) developed the first commercially available restraint, consisting of a knitted polyester mesh to be fitted around both ventricles. In both canine[29,30] and ovine[28,31] MI models, the Acorn CorCap restraint decreased LV volumes and improved cardiac function, leading to a clinical trial encompassing 300 patients; use of CorCap decreased LV volumes, improved quality of life, and decreased the likelihood of additional cardiac procedures.[17] More recently, Paracor Medical (Sunnyvale, California) developed a HeartNet LV support system made from a nitinol mesh that is placed around the ventricles. Similarly, both canine[32] and ovine[33] MI models showed decreased LV end diastolic volume (EDV) and end systolic volume (ESV). An initial clinical feasibility study suggested a functional and clinical benefit from the Paracore HeartNet device,[16] leading to a phase II clinical trial (PEERLESS-HF).[34] However, the clinical trial was stopped prematurely due to limited effects on certain functional end points.[35] Ultimately, despite promising results in both large animal models and clinical trials, a major limitation of polymeric LV restraints is the highly invasive surgical procedure necessary for implantation.

2.2. Injectable hydrogels as bulking agents

Similar to infarct restraints, acellular injectable biomaterials can also limit infarct expansion by bulking the damaged myocardial wall to provide mechanical stabilization.[19] Infarcts naturally stiffen over time as wound healing progresses and collagen

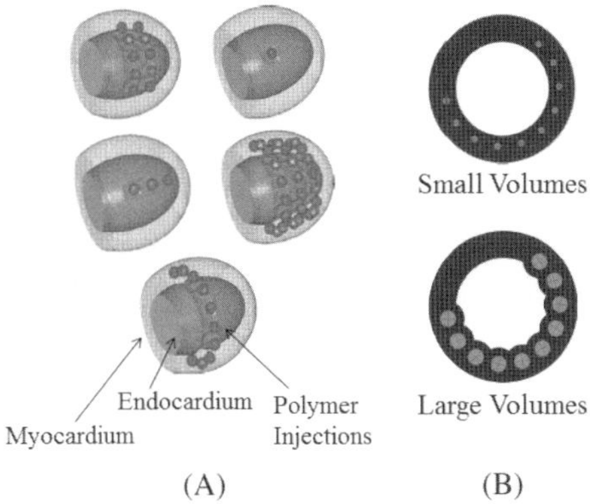

Fig. 3. FE model of simulated polymer injections, demonstrating injected materials and the importance of injection pattern (A) and volume of material dispersion (B) in decreasing myocardial stress. Figure adapted from (38).

is deposited. Therefore, the goal of injectable hydrogels is to modify the material properties of the infarct region before the body compensates for the remodeling process; in doing so, infarct expansion and remodeling post-MI can be limited.[19] Referring back to the Law of Laplace (Eq. 1), the theory behind the use of injectable hydrogels as bulking agents is that they increase the wall thickness (h) to decrease LV dilatation (R) and in turn, decrease wall stress (T). Theoretical finite element (FE) models have been used to evaluate the ability of injectable biomaterials to attenuate remodeling (Fig. 3).[36–38] Simulated injections of a non-contractile material into an infarcted LV were shown to decrease LV dilatation and reduce elevated myofiber stresses.[36] In a subsequent study, injection of a tissue filler into the infarct reduced stress in the LV and improved ejection fraction (EF) in an ovine infarct model.[37] For the first time, the use of injectable biomaterials presents the real possibility of limiting adverse LV remodeling using minimally invasive delivery.

2.2.1. *Natural acellular hydrogels*

Hydrogels can be grouped into either natural or synthetic materials. Natural materials offer several advantages due to their inherent biological properties, including presentation of receptor-binding ligands and susceptibility to cell-triggered proteolytic degradation[39,40] For cardiac applications where the goal is to replace or repair the damaged ECM, natural biomaterials more closely mimic features of the native ECM, enabling them to increase cell adhesion, activate cell proliferation, and maintain cell differentiation.[41] Commonly used natural, injectable materials for cardiac repair include fibrin, alginate, collagen, Matrigel, chitosan, hyaluronic acid (HA), keratin, and decellularized matrices (see Table 1).

Table 1. Summary of acellular hydrogel systems and their assessment in animal MI models.

Material	Animal/MI model	Injection time (post-MI)	Injection volume	Evaluation time (post-delivery)	Reference
Alginate	Rat: Proximal LCA ligation	1 week and 2 months	100–150 μL	2 months	50
	Swine: Transient LAD occlusion	4 days	1, 2, and 4 mL	2 months	49
	Rat: Proximal LCA ligation	1 week	130 μL	2 months	53
Collagen	Rat: Proximal LCA ligation	1 week	100 μL	6 weeks	57
Matrigel	Rat: LAD ligation	Immediate	50 μL × 5 sites	4 weeks	60
HA	Rat: LCX ligation	2 weeks	50 μL	4 weeks	63
	Sheep: LAD and D2 ligation	30 min	200 μL × 20 sites	2 and 8 weeks	64
	Sheep: LAD and D2 ligation	30 min	200 μL × 20 sites	2 and 8 weeks	65
Keratin	Rat: LAD ligation	Immediate	100 μL	4 and 8 weeks	66
Decellularized ventricular ECM	Rat: LAD ligation/ reperfusion	2 weeks	75 μL	1 and 4 weeks	67
	Swine: LAD embolization	2 weeks	250 μL × 14–15 sites	3 months	68
Decellularized SIS ECM	Mouse: LAD ligation	Immediate	20 μL	2 and 6 weeks	69
Poly(NIPAAm-co-AAc-co-HEMAPTMC)	Rat: LAD ligation	2 weeks	100 μL × 5 sites	4 and 8 weeks	70
Dex-PCL-HEMA/ PNIPAAm	Rabbit: Proximal LCA ligation	4 days	50 μL × 4 sites	1 month	71
αCD/MPEG-PCL-MPEG	Rabbit: LAD ligation	1 week	50 μL × 4 sites	4 weeks	72
PEG-VS	Rat: LAD ligation	2 min	100 μL (via 2–3 sites)	4 and 13 weeks	73
	Rat: LCA ligation	9 days	100 μL	7 and 50 days	74

* SIS = small intestinal submucosa; LAD = left anterior descending coronary artery; LCX = left circumflex coronary artery; LCA = left coronary artery; OM1 = obtuse marginal I branch; OM2 = obtuse marginal II branch; D2 = second diagonal branch.

Fibrin: Fibrin, a fibrillar protein consisting of fibrinogen and thrombin, is natively a critical blood component involved in the clotting cascade. A two component system is required to form fibrin hydrogels, one consisting of fibrinogen and aprotinin, a fibrinolysis inhibitor, and the other containing thrombin, factor XIIIa, and calcium chloride ($CaCl_2$). When these solutions are combined, fibrinogen converts to fibrin and self-assembles to form a mesh crosslinked by factor XIIIa. Fibrin naturally has binding domains for soluble growth factors and cellular integrin receptors, making it an ideal candidate for wound healing applications.[42] Fibrin hydrogels have been used for MI therapy[43–46] due to favorable properties such as biodegradability, mechanical strength, and angiogenic activity.[47]

Christman *et al.* first injected fibrin and showed positive outcomes on cardiac function.[43–45] Specifically, MI was induced in rats (temporary coronary occlusion) followed by injection of fibrin glue 10 days later. Assessment 5 weeks after delivery revealed preserved infarct wall thickness and fractional shortening (FS) as compared to bovine serum albumin (BSA) controls.[43] In a subsequent study, Christman *et al.* showed that injectable fibrin alone significantly decreased infarct area and increased arteriole density in the infarct region as compared to control BSA injections.[44] The ability of injectable fibrin to induce neovascularization in the infarct region was later confirmed by Huang *et al.*[46]

Alginate: Alginate is a naturally occurring, linear polysaccharide derived from seaweed. Its tailorability and non-thrombogenicity make it an attractive cardiac repair candidate.[48–50] It consists of linked *ß*-D-mannuronate and α-L-guluronate residues that crosslink into hydrogels in the presence of calcium.[51] Alginate's unique gelation enables delivery using minimally invasive, catheter-based intracoronary infusion, which relies on the infarct myocardium's leaky vasculature. Due to inadequate levels of calcium in the vasculature, alginate does not crosslink until it reaches the infarct, where high levels of calcium induce gelation.[49]

Landa *et al.* delivered alginate to rats at 7 and 60 days post-MI (permanent coronary occlusion) for acute and chronic remodeling, respectively.[50] Around 8 weeks after injection, delivery into recent or old infarcts increased scar thickness and attenuated LV systolic and diastolic dilatation and dysfunction. However, FS was not improved in either group.[50] These results provide insight into the timing for hydrogel delivery post-MI. In contrast, Leor *et al.* delivered alginate hydrogels to a clinically relevant, porcine MI model using a coronary catheter.[49] To investigate the effects of delivering different hydrogel volumes, alginate solutions at incremental volumes (1, 2, or 4 mL) were delivered 4 days post-MI (temporary occlusion). At 60 days after delivery, intracoronary injection of alginate (2 and 4 mL) prevented and even reversed LV systolic and diastolic dilatation. In addition, the 2 mL alginate injection led to myofibroblast infiltration and increased scar thickness as compared to saline controls.[49] Similar to the study by Landa *et al.*, there were no significant FS improvements.

Due to the positive results of alginate delivery in both small[50] and large[49] animal models, it has become the first injectable acellular biomaterial to be explored in a clinical trial. The injectable system of sodium alginate and calcium gluconate was first developed and tested by BioLineRx as BL-1040. In a phase I/II pilot clinical trial, the safety and feasibility of BL-1040 was demonstrated in 27 patients, including improved survival, preserved LV, EDV, and ESV, prevention of FS deterioration, improved mitral regurgitation, and limitation of systolic wall thinning (NCT00557531). Recently, BioLine out-licensed BL-1040 to Ikaria Holdings, Inc., which is currently performing a placebo-controlled, double-blind phase II clinical trial with 300 patients (NCT01226563).

Several studies have studied the effect of modifying normally non-adhesive alginate with adhesive peptides such as Arg–Gly–Asp (RGD) or Tyr–Ile–Gly–Ser–Arg (YIGSR).[52,53] Yu *et al.* delivered modified and unmodified alginate to rats 5 weeks post-MI (temporary occlusion).[52] About 5 weeks after injection, delivery of either RGD modified or unmodified alginate led to decreased EDV and ESV, improved FS, significantly increased myocardial wall thickness, and significantly increased arteriogenesis as compared with BSA controls. Although both groups led to increased arterioles in the infarct, the RGD modified alginate group had the highest number of arterioles,[52] demonstrating that surface modification can influence the infarcted myocardial microenvironment. Further, Tsur-Gang *et al.* delivered unmodified alginate, alginate modified with adhesion peptides RGD and YIGSR, and alginate modified with a non-specific peptide, RGE, to rats 7 days post-MI (permanent occlusion).[53] In this case, unmodified alginate led to the best improvements in infarct thickness, LV geometry, and FS 60 days after injection. However, no significant differences were seen in relative infarct thickness, cell infiltration, or blood vessel density between the groups.[53] Differences in material viscosity may explain some of these findings.

Due to the positive benefits of fibrin and alginate alone, Mukherjee *et al.* delivered a fibrin-alginate composite to pigs 1 week post-MI (permanent occlusion).[54] Injections were performed using a double-barrel syringe; one component was composed of fibrinogen, fibronectin, factor XIII, plasminogen, and gelatin-grafted alginate dissolved in an aprotinin solution while the other consisted of thrombin and $CaCl_2$. Assessment performed at 1, 2, and 3 weeks after delivery demonstrated significantly increased wall thickness at 1 week and limited LV expansion at 2 and 3 weeks post-injection. However, no functional improvements were observed. In addition, the composite led to significantly reduced soluble collagen vulnerable to protease degradation and decreased protease levels (e.g., MMP-2) in the infarct as compared to saline controls, providing a potential mechanism for attenuated infarct expansion.[54]

Collagen: Collagen is the most abundant protein in the human body and the predominant ECM protein. Due to extensive ECM damage post-MI, delivery of injectable collagen is well-suited for cardiac repair.[55] Collagen can be prepared in an injectable form which self-assembles at physiological pH and temperature into a fibrous hydrogel.[56] In addition, collagen has been shown to promote angiogenesis and myoblast infiltration in the ischemic myocardium.[46] Dai *et al.* delivered commercially available Zyderm (INAMED Corp., Santa Barbara, California), a bovine dermal collagen (95% collagen type I, 5% collagen type III) to rats 1 week post-MI (permanent occlusion).[57] Around 6 weeks after delivery, collagen treated animals had higher stroke volume (SV), increased EF, increased wall thickness, and absence of paradoxical LV systolic bulging as compared to controls.[57] Unlike other natural hydrogels for cardiac repair, this study did not show increased angiogenesis or cell infiltration, potentially due to differences in collagen types or concentration injected, limiting cell migration into the infarct.

Matrigel: As an alternative to pure collagen, Matrigel (BD Biosciences) is a commercially available, purified matrix isolated from a mouse sacrcoma line that encompasses collagen as well as basement membrane proteins, including laminin and entactin, and growth factors.[58,59] Similar to collagen, it undergoes a phase transition at physiological temperature to generate hydrogels. Ou *et al.* delivered Matrigel to rats immediately post-MI (permanent occlusion) and demonstrated significantly increased vessel density, enhanced recruitment of cardiac progenitor cells (CPCs), improved LV peak rate of pressure rise, increased wall thickness, and enhanced EF 4 weeks post-MI.[60] In contrast, Kofidis *et al.* delivered growth-factor free Matrigel to mice immediately post-MI (permanent occlusion) but did not demonstrate significantly increased FS at 2 weeks post-MI as compared to controls but did show trends towards increased scar thickening and improved function.[61] In another study, Kofidis *et al.* delivered Matrigel to a rat MI model and saw trends towards increased wall thickness and decreased end diastolic diameter (EDD) compared to controls 2 weeks post-MI.[62]

Chitosan: Chitosan is a non-mammalian, linear polysaccharide derived by deacetylating chitin, the main component of crustacean exoskeletons. Chitosan undergoes a temperature-induced phase transition to form hydrogels at physiological temperature when mixed with glycerol phosphate and a glyoxal salt solution.[75,76] Gelation temperature can be altered by changing the glyoxal concentration and degradation kinetics can be controlled by the degree of deacetylation.[77–79] Lu *et al.* delivered a previously developed chitosan hydrogel[77] to rats 1 week post-MI (permanent occlusion). About 4 weeks after injection, the myocardial wall thickness was significantly increased, LV EDD and end systolic diameter (ESD) significantly decreased, EF and FS enhanced, infarct size decreased, and microvessels within the infarct increased for animals treated with chitosan as compared to saline controls.[75] These results were further confirmed by Wang *et al.*[80] and Liu *et al.*[76] where chitosan alone resulted in significant improvements in EF, FS, infarct size, wall thickness, and vessel density.

HA: HA is a linear polysaccharide abundant in the body that serves a role in processes such as scar reduction, cell migration, and angiogenesis.[81] Even though HA is naturally derived, it is often chemically modified to tune its material properties and enable hydrogel formation.[63–65] Yoon *et al.* developed HA hydrogels consisting of acrylated HA with thiol-terminated poly(ethylene glycol) (PEG) crosslinkers ($PEG\text{-}SH_4$). Injection into rats 2 weeks post-MI (permanent occlusion) led to increased vessel formation, decreased apoptosis, reduced infarct area, and significantly enhanced EF and SV as compared to infarct controls at 4 weeks after delivery. Interestingly, results showed increased wall thickness while histology revealed complete hydrogel degradation at 4 weeks.[63]

Ifkovits *et al.* developed methacrylated HA (MeHA) hydrogels crosslinked with redox radical initiators, ammonium persulfate (APS) and *N, N, N′,N′*-tetramethylethylenediamine (TEMED).[64] Two formulations with different initial mechanics but similar degradation were injected into sheep 30 minutes post-MI (permanent occlusion). Both hydrogels significantly increased wall thickness relative to infarct controls 8 weeks post-MI (Fig. 4). However, only the higher modulus gels significantly reduced infarct expansion with trends toward decreased EDV and ESV and increased EF.[64] In a subsequent study, Tous *et al.* compared stable MeHA hydrogels with hydrolytically degradable HA hydrogels and demonstrated that stable hydrogels improved wall thickness and LV volumes more efficiently than degradable hydrogels.[65] These studies highlight the importance of material properties (e.g., mechanics, degradation) in cardiac repair.

Keratin: Keratin is another injectable, natural hydrogel being explored for cardiac applications. It can be extracted from human hair or animal wool and processed into an injectable material that self-assembles into a fibrous hydrogel.[82] Shen *et al.*

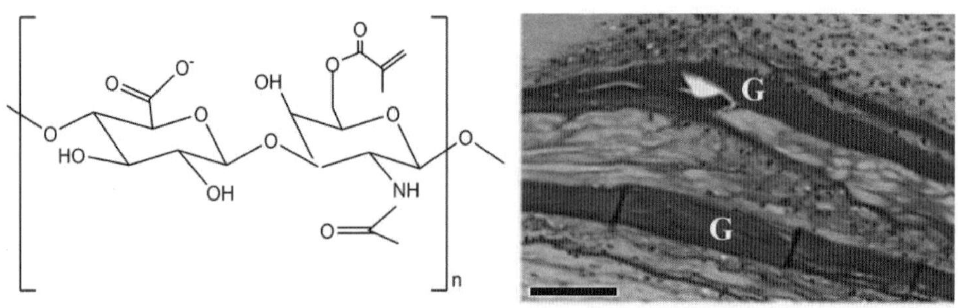

(A) (B)

Fig. 4. Natural, injectable hydrogel for acellular cardiac repair. Chemical structure of methacrylated HA (A) and representative H&E stain of the hydrogel at 8 weeks post-MI (B). G = Gel, Scale bar = 50 mm. Figure adapted from (64).

delivered keratin-based hydrogels to rats immediately post-MI (permanent occlusion) and revealed increased angiogenesis, increased wall thickness, decreased infarct size, and enhanced EF and FS as compared to saline controls at 4 and 8 weeks post-MI.[66] Based on these results, keratin-based hydrogels are a promising acellular therapy for cardiac repair.

Decellularized ECM: Decellularization of tissues is appealing for tissue engineering because tissue-specific materials can be generated from the same tissue being treated.[83] Decellularized ECM more closely mimics the unique ECM composition of each tissue, which includes specific fibrous proteins (collagen, fibrin, elastin, etc.), proteoglycans and glycosaminoglycans. In addition, its natural thermoresponsive behavior enables it to be a liquid hydrogel precursor at 25°C and then self-assemble to form nanofibrous hydrogels at 37°C.[67–69,84]

Singelyn *et al.* utilized a decellularized porcine ECM that self-assembles to form a myocardial matrix at physiological temperature.[67,68,84] When injected into the LV of healthy rats, it induced significant increase in arteriole formation 11 days after injection.[84] Injection into infarcted rats 2 weeks post-MI (temporary occlusion) led to preserved cardiac function and structure at 1 and 4 weeks after delivery, as evidenced by maintained EF, EDV, and ESV. Histology revealed increased proliferating cells and recruitment of endogenous CPCs to the infarct. A feasibility study was then performed to assess the efficacy of catheter delivery in a porcine model; histology confirmed hydrogel retention and distribution. This was the first study to successfully demonstrate transendocardial injections of an *in situ* gelling material in a clinically relevant, large animal model.[67] In a subsequent study, Seif-Naraghi *et al.* delivered this porcine ECM hydrogel to pigs 2 weeks post-MI (permanent occlusion) using a NOGA-guided MyoStar catheter (Biologics Delivery Systems, Irwindale, California). At 3 months after delivery, echocardiography revealed increased cardiac function, decreased LV volumes, and improved global wall motion scores.[68]

Okada *et al.* developed injectable hydrogels from decellularized small intestinal submucosa (SIS) and injected them into mice immediately post-MI (permanent occlusion).[69] Two hydrogel forms, SIS-B and SIS-C, were evaluated for their ability to provide a therapeutic effect post-MI. SIS-B was characterized to have increased mechanical properties and significantly more basic fibroblast growth factor (bFGF) as compared to SIS-C. Assessment at 2 and 6 weeks post-MI revealed SIS-B delivery led to reduced infarct size, increased capillary density, preserved LV end systolic geometry, and improved cardiac contractility whereas SIS-C had minimal to no benefits as compared to saline controls.[69] Since no further information was provided on the two SDS forms, it can be hypothesized that the outcomes were due to differences in bFGF concentration.

2.2.2. *Synthetic acellular hydrogels*

Despite several advantageous properties, natural materials are generally limited in the ability to tailor their properties. In contrast, synthetic materials have defined material properties, including molecular weight, gelation, hydrophilic/hydrophobic properties, degradation, and mechanics, without batch-to-batch variations.[40] Despite their inherent lack of bioactivity, synthetic materials can be modified with cell binding sites or adhesive ligands to encourage cell interaction.[85] For cardiac applications, synthetic materials can be designed to trigger cellular events, minimize the immune response, and degrade at a controlled rate, thus enabling cell integration and potentially, improvements in cardiac function long-term.[39] Various synthetic materials have been explored for cardiac repair therapy, including poly(N-isopropylacrylamide) (PNIPAAm) and PEG-based hydrogels (see Table 1).

PNIPAAm: Several studies have focused on the use of PNIPAAm-based hydrogels due to their thermoresponsive properties.[70,71,86,87] Fujimoto *et al.* developed a hydrogel from the copolymerization of NIPAAm, acrylic acid (AAc), and hydroxylethyl methacrylate-poly(trimethylene carbonate) (HEMAPTMC) to generate poly(NIPAAm-co-AAc-co-HEMAPTMC), which gels at 37°C and degrades over 5 months.[70] Injection into rats 2 weeks post-MI (permanent occlusion) led to significantly increased wall thickness, maintained LV end diastolic area, and preserved percent LV fractional area change compared to saline controls at 4 and 8 weeks post-MI. There was also enhanced capillary density and positive staining for α-smooth muscle actin in the infarct, providing insight to the mechanism of functional repair.[70]

Similarly, Wang *et al.* developed a thermosensitive hydrogel consisting of biodegradable dextran grafted to hydrophobic poly(ε-caprolactone)-2-hydroxylethyl methacrylate (PCL-HEMA) and copolymerized with thermoresponsive PNIPAAm to form Dex-PCL-HEMA/PNIPAAm.[71] Injection into rabbits 4 days post-MI (permanent occlusion) led to significantly increased EF, reduced EDD and ESD, increased ESP, decreased EDP, increased cardiac contractility, and reduced infarct size as compared to saline controls at 1 month after delivery. Hydrogel treatment led to significantly increased infarct thickness despite no histological evidence of the material remaining at 30 days.[71]

Wall *et al.* developed semi-interpenetrating polymer network (sIPN) hydrogels with RGD adhesion sites and MMP degradable crosslinkers.[86] Thermoresponsive poly(*N*-isopropylacrylamide-*co*-acrylic acid) [p(NIPAAm-*co*-AAc)] was gelled using free radical polymerization induced by APS and TEMED. Injection into mice immediately post-MI (permanent ligation) led to significantly increased wall thickness and increased FS as compared to saline controls with a trend towards improved EF at 6 weeks post-MI. In addition, there were no statistical differences in FS or EF between animals treated with sIPN injection as compared to sham-operated, uninfarcted mice by 6 weeks.[86]

PEG: PEG and its derivatives have been widely studied as injectable hydrogels for MI treatment.[72–74,88–90] Jiang *et al.* developed a novel hydrogel consisting of α-cyclodextrin (α-CD) and linear triblock copolymer, poly(ethylene glycol)-*b*-polycaprolactone-(dodecanedioic acid)-polycaprolactone-poly(ethylene glycol) (MPEG-PCL-MPEG), where degradation is controlled by the triblock copolymer.[72] Injection into rabbits 1 week post-MI (permanent occlusion) resulted in significantly improved LV EF, decreased EDD and ESD, reduced scar thickness, and decreased infarct size at 4 weeks after delivery. In a subsequent study, Wang *et al.* applied the same α-CD/MPEG-PCL-MPEG hydrogel to rats 5 min post-MI (permanent occlusion) and demonstrated significantly improved FS, decreased LV EDD and ESD, increased ESP, and decreased infarct size as compared to saline controls 1 month after delivery.[88] In both studies, no significant improvements were observed in neovascularization in the infarct.

Wu *et al.* developed a thermoresponsive PEG-based hydrogel consisting of an aliphatic, triblock copolymer, poly (δ-valerolactone)-*block*-poly (ethylene glycol)-*block*-poly (δ-valerolactone) (PVL-*b*-PEG-*b*-PVL).[89] This polymer dissolves in water at room temperature, but gels at 37°C. Injection into rats 1 week post-MI (permanent occlusion) led to significantly increased EF, improved end systolic elastance, improved preload recruitable strong work, reduced ESV and EDV, decreased infarct area, increased wall thickness, and increased blood vessel density 5 weeks after injection.[89] Similarly, Kraehenbuehl *et al.* delivered a MMP-responsive, PEG-vinylsulfone (PEG-VS) hydrogel to rats 1 h post-MI (permanent occlusion) and showed significantly increased EF and decreased ESV with trends toward decreased EDV and reduced infarct size as compared to saline controls at 6 weeks post-MI.[90]

In contrast to degradable PEG hydrogels, Dobner *et al.* delivered a non-degradable PEG-VS hydrogel polymerized with dithiothreitol (DTT) to rats two minutes post-MI (permanent occlusion).[73] At 4 weeks post-MI, the PEG-VS hydrogel significantly increased wall thickness and decreased LV EDD. However, at 13 weeks, the pathological progression of the treatment animals was similar to that of saline controls. FS was not improved in hydrogel animals at either time point.[73] Rane *et al.* applied the same bioinert, non-degradable PEG-VS hydrogel to rats 9 days post-MI (permanent occlusion) and demonstrated significant increases in wall thickness as compared to saline controls at 50 days post-MI.[74] Despite this, there was a steady decline in cardiac function and expansion of LV volumes in both the hydrogel and control animals. In addition, there were no significant differences in neovascularization or cell infiltration between hydrogel and control groups.[74]

Similar to Dobner *et al.*, Rane *et al.* found that increased wall thickness alone was insufficient to cause positive structural and functional outcomes. These studies suggest that the bioactivity and ability for cells to infiltrate degradable biomaterials are important to preserving cardiac function post-MI. In addition, prolonged material presence alone may be insufficient to attenuate adverse LV remodeling.

3. Cellular Approaches

3.1. Cell delivery

MI results in the loss of over one billion cardiomyocytes in the infarct region.[91] It was previously believed that after such a massive cardiomyocyte loss, the adult myocardium lacked an intrinsic regenerative capacity to repair itself.[92] Recently, studies have confirmed the activation of resident cardiac stem/progenitor cells and/or recruitment of endogenous stem cells post-MI.[93] However, this innate regenerative capacity is not sufficient to compensate for the large-scale tissue loss post-MI.[94] As a result, the delivery of exogenous cells has emerged as a therapy for cardiac MI repair.[95] A variety of cell types have been shown to be effective for cardiac repair, including fetal or neonatal cardiomyocytes, embryonic stem cells (ESCs), skeletal myoblasts, bone marrow-derived stem cells (BSCs), adipose-derived stem cells, and cardiac stem cells (CSCs).[96,97]

ESCs offer the advantage of differentiating into both cardiomyocyte and vascular lineages but are limited due to their immunogenicity, risk of tumor development, and ethical concerns.[98] Skeletal myoblasts have fewer ethical concerns and offer the advantages of being autologous, easy to obtain and expand, and resistant to ischemia.[99] However, they cannot differentiate into cardiomyocytes[100] and, even though they are naturally contractile, they have a limited ability to contract synchronously with host cardiomyocytes.[101] BSCs are another autologous option that can be readily isolated, expanded rapidly in culture, and maintain or modify their phenotype to complement the host tissue.[102] The bone marrow is a common source of adult stem/progenitor cells, including mesenchymal stem cells (MSCs), endothelial progenitor cells (EPCs), and hematopoietic stem cells (HSCs).[103] However, it remains unclear if BSCs can truly differentiate into cardiomyocytes *in vivo*.[100] In contrast, resident CSCs are multipotent and can differentiate into cardiomyocytes *in vivo*,[93] but there are concerns with scalability and autologous isolation from patients with recent MI.[104] Therefore, each cell type offers its own set of advantages and limitations, making it challenging to determine the ideal cell source for cardiac repair.

Cell delivery, or cellular cardiomyoplasty, involves directly delivering cells to the infarcted myocardium using intramyocardial injection, intracoronary infusion, or interstitial retrograde coronary venous delivery. The feasibility of delivering cells alone to the heart has been confirmed for nearly two decades.[105] While both animal models[97] and clinical studies[96] demonstrate significant improvements in cardiac function, this improvement is often not clinically relevant and, for the most part, transient.[106] This transiency is due to unsatisfactory cell retention, survival, and engraftment.[95] Regardless of the delivery method, there is lack of cell retention at the injection site due to the low viscosity of a cell–saline suspension, causing the cells to leak out during injection.[107] Of the small percentage of cells that do engraft, they often lack appropriate functionality to have a therapeutic effect.[108] Hofmann *et al.* and Hou *et al.* showed that less than 10% of BMCs delivered are detected 2h after injection.[109,110] Of the retained cells, approximately 90% die within the first week due to the myriad of stresses encountered upon injection into the infarct, including physical stress, ischemia due to microvasculature obstruction, myocardial hypoxia, inflammation, and release of cytokines and reactive oxygen species.[111] In addition, due to ECM breakdown post-MI, the retained cells lack an anchoring matrix which may result in anoikis, a form of programmed cell death due to loss of normal cell-matrix interactions.[112] Ultimately, in spite of enthusiasm for cell delivery, regeneration of functional, integrated myocardium in the infarct has yet to be demonstrated in large animal models or patients.

Due to these challenges, biomaterials have been explored to enhance cell retention and engraftment for cardiac repair by improving cell attachment, migration, and survival upon delivery.[46] In addition, biomaterials can enhance tissue angiogenesis,[44,46] activate cells to produce cytokines,[46] enhance cell retention,[44] and influence cell differentiation.[113] Polymers can be designed to influence integrin binding sites with a range of matrix pore sizes, substrate topography, and substrate rigidity.[114] Biomaterial approaches for cell delivery include *in vitro* engineered cardiac muscle patches and injectable biomaterials.

3.2. Cardiac patches

Cardiac patches may provide structural support to limit LV dilation post-MI, similar to LV restraints, but with the added benefit of cell delivery. They are constructed by seeding cells onto a biomaterial scaffold *in vitro* for subsequent *in vivo* implantation onto the infarcted myocardium.[115,116] Theoretically, cardiac function improves through the combination of cell-mediated remodeling[117] and material-based mechanical stabilization which attenuates myocardial dilatation.[118] Because the cells are anchored to the scaffold, cardiac patches lead to better engraftment when compared to cell delivery alone.[118]

Li *et al.* demonstrated the use of a cardiac patch to deliver cells for cardiac repair.[119] Fetal cardiomyocytes were seeded on a gelatin mesh *in vitro* and subsequently implanted onto the infarct of a rat MI model, leading to prolonged cell survival over 5 weeks and a trend towards improved LV function.[119] Since this early study, a variety of natural and synthetic cardiac patches have led to positive functional outcomes *in vivo* as compared to injection of cells alone.[115,116] Hamdi *et al.* compared delivery of skeletal myoblasts through intramyocardial injection, formation as a cell sheet, or using a collagen-based cardiac patch at 4 weeks post-MI.[120] Animals that received the cell sheet or cell seeded scaffold had significantly improved EF and increased vessel density 1 month after delivery as compared to culture medium controls.[120] The promising results in animal models have led to clinical trials involving cardiac patches for MI repair. In the MAGNUM phase I clinical trial, BMCs were seeded on type I collagen cardiac patches and delivered to 10 patients with coronary artery bypass grafts and intrainfarct implantation of autologous bone marrow cells. The combined cell injection and cardiac patch approach led to improved LV EDV and increased scar thickness as compared to cell injection alone.[121] Additional studies using three-dimensional (3D) fibrin cardiac patches have shown similar results.[122–124]

Despite success with cardiac patches, major drawbacks still remain. Although the purpose is to improve cell engraftment, the reality is, there is low engraftment within patches when implanted *in vivo* due to inadequate angiogenesis.[118,122,123] Low vascularization limits graft thickness due to an inability to support the high oxygen and nutrient demands of cells at depths greater than ~100 μm.[125] In addition, the patches must be sutured to the infarct,[48] a technique which may be useful to prevent LV dilatation but is unlikely to directly contribute to myocardial contractility and function. The location of the patch on the epicardial surface of the heart also limits the region of therapeutic benefit. Most importantly, the invasive surgical procedure necessary for implantation limits clinical translation.

3.3. Injectable hydrogels for cell delivery

Injectable hydrogels for cell delivery offer the advantage of enhancing cell retention, survival, and engraftment with the potential for minimally-invasive, percutaneous catheter-based delivery,[49,67,68] which may enhance clinical translation.[50,59,75,126] As compared to rigid cardiac patches with limited radial engraftment, injectable hydrogels can form *in situ*, integrate into the host tissue, and adapt to the surrounding contracting myocardium.[127] One of their major benefits is that hydrogel crosslinking typically occurs after the addition of cells, enabling complete cell encapsulation into a viscous but flowable delivery matrix that solidifies upon injection into the myocardium. Hydrogel shape, stiffness, and thickness can be controlled to ensure formation into the geometry of the injection site.[128] Hydrogels not only provide 3D support to enhance cell retention, but also permit control over the biophysical and biochemical microenvironment of the transplanted cells to prolong survival and engraftment.[129] In turn, encapsulated cells are allowed to produce their own ECM and remodel the construct into a more suitable environment.

Delivery of cells using injectable biomaterials may provide additional benefit over acellular materials. However, many studies show similar functional benefits with delivery of hydrogels alone. As previously discussed in Sec. 2.2, injection of hydrogels into the infarct region results in improved passive myocardial properties, mechanical stabilization due to LV geometry alterations, and ultimately, reduced myocardial wall stress.[36] However, to improve cardiac function long-term, the therapy should also contribute to active myocardial regeneration, including the introduction of viable, contractile tissue.[130] This can potentially be achieved by combining cell approaches with injectable hydrogels. Ultimately, hydrogels provide a platform for cell delivery, a structural matrix to direct tissue formation, and an improved microenvironment for myocardial repair.[43,85] Similar to acellular approaches both natural and synthetic hydrogels have been utilized to deliver cells for myocardial repair (see Table 2).

3.3.1. *Natural hydrogels for cell delivery*

Natural materials are a popular choice for cell delivery due to their inherent ability to be recognized by cell-surface receptors and potential for cell-mediated remodeling, thus facilitating cell-biomaterial interactions. Similar to acellular approaches, commonly used naturally derived hydrogels for cell delivery include fibrin, alginate, collagen, Matrigel, chitosan, and self-assembling peptides.

Fibrin: Injectable fibrin has been used to deliver a variety of cell types, including skeletal myoblasts,[43,44] BSCs,[131] bone marrow mononuclear cells (BMNCs),[132,133] bone marrow-derived CSCs,[124] MSCs,[134] endothelial cells,[135] and adipose-derived stem cells.[136,137] Regardless of the cell source, delivery of cells encapsulated in fibrin leads to enhanced neovascularization, improved LV function, and attenuated LV remodeling.[138] Christman *et al.* demonstrated the benefits of injectable hydrogels for cell delivery over traditional cellular cardiomyoplasty.[44] Injection of skeletal myoblasts encapsulated in fibrin glue in rats 1 week post-MI (temporary occlusion) led to a significantly greater cell area relative to myoblasts injected in BSA 5 weeks after delivery, indicating that fibrin increases cell survival by potentially serving as a temporary ECM for the injected cells.[44] Similar to previous studies,[43] fibrin glue, with or without cells, induced neovascularization in the ischemic myocardium and reduced infarct expansion.

Table 2. Summary of cellular hydrogel systems and their assessment in animal MI models.

Material	Animal/ MI model	Cells delivered	Injection time (post-MI)	Injection volume	Evaluation time (post-delivery)	Reference
Fibrin	Rat: LAD ligation/ reperfusion	Skeletal myoblasts (5×10^6)	10 days[43] or 1 week[44]	50 μL	5 weeks	43,44
	Rat: cryoinjury	BMNCs (2×10^7)	3 weeks	100 μL × 2 sites	8 weeks	133
	Sheep; LCX constriction	Endothelial cells	4 weeks	800 μL × 3 sites	2 months	135
	Rat: LAD ligation	MSCs (5×10^5)	Immediate	150 μL × 2 sites	90 min	134
Collagen	Rat: LAD ligation	MSCs (2×10^6)	1 week	70 μL	4 weeks	140
Matrigel	Mouse: LAD ligation	ESCs (1×10^6)	Immediate	50 μL	2 weeks	61
	Rat: LAD ligation/ heterotopic transplant	ESCs (2×10^6)	Immediate	125 μL	2 weeks	142
Chitosan	Rat: LAD ligation	ESCs (1×10^7)	1 week	100 μL	4 weeks	75
	Rat: LAD ligation	Adipose-derived MSCs (4×10^6)	1 week	100 μL (via 3 sites)	4 weeks	76
Self-assembling peptides	Minipigs: Mid-LAD ligation	BMNCs (1×10^8)	Immediate	50 μL × 40 sites	4 weeks	143
Dex-PCL-HEMA/ PNIPAAm	Rabbit: Proximal LCA ligation	BMNCs (1×10^7)	1 week	50 μL × 4 sites	1 month	87
αCD/MPEG-PCL-MPEG	Rat: LCA ligation	BMNCs (2×10^7)	1 week	50 μL × 4 sites	4 weeks	144
Oligo(PEG fumarate)	Rat: LAD ligation	ESCs (1×10^6)	1 week	100 μL (via 3 sites)	4 weeks	145
PEGylated-fibrinogen	Rat: Proximal LAD ligation	Neonatal cardio-myocytes (3×10^6)	1 week	300 μL	30 days	146

* BMNCs = bone marrow-derived mononuclear cells; MSCs = mesenchymal stem cells; ESCs = embryonic stem cells; CPCs = cardiac progenitor cells; ELC = endothelial-like cells; SMLC = smooth muscle-like cells (see Table 1 for additional abbreviations).

Ryu *et al.* delivered BMNCs with and without fibrin encapsulation to rats 3 weeks post-MI (cryoinjury).[133] About 8 weeks after delivery, histology revealed increased neovascularization when cells were delivered using fibrin hydrogels, including significantly increased microvessel density and larger internal vessel diameters as compared to cells alone or culture medium controls.[133] However, it is unclear if the improvements were due to fibrin alone or BMMC delivery due to lack of a fibrin control. Chekanov *et al.* reported similar results when delivering endothelial cells with and without a fibrin matrix to an ovine chronic ischemia cardiomyopathy model.[135] Finally, Martens *et al.* delivered labeled MSCs encapsulated in fibrin to rats immediately post-MI (permanent occlusion) using a minimally invasive, percutaneous catheter.[134] Approximately 90 minutes after delivery, cells delivered using fibrin had significantly increased retention in the heart and reduction of redistribution to the liver and kidney as compared to cells delivered alone.[134]

Alginate: Alginate hydrogels have also been utilized for cell delivery for cardiac repair. Yu *et al.* used injectable, RGD modified, alginate hydrogel microspheres to deliver encapsulated MSCs (temporary occlusion).[139] Alginate encapsulation improved cell attachment and growth and increased angiogenic growth factor expression *in vitro*. Injection of alginate, with or without cells, into rats 1 week post-MI (temporary occlusion) led to preserved wall thickness and significantly improved FS at 10 weeks after delivery as compared to baseline. Histology demonstrated that each treatment group (alginate alone, cells alone, and alginate with cells) reduced infarct area and increased arteriole density indicative of angiogenesis, with no significant differences between the groups. Despite comparable functional outcomes between the groups, the use of alginate significantly increased cell survival over cells alone or saline controls.[139]

Collagen: Injectable collagen has also been used to deliver cells for cardiac repair. Dai *et al.* delivered nanoparticle labeled MSCs encapsulated in a commercially available collagen type I matrix to rats 1 week post-MI (permanent occlusion).[140] After 4 weeks following injection, collagen encapsulation significantly reduced MSC relocation to regions of non-infarcted myocardium and remote organs relative to delivery of cells alone. However, cell delivery using collagen hydrogels did not significantly improve LV EF as compared to saline controls. EF was significantly increased in animals that received collagen alone, consistent with a previous study,[57] as well as cells alone as compared to saline.[140] To assess the translational potential of collagen-cell systems, Thompson *et al.* injected autologous BMCs encapsulated in a collagen hydrogel into the myocardium of pigs using a composite catheter system (TransAccess).[141] Histology revealed hydrogel retention in the myocardium immediately after injection and at 14 and 28 days following injection, indicating the efficacy of intramyocardial catheter delivery. However, since this procedure was only performed on healthy, non-infarcted hearts, it is difficult to assess the effects of this therapy on LV remodeling.[141]

Matrigel: Several cell types have been delivered using Matrigel hydrogels, including ESCs[61,142] and ESC-derived cardiomyocytes.[59] Laflamme *et al.* delivered human ESC-derived cardiomyocytes encapsulated in Matrigel to rats 4 days post-MI (temporary occlusion) and saw preserved cell retention in the infarct as compared to cell delivery in serum free medium at 1 week after injection.[59] Kofidis *et al.* delivered labeled mouse ESCs encapsulated in Matrigel to a rat heterotopic heart transplant model immediately post-MI (permanent occlusion).[142] About 2 weeks later, delivery of Matrigel encapsulated ESCs led to significantly increased FS, increased thickness in the posterior and septal walls, and reduced EDD as compared with controls (infarcted, Matrigel alone, ESCs alone).[142] These functional and structural results were confirmed in a subsequent study using a mouse MI model. In this study, the Matrigel encapsulated ESCs showed expression of gap junctional protein connexin 43 at intercellular sites, indicating possible connections with the host myocardium.[61]

Composites of collagen and Matrigel have also been used for cardiac repair. The Zimmerman group developed a novel cardiac muscle construct known as engineered heart tissue (EHT), which consists of a combination of neonatal cardiomyocytes with an artificial ECM made from collagen type I, Matrigel and cell growth medium.[147] Due to the properties of collagen and Matrigel, EHT remains a liquid at room temperature but solidifies at 37°C. Zhang *et al.* delivered liquid EHT as an injectable hydrogel to rats 3 weeks post-MI (permanent occlusion). About 1 month after delivery, collagen-Matrigel encapsulated cardiomyocytes led to increased wall thickness, decreased LV ESD, and enhanced FS as compared to matrix alone, cells alone, or medium controls.[148]

Chitosan: Chitosan hydrogels have also been used for cell delivery to attenuate remodeling post-MI. Lu *et al.* delivered chitosan encapsulated ESCs to rats 1 week post-MI (permanent occlusion) and saw significantly improved 24-h cell retention and 4-week graft size as compared to delivery of ESCs alone.[75] In addition, delivery of ESCs encapsulated in chitosan led to significantly increased microvasculature density in the infarct, increased wall thickness, reduced infarct area, decreased EDD and ESD, and improved FS and EF at 4 weeks after injection as compared to chitosan alone, ESCs alone, or saline controls.[75] These results were later confirmed by delivering a chitosan hydrogel with encapsulated adipose-derived MSCs to a rat MI model. Even though injection of chitosan or cells alone improved cardiac structure and function as compared to saline controls, the combination of cells with chitosan further improved outcomes.[76]

Self-assembling peptides: Self-assembling peptides consist of alternating hydrophilic and hydrophobic amino acids that can form stable β-sheets. At physiological osmolarity and pH, these peptides self-assemble to form nanofibrous hydrogels. Due to their ability to promote cell organization through their structure, these biologically inspired materials have been used for cell delivery in both small[149,150] and large[143] animal models. Various cells have been delivered, including cardiomyocytes,[149,151] CPCs,[152] skeletal myoblasts,[150] and BMNCs.[143] Dubois *et al.* injected skeletal myoblasts with self-assembling peptides into rats 2 weeks post-MI (permanent ligation).[150] Nearly 1 month after injection, peptide nanofibers alone significantly increased angiogenesis, but did not improve survival of the encapsulated skeletal myoblasts. Consistent with the decreased cell survival, no functional improvements were seen in any of the therapy groups relative to culture medium controls.[150] In contrast, Lin *et al.* used self-assembling peptides to deliver cells to a large animal MI model.[143] Injection of BMNCs and self-assembling peptides to minipigs immediately post-MI (permanent occlusion) led to significantly increased EF as compared to all other groups (cells alone, peptides alone, saline controls) and increased interventricular septum (IVS) systolic and diastolic thickness relative to saline controls at 4 weeks after delivery. In addition, peptide nanofibers alone significantly enhanced diastolic function as evidenced by increased IVS diastolic thickness relative to BMNCs or saline controls, whereas BMNCs alone significantly improved EF and increased IVS systolic thickness relative to saline controls. Overall, the combination of cells and nanofibers led to the greatest improvements in diastolic and systolic function.[143]

3.3.2. *Synthetic hydrogels for cell delivery*

Synthetic hydrogels can also be used to deliver cells for cardiac repair. Due to their tunable properties, small peptides can be added to promote cell interactions or enable degradation. Similar to acellular hydrogels, the primary synthetic materials used for cell delivery are PNIPAAm and PEG-based hydrogels.

PIPAAm: Thermoresponsive PIPAAm hydrogels have successfully encapsulated and delivered cells post-MI in several studies.[86,87] For instance, Li *et al.* delivered BMNCs encapsulated in a thermoresponsive Dex-PCL-HEMA/PNIPAAm hydrogel to rabbits 1 week post-MI (permanent occlusion), as previously discussed in Sec. 2.2.2.[87] Histology performed 48h after injection showed significantly increased cell retention in the infarct area with the presence of the hydrogel as compared to BMNCs alone. After 30 days following injection, animals that received BMNCs in hydrogels had significantly decreased infarct size, increased neovascularization, reduced LV EDD and ESD, and increased EF as compared to cells alone, hydrogel alone, or culture medium controls. Even though material alone led to improved structural and functional cardiac outcomes as compared to the medium control, there were no improvements in neovascularization, providing motivation for cell encapsulation strategies to repair the damaged myocardium.[87] Similarly, Wall *et al.* injected bone marrow-derived MSCs encapsulated in p(NIPAAm-*co*-AAc) hydrogels into mice immediately post-MI (permanent occlusion) and saw trends towards increased FS and EF and decreased EDV at 2 weeks as compared to saline controls.[86]

PEG: PEG hydrogels have delivered several cell types, including ESC-derived vascular cells,[90] BMNCs,[123,144] and ESCs.[145] Wang *et al.* delivered BMNCs encapsulated in an α-CD/MPEG-PCL-MPEG hydrogel to rabbits 7 days post-MI (permanent occlusion), as previously discussed in Sec. 2.2.2.[144] About 4 weeks after delivery, hydrogel encapsulated BMNCs led to significantly increased cell retention and neovascularization in the infarct, decreased LV EDD, reduced infarct size, and increased EF as compared to cells alone or saline controls.[144] Similarly, Wang *et al.* delivered ESCs encapsulated in a biodegradable oligo(poly[ethylene glycol] fumarate) (OPF) hydrogel to rats 1 week post-MI (permanent occlusion) and showed significantly improved 24-h cell retention and 4-week graft size as compared to delivery of ESCs alone.[145] At 4 weeks after injection, delivery of OPF encapsulated ESCs led to significantly decreased infarct area, reduced collagen deposition in the infarct (fibrosis), increased infarct neovascularization, decreased LV EDV and ESV, and increased FS as compared to ESCs alone, OPF alone, or saline controls. Even though outcomes were most pronounced with the cell-hydrogel system, delivery of OPF alone or ESCs alone led to significantly reduced infarct area, decreased fibrosis, reduced LV volumes and enhanced FS as compared to saline controls. In addition, each treatment group showed significantly decreased protease expression (i.e., MMP-2, MMP-9) relative to saline controls, with the cell-hydrogel system leading to the greatest decrease, which could favor infarct stiffening at later time points.[145]

Building on the benefits of synthetic and natural systems, Habib *et al.* used a photopolymerizable, biodegradable PEGylated-fibrinogen hydrogel to deliver neonatal rat cardiomyocytes to rats 1 week post-MI (permanent occlusion).[146] After 30 days following injection, delivery of hydrogel encapsulated neonatal cardiomyocytes encapsulated led to significantly increased FS as compared to all other groups (cells alone, hydrogel alone, saline controls), decreased wall motion score index (measure of regional systolic function) relative to saline controls, and preserved LV EDD as compared to hydrogel alone or saline. Histology revealed that co-injection of the cells and hydrogel significantly increased the cell-graft area and wall thickness as compared to all other groups (Fig. 5).[146]

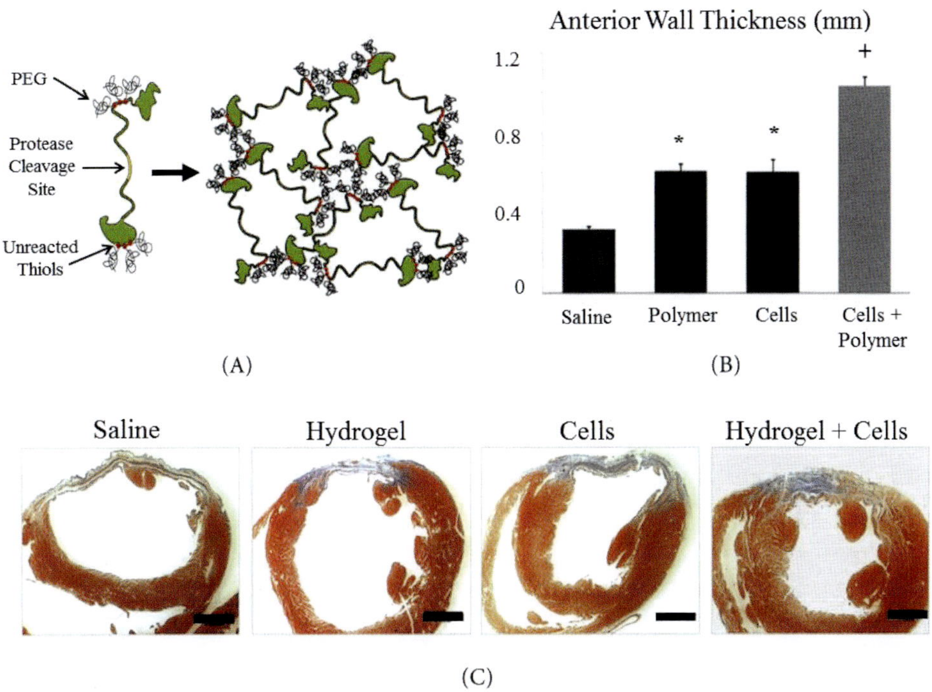

Fig. 5. Synthetic hydrogel for cell delivery post-MI. Schematic of PEG-fibrinogen hydrogel assembly (A). Infarct thickness (B) and corresponding Masson's Trichrome staining (C) of infarcted hearts injected with saline, PEG hydrogel alone, rat cardiomoyctes alone, or combined cell-hydrogel therapy at 30 days post-MI. Scale bar = 3 mm.

*$p < 0.05$ Versus saline controls, †<0.005 Versus all other groups. Figure adapted from (146).

4. Injectable Hydrogels for Molecule Delivery

In addition to providing mechanical stabilization to the injured myocardium and serving as an exogenous cell delivery vehicle, injectable hydrogels can encapsulate and deliver therapeutic molecules to the myocardium post-MI. Tissue repair is a complex process controlled in part by growth factors and cytokines that regulate the pathways of wound healing and tissue generation. Therefore, delivery of exogenous molecules, such as growth factors, cytokines, and stem cell mobilizing factors, can modulate endogenous biological responses post-MI.[153] Delivery of therapeutic molecules alone, either by direct myocardial injection or systemic intravenous circulation, has helped restore cardiac function in some animal models.[154] However, clinical trials have been less conclusive and there are several limitations with molecule delivery alone, such as the need for high concentrations or repeated doses due to high diffusion rates and short half-lives.[155] In addition, systemic delivery introduces numerous safety concerns and off-target effects, such as increased incidence of restenosis, elevated blood

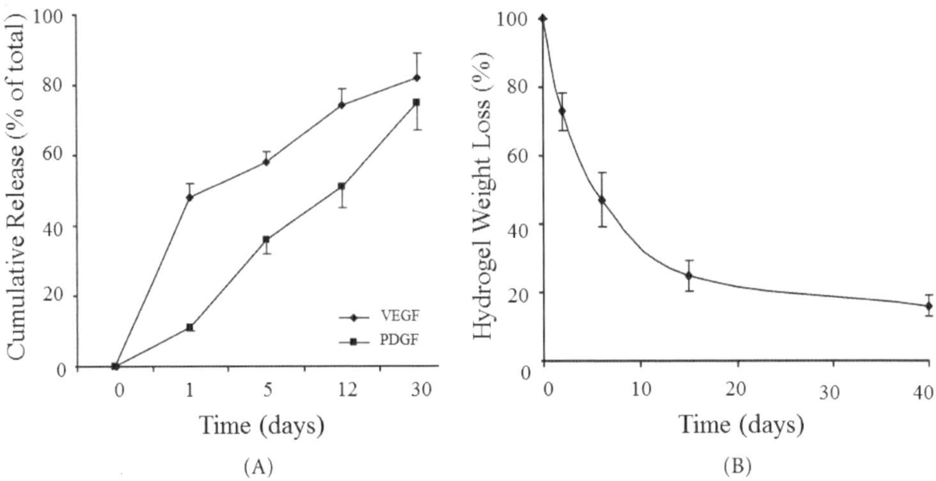

Fig. 6. Molecule release from injectable, natural hydrogels. Dual growth factor release of VEGF and PDGF (A) tracks with alginate hydrogel degradation (B) *in vitro*. Figure adapted from (161).

pressure and viscosity, and thrombolytic events in congestive heart failure patients.[156] Due to limitations in molecule delivery alone, injectable hydrogels have been used as delivery vehicles to localize molecules and tailor release kinetics through changes in polymer-molecule interactions, polymer hydrophobicity, and hydrogel degradation (Fig. 6).[157,158] Hydrogels can both sustain local molecule release and prolong molecule bioactivity.[159] For cardiac applications, injectable hydrogels deliver anti-apoptotic molecules, angiogoenic factors, or chemoatttractants to attenuate remodeling post-MI (see Table 3).

4.1. Anti-apoptotic factors

Hydrogels have been used to deliver anti-apoptotic factors to attenuate further loss of cardiomyocytes post-MI. Ruvinov *et al.* used alginate hydrogel microparticles to deliver both insulin-like growth factor 1 (IGF-1) and hepatocyte growth factor (HGF) to rats 6 days post-MI (permanent occlusion).[160] This growth factor combination was selected due to the cytoprotective properties of IGF-1 and pro-angiogenic and anti-fibrotic effects of HGF. To increase growth factor affinity to the hydrogel, alginate was chemically modified with sulfate groups; due to differences in sulfate affinity between IGF-1 and HGF, there was sequential release of IGF-1 followed by HGF. Around 4 weeks following injection, the dual IGF-1/HGF affinity-bound alginate hydrogel significantly decreased fibrosis, reduced apoptosis, increased vessel density and area, and increased cardiomyocyte proliferation as compared to IGF/HGF alone and hydrogel alone. In addition, the dual growth factor hydrogel system led to significantly

Table 3. Summary of hydrogels for molecule delivery and their assessment in animal MI models.

Material	Animal/MI model	Molecules delivered	Injection time (post-MI)	Injection volume	Evaluation time (post-delivery)	Reference
Alginate	Rat: LAD ligation	IGF-1 (250 ng), HGF (250 ng)	6 days	120 μL	4 weeks	160
	Rat: LAD ligation	VEGF (3 μg), PDGF(3 μg)	1 week	100 μL	4 weeks	161
Chitosan	Rat: LAD ligation	FGF (2 μg)	1 week	100 μL (via 3 sites)	4 weeks	80
HA	Mouse: cryoinjury	SDF-1α (0.25 μg)	Immediate	25 μL	1 week	162
Gelatin	Rat: LAD ligation	FGF (120 μg)	Immediate	57 μL × 6 sites	2 and 4 weeks	163
	Dog: LAD ligation	FGF (100 μg)	Immediate	1 mL × 5 sites	10 and 17 days	164
Self-assembling peptides	Rat: LCA ligation	FGF (40 ng), PDGF (160 ng)	Immediate	50 μL (via 3 sites)	4 and 8 weeks	165
	Rat: LCA ligation	SDF-1α (30 nmol/L)	Immediate	80 μL (via 3 sites)	4 weeks	166
α-CD/ MPEG-PCL-MPEG	Rat: LCA ligation	EPO (600 U)	5 minutes	100 μL × 3 sites	1 month	88
PVL-*b*-PEG-*b*-PVL	Rat: LAD ligation	VEGF (40 ng)	1 week	25 μL × 4 sites	1, 2, 3, and 5 weeks	89

* IGF-1 = insulin-like growth factor 1; HGF = hepatocyte growth factor; VEGF = vascular endothelial growth factor; PDGF = platelet derived growth factor; FGF = fibroblast growth factor; SDF-1α = stromal cell derived factor-1 alpha; EPO = erythropoietin (see Tables 1 and 2 for additional abbreviations).

increased scar thickness, limited infarct expansion and reduced fibrosis as compared to saline controls. There was also evidence of cardiogenesis (GATA-4-positive cells), suggesting the potential for endogenous cardiac muscle regeneration.[160]

Davis *et al.* used self-assembling peptide nanofibers to deliver IGF-1 to rats immediately post-MI (permanent occlusion).[149] To sustain IGF-1 release, biotinylated IGF-1 was tethered to biotinylated self-assembling peptides using a streptavidin linker complexed to the IGF-1. This attachment strategy enables IGF-1 binding but does not prevent peptide self-assembly into nanofibrous hydrogels. A preliminary study in healthy rats demonstrated that tethering of IGF-1 to peptide nanofibers led to significant retention in the myocardium for 28 days and preserved bioactivity for 14 days as compared to untethered IGF-1 or IGF alone. However, when delivered to the rat MI model, no differences in cardiac function were observed between treatment groups at 1 or 21 days post-MI.[149] In contrast, Kraehenbuehl *et al.* demonstrated improvements in cardiac function with hydrogel-based growth factor delivery. Thymosin beta-4 (Tβ4), a pro-angiogenic and pro-survival factor, was encapsulated in an MMP degradable PEG-VS hydrogel and injected into rats 1h post-MI (permanent occlusion).[90] About 6 weeks later, PEG hydrogels with or without Tβ4 led to significantly decreased LV ESV and increased EF as compared to saline controls, with a trend towards decreased infarct size and reduced LV EDV. However, outcomes were more pronounced with Tβ4. Interestingly, delivery of *Tβ4* using PEG hydrogels led to a significantly smaller LV ESV as compared to hydrogels alone, demonstrating the beneficial effects of growth factor delivery.[90]

4.2. Angiogenic factors

Hydrogels can also deliver pro-angiogenic factors to increase vascularization to the ischemic myocardium and generate an environment more conducive to cell infiltration for repair. Basic fibroblast growth factor (bFGF) is a powerful angiogenic molecule that has been delivered using hydrogels to sustain its release and prolong its activity.[78,80,163–165,167–170] For instance, bFGF encapsulation in gelatin hydrogel microspheres increases its *in vivo* half-life from days to weeks.[167–169] Hence, gelatin has been used to deliver FGF in both small[163] and large[164] animal models. Shao *et al.* delivered gelatin encapsulated bFGF to rats immediately post-MI (permanent occlusion) and saw significantly increased capillary and arteriolar density in the BZ, reduced cardiomyocyte apoptosis, increased infarct thickness, and reduced infarct expansion as compared to saline controls at 4 weeks after delivery.[163] Similarly, Liu *et al.* delivered gelatin encapsulated bFGF microspheres to dogs immediately post-MI (permanent occlusion) and found significantly increased microvessel density and improved EF as compared to controls of gelatin microspheres in saline at 10 and 17 days post-MI.[164]

Similar improvements in vascularization and cardiac function have been shown for bFGF delivery from chitosan hydrogels post-MI.[80,170]

Despite increased neovascularization with pro-angiogenic factor delivery, there is concern that the newly formed vessels lack stability and connectivity.[171] To enhance vessel stability, multiple pro-angiogenic factors can be simultaneously or sequentially released from hydrogels.[161,165,172] Kim *et al.* delivered both bFGF and platelet derived growth factor (PDGF) using self-assembling peptides to rats immediately post-MI (permanent occlusion).[165] The goal is to first stimulate vessel formation using bFGF, then recruit smooth muscle cells to form stable vasculature using PDGF. In addition, self-assembling peptide nanofibers alone recruit endothelial cells and vascular smooth muscle cells,[151] where both cell types are necessary for formation of stable vasculature. Dual bFGF/PDGF delivery from peptide nanofibers significantly reduced infarct size, increased wall thickness, decreased cardiomyocyte apoptosis, and increased capillary and arteriole density at four and week weeks after delivery and significantly improved cardiac function and contractility (LV EDP, +dP/dt, −dP/dt) after 4 weeks relative to sucrose controls. Similarly, Hao *et al.* sequentially delivered vascular endothelial growth factor (VEGF) then PDGF from alginate hydrogels (Fig. 6) to rats 1 week post-MI (permanent occlusion).[161] About 4 weeks later, sequential release of VEGF then PDGF significantly increased α-actin positive vessels density, indicating smooth muscle cell vessel stabilization, and increased the systolic velocity-time integral, a marker of myocardial function, as compared to either growth factor alone.[161]

To further control angiogenic factor delivery, Wu *et al.* conjugated VEGF to a PVL-*b*-PEG-*b*-PVL hydrogel (discussed previously in Sec. 2.2.2) and injected the system into rats 1 week post-MI (permanent occlusion).[89] Around 5 weeks after delivery, animals that received VEGF tethered to the hydrogel had improved EF and FE, increased contractility (+dP/dt, −dP/dt), preserved EDV and ESV, increased wall thickness, and increased vessel density as compared to VEGF mixed with the hydrogel, hydrogel alone, or saline controls.[89] Similar to previous studies,[149,160] conjugating biomolecules to injectable hydrogels prolongs their release and biological effects.

4.3. Chemoattractants

In addition to promoting cell survival and angiogenesis, therapeutic molecule delivery from injectable hydrogels can also focus on recruiting endogenous cells to repair the damaged myocardium. Stromal cell derived factor-1 alpha (SDF-1α) is a chemokine shown to selectively recruit stem and progenitor cells.[162,166] Segers *et al.* locally delivered SDF-1α to rats immediately post-MI (permanent occlusion) using self-assembling peptide nanofibers.[166] SDF-1α release was tailored by tethering it to the hydrogel using a peptide sequence susceptible to endogenous protease cleavage. After 4 weeks following injection, delivery of SDF-1α tethered to peptide nanofibers significantly increased recruitment of cells expressing the SDF-1α receptor (CXCR4+/c-Kit+ cells), increased capillary and arteriole density in the infarct, and improved EF relative to infarct controls.[166] Similar to self-assembling peptides being well-suited for molecule delivery due to their ability to tether proteins,[153] HA has been shown to cooperate with SDF-*1*α mediated BMC homing[173] and thus, may be well-suited for SDF-1α delivery (Fig. 4). In addition to their native binding affinity, both SDF-1α and HA stimulate BMC chemotaxis *in vitro* through their CXCR4 and CD44 receptors, respectively.[162] Purcell *et al.* delivered SDF-1α encapsulated in a degradable HA hydrogel to mice immediately post-MI (cryoinjury) to assess their synergistic effects on cell homing.[162] A week after delivery, HA hydrogel encapsulation of SDF-1α led to significantly increased endogenous BMC homing and engraftment in the infarct as compared to SDF-1α alone or hydrogel alone.[162] In a subsequent study, MacArthur *et al.* demonstrated that sustained release of an engineered SDF-1α analog (ESA) from degradable HA hydrogels causes both endogenous EPC recruitment to the myocardium and preserved LV function.[174]

5. Combined Hydrogel Delivery of Cells and Molecules

Building from the benefits of delivery of hydrogels alone and hydrogels with encapsulated cells or therapeutic molecules, optimal therapies may rely on the combination of all three components. The addition of growth and differentiating factors can further improve the survival and differentiation of encapsulated cells, and impact the immediate surrounding environment, thus enabling better graft integration. Hydrogels are advantageous over other material-based delivery systems due to their ability to mechanically support the myocardium and simultaneously serve as a controlled delivery platform. Ultimately, the goal is to improve cell engraftment and long-term cell survival to facilitate myocardial tissue reconstruction and/or stem cell mediated cardiac regeneration.

Both natural and synthetic hydrogels have been used to simultaneously deliver cells and anti-apoptotic or angiogenic therapeutic molecules to the infarct post-MI (see Table 4). Laflamme *et al.* delivered Matrigel encapsulated human ESC-derived cardiomyocytes with pro-survival growth factors, including IGF-1, cyclosporine A, anti-apoptotic peptide Bcl-XL BH4, caspase inhibitor ZVAD-fmk and vasodilator pinacidil, to rats 4 days post-MI (temporary occlusion).[59] Delivery of Matrigel with both cells and growth factors led to significantly increased cell engraftment in the infarct as compared to delivery of cells with Matrigel alone 1 week following injection. At 4 weeks after delivery, the combined hydrogel-cell-growth factor system significantly increased FS and preserved LV EDD as compared to non-cardiac cells

Table 4. Summary of hydrogels for cell and molecule delivery and their *in vivo* assessment.

Material	Animal/MI model	Cells and molecules delivered	Injection time (post-MI)	Injection volume	Evaluation time (post-delivery)	Reference
Fibrin with PLGA nano-spheres	Rat: LCA ligation	BMNCs (2×10^7), TGF-β1 (2 ng)	1 week	100 μL	4 weeks	132
Matrigel	Rat: LAD ligation/ reperfusion	ESC-derived cardio-myocytes (1×10^7), IGF-1 (7 ng), pinacidil (50 μM), cyclosporine A (0.2 μM), Bcl-X_L BH4 (50 nM), ZVAD (100 μM)	4 days[61] or 1 month[128]	70 μL	1and 4 weeks[61] or 1, 2, and 3 months[128]	59,126
Self-assembling peptides	Rat: LCA ligation	Skeletal myoblasts (5×10^6), PDGF (100 ng)	2 weeks	200 μL (via 3–4 sites)	1 month	150
	Rat: LCA ligation	Neonatal cardiomyocyte (5×10^5), IGF-1 (25 ng)	Immediate	80 μL	1, 14, 21, and 28 days	149
	Rat: LCA ligation	CPCs (1×10^5), IGF-1 (1 ng)	Immediate	5 μL/ injection (multiple sites)	1 month	152
PEG-VS	Rat: LAD ligation	ESC-derived ELCs (6.6×10^5) and SMLCs (3.3×10^5), Tβ4 (2.5 μg)	1h	60 μL	3 days and 6 weeks	90
PEGylated-fibrinogen	Mouse: LAD ligation	BMNCs (5×10^5), HGF (100 ng)	Immediate	50 μL	2 and 4 weeks	123

* PLGA = poly(lactic-*co*-glycolic acid); CPCs = cardiac progenitor cells; ELC = endothelial-like cells; SMLC = smooth muscle-like cells; TGF-β1 = transforming growth factor-beta 1; Bcl-XL BH4 = cell-permeant TAT peptide; ZVAD = benzyloxycarbonyl-Val-Ala-Asp(O-methyl)-fluoromethyl ketone; FGF = fibroblast growth factor; Tβ4 = thymosin beta-4 (see Tables 1, 2, and 3 for additional abbreviations).

with growth factors, growth factors alone, or culture media controls.[59] In a subsequent study, Fernandes *et al.* delivered the same system to rats 1 month post-MI (temporary occlusion) and saw successful cell engraftment into the infarct 3 months after injection. Despite improved cell engraftment, there were no significant differences in LV EDD, EDV and ESV, SV, or FS between any of the groups at any time point.[126]

Davis *et al.* delivered neonatal cardiomyocytes and IGF-1 tethered to self-assembling peptide nanofibers to rats immediately post-MI (permanent occlusion).[149] A preliminary study in healthy rats demonstrated that peptide tethered IGF-1 delivery significantly decreased activation of caspase-3, an apoptosis marker, in the encapsulated cardiomyocytes and increased transplanted cell area, as compared to untethered IGF-1 or no IGF-1 controls 2 weeks after delivery. Around 3 weeks after delivery to the rat MI model, the tethered IGF-1 cell therapy significantly decreased apoptosis of the encapsulated cardiomyocytes compared with peptides alone or untethered IGF-1 and overall, significantly decreased LV dilatation and increased FS as compared with all other groups.[149] In another study by the same group, Padin-Iruegas *et al.* delivered rat CPCs with IGF-1-modified self-assembling peptide nanofibers to rats immediately post-MI (permanent occlusion).[152] About 1 month following delivery, CPCs encapsulated in IGF-1 tethered peptides significantly enhanced myocardial regeneration (as indicated by increased cardiomyocyte mass, volume, and number), decreased infarct size, decreased wall stress, enhanced contractility (+dP/dt, −dP/dt), and increased EF as compared to CPCs or IGF-1 tethered peptides alone.[152] In contrast to previous studies,[175] Dubois *et al.* delivered skeletal myoblasts encapsulated in PDGF loaded self-assembling peptide nanofibers to rats 2 weeks post-MI (permanent occlusion) and saw no improvements in LV EF or encapsulated cell survival despite increases in angiogenesis as compared to culture medium controls at 1 month after injection.[150]

Synthetic hydrogels have also been used to simultaneously deliver cells and biomolecules for cardiac repair. Kraehenbuehl *et al.* delivered human ESC-derived endothelial- and smooth muscle like cells (ELCs, SMLCs) encapsulated in an MMP degradable, PEG-VS hydrogel loaded with Tβ4 to rats 1h post-MI (permanent ligation).[90] About 6 weeks post-MI, delivery of ELCs/SMLCs encapsulated in a Tβ4-loaded hydrogel significantly increased neovascularization in the infarct, decreased infarct size, reduced LV, EDV, ESV, and increased EF relative to saline controls. In addition, the combined cell-Tβ4 delivery from PEG hydrogels significantly increased EF compared to PEG hydrogels with Tβ4 or PEG hydrogels alone.[90] Zhang *et al.* delivered murine BMNCs and HGF covalently bound to PEGylated-fibrinogen hydrogels to mice immediately post-MI (permanent occlusion) and saw significantly decreased apoptosis in the BZ and infarct, increased angiogenesis, decreased fibrosis, decreased LV EDV and ESV, and increased EF as compared to saline controls at 4 weeks post-MI.[123] Delivery of BMNCs alone led to significantly decreased apoptosis in BZ regions, increased angiogenesis, and decreased fibrosis whereas delivery of HGF-bound hydrogel led to significantly decreased apoptosis and increased angiogenesis in the infarct relative to saline controls. In addition, BMNCs delivery using a HGF-loaded hydrogel significantly increased cell engraftment in the infarct 15-fold relative to delivery of BMNCs alone. All treatment groups (BMSCs, hydrogel+HGF, hydrogel+HGF+BMNCs) improved LV volumes and FS relative to saline controls, with no statistical difference in FS between the groups.[123]

6. Future Directions and Conclusions

As discussed throughout this chapter, a range of injectable hydrogels, cell types, and molecules have been delivered with the intent of attenuating LV remodeling post-MI. Although many hydrogels have shown positive outcomes in animal models, only one has progressed to clinical trials (NCT01226563). Therefore, several areas must be explored prior to clinical translation. For example, the exact mechanism by which injectable biomaterials lead to positive functional outcomes is not fully understand and is likely quite complex, involving both mechanical (e.g., stress reduction) and biological (e.g., exogenous cell delivery, endogenous cell recruitment, angiogenesis, inflammatory response) interactions. It is evident that increased myocardial wall thickness,[45,50,54,73,74] decreased infarct size,[45] and/or increased angiogenesis[57,72,144] do not directly correlate with improved cardiac function; thus, finding specific correlations may improve future therapy design. Theoretical FE models provide some insight into the mechanisms by which injectable biomaterials impact LV remodeling[36–38] and further development of such models may improve our understanding of material injection post-MI.

From a translational perspective, it is important to elucidate the effects of hydrogel properties (e.g., mechanics, degradation), injection parameters (e.g., volume, number of injection), mode of delivery (e.g., direction injection Versus catheter delivery) (Fig. 7), and timing of delivery (e.g., acute Versus chronic MI) on LV remodeling. Much of this can be explored through both experimental and modeling approaches and by selecting the most appropriate mimic of clinical scenarios to enhance our understanding of outcomes. Future studies should further investigate the mechanisms by which hydrogels act on the heart, including both biological and mechanical effects, and focus on clinically relevant parameters, such as the mode of delivery. This information has considerable clinical impact with respect to improved therapies to attenuate LV remodeling and ultimately, to aid patients by preventing progression to heart failure.

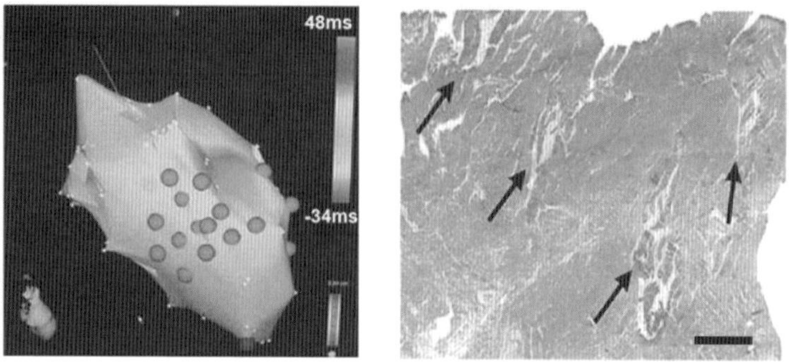

(A) (B)

Fig. 7. Percutaneous catheter delivery of decellularized ventricular ECM to healthy rats. NOGA maps representing injection sites (A) and representative H&E stain indicating presence of hydrogel within the myocardium (B). Scale bar = 1 mm. Figure adapted from (67).

Acknowledgments

This work was supported by the National Institutes of Health (R01 HL111090, T32 HL007954).

References

1. A. L. Bui, T. B. Horwich, and G. C. Fonarow, *Nat. Rev. Cardiol.*, **8** (2011).
2. V. L. Roger, A. S. Go, D. M. Lloyd-Jones, R. J. Adams, J. D. Berry, T. M. Brown, M. R. Carnethon, S. Dai, G. de Simone, E. S. Ford, C. S. Fox, H. J. Fullerton, C. Gillespie, K. J. Greenlund, S. M. Hailpern, J. A. Heit, P. M. Ho, V. J. Howard, B. M. Kissela, S. J. Kittner, D. T. Lackland, J. H. Lichtman, L. D. Lisabeth, D. M. Makuc, G. M. Marcus, A. Marelli, D. B. Matchar, M. M. McDermott, J. B. Meigs, C. S. Moy, D. Mozaffarian, M. E. Mussolino, G. Nichol, N. P. Paynter, W. D. Rosamond, P. D. Sorlie, R. S. Stafford, T. N. Turan, M. B. Turner, N. D. Wong, J. Wylie-Rosett, and American Heart Association Statistics Committee and Stroke Statistics Subcommittee, *Circulation*, **123** (2011).
3. M. Gheorghiade and R. O. Bonow, *Circulation*, **97** (1998).
4. J. P. Cleutjens and E. E. Creemers, *J. Card. Fail.*, **8** (2002).
5. J. W. Holmes, T. K. Borg, and J. W. Covell, *Annu. Rev. Biomed. Eng.*, **7** (2005).
6. M. Dobaczewski, C. Gonzalez-Quesada, and N. G. Frangogiannis, *J. Mol. Cell. Cardiol.*, **48** (2010).
7. F. G. Spinale, *Physiol. Rev.*, **87** (2007).
8. G. D. Buckberg, *Eur. J. Cardiothorac. Surg.*, **27** (2005).
9. M. Nahrendorf, *Circ. Cardiovasc. Imaging*, **4** (2011).
10. C. M. Kramer, J. A. Lima, N. Reichek, V. A. Ferrari, M. R. Llaneras, L. C. Palmon, I. T. Yeh, B. Tallant, and L. Axel, *Circulation*, **88** (1993).

11. B. M. Jackson, J. H. Gorman III, I. S. Salgo, S. L. Moainie, T. Plappert, M. St. John-Sutton, L. H. Edmunds Jr., and R. C. Gorman, *Am. J. Physiol. Heart Circ. Physiol.*, **284** (2003).
12. F. H. Epstein, Z. Yang, W. D. Gilson, S. S. Berr, C. M. Kramer, and B. A. French, *Magn. Reson. Med.*, **48** (2002).
13. J. J. Pilla, A. S. Blom, J. H. Gorman III, D. J. Brockman, J. Affuso, L. M. Parish, H. Sakamoto, B. M. Jackson, M. A. Acker, and R. C. Gorman, *Ann. Thorac. Surg.*, **80** (2005).
14. R. Batista, *Eur. J. Cardiothorac. Surg.*, **15** (Suppl 1) (1999).
15. S. T. Kelley, R. Malekan, J. H. Gorman III, B. M. Jackson, R. C. Gorman, Y. Suzuki, T. Plappert, D. K. Bogen, M. G. Sutton, and L. H. Edmunds Jr., *Circulation*, **99** (1999).
16. C. T. Klodell Jr., J. M. Aranda Jr., D. C. McGiffin, B. K. Rayburn, B. Sun, W. T. Abraham, W. E. Pae Jr., J. P. Boehmer, H. Klein, and C. Huth, *J. Thorac. Cardiovasc. Surg.*, **135** (2008).
17. R. C. Starling, M. Jessup, J. K. Oh, H. N. Sabbah, M. A. Acker, D. L. Mann, and S. H. Kubo, *Ann. Thorac. Surg.*, **84** (2007).
18. D. M. Nelson, Z. Ma, K. L. Fujimoto, R. Hashizume, and W. R. Wagner, *Acta Biomater.*, **7** (2011).
19. E. Tous, B. Purcell, J. L. Ifkovits, and J. A. Burdick, *J. Cardiovasc. Transl. Res.*, **4** (2011).
20. J. L. Drury and D. J. Mooney, *Biomaterials*, **24** (2003).
21. E. Ruel-Gariepy and J. C. Leroux, *Eur. J. Pharm. Biopharm.*, **58** (2004).
22. K. Y. Lee and D. J. Mooney, *Chem. Rev.*, **101** (2001).
23. K. Gupta, M. Ratcliffe, M. Fallert, L. Edmunds, and D. Bogen, *Circulation*, **89** (1994).
24. Y. Enomoto, J. H. Gorman III, S. L. Moainie, B. M. Jackson, L. M. Parish, T. Plappert, A. Zeeshan, M. G. St. John-Sutton, and R. C. Gorman, *Ann. Thorac. Surg.*, **79** (2005).
25. S. L. Moainie, T. S. Guy, J. H. Gorman III, T. Plappert, B. M. Jackson, M. G. St. John-Sutton, L. H. Edmunds Jr., and R. C. Gorman, *Ann. Thorac. Surg.*, **74** (2002).
26. D. Mann, *Circulation*, **100** (1999).
27. F. W. Bowen, S. C. Jones, N. Narula, M. G. St. John Sutton, T. Plappert, L. H. Edmunds Jr., and I. M. Dixon, *Ann. Thorac. Surg.*, **72** (2001).
28. A. Blom, R. Mukherjee, J. Pilla, A. Lowry, W. Yarbrough, J. Mingoia, J. Hendrick, R. Stroud, J. McLean, J. Affuso, R. Gorman, J. Gorman, M. Acker, and F. Spinale, *Circulation*, **112** (2005).
29. P. A. Chaudhry, T. Mishima, V. G. Sharov, J. Hawkins, C. Alferness, G. Paone, and H. N. Sabbah, *Ann. Thorac. Surg.*, **70** (2000).
30. H. N. Sabbah, *Ann. Thorac. Surg.*, **75** (2003).
31. J. J. Pilla, A. S. Blom, D. J. Brockman, F. Bowen, Q. Yuan, J. Giammarco, V. A. Ferrari, J. H. Gorman III, R. C. Gorman, and M. A. Acker, *Circulation*, **106** (2002).
32. I. George, Y. Cheng, G. Yi, K. He, X. Li, M. C. Oz, J. Holmes, and J. Wang, *Eur. J. Cardiothorac. Surg.*, **31** (2007).
33. J. Magovern, L. Teekell-Taylor, S. Mankad, U. Dasika, W. McGregor, R. Biederman, J. Yamrozik, and D. Trumble, *Asaio J.*, **52** (2006).
34. W. T. Abraham, I. Anand, J. M. Aranda Jr., J. Boehmer, M. R. Costanzo, T. DeMarco, R. Holcomb, R. Ivanhoe, M. Kolber, and B. Rayburn, *Am. Heart J.*, **164** (2012).

35. M. R. Costanzo, R. J. Ivanhoe, A. Kao, I. S. Anand, A. Bank, J. Boehmer, T. Demarco, C. M. Hergert, R. G. Holcomb, S. Maybaum, B. Sun, T. A. Vassiliades Jr., B. K. Rayburn, and W. T. Abraham, *J. Card. Fail.*, **18** (2012).
36. S. T. Wall, J. C. Walker, K. E. Healy, M. B. Ratcliffe, and J. M. Guccione, *Circulation*, **114** (2006).
37. J. F. Wenk, P. Eslami, Z. Zhang, C. Xu, E. Kuhl, J. H. Gorman III, J. D. Robb, M. B. Ratcliffe, R. C. Gorman, and J. M. Guccione, *Ann. Thorac. Surg.*, **92** (2011).
38. J. F. Wenk, S. T. Wall, R. C. Peterson, S. L. Helgerson, H. N. Sabbah, M. Burger, N. Stander, M. B. Ratcliffe, and J. M. Guccione, *J. Biomech. Eng.*, **131** (2009).
39. J. P. Karam, C. Muscari, and C. N. Montero-Menei, *Biomaterials*, **33** (2012).
40. M. Lutolf and J. Hubbell, *Nat. Biotechnol.*, **23** (2005).
41. R. Sui, X. Liao, X. Zhou, and Q. Tan, *Stem Cell. Rev.*, **7** (2011).
42. M. C. Barsotti, F. Felice, A. Balbarini, and R. Di Stefano, *Biotechnol. Appl. Biochem.*, **58** (2011).
43. K. L. Christman, H. H. Fok, R. E. Sievers, Q. Fang, and R. J. Lee, *Tissue Eng.*, **10** (2004).
44. K. L. Christman, A. J. Vardanian, Q. Fang, R. E. Sievers, H. H. Fok, and R. J. Lee, *J. Am. Coll. Cardiol.*, **44** (2004).
45. J. Yu, K. L. Christman, E. Chin, R. E. Sievers, M. Saeed, and R. J. Lee, *J. Thorac. Cardiovasc. Surg.*, **137** (2009).
46. N. F. Huang, J. Yu, R. Sievers, S. Li, and R. J. Lee, *Tissue Eng.*, **11** (2005).
47. W. D. Thompson, E. B. Smith, C. M. Stirk, F. I. Marshall, A. J. Stout, and A. Kocchar, *J. Pathol.*, **168** (1992).
48. J. Leor, S. Aboulafia-Etzion, A. Dar, L. Shapiro, I. M. Barbash, A. Battler, Y. Granot, and S. Cohen, *Circulation*, **102** (2000).
49. J. Leor, S. Tuvia, V. Guetta, F. Manczur, D. Castel, U. Willenz, O. Petnehazy, N. Landa, M. S. Feinberg, E. Konen, O. Goitein, O. Tsur-Gang, M. Shaul, L. Klapper, and S. Cohen, *J. Am. Coll. Cardiol.*, **54** (2009).
50. N. Landa, L. Miller, M. S. Feinberg, R. Holbova, M. Shachar, I. Freeman, S. Cohen, and J. Leor, *Circulation*, **117** (2008).
51. J. Rowley, G. Madlambayan, and D. Mooney, *Biomaterials*, **20** (1999).
52. J. Yu, Y. Gu, K. T. Du, S. Mihardja, R. E. Sievers, and R. J. Lee, *Biomaterials*, **30** (2009).
53. O. Tsur-Gang, E. Ruvinov, N. Landa, R. Holbova, M. S. Feinberg, J. Leor, and S. Cohen, *Biomaterials*, **30** (2009).
54. R. Mukherjee, J. A. Zavadzkas, S. M. Saunders, J. E. McLean, L. B. Jeffords, C. Beck, R. E. Stroud, A. M. Leone, C. N. Koval, W. T. Rivers, S. Basu, A. Sheehy, G. Michal, and F. G. Spinale, *Ann. Thorac. Surg.*, **86** (2008).
55. R. Parenteau-Bareil, R. Gauvin, and F. Berthod, *Materials*, **3** (2010).
56. J. Rosenblatt, B. Devereux, and D. Wallace, *Biomaterials*, **15** (1994).
57. W. Dai, L. E. Wold, J. S. Dow, and R. A. Kloner, *J. Am. Coll. Cardiol.*, **46** (2005).
58. C. S. Hughes, L. M. Postovit, and G. A. Lajoie, *Proteomics*, **10** (2010).
59. M. A. Laflamme, K. Y. Chen, A. V. Naumova, V. Muskheli, J. A. Fugate, S. K. Dupras, H. Reinecke, C. Xu, M. Hassanipour, S. Police, C. O'Sullivan, L. Collins, Y. Chen,

E. Minami, E. A. Gill, S. Ueno, C. Yuan, J. Gold, and C. E. Murry, *Nat. Biotechnol.*, **25** (2007).
60. L. Ou, W. Li, Y. Zhang, W. Wang, J. Liu, H. Sorg, D. Furlani, R. Gaebel, P. Mark, C. Klopsch, L. Wang, K. Luetzow, A. Lendlein, K. Wagner, D. Klee, A. Liebold, R. Li, D. Kong, G. Steinhoff, and N. Ma, *J. Cell. Mol. Med.*, **15** (2011).
61. T. Kofidis, D. R. Lebl, E. C. Martinez, G. Hoyt, M. Tanaka, and R. C. Robbins, *Circulation*, **112** (2005).
62. T. Kofidis, J. L. de Bruin, G. Hoyt, Y. Ho, M. Tanaka, T. Yamane, D. R. Lebl, R. J. Swijnenburg, C. P. Chang, T. Quertermous, and R. C. Robbins, *J. Heart Lung Transplant.*, **24** (2005).
63. S. J. Yoon, Y. H. Fang, C. H. Lim, B. S. Kim, H. S. Son, Y. Park, and K. Sun, *J. Biomed. Mater. Res. B Appl. Biomater.*, **91** (2009).
64. J. L. Ifkovits, E. Tous, M. Minakawa, M. Morita, J. D. Robb, K. J. Koomalsingh, J. H. Gorman III, R. C. Gorman, and J. A. Burdick, *Proc. Natl. Acad. Sci. USA*, **107** (2010).
65. E. Tous, J. L. Ifkovits, K. J. Koomalsingh, T. Shuto, T. Soeda, N. Kondo, J. H. Gorman III, R. C. Gorman, and J. A. Burdick, *Biomacromolecules*, **12** (2011).
66. D. Shen, X. Wang, L. Zhang, X. Zhao, J. Li, K. Cheng, and J. Zhang, *Biomaterials*, **32** (2011).
67. J. M. Singelyn, P. Sundaramurthy, T. D. Johnson, P. J. Schup-Magoffin, D. P. Hu, D. M. Faulk, J. Wang, K. M. Mayle, K. Bartels, M. Salvatore, A. M. Kinsey, A. N. DeMaria, N. Dib, and K. L. Christman, *J. Am. Coll. Cardiol.*, **59** (2012).
68. S. B. Seif-Naraghi, J. M. Singelyn, M. A. Salvatore, K. G. Osborn, J. J. Wang, U. Sampat, O. L. Kwan, G. M. Strachan, J. Wong, P. J. Schup-Magoffin, R. L. Braden, K. Bartels, J. A. DeQuach, M. Preul, A. M. Kinsey, A. N. DeMaria, N. Dib, and K. L. Christman, *Sci. Transl. Med.*, **5** (2013).
69. M. Okada, T. R. Payne, H. Oshima, N. Momoi, K. Tobita, and J. Huard, *Biomaterials*, **31** (2010).
70. K. L. Fujimoto, Z. Ma, D. M. Nelson, R. Hashizume, J. Guan, K. Tobita, and W. R. Wagner, *Biomaterials*, **30** (2009).
71. T. Wang, D. Q. Wu, X. J. Jiang, X. Z. Zhang, X. Y. Li, J. F. Zhang, Z. B. Zheng, R. Zhuo, H. Jiang, and C. Huang, *Eur. J. Heart Fail.*, **11** (2009).
72. X. Jiang, T. Wang, X. Li, D. Wu, Z. Zheng, J. Zhang, J. Chen, B. Peng, H. Jiang, C. Huang and X. Zhang, *J. Biomed. Mater. Res. A*, **90A** (2009).
73. S. Dobner, D. Bezuidenhout, P. Govender, P. Zilla, and N. Davies, *J. Card. Fail.*, **15** (2009).
74. A. A. Rane and K. L. Christman, *J. Am. Coll. Cardiol.*, **58** (2011).
75. W. N. Lu, S. H. Lu, H. B. Wang, D. X. Li, C. M. Duan, Z. Q. Liu, T. Hao, W. J. He, B. Xu, Q. Fu, Y. C. Song, X. H. Xie, and C. Y. Wang, *Tissue Eng. Pt. A*, **15** (2009).
76. Z. Liu, H. Wang, Y. Wang, Q. Lin, A. Yao, F. Cao, D. Li, J. Zhou, C. Duan, Z. Du, Y. Wang, and C. Wang, *Biomaterials*, **33** (2012).
77. C. D. Hoemann, A. Chenite, J. Sun, M. Hurtig, A. Serreqi, Z. Lu, E. Rossomacha, and M. D. Buschmann, *J. Biomed. Mater. Res. A*, **83A** (2007).

78. A. Chenite, C. Chaput, D. Wang, C. Combes, M. Buschmann, C. Hoemann, J. Leroux, B. Atkinson, F. Binette, and A. Selmani, *Biomaterials*, **21** (2000).
79. E. Ruel-Gariepy, M. Shive, A. Bichara, M. Berrada, D. Le Garrec, A. Chenite, and J. Leroux, *Eur. J. Pharm. Biopharm.*, **57** (2004).
80. H. Wang, X. Zhang, Y. Li, Y. Ma, Y. Zhang, Z. Liu, J. Zhou, Q. Lin, Y. Wang, C. Duan, and C. Wang, *J. Heart Lung Transplant.*, **29** (2010).
81. B. P. Toole, *Nat. Rev. Cancer.*, **4** (2004).
82. J. G. Rouse and M. E. Van Dyke, *Materials*, **3** (2010).
83. P. M. Crapo, T. W. Gilbert, and S. F. Badylak, *Biomaterials*, **32** (2011).
84. J. M. Singelyn, J. A. DeQuach, S. B. Seif-Naraghi, R. B. Littlefield, P. J. Schup-Magoffin, and K. L. Christman, *Biomaterials*, **30** (2009).
85. M. E. Davis, P. C. Hsieh, A. J. Grodzinsky, and R. T. Lee, *Circ. Res.*, **97** (2005).
86. S. T. Wall, C. Yeh, R. Y. K. Tu, M. J. Mann, and K. E. Healy, *J. Biomed. Mater. Res. A*, **95A** (2010).
87. X. Li, T. Wang, X. Jiang, T. Lin, D. Wu, X. Zhang, E. Okello, H. Xu, and M. Yuan, *Cardiology*, **115** (2010).
88. T. Wang, X. J. Jiang, T. Lin, S. Ren, X. Li, X. Zhang, and Q. Tang, *Biomaterials*, **30** (2009).
89. J. Wu, F. Zeng, X. Huang, J. C. Chung, F. Konecny, R. D. Weisel, and R. Li, *Biomaterials*, **32** (2011).
90. T. P. Kraehenbuehl, L. S. Ferreira, A. M. Hayward, M. Nahrendorf, A. J. van der Vlies, E. Vasile, R. Weissleder, R. Langer, and J. A. Hubbell, *Biomaterials*, **32** (2011).
91. C. A. Beltrami, N. Finato, M. Rocco, G. A. Feruglio, C. Puricelli, E. Cigola, F. Quaini, E. H. Sonnenblick, G. Olivetti, and P. Anversa, *Circulation*, **89** (1994).
92. M. H. Soonpaa and L. J. Field, *Circ. Res.*, **83** (1998).
93. A. P. Beltrami, L. Barlucchi, D. Torella, M. Baker, F. Limana, S. Chimenti, H. Kasahara, M. Rota, E. Musso, K. Urbanek, A. Leri, J. Kajstura, B. Nadal-Ginard, and P. Anversa, *Cell*, **114** (2003).
94. A. G. Zenovich and D. A. Taylor, *Front. Biosci.*, **13** (2008).
95. H. Reinecke and C. E. Murry, *J. Mol. Cell. Cardiol.*, **34** (2002).
96. P. Menasche, *Thromb. Haemost.*, **94** (2005).
97. V. F. Segers and R. T. Lee, *Nature*, **451** (2008).
98. W. H. Zimmermann, *J. Mol. Cell. Cardiol.*, **50** (2011).
99. P. Menasche, *J. Mol. Cell. Cardiol.*, **45** (2008).
100. C. E. Murry, M. H. Soonpaa, H. Reinecke, H. Nakajima, H. O. Nakajima, M. Rubart, K. B. Pasumarthi, J. I. Virag, S. H. Bartelmez, V. Poppa, G. Bradford, J. D. Dowell, D. A. Williams, and L. J. Field, *Nature*, **428** (2004).
101. C. E. Murry, M. L. Whitney, and H. Reinecke, *J. Card. Fail.*, **8** (2002).
102. K. Le Blanc and M. Pittenger, *Cytotherapy*, **7** (2005).
103. D. Orlic, *Cytotherapy*, **7** (2005).
104. A. M. Smits, L. W. van Laake, K. den Ouden, C. Schreurs, K. Szuhai, C. J. van Echteld, C. L. Mummery, P. A. Doevendans, and M. J. Goumans, *Cardiovasc. Res.*, **83** (2009).
105. G. Y. Koh, M. G. Klug, M. H. Soonpaa, and L. J. Field, *J. Clin. Invest.*, **92** (1993).

106. D. A. D'Alessandro and R. E. Michler, *Curr. Treat. Options Cardiovasc. Med.*, **12** (2010).
107. J. V. Terrovitis, R. R. Smith, and E. Marban, *Circ. Res.*, **106** (2010).
108. Y. S. Choi, G. J. Dusting, S. Stubbs, S. Arunothayaraj, X. L. Han, P. Collas, W. A. Morrison, and R. J. Dilley, *J. Cell. Mol. Med.*, **14** (2010).
109. M. Hofmann, K. C. Wollert, G. P. Meyer, A. Menke, L. Arseniev, B. Hertenstein, A. Ganser, W. H. Knapp, and H. Drexler, *Circulation*, **111** (2005).
110. D. Hou, E. A. Youssef, T. J. Brinton, P. Zhang, P. Rogers, E. T. Price, A. C. Yeung, B. H. Johnstone, P. G. Yock, and K. L. March, *Circulation*, **112** (2005).
111. T. E. Robey, M. K. Saiget, H. Reinecke, and C. E. Murry, *J. Mol. Cell. Cardiol.*, **45** (2008).
112. I. Zvibel, F. Smets, and H. Soriano, *Cell Transplant.*, **11** (2002).
113. T. Matsumoto and D. J. Mooney, *Adv. Biochem. Eng. Biotechnol.*, **102** (2006).
114. S. Li, N. F. Huang and S. Hsu, *J. Cell. Biochem.*, **96** (2005).
115. S. Cohen and J. Leor, *Sci. Am.*, **291** (2004).
116. N. Tran, Y. Li, S. Bertrand, S. Bangratz, J. P. Carteaux, J. F. Stoltz, and J. P. Villemot, *Biorheology*, **40** (2003).
117. V. Schachinger, B. Assmus, S. Erbs, A. Elsasser, W. Haberbosch, R. Hambrecht, J. Yu, R. Corti, D. G. Mathey, C. W. Hamm, T. Tonn, S. Dimmeler, A. M. Zeiher, and REPAIR-AMI investigators, *Eur. J. Heart Fail.*, **11** (2009).
118. D. Simpson, H. Liu, T. H. Fan, R. Nerem, and S. C. Dudley Jr., *Stem Cells*, **25** (2007).
119. R. K. Li, Z. Q. Jia, R. D. Weisel, D. A. Mickle, A. Choi, and T. M. Yau, *Circulation*, **100** (1999).
120. H. Hamdi, A. Furuta, V. Bellamy, A. Bel, E. Puymirat, S. Peyrard, O. Agbulut, and P. Menasche, *Ann. Thorac. Surg.*, **87** (2009).
121. J. C. Chachques, J. C. Trainini, N. Lago, M. Cortes-Morichetti, O. Schussler, and A. Carpentier, *Ann. Thorac. Surg.*, **85** (2008).
122. G. Zhang, X. Wang, Z. Wang, J. Zhang, and L. Suggs, *Tissue Eng.*, **12** (2006).
123. G. Zhang, Q. Hu, E. A. Braunlin, L. J. Suggs, and J. Zhang, *Tissue Eng. Pt. A*, **14** (2008).
124. H. D. Guo, H. J. Wang, Y. Z. Tan, and J. H. Wu, *Tissue Eng. Part A*, **17** (2011).
125. H. C. Ott, T. S. Matthiesen, S. K. Goh, L. D. Black, S. M. Kren, T. I. Netoff, and D. A. Taylor, *Nat. Med.*, **14** (2008).
126. S. Fernandes, A. V. Naumova, W. Z. Zhu, M. A. Laflamme, J. Gold, and C. E. Murry, *J. Mol. Cell. Cardiol.*, **49** (2010).
127. J. H. Traverse, *J. Cardiovasc. Transl. Res.*, **5** (2012).
128. G. D. Nicodemus and S. J. Bryant, *Tissue Eng. Part B Rev.*, **14** (2008).
129. W. Bian, B. Liau, N. Badie, and N. Bursac, *Nat. Protoc.*, **4** (2009).
130. G. R. Gaudette and I. S. Cohen, *Circulation*, **114** (2006).
131. J. S. Nakamuta, M. E. Danoviz, F. L. N. Marques, L. dos Santos, C. Becker, G. A. Goncalves, P. F. Vassallo, I. T. Schettert, P. J. F. Tucci, and J. E. Krieger, *Plos One*, **4** (2009).
132. H. S. Yang, S. H. Bhang, I. Kim, T. Lee, J. Kang, D. Lee, S. Lee, K. Hwang, and B. Kim, *Cell Transplant.*, **21** (2012).

133. J. H. Ryu, I. K. Kim, S. W. Cho, M. C. Cho, K. K. Hwang, H. Piao, S. Piao, S. H. Lim, Y. S. Hong, C. Y. Choi, K. J. Yoo, and B. S. Kim, *Biomaterials*, **26** (2005).
134. T. P. Martens, A. F. G. Godier, J. J. Parks, L. Q. Wan, M. S. Koeckert, G. M. Eng, B. I. Hudson, W. Sherman, and G. Vunjak-Novakovic, *Cell Transplant.*, **18** (2009).
135. V. Chekanov, M. Akhtar, G. Tchekanov, G. Dangas, M. Z. Shehzad, F. Tio, M. Adamian, A. Colombo, G. Roubin, M. B. Leon, J. W. Moses, and N. N. Kipshidze, *Pacing Clin. Electrophysiol.*, **26** (2003).
136. M. E. Danoviz, J. S. Nakamuta, F. L. N. Marques, L. dos Santos, E. C. Alvarenga, A. A. dos Santos, E. L. Antonio, I. T. Schettert, P. J. Tucci, and J. E. Krieger, *Plos One*, **5** (2010).
137. X. Zhang, H. Wang, X. Ma, A. Adila, B. Wang, F. Liu, B. Chen, C. Wang, and Y. Ma, *Exp. Biol. Med. (Maywood)*, **235** (2010).
138. H. Jawad, N. N. Ali, A. R. Lyon, Q. Z. Chen, S. E. Harding, and A. R. Boccaccini, *J. Tissue Eng. Regen. Med.*, **1** (2007).
139. J. Yu, K. T. Du, Q. Fang, Y. Gu, S. S. Mihardja, R. E. Sievers, J. C. Wu, and R. J. Lee, *Biomaterials*, **31** (2010).
140. W. Dai, S. L. Hale, G. L. Kay, A. J. Jyrala, and R. A. Kloner, *Regen. Med.*, **4** (2009).
141. C. A. Thompson, B. A. Nasseri, J. Makower, S. Houser, M. McGarry, T. Lamson, I. Pomerantseva, J. Y. Chang, H. K. Gold, J. P. Vacanti, and S. N. Oesterle, *J. Am. Coll. Cardiol.*, **41** (2003).
142. T. Kofidis, J. L. de Bruin, G. Hoyt, D. R. Lebl, M. Tanaka, T. Yamane, C. P. Chang, and R. C. Robbins, *J. Thorac. Cardiovasc. Surg.*, **128** (2004).
143. Y. Lin, M. Yeh, Y. Yang, D. Tsai, T. Chu, Y. Shih, M. Chang, Y. Liu, A. C. L. Tang, T. Chen, C. Luo, K. Chang, J. Chen, H. Wu, T. Hung, and P. C. H. Hsieh, *Circulation*, **122** (2010).
144. T. Wang, X. Jiang, Q. Tang, X. Li, T. Lin, D. Wu, X. Zhang, and E. Okello, *Acta Biomaterialia*, **5** (2009).
145. H. Wang, Z. Liu, D. Li, X. Guo, F. K. Kasper, C. Duan, J. Zhou, A. G. Mikos, and C. Wang, *J. Cell. Mol. Med.*, **16** (2012).
146. M. Habib, K. Shapira-Schweitzer, O. Caspi, A. Gepstein, G. Arbel, D. Aronson, D. Seliktar, and L. Gepstein, *Biomaterials*, **32** (2011).
147. W. H. Zimmermann, C. Fink, D. Kralisch, U. Remmers, J. Weil, and T. Eschenhagen, *Biotechnol. Bioeng.*, **68** (2000).
148. P. Zhang, H. Zhang, H. Wang, Y. Wei, and S. Hu, *Artif. Organs*, **30** (2006).
149. M. E. Davis, P. C. Hsieh, T. Takahashi, Q. Song, S. Zhang, R. D. Kamm, A. J. Grodzinsky, P. Anversa, and R. T. Lee, *Proc. Natl. Acad. Sci. USA*, **103** (2006).
150. G. Dubois, V. F. M. Segers, V. Bellamy, L. Sabbah, S. Peyrard, P. Bruneval, A. A. Hagege, R. T. Lee, and P. Menasche, *J. Biomed. Mater. Res. Part B Appl. Biomater.*, **87B** (2008).
151. M. E. Davis, J. P. Motion, D. A. Narmoneva, T. Takahashi, D. Hakuno, R. D. Kamm, S. Zhang, and R. T. Lee, *Circulation*, **111** (2005).
152. M. E. Padin-Iruegas, Y. Misao, M. E. Davis, V. F. Segers, G. Esposito, T. Tokunou, K. Urbanek, T. Hosoda, M. Rota, P. Anversa, A. Leri, R. T. Lee, and J. Kajstura, *Circulation*, **120** (2009).

153. V. F. Segers and R. T. Lee, *J. Cardiovasc. Transl. Res.*, **3** (2010).
154. K. Urbanek, M. Rota, S. Cascapera, C. Bearzi, A. Nascimbene, A. De Angelis, T. Hosoda, S. Chimenti, M. Baker, F. Limana, D. Nurzynska, D. Torella, F. Rotatori, R. Rastaldo, E. Musso, F. Quaini, A. Leri, J. Kajstura, and P. Anversa, *Circ. Res.*, **97** (2005).
155. S. Epstein, S. Fuchs, Y. Zhou, R. Baffour, and R. Kornowski, *Cardiovasc. Res.*, **49** (2001).
156. S. Vandervelde, M. J. van Luyn, R. A. Tio, and M. C. Harmsen, *J. Mol. Cell. Cardiol.*, **39** (2005).
157. J. D. Kretlow, L. Klouda, and A. G. Mikos, *Adv. Drug Deliv. Rev.*, **59** (2007).
158. R. R. Chen and D. J. Mooney, *Pharm. Res.*, **20** (2003).
159. R. Langer and J. Folkman, *Nature*, **263** (1976).
160. E. Ruvinov, J. Leor, and S. Cohen, *Biomaterials*, **32** (2011).
161. X. Hao, E. A. Silva, A. Mansson-Broberg, K. H. Grinnemo, A. J. Siddiqui, G. Dellgren, E. Wardell, L. A. Brodin, D. J. Mooney, and C. Sylven, *Cardiovasc. Res.*, **75** (2007).
162. B. P. Purcell, J. A. Elser, A. Mu, K. B. Margulies, and J. A. Burdick, *Biomaterials*, **33** (2012).
163. Z. Q. Shao, K. Takaji, Y. Katayama, R. Kunitomo, H. Sakaguchi, Z. F. Lai, and M. Kawasuji, *Circ. J.*, **70** (2006).
164. Y. Liu, L. Sun, Y. Huan, H. Zhao, and J. Deng, *Eur. J. Cardiothorac. Surg.*, **30** (2006).
165. J. H. Kim, Y. Jung, S. Kim, K. Sun, J. Choi, H. C. Kim, Y. Park, and S. H. Kim, *Biomaterials*, **32** (2011).
166. V. F. Segers, T. Tokunou, L. J. Higgins, C. MacGillivray, J. Gannon, and R. T. Lee, *Circulation*, **116** (2007).
167. Y. Tabata, S. Hijikata, and Y. Ikada, *J. Control. Release*, **31** (1994).
168. A. Iwakura, M. Fujita, K. Kataoka, K. Tambara, Y. Sakakibara, M. Komeda, and Y. Tabata, *Heart Vessels*, **18** (2003).
169. Y. Sakakibara, M. Yamamoto, K. Nishimura, T. Nishina, S. Miwa, N. Handa, T. Nomoto, K. Tanbara, Y. Tabata, and M. Komeda, *Circulation*, **102** (2000).
170. M. Fujita, M. Ishihara, Y. Morimoto, M. Simizu, Y. Saito, H. Yura, T. Matsui, B. Takase, H. Hattori, Y. Kanatani, M. Kikuchi, and T. Maehara, *J. Surg. Res.*, **126** (2005).
171. G. Yancopoulos, S. Davis, N. Gale, J. Rudge, S. Wiegand, and J. Holash, *Nature*, **407** (2000).
172. S. Banquet, E. Gomez, L. Nicol, F. Edwards-Levy, J. P. Henry, R. Cao, D. Schapman, B. Dautreaux, F. Lallemand, F. Bauer, Y. Cao, C. Thuillez, P. Mulder, V. Richard, and E. Brakenhielm, *Circulation*, **124** (2011).
173. A. Avigdor, P. Goichberg, S. Shivtiel, A. Dar, A. Peled, S. Samira, O. Kollet, R. Hershkoviz, R. Alon, I. Hardan, H. Ben-Hur, D. Naor, A. Nagler, and T. Lapidot, *Blood*, **103** (2004).
174. J. W. MacArthur, B. P. Purcell, Y. Shudo, J. E. Cohen, A. Fairman, A. Trubelja, J. Patel, P. Hsiao, E. Yang, W. Hiesinger, P. Atluri, J. A. Burdick, and Y. J. Woo., *Circulation*, (in press).
175. P. C. Hsieh, M. E. Davis, J. Gannon, C. MacGillivray, and R. T. Lee, *J. Clin. Invest.*, **116** (2006).

Chapter 12

Application of Hydrogels in Heart Valve Tissue Engineering

Jesper Hjortnaes[*,†,‡] **and Frederick J. Schoen**[§,¶]
[*]*Department of Cardiothoracic Surgery, Division of Heart and Lungs*
University Medical Center Utrecht
Utrecht, The Netherlands
[†]*Center of Excellence for Vascular Biology, Department of Medicine*
Brigham and Women's Hospital, Harvard Medical School,
Boston, MA, USA
[‡]*Division of Biomedical Engineering*
Brigham and Women's Hospital
Harvard Medical School
Boston, MA, USA
[§]*Department of Pathology*
Brigham and Women's Hospital
Harvard Medical School
Boston, MA, USA
[¶]*fschoen@partners.org*

1. Introduction

Heart valve disease is a leading cause of mortality and morbidity in the Western world.[1] Nearly 300,000 valve repair and replacement surgeries are performed each year worldwide, and this number is expected to increase sharply as life expectancy rises, and the aging population increases over the next 30 years.[2] Commercially available mechanical and bioprosthetic heart valves, and cryopreserved homograft valves are the standard of practice in surgical valve replacement.[3] However, contemporary valve substitutes are associated with significant shortcomings.

Firstly, mechanical valves generally have excellent durability, but are thrombogenic, which necessitates life-long anti-coagulation therapy.[4] Conversely, bioprosthetic valves do not generally require anti-coagulation treatment. However, these valves have limited durability (on average approximately 15 years) due to valve leaflet mineralization and structural deterioration over time,[5] necessitating reoperation. In addition, both mechanical and bioprosthetic valves are obstructive relative to native valves and both types are susceptible to infection, which often requires another replacement surgery as this complication may be difficult to treat medically. Cryopreserved homografts, which are valves derived from human cadavers and non-viable when implanted, share similar durability limitations with bioprosthetic valves, in addition to being of very limited supply.[6] Finally, all existing valve substitutes are particularly unsuitable for congenital valve disease. Not only are very small valve sizes required, and bioprosthetic valves are subject to accelerated calcification in the young but also because growth and enlargement of a valve replacement, essential to eliminate the need for reoperations in the rapidly growing pediatric patient, cannot occur with conventional valve replacement.[7]

The development of tissue engineered heart valves (TEHV) aims to overcome these limitations by permitting living heart valve substitutes that (i) are non-obstructive, (ii) are non-thrombogenic, (iii) provide for ongoing remodeling to maintain tissue homeostasis, (iv) repair any injury occurring during function, and importantly, (v) grow with the patient.[8,9]

Various approaches have been utilized in the pursuit of developing a TEHV. A widely studied paradigm is to pre-seed autologous cells on synthetic biodegradable polymer scaffolds, which have been fabricated in the shape of a trileaflet valve. These are then conditioned in an *in vitro* bioreactor, which simulates the physiological mechanical and hemodynamic environment of a native heart valve.[7,10] The resulting construct is implanted as a valve *in vivo*. This approach stimulates tissue growth *in vitro* prior to surgical implantation, followed by remodeling *in vivo* after implantation. TEHVs grown from autologous cells seeded on biodegradable synthetic polymers have typically been implanted in the pulmonary circulation of growing lambs and evaluated in experimental models for up to 5 months. In some cases, the constructs seem to have evolved into functionally appropriate structures that mimic the complex architecture of native semilunar valves.[11–14] However, long-term evaluation of this approach is needed, before clinical trials can be done.

Other approaches to engineer TEHV are also being investigated; these approaches include '*in situ* tissue regeneration',[13] in which an implanted scaffold of degradable biomaterial or a decellularized non-aldehyde-fixed or otherwise chemically preserved valve is either seeded with cells *in vitro*, or is functionalized, in an effort to attract endogenous circulating precursor cells as the source of cells. Decellularized tissue scaffolds derived from porcine valves have been successfully used in clinical studies in the pulmonary position[15] but have suffered mechanical failure in the systemic circulation.[16] Decellularized valves pre-seeded with autologous vascular endothelial cells prior to orthotopic implantation and implanted into the aortic valve position in a growing lamb model demonstrated favorable functionality compared to cryopreserved allografts for up to 9 months. However, in some cases predisposition to mineralization of the tissue engineered constructs were observed in these animal models.[17] Moreover, and importantly, these studies have not shown that pre-seeding is either necessary or advantageous to function or that appropriate cellular repopulation occurs *in vivo*. Thus, despite the exciting possibility of TEHV, many significant challenges remain before this form of therapy can be validated, and shown to be sufficiently safe and effective to warrant successful translation to clinical use.[18]

The scaffold material is a paramount consideration in fabricating a valve substitute by tissue engineering approaches. Moreover, an emerging area of biomaterials research is focused on fusing techniques of engineering with developmental heart valve biology. Specifically, there has been increased focus on finding materials that simulate the natural physiological environment that promotes fetal heart valve development.[19] This could lead to better understanding and harnessing of the environmental signals that control heart valve development and regeneration. Indeed, focusing on re-creating the natural valve micro-environment has helped to understand and potentially facilitate valve tissue remodeling, and hydrogel micro-engineering holds promise as a new approach to engineer heart valve substitutes.

Hydrogels are colloidal polymers composed of particles and water. In the present context hydrogels are defined as 'three-dimensional (3D) *in vitro* culture platforms that enable cell growth in or on user-defined microenvironments that mimic critical aspects of native tissue'.[20] Chemical and mechanical properties of hydrogels can be tuned by engineering methodology. This allows for a predictable functionality which directly influences cell fate. Theoretically, by mimicking the native cellular microenvironment or niche within the native, developing and mature, healthy heart valve, tissue heart valve regeneration can be understood and mimicked, and thus provides a potential means to engineer a heart valve substitute.

This chapter summarizes the application of hydrogels in heart valve tissue engineering and discusses how the imaginative use of hydrogels can further the development of a successful tissue engineered therapeutic option.

2. Heart Valve Structure, Function, and Development: Relevance to Engineering

2.1. Aortic valve structure and function

We provide a brief overview of valve function and structure as a scientific context for TEHV. Normal heart valves ensure unidirectional blood flow through the heart. The semilunar valves — the pulmonary and the aortic valves — prevent backflow into the ventricles during diastole (when they are closed), whereas the atrioventricular valves — the mitral and the tricuspid valves — prevent reverse flow from the ventricles into the atria during systole (when these valves are closed and the semilunar valves are open). Human heart valves open and close approximately 40 million times annually and thus 3 billion times over a lifetime. The cyclical opening and closing of the valves with each cardiac cycle is facilitated by a synergistic process of mechanical forces exerted by the surrounding blood and heart onto the mobile part of the valves, termed their cusps (semilunar valves) or leaflets (for atrioventricular valves).[7] The ability of valves to maintain structural and mechanical integrity is largely attributed to the unique and complex organization of valve in which both the extracellular matrix (ECM), and cells are highly specialized (Fig. 1).

Heart valve leaflets demonstrate a layered ECM architecture: (i) near the outflow side of the valve, there is a layer with densely packed collagen, giving the leaflets their strength, called the *fibrosa*; (ii) a middle layer of loose connective tissue enriched in glycosaminoglycans (GAGs), named the *spongiosa*; and (iii) a layer below the inflow surface of the leaflet rich in elastin, known as the *ventricularis*.[21] This unique organization of ECM proteins contributes to the structural and functional integrity of the heart valve. Collagen is the major stress-bearing component of the heart valve ECM. It maintains apposition of the cusps during the cardiac cycle in addition to transferring mechanical load from the cusps to the aortic wall when the valve is closed and needs to withstand the pressure of backflow. Since collagen fibrils cannot be stretched, the changes of cusp shape and size during the cardiac cycle are attributed to rearrangements in collagen structure such as crimping in systole and straightening of the crimp in diastole (similar to an accordion).[22] These rearrangements of collagen during opening and closing are facilitated by the pull of the elastin in the *ventricularis* that was stretched during valve closure. Finally, GAGs in the *spongiosa* function as a shock absorber to the mechanical forces and a lubricant for internal tissue reaugmentation that occurs during the cardiac cycle.[21]

2.2. Valvular cell population

The surfaces of the valve are covered with an endothelial monolayer comprised valvular endothelial cells (VECs) and the entire internal valve cross section is composed of valvular interstitial cells (VICs) embedded in ECM. VECs contribute to the non-thrombogenic properties of the valve. Moreover, VECs have shown to be able to undergo endothelial to mesenchymal transformation (EMT) in response to pathological levels of mechanical strain, increased TGF-β signaling, or Notch-1 activation,[23] which in turn may play a role in valve morphogenesis and development, adaptive remodeling, and sometimes pathology.[21,24] VICs are crucial to maintaining valvular function. VICs continuously repair and remodel the valvular

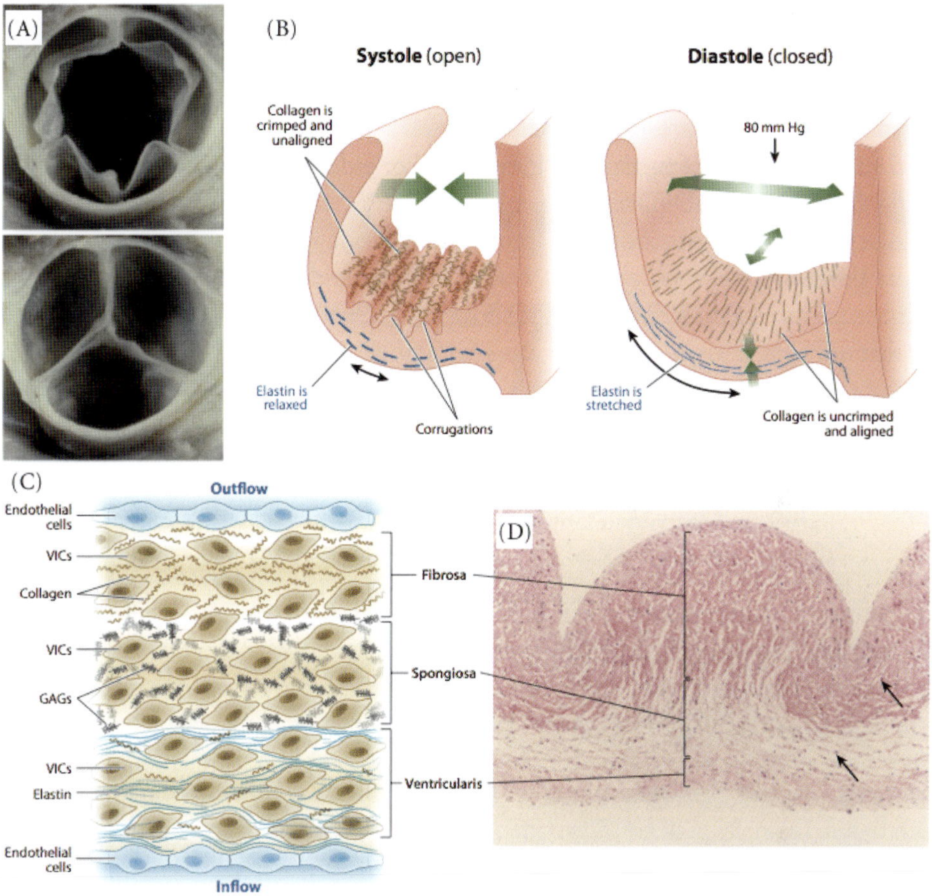

Fig. 1. Aortic valve functional structure at macroscopic and microscopic levels. (A) Outflow aspect of aortic valve in open (*top*) and closed (*bottom*) configurations, corresponding to systole and diastole, respectively. (B) Schematic representation of architecture and configuration of aortic valve cusp in cross section and of collagen and elastin in systole and diastole. Modified from Ref. 11. (C) Schematic diagram of the detailed cellular and extracellular matrix architecture of a normal aortic valve. Modified with permission from Ref. 12. (D) Tissue architecture, shown as low-magnification photomicrograph of cross-section cuspal configuration in the non-distended state (corresponding to systole), emphasizing three major layers: Ventricularis, spongiosa, and fibrosa. Valvular interstitial cells (VICs) are denoted by arrows. The outflow surface is at the top. Original magnification: 100×. Hematoxylin and eosin stain. (E) Transmission electron photomicrograph of relaxed fresh porcine aortic valve (characteristic of the systolic configuration), demonstrating the fibroblast morphology of VICs (*arrow*); the dense, surrounding closely apposed collagen (*asterisk*) with wavy crimp; and the potential for VIC–collagen and VIC–VIC interactions. Reproduced with permission from Ref. 2. (F) Transmission electron photomicrograph of surface of aortic valve demonstrating valvular endothelial cell (VEC; *arrow*) and proximity of deeper VICs (*double arrows*) and potential for VEC–VIC interactions. Reproduced with permission from Ref. 14. Abbreviation: GAG, glycosoaminoglycan.

Source: Reproduced with permission from *Annu. Rev. Pathol. Mech. Dis.*, **7**, 161–183.

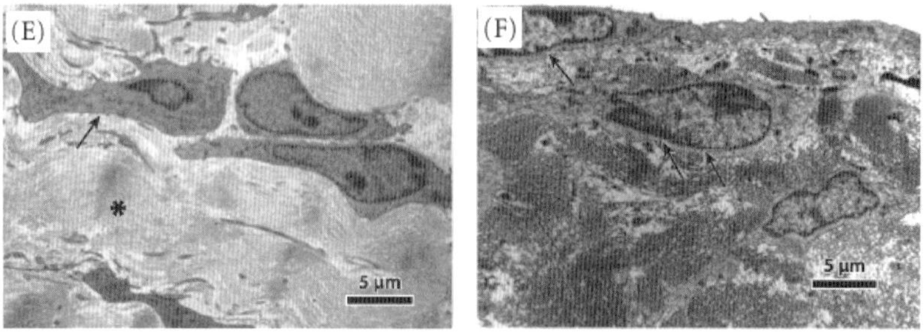

Fig. 1. (*Continued*)

ECM, by synthesizing ECM proteins, expressing matrix degrading enzymes (e.g., (MMPs) and tissue inhibitors of metalloproteinases (TIMPs)).[21,25,26] VICs are a very heterogeneous cell population characterized by five specifically identifiable phenotypes: Embryonic progenitor endothelial VICs, adult progenitor VICs, quiescent VICs, activated VICs and osteoblastic VICs.[27] Valvular homeostasis and the ability to repair/remodel as a consequence of injury is considered to be mediated primarily by active VIC phenotype transition primarily driven by environmental cues.[21,28] It is believed that the transition of quiescent fibroblast-like cells to activated myofibroblast-like cells is key to maintaining valvular homeostasis. For example, VICs are activated during embryonic development and valvular maturation.[26] In addition, mechanical stress changes or transforming growth factor (TGF)-β increase VIC activation and ECM protein synthesis by isolated VICs *in vitro.*[29–32] VIC activation, marked by expression of alpha smooth muscle actin (α-SMA), increases stress fiber formation causing enhanced contractility of cells onto the surrounding ECM. In turn, this may promote physiological and pathological remodeling of the valve.[28] In addition, although VECs may interact with VICs to contribute to valvular tissue integrity, the exact mechanisms of the functional interactions between VECs and VICs are uncertain.

Consistent with the rapidly evolving view that the mechanical environment is a major determinant of stem and cell differentiated cell phenotype, the interplay between VIC phenotypical transitions and the surrounding ECM, vital to the understanding of valvular physiology and even pathology, is under active investigation. More specifically, in the pursuit of engineering a valve substitute that is able to grow, remodel and repair, an understanding of both the effects of VICs on their environment in addition to the VIC response to their environment, is crucial.

2.3. Cardiac valve development

Since creating a TEHV may in large measure recapitulate normal cardiac morphogenesis, mechanisms of embryonic and foetal heart valve development are important to consider. Valves develop within the atrioventricular canal and ventricular outflow tracts, and mature in the fetus.[33,34] During normal development of the heart, the heart tube consists of endocardium and myocardium separated by a primitive acellular amorphous ECM called the cardiac jelly. After the completion of heart looping, the valve cusps/leaflets originate from endocardial cushions, the precursors of valves and the cardiac septums.[35] A subset of endothelial cells of the endocardial cushion, driven by signals from the underlying myocardium, transform to mesenchymal cells and migrate into the cardiac jelly to form VICs (through EMT as described earlier). Experimental studies have suggested that VICs in adult valves may be also replenished via circulating endothelial or mesenchymal cell precursors derived from the bone marrow[36] and subsequently undergo EMT.[37] Even though the role of specific ECM components in remodeling of the endocardial cushions into heart valves is poorly understood, it is clear that the temporal and spatial patterns of collagen expression are vital to function.[38] In addition, the glycosaminoglycan hyaluronic acid (HA) is well documented to have important function in EMT and subsequent valve development.[39]

3. Biomaterials in Heart Valve Tissue Engineering

Investigators in the field of heart valve tissue engineering have focused on either (i) shaping a valve and leaflet geometries out of synthetic polymer scaffolds and seeding these with cells prior to implantation into animal models, (ii) using decellularized natural valve materials, or (iii) using fabricated ECM with or without embedded cells. The biggest challenge to overcome before TEHV can successfully be translated to patients is to exactly formulate the optimal components and conditions (e.g., scaffold material, cell source, *in vitro* manipulation, animal models) for tissue engineering to yield progress toward an adequate valve substitute in a bioreactor. Other key challenges have been discussed.[10,18]

As a cornerstone in tissue engineering over the last three decades, biomaterials science has focused on how to create scaffolds to promote natural tissue regeneration.[40] A wide range of synthetic polymers have been developed and used in heart valve tissue engineering. Widely used polymers include poly(lactic acid) (PLA), poly(glycoclic acid) (PGA) and poly(lactic-*co*-glycolic acid) (PLGA), which are clinically used as biodegradable sutures.[41–43] Because they had been already Food and Drug Administration (FDA)-approved and accepted as biocompatible and biodegradable polymers over a reasonable interval, initial heart valve tissue engineering research focused on these materials. However, the early studies into using PLA: PGA scaffolds revealed inadequate mechanical properties suitable for TEHV, which lead to the addition of other polymers to the PGA-based scaffolds. The most common modification used was the addition of poly-4-hydroxybutyrate (P4HB), a bioresorbable thermoplastic that can be molded into the complex shape of a heart valve and sealed without the use of sutures.[44] One study seeded these scaffolds with amniotic fluid-derived stem cells and conditioned them in an *in vitro* pulsatile system,[45] resulting in endothelialization, cell proliferation, and ECM synthesis. Upon evaluation in a lamb model, PGA:P4HB scaffolds seeded with ovine VICs and VECs, yielded ECM protein synthesis and mechanical properties similar to the native valve after 20 weeks of implantation.[41]

Despite what seems like initial short-term success, long-term function remains to be evaluated and reliably reproduced. Using these synthetic polymers for TEHV has yet to produce a viable tissue engineered valve substitute that is close to human trials. In addition, these materials possess certain limitations. While these materials exhibit good mechanical properties, they usually possess high initial stiffness compared with native tissues. In addition, they require considerable time for the polymer to undergo degradation by hydrolysis and resorption.[46] Moreover, these materials do not possess any functional biological moieties that could direct cell signaling, ECM production or enzymatic material degradation. Some of these materials can be functionalized with such tissue guiding motifs; however, there has been no control over functional molecule localization.[47,48] Moreover, this approach does not uniformly lead to a physiologically relevant environment that can be invaded and remodeled by cells. To this end, alternative approaches are being pursued to controllably provide requisite environmental cues to promote and control cell growth and provide functional ECM.

4. Hydrogels: Tunable, Controllable, Bio-mimetic Scaffolds

4.1. Hydrogels

Cells often behave very differently when they are isolated from the complex architecture of their native tissues and cultured in Petri dishes. In particular, VICs in two-dimensional culture will undergo spontaneous myofibroblast-like differentiation (aVICs), whereas *in situ* they are mostly quiescent fibroblast-like cells (qVICs).[27] This difference in cell behavior has constituted a major obstacle for tissue engineers. However, in the past 15 years the field has made progress in creating successful 3D cellular microenvironments with hydrogels that possess elasticity similar to that of natural tissues.[49] The structure and composition of these hydrogels can be tailored to deliver the appropriate chemical, biological, and physical cues that encourage the development of tissue-like structures *in vitro*. Hydrogels can be fabricated from synthetic and from natural biomaterials. Both have been applied in the field of heart valve tissue engineering.

4.2. Synthethic hydrogel: Poly(ethylene glycol) (PEG)

One of the most common polymers used in hydrogels for heart valve tissue engineering is poly(ethylene glycol) (PEG).[50–52] PEG hydrogels are intrinsically biocompatible due to biologically neutral base gels, resistance to protein adsorption, no release of acidic products during degradation, and allow for polymer crosslinking with low toxicity. In turn, this allows for high-density cellular encapsulation. PEG hydrogels can be chemically tuned by crosslinking bioactive moieties such as peptides, GAG, or growth factors. More specifically, peptides attached to the surface or bulk of PEG will retain their bioactivity.[53] Mechanical properties can also be altered by changing the amount of polymer crosslinking, or copolymerization. As such, the properties of PEG hydrogels can be modified to mimic tissue function or even a specific pericellular environment. Changing the molecular weight or the weight fraction in solution of the macromere or altering crosslinking time can control physical properties of PEG hydrogels.[54,55] Hence, dynamically tunable PEG-based hydrogel platforms have been developed to understand valve cellular biology.[56] A photosensitive PEG hydrogel was functionalized with an MMP-sensitive peptide to study the response of VICs.[50] A similar approach was used to tune mechanical properties of PEG-based hydrogels. Interestingly, this work showed that VIC differentiation can be directed by gradients in substrate stiffness. Moreover, whereas it was previously thought that stiff environments would favor activation of VICs, this work showed that soft substrates could also facilitate activation of VICs.[57,58] However, PEG synthetic hydrogels have the significant drawback of being devoid of cell binding motifs. Functionalization with cell-adhesive peptides is thus necessary to facilitate cell attachment. The potential of using this material thus mostly lies in the tunability of such a platform to a specific purpose.

4.3. 'Natural' hydrogels

Increased interest has risen to use so-called 'natural materials' which are based on valvular ECM components. Although they have less flexibility in functionalization, they do provide mechanical properties comparable to native tissue and the possibility of biological cues needed to direct and control cell growth. Most importantly, natural materials allow for the fabrication of a more *in situ*-like environment analogous to the native heart valve. Cells surrounded by a native-like environment mimicking their natural state may be presented with proteins similar to those of fetal and adult heart valves, which could promote *in vitro* and *in vivo* remodeling. In turn, this may be beneficial in creating a viable heart valve replacement. In addition, such an approach can be used *in vitro* to controllably study the mechanism of valvular physiology and disease as it may occur *in vivo*.

Native valve ECM components such as collagen, HA, fibronectin, and elastin have been used as biomaterials to fabricate hydrogels. The two most studied natural biomaterials for hydrogel-based heart valve tissue engineering are HA and collagen.[59–63] Both collagen and HA are abundant in adult heart valve ECM, and also vital for valve maturation.

4.4. HA

As mentioned before, the endocardial cushions from which heart valves develop contain a significant amount of HA, playing a key role in regulating ECM production and control of EMT by endocardial cells. From an engineering point of view, HA hydrogels are promising scaffolds for *in vitro* and *in vivo* remodeling, due to their inherent slow biodegradation by VIC-secreted hyaluronases. The *spongiosa* of native valves contain more than 90% HA. HA is also a non-thrombogenic and non-immunogenic material. Using HA in hydrogel scaffolds also seems to provide for biological cues to promote specific ECM production. For example, in a study where neonatal rat smooth muscle cells were cultured on HA hydrogels, results demonstrated increased elastin production compared to cells grown on a Petri dish.[60] In addition, photo-crosslinkable HA hydrogels modified with methacrylate groups showed that ECM protein production of encapsulated VICs is based on the molecular weight of the HA degradation products.[64] This is by itself very promising, since lack of elastin production seems to hinder the successful long-term performance of synthetic based heart valves.

4.5. Collagen

Collagen is the main ECM protein produced in developing valves and maintained in mature valves through a homeostatic process described earlier. Collagen is largely responsible for the mechanical strength of heart valves, and seems logical to be considered a scaffold.[65] However, on its own, collagen lacks the necessary mechanical properties to fabricate a durable heart valve substitute. It is thus often used as one component of a composite material. Collagen has been used in combination with photocrosslinkable methacrylated HA.[66] Interestingly, such a composite material which provides for a similar fibrous structure of native valve leaflets, through collagen, together with increased mechanical properties of HA, has shown to promote proliferation of fibroblasts, analogous to quiescent VICs, predominantly present in healthy heart valves.[67] By adding HA, the formation of a collagen fiber network is altered, meaning that cells cannot contract the composite gels, compared with collagen gels alone. Collagen has also been combined with the GAG chondroitin sulfate to create valve replacements.[61] This composite scaffold demonstrated *in vitro* development into similar architecture as native valves when seeded with porcine VICs and VECs.

Alternatively, a technique has been developed to create collagen-based valve scaffolds *in vivo* using fibrous encapsulation of an implanted foreign body by an organism.[68,69] First, silicone rods were shaped into valve molds and wrapped in polyurethane or autologous connective tissue from white rabbits. Then, these constructs were implanted subcutaneously in the same animals. After 4 weeks, the host response to the implant was to secrete connective tissue primarily in the form of collagen, which would coat the valve-shaped mold. Results demonstrated a tri-leaflet valve composed primarily of collagen, which performed well in bioreactors simulating the pulmonary circulation.[69] Future studies are ongoing on using this approach to create an autologous valve substitute.

Another approach using collagen-based hydrogels is to use its hydrolyzed form, gelatin. Gelatin is synthesized upon the partial breakdown of the natural triple-helical structure of collagen and is considered safe by the FDA, which made it an attractive material towards clinical applications. It maintains similar bioactivity as collagen, however unmodified gelatin as a hydrogel platform is mechanically very weak and the gelation temperature is below 37°C, the physiological temperature. To overcome its mechanical shortcomings, gelatin can be altered with methacrylate groups (GelMA), which when exposed to light in the presence of a photo-initiatior, polymerizes to form a covalently and physically crosslinked hydrogel. Changing macromer molecular weight or percentage per volume or crosslinking time can alter its physical mechanical properties, making it an attractive scaffold. Encapsulating VICs into GelMA hydrogels resulted in high viability, myofibroblast differentiation and matrix remodeling in 3D.[70] However, similar to collagen, GelMA is an intrinsically weak material, which is not at par with the mechanical properties required for a tissue engineered valve substitute. To what extent GelMA can be explored in combination with other materials to address this limitation remains to be studied.

4.6. Fibrin

Fibrin-based hydrogels present another viable alternative for TEHVs, since they can be derived from the patient. This would eliminate the worry of an adverse immunogenic response. Although studies of fibroblasts encapsulated in fibrin gels show increased collagen production and anisotropic mechanical properties, these materials are not strong enough to be considered sustainable heart valve scaffold materials.[71,72] However, when mechanical conditioning in the form of cyclic distension is performed, fibrin-based scaffolds with porcine VICs or firoblasts showed improved mechanical properties and collagen production compared to statically grown control samples.[73] This could mean that fibrin-based scaffolds may facilitate increased remodeling and thus eventually aid in generating mechanical properties needed for *in vivo* implantation.

4.7. 3D printing

A technique that has recently gained a great deal of attention, particularly in organ tissue engineering is 3D printing. Considering hydrogels can be designed to mimic mechanical and biological properties of soft tissues, they are very suitable to be used and controlled for 3D printing. The technology that can print clinically sized, geometrically, complex constructs with hydrogels exists. This technique has primarily been demonstrated for hard tissues such as bone and cartilage.[74] However, a recent study used this approach to print aortic valves (Fig. 2).[75]

By using PEG-diacrylate (PEGDA) as a base polymer hydrogel, aortic valves including leaflets, coronary sinus, and root geometry at different scales were printed. Briefly, the valves were fabricated by simultaneously printing and crosslinking small hydrogel paths by exposing the hydrogel paths to light. Each small path would then brick-wise be printed to form an aortic valve. These constructs exhibited tunable mechanical properties, and adequate cytocompatibility of seeded porcine VICs.

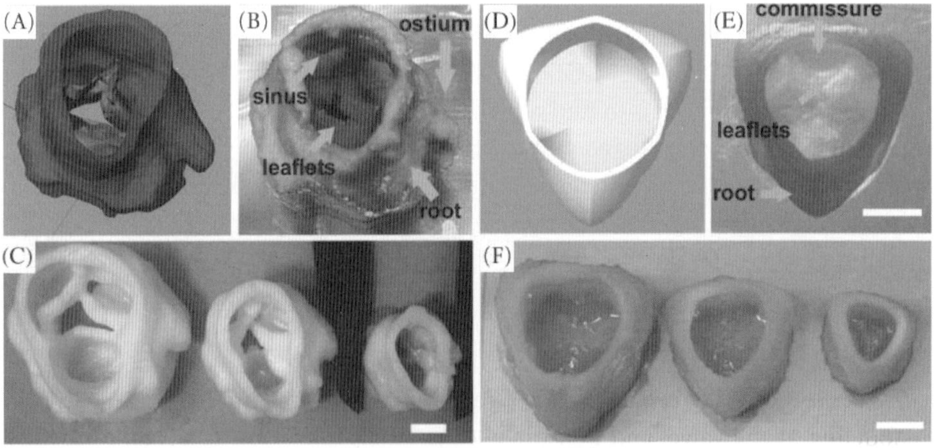

Fig. 2. Printing heterogeneous valve constructs. (A) Porcine aortic valve model was (B) Printed, where root was formed with 700 MW PEG-DA hydrogel while the leaflets were formed with 700/8000 MW PEG-DA hydrogels. Key features such as the coronary ostium and sinuses were present. (C) Scaffolds were printed with 700 MW PEG-DA at different scale for fidelity analysis, where the inner diameters (ID) were 22, 17, and 12 mm. (D) Axisymmetric valve model was (E) Printed with two blends of hydrogels (F) And at 22, 17, and 12 mm ID. Scale bar = 1 cm.

Source: With permission from Hockaday *et al. Biofabrication* (2013).

Although this approach is in its infancy, this use of hydrogels could prove an enabling technique for heart valve tissue engineering, because the 3D printed heart valve can be sized and shaped according to specific patient's needs. In essence, this is personalized medicine applied to cardiac valve replacement. In addition, 3D printing hydrogels has been utilized in creating vascularized constructs using thermal gelation after printing. Using different biomaterial inks and print-heads simultaneously, vascularized multicellular heterogeneous tissue constructs were printed, demonstrating good viability and little to no damage to encapsulated cells.[76] This approach may open up new avenues for research on angiogenesis, drug screening and stem cell niches.

5. Hydrogels as Model Systems to Study Biology and Pathophysiology of VICs in Health and Disease

The chemical and mechanical tunability of hydrogels — as described previously — does not only harbor potential in creating functional tissue engineered valve substitutes, but also facilitates the ability to mimic the native cellular niche as it may occur in the heart valve. More specifically, hydrogels provide for a 3D *in vitro* platform to study the biology and pathophysiology of valvular cells, in a model which may be much more physiological than standard culture (Fig. 3).

Indeed, this is an active area of investigation.[77–79] Of special interest is the pursuit of understanding the mechanisms of calcific aortic valve disease (CAVD). CAVD is the most common valve disease in the Western world[2] and is characterized by progressive fibrosis and mineralization of the aortic valve cusps, causing aortic valve stenosis.[21] Traditionally viewed as a degenerative disease, CAVD is now seen as a dynamic process involving phenotypic modulation of resident VICs.[27,80] As described before, VICs are responsible for maintaining valve tissue homeostasis. However, when repeated valve injury or chronic exposure to exogenous pathogenic stimuli such as inflammation or excessive mechanical stress occur, regulation of healthy VIC phenotypes can be altered. One of the postulated mechanisms is that with changes in valve architecture that occur with age and upon persistent VIC myofibroblastic activation, VICs can increase valve stiffness through excess remodeling of the valve, and eventually cause VICs to express genes and proteins associated with osteogenesis.[81] This process of osteogenic activation is subject to several factors amongst which the potent cytokine transforming growth factor b1 (TGFb1) has been well documented.[82] Because symptoms do not occur until there is a significant aortic valve stenosis caused by severe calcification with CAVD, clinical specimens obtained at surgery have end-stage calcific valve disease in humans. In addition, animal models have failed to truly recapitulate the human disease, and

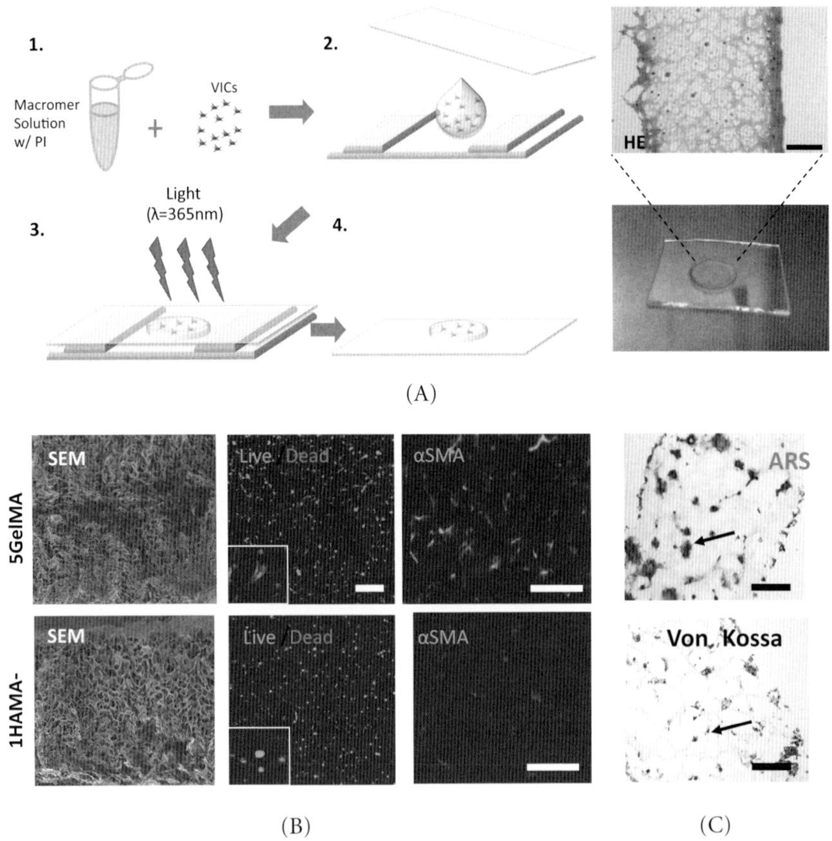

Fig. 3. Using hydrogels as model systems for valve pathobiology. (A) Typical fabrication technique of photo-crosslinkable hydrogels in which VICs are resuspended in a macromer solution whichin the presence of PI solution can crosslink when exposed to UV light. (B) Changing material composition affects VIC behavior. When HAMA was added to GelMA it facilitated a quiescent VIC phenotype, contrary to VICs encapsulated in GelMA alone, which demonstrate spreading and myofiboblast-like differentation. (C) When exposed to osteogenic enviroment VICs encaspulated in hybrid hydrogels undergo osteoblastic differentiation and deposit calcium as visualized by Alizarin Red S and Von Kossa staining. Arrows indicate positive staining Bar 50 um.

in vitro standard culture experiments are arguably unable to simulate the cellular processes that occur *in vivo*. As such, hydrogels have been explored to simulate specific biological and pathophysiological processes to gain insight into underlying mechanisms. A key advance was on seeding VICs onto HA-based hydrogels, demonstrating the feasibility of utilizing natural material-based hydrogel microengineering as an *in vitro* culture platform (Fig. 3).[59] GelMA-based hydrogel platforms have also been widely explored as hydrogel culture systems as a tool to understand and direct VIC phenotypic fate,[20,70] in addition to synthetic polymers such as PEG. After the feasibility of using various materials to create relevant 3D cellular micro-environments was demonstrated, investigators started using these systems to understand the phenotypic change of VICs that may occur in heart valve biology and pathophysiology. Understanding the phenotypic switch from quiescent fibroblast-like VIC to activated myofibroblastic VIC is of vital importance in unraveling mechanisms of disease. For instance, it became clear that matrix elasticity is able to regulate proliferation and differentiation of quiescent fibroblastic VIC[57,79]; particularly stiffened matrix in fibrotic lesions are thought to promote pathogenic myofibroblast activation.[81] PEG-based hydrogels have been modulated to simulate physiological matrix elasticities, demonstrating that soft hydrogels are able to preserve the quiescent fibroblast phenotype of VICs much better than stiff surfaces.[83,84] Decreasing matrix elasticity during culture even facilitated a phenotypic change from activated myofibroblasts to quiescent fibroblasts.[58,77,78] In addition to controlling the mechanical characteristics of hydrogels, the chemical composition of matrices and its effect on VIC phenotype can also be modulated. Indeed, hydrogels have been functionalized with other valve matrix molecules such as fibronectin and fibrin, which revealed that fibronectin had an inhibitory effect on calcification markers of VICs, whereas fibrin seemed to enhance them.[72] Hydrogels can also be used as a platform to test the effect of exogenous stimuli such as cytokines on VIC phenotype. The addition of TGFb1 revealed an increased myofibroblastic differentiation of VICs encapsulated in hydrogels,[29] whereas FGF-2 has demonstrated an inhibitory effect on myofibroblastic differentiation of VICS. All in all, hydrogels are dynamic substrates which hold great potential in deciphering the complex role of matrix composition and cues on the phenotypic plasticity of VICs and the role of this most influential cell type in valve homeostasis and disease.

6. Conclusion

The success of TEHV depends on living functional tissue that is able to adapt to a viable and durable homeostatic state. In addition, there must be an intricate ability to repair and remodel when needed, whether it is as a response to injury or growth. Valvular homeostasis is a complex interaction between the endothelium, VIC phenotypes, the ECM, and biomechanics. Using natural materials to fabricate hydrogels and generate living heart valves is a relatively new approach in heart valve tissue engineering. Even though hydrogels have substantial potential due to their tunable, controllable, and biomimetic nature caused by using the body's own ECM proteins, significant challenges need to be overcome. Up until this point the use of hydrogels in TEHV has mostly been *in vitro* and used to understand cell-material relationships in 3D platforms. In addition, even with the advent of tissue controlling natural materials, biocompatibility alone will not be enough to create the 'perfect' heart valve replacement. The valve substitute must have sufficient mechanical properties to withstand the hemodynamic demands of a beating heart. The combination of cell-instructive biomaterials with other techniques that introduce mechanical strength could create such a scaffold system. Although considerable progress has been made in developing, characterizing, and using hydrogel-based tissue engineering for heart valve replacement, the approach is still in its rudimentary stages and far short of clinical implementation. Nevertheless, considering the premise that hydrogel engineering can harness the body's own cells and ECM to stimulate a living, durable valve substitute, hydrogels hold exciting potential. Moreover, hydrogel systems can be used to explore mechanisms of heart valve health and disease.

References

1. C. M. Otto, B. K. Lind, D. W. Kitzman, B. J. Gersh, and D. S. Siscovick, Association of aortic-valve sclerosis with cardiovascular mortality and morbidity in the elderly, *New Engl. J. Med.*, **341**(3), 142–147 (1999).
2. J. J. Takkenberg, N. M. Rajamannan, R. Rosenhek, A. S. Kumar, J. R. Carapetis, and M. H. Yacoub, The need for a global perspective on heart valve disease epidemiology. The SHVD working group on epidemiology of heart valve disease founding statement. *J. Heart Valve Dis.*, **17**(1), 135–139 (2008).
3. S. H. Rahimtoola, Choice of prosthetic heart valve in adults an update. *J. Am. Coll. Cardiol.*, **55**(22), 2413–2426 (2010).
4. K. Hammermeister, G. K. Sethi, W. G. Henderson, F. L. Grover, C. Oprian, and S. H. Rahimtoola, Outcomes 15 years after valve replacement with a mechanical versus a bioprosthetic valve: Final report of the Veterans Affairs randomized trial. *J. Am. Coll. Cardiol.*, **36**(4), 1152–1158.
5. F. J. Schoen and R. J. Levy, Calcification of tissue heart valve substitutes: Progress toward understanding and prevention. *Ann. Thorac. Surg.*, **79**(3), 1072–1080 (2005).
6. R. N. Mitchell, R. A. Jonas, and F. J. Schoen, Pathology of explanted cryopreserved allograft heart valves: Comparison with aortic valves from orthotopic heart transplants. *J. Thorac. Cardiovasc. Surg.*, **115**(1), 118–127.
7. M. S. Sacks, F. J. Schoen, and J. E. Mayer, Bioengineering challenges for heart valve tissue engineering. *Annu. Rev. Biomed. Eng.*, **11**, 289–313 (2009).
8. I. Vesely, Heart valve tissue engineering. *Circ. Res.*, **97**(8), 743–755 (2004).
9. D. E. Harken, Heart valves: Ten commandments and still counting. *Ann. Thorac. Surg.*, **48**(3 Suppl), S18–S19 (1989).
10. F. J. Schoen, Heart valve tissue engineering: Quo vadis? *Curr. Opin. Biotechnol.*, **22**(5), 698–705 (2011).
11. F. W. Sutherland, T. E. Perry, Y. Yu, M. C. Sherwood, E. Rabkin, Y. Masuda *et al.*, From stem cells to viable autologous semilunar heart valve. *Circulation*, **111**(21), 2783–2791 (2005).

12. S. P. Hoerstrup, A. Kadner, S. Melnitchouk, A. Trojan, K. Eid, J. Tracy *et al.*, Tissue engineering of functional trileaflet heart valves from human marrow stromal cells. *Circulation.* **106**(12 Suppl 1), I143–I150.
13. D. Gottlieb, T. Kunal, S. Emani, E. Aikawa, D. W. Brown, A. J. Powell *et al.*, *In vivo* monitoring of function of autologous engineered pulmonary valve. *J. Thorac. Cardiovasc. Surg.*, **139**(3), 723–731 (2010).
14. N. Hibino, E. McGillicuddy, G. Matsumura, Y. Ichihara, Y. Naito, C. Breuer *et al.*, Late-term results of tissue-engineered vascular grafts in humans. *J. Thorac. Cardiovasc. Surg.*, **139**(2), 431–436, 436e1–436e2 (2010).
15. P. M. Dohmen, A. Lembcke, S. Holinski, D. Kivelitz, J. P. Braun, A. Pruss *et al.*, Mid-term clinical results using a tissue-engineered pulmonary valve to reconstruct the right ventricular outflow tract during the Ross procedure. *Ann. Thorac. Surg.*, **84**(3), 729–736 (2007).
16. P. Simon, M. T. Kasimir, G. Seebacher, G. Weigel, R. Ullrich, U. Salzer-Muhar *et al.*, Early failure of the tissue engineered porcine heart valve SYNERGRAFT in pediatric patients. *Eur. J. Cardiothorac. Surg.*, **23**(6), 1002–1006; discussion 6 (2003).
17. I. Tudorache, A. Calistru, H. Baraki, T. Meyer, K. Hoffler, S. Sarikouch *et al.*, Orthotopic replacement of aortic heart valves with tissue-engineered grafts. *Tissue Eng. Pt. A*, **19**(15–16), 1686–1694.
18. J. Hjortnaes, C. V. Bouten, L. A. Van Herwerden, P. F. Grundeman, and J. Kluin, Translating autologous heart valve tissue engineering from bench to bed. *Tissue Eng. Pt. B, Rev.*, **15**(3), 307–317 (2009).
19. M. K. Sewell-Loftin, Y. W. Chun, A. Khademhosseini, and W. D. Merryman, EMT-inducing biomaterials for heart valve engineering: Taking cues from developmental biology. *J. Cardiovasc. Transl. Res.*, **4**(5), 658–671 (2011).
20. C. A. DeForest and K. S. Anseth, Advances in bioactive hydrogels to probe and direct cell fate. *Annu. Rev. Chem. Biomol. Eng.*, **3**, 421–444 (2012).
21. F. J. Schoen, Evolving concepts of cardiac valve dynamics: The continuum of development, functional structure, pathobiology, and tissue engineering. *Circulation*, **118**(18), 1864–1880 (2008).
22. E. M. Joyce, J. Liao, F. J. Schoen, Jr. J. E. Mayer, and M. S. Sacks, Functional collagen fiber architecture of the pulmonary heart valve cusp. *Ann. Thorac. Surg.*, **87**(4), 1240–1249 (2009).
23. G. Paranya, S. Vineberg, E. Dvorin, S. Kaushal, R. J. Roth, E. Rabkin *et al.*, Aortic valve endothelial cells undergo transforming growth factor-beta-mediated and non-transforming growth factor-beta-mediated transdifferentiation *in vitro*. *Am. J. Pathol.*, **159**(4), 1335–1343 (2001).
24. J. Bischoff and E. Aikawa, Progenitor cells confer plasticity to cardiac valve endothelium. *J. Cardiovasc. Transl. Res.*, **4**(6), 710–719 (2011).
25. E. Rabkin, M. Aikawa, J. R. Stone, Y. Fukumoto, P. Libby, and F. J. Schoen, Activated interstitial myofibroblasts express catabolic enzymes and mediate matrix remodeling in myxomatous heart valves. *Circulation*, **104**(21), 2525–2532.

26. E. Aikawa, P. Whittaker, M. Farber, K. Mendelson, R. F. Padera, M. Aikawa *et al.*, Human semilunar cardiac valve remodeling by activated cells from fetus to adult: Implications for postnatal adaptation, pathology, and tissue engineering. *Circulation*, **113**(10), 1344–1352 (2006).
27. A. C. Liu, V. R. Joag, and A. I. Gotlieb, The emerging role of valve interstitial cell phenotypes in regulating heart valve pathobiology. *Am. J. Pathol.*, **171**(5), 1407–1418 (2007).
28. G. A. Walker, K. S. Masters, D. N. Shah, K. S. Anseth, and L. A. Leinwand, Valvular myofibroblast activation by transforming growth factor-beta: Implications for pathological extracellular matrix remodeling in heart valve disease. *Circ. Res.*, **95**(3), 253–260 (2004).
29. M. C. Cushing, J. T. Liao, and K. S. Anseth, Activation of valvular interstitial cells is mediated by transforming growth factor-beta1 interactions with matrix molecules. *Matrix Biol.*, **24**(6), 428–437 (2005).
30. A. C. Liu and A. I. Gotlieb, Transforming growth factor-beta regulates *in vitro* heart valve repair by activated valve interstitial cells. *Am. J. Pathol.*, **173**(5), 1275–1285 (2008).
31. W. D. Merryman, H. D. Lukoff, R. A. Long, G. C. Engelmayr, Jr., R. A. Hopkins, and M. S. Sacks, Synergistic effects of cyclic tension and transforming growth factor-beta1 on the aortic valve myofibroblast. *Cardiovasc. Pathol.*, **16**(5), 268–276 (2007).
32. P. Thayer, K. Balachandran, S. Rathan, C. H. Yap, S. Arjunon, H. Jo *et al.*, The effects of combined cyclic stretch and pressure on the aortic valve interstitial cell phenotype. *Ann. Biomed. Eng.*, **39**(6), 1654–1667 (2001).
33. M. D. Combs and K. E. Yutzey, Heart valve development: Regulatory networks in development and disease. *Circ. Res.*, **105**(5), 408–421 (2009).
34. P. W. Riem Vis, J. Kluin, J. P. Sluijter, L. A. van Herwerden, and C. V. Bouten, Environmental regulation of valvulogenesis: Implications for tissue engineering. *Eur. J. Cardiothorac. Surg.*, **39**(1), 8–17 (2011).
35. E. J. Armstrong and J. Bischoff, Heart valve development: Endothelial cell signaling and differentiation. *Circ. Res.*, **95**(5), 459–470 (2004).
36. R. P. Visconti, Y. Ebihara, A. C. LaRue, P. A. Fleming, T. C. McQuinn, M. Masuya *et al.*, An *in vivo* analysis of hematopoietic stem cell potential: Hematopoietic origin of cardiac valve interstitial cells. *Circ. Res.*, **98**(5), 690–696 (2006).
37. A. H. Kumar, P. Metharom, J. Schmeckpeper, S. Weiss, K. Martin, and N. M. Caplice, Bone marrow-derived CX3CR1 progenitors contribute to neointimal smooth muscle cells via fractalkine CX3CR1 interaction. *FASEB J.*, **24**(1), 81–92 (2010).
38. J. D. Peacock, Y. Lu, M. Koch, K. E. Kadler, and J. Lincoln, Temporal and spatial expression of collagens during murine atrioventricular heart valve development and maintenance. *Dev. Dynam.*, **237**(10), 3051–3058 (2008).
39. T. D. Camenisch, J. A. Schroeder, J. Bradley, S. E. Klewer, and J. A. McDonald, Heart-valve mesenchyme formation is dependent on hyaluronan-augmented activation of ErbB2-ErbB3 receptors. *Nat. Med.*, **8**(8), 850–855.
40. R. Langer and J. P. Vacanti, Tissue engineering, *Science*, **260**(5110), 920–926 (1993).

41. S. P. Hoerstrup, R. Sodian, S. Daebritz, J. Wang, E. A. Bacha, D. P. Martin *et al.*, Functional living trileaflet heart valves grown *in vitro*. *Circulation*, **102**(19 Suppl 3), II44–II49 (2000).
42. T. Shinoka, C. K. Breuer, R. E. Tanel, G. Zund, T. Miura, P. X. Ma *et al.*, Tissue engineering heart valves: Valve leaflet replacement study in a lamb model. *Ann. Thorac. Surg.*, **60**(6 Suppl), S513–S516 (1995).
43. G. Zund, C. K. Breuer, T. Shinoka, P. X. Ma, R. Langer, J. E. Mayer *et al.*, The *in vitro* construction of a tissue engineered bioprosthetic heart valve. *Eur. J. Cardiothorac. Surg.*, **11**(3), 493–497 (1997).
44. R. Sodian, S. P. Hoerstrup, J. S. Sperling, D. P. Martin, S. Daebritz, Jr. J. E. Mayer *et al.*, Evaluation of biodegradable, three-dimensional matrices for tissue engineering of heart valves. *ASAIO J.*, **46**(1), 107–110 (2000).
45. D. Schmidt, J. Achermann, B. Odermatt, C. Breymann, A. Mol, M. Genoni *et al.*, Prenatally fabricated autologous human living heart valves based on amniotic fluid derived progenitor cells as single cell source. *Circulation*, **116**(11 Suppl), I64–I70 (2007).
46. G. C. Engelmayr, Jr. D. K. Hildebrand, F. W. Sutherland, J. E. Mayer, and Jr. M. S. Sacks, A novel bioreactor for the dynamic flexural stimulation of tissue engineered heart valve biomaterials. *Biomaterials*, **24**(14), 2523–2532 (2003).
47. S. Nemir, H. N. Hayenga, and J. E. West, PEGDA hydrogels with patterned elasticity: Novel tools for the study of cell response to substrate rigidity. Biotechnology and bioengineering, **105**(3), 636–644 (2010).
48. S. Nemir and J. L. West, Synthetic materials in the study of cell response to substrate rigidity. *Ann. Biomed. Eng.*, **38**(1), 2–20 (2010).
49. M. C. Cushing and K. S. Anseth, Materials science. Hydrogel cell cultures. *Science*, **316**(5828), 1133–1134 (2007).
50. J. A. Benton, B. D. Fairbanks, and K. S. Anseth, Characterization of valvular interstitial cell function in three dimensional matrix metalloproteinase degradable PEG hydrogels. *Biomaterials*, **30**(34), 6593–6603 (2009).
51. J. A. Benton, H. B. Kern, and K. S. Anseth, Substrate properties influence calcification in valvular interstitial cell culture. *J. Heart Valve Dis.*, **7**(6), 689–699 (2008).
52. A. M. Kloxin, A. M. Kasko, C. N. Salinas, and K. S. Anseth, Photodegradable hydrogels for dynamic tuning of physical and chemical properties. *Science*, **324**(5923), 59–63 (2009).
53. L. M. Weber and K. S. Anseth, Hydrogel encapsulation environments functionalized with extracellular matrix interactions increase islet insulin secretion. *Matrix Biol.*, **27**(8), 667–673 (2008).
54. M. P. Cuchiara, A. C. Allen, T. M. Chen, J. S. Miller, and J. S. West, Multilayer microfluidic PEGDA hydrogels. *Biomaterials*, **31**(21), 5491–5497 (2010).
55. J. S. Temenoff, K. A. Athanasiou, R. G. LeBaron, and A. G. Mikos, Effect of poly (ethylene glycol) molecular weight on tensile and swelling properties of oligo(poly(ethylene glycol) fumarate) hydrogels for cartilage tissue engineering. *Journal of biomedical materials research*, **59**(3), 429–437 (2002).
56. M. S. Cushing, J. T. Liao, M. P. Jaeggli, and K. S. Anseth, Material-based regulation of the myofibroblast phenotype. *Biomaterials*, **28**(23), 3378–3387 (2007).

57. A. M. Kloxin, J. A. Benton, and K. S. Anseth, *In situ* elasticity modulation with dynamic substrates to direct cell phenotype. *Biomaterials*, **31**(1), 1–8 (2010).
58. A. M. Kloxin, M. W. Tibbitt, A. M. Kasko, J. A. Fairbairn, and K. S. Anseth, Tunable hydrogels for external manipulation of cellular microenvironments through controlled photodegradation. *Adv. Mater.*, **22**(1), 61–66 (2010).
59. K. S. Masters, D. N. Shah, L. A. Leinwand, and K. S. Anseth, Crosslinked hyaluronan scaffolds as a biologically active carrier for valvular interstitial cells. *Biomaterials*, **26**(15), 2517–2525 (2005).
60. A. Ramamurthi and I. Vesely, Evaluation of the matrix-synthesis potential of crosslinked hyaluronan gels for tissue engineering of aortic heart valves. *Biomaterials*, **26**(9), 999–1010 (2005).
61. T. C. Flanagan, B. Wilkins, A. Black, S. Jockenhoevel, T. J. Smith, A. S. Pandit, A collagen-glycosaminoglycan co-culture model for heart valve tissue engineering applications. *Biomaterials*, **27**(10), 2233–2246 (2006).
62. K. S. Masters, D. N. Shah, G. Walker, L. A. Leinwand, and K. S. Anseth, Designing scaffolds for valvular interstitial cells: Cell adhesion and function on naturally derived materials. *J. Biomed. Mater. Res. Pt. A*, **71**(1), 172–180 (2004).
63. S. Suri and C. E. Schmidt, Photopatterned collagen-hyaluronic acid interpenetrating polymer network hydrogels. *Acta Biomater.*, **5**(7), 2385–2397 (2009).
64. K. J. Rodriguez, L. M. Piechura, and K. S. Masters, Regulation of valvular interstitial cell phenotype and function by hyaluronic acid in 2-D and 3-D culture environments. *Matrix Biol.*, **30**(1), 70–82 (2011).
65. M. E. Tedder, J. Liao, B. Weed, C. Stabler, H. Zhang, A. Simionescu *et al.*, Stabilized collagen scaffolds for heart valve tissue engineering. *Tissue Eng. Pt. A*, **15**(6), 1257–1268 (2009).
66. M. D. Brigham, A. Bick, E. Lo, A. Bendali, J. A. Burdick, A. Khademhosseini, Mechanically robust and bioadhesive collagen and photocrosslinkable hyaluronic acid semi-interpenetrating networks. *Tissue Eng. Pt. A*, **15**(7), 1645–1653 (2009).
67. S. T. Kreger and S. L. Voytik-Harbin, Hyaluronan concentration within a 3D collagen matrix modulates matrix viscoelasticity, but not fibroblast response. *Matrix Biol.*, **28**(6), 336–346 (2009).
68. K. Hayashida, K. Kanda, H. Yaku, J. Ando, and Y. Nakayama. Development of an *in vivo* tissue-engineered, autologous heart valve (the biovalve): Preparation of a prototype model. *J. Thorac. Cardiovasc. Surg.*, **134**(1), 152–159 (2007).
69. M. Yamanami, Y. Yahata, M. Uechi, M. Fujiwara, H. Ishibashi-Ueda, K. Kanda *et al.*, Development of a completely autologous valved conduit with the sinus of Valsalva using in-body tissue architecture technology: A pilot study in pulmonary valve replacement in a beagle model. *Circulation*, **122**(11 Suppl), S100–S106.
70. J. A. Benton, C. A. DeForest, V. Vivekanandan, and K. S. Anseth, Photocrosslinking of gelatin macromers to synthesize porous hydrogels that promote valvular interstitial cell function. *Tissue Eng. Pt. A*, **15**(11), 3221–3230 (2009).
71. Q. Ye, G. Zund, P. Benedikt, S. Jockenhoevel, S. P. Hoerstrup, S. Sakyama *et al.*, Fibrin gel as a three dimensional matrix in cardiovascular tissue engineering. *Eur. J. Cardiothorac. Surg.*, **17**(5), 587–591 (2000).

72. S. Jockenhoevel, G. Zund, S. P. Hoerstrup, K. Chalabi, J. S. Sachweh, L. Demircan *et al.*, Fibrin gel — advantages of a new scaffold in cardiovascular tissue engineering. *Eur. J. Cardiothorac. Surg.*, **19**(4) 424–430 (2001).
73. Z. H. Syedain, J. S. Weinberg, and R. T. Tranquillo, Cyclic distension of fibrin-based tissue constructs: Evidence of adaptation during growth of engineered connective tissue. *Proceedings of the National Academy of Sciences of the United States of America*, **105**(18), 6537–6542 (2008).
74. S. J. Hollister, Porous scaffold design for tissue engineering. *Nat. Mater.*, **4**(7), 518–524 (2005).
75. B. Duan, L. A. Hockaday, K. H. Kang, J. T. Butcher, 3D bioprinting of heterogeneous aortic valve conduits with alginate/gelatin hydrogels. *J. Biomed. Mater. Res. Pt. A*, **101**(5), 1255–1264 (2013).
76. D. B. Kolesky, R. L. Truby, A. S. Gladman, T. A. Busbee, K. A. Homan, and J. A. Lewis, 3D Bioprinting of Vascularized, Heterogeneous Cell-Laden Tissue Constructs. *Adv. Mater.*, **26**(19), 3124–3130 (2014).
77. C. M. Kirschner, D. L. Alge, S. T. Gould, and K. S. Anseth, Clickable, photodegradable hydrogels to dynamically modulate valvular interstitial cell phenotype. *Adv. Health. Mater.*, **3**(5), 649–657 (2014).
78. A. M. Quinlan and K. L. Billiar, Investigating the role of substrate stiffness in the persistence of valvular interstitial cell activation. *J. Biomed. Mater. Res. Pt. A*, **100**(9), 2474–2482 (2012).
79. S. T. Gould, E. E. Matherly, J. N. Smith, D. D. Heistad, and K. S. Anseth, The role of valvular endothelial cell paracrine signaling and matrix elasticity on valvular interstitial cell activation. *Biomaterials*, **35**(11), 3596–3606 (2014).
80. N. M. Rajamannan, F. J. Evans, E. Aikawa, K. J. Grande-Allen, L. L. Demer, D. D. Heistad *et al.*, Calcific aortic valve disease: Not simply a degenerative process: A review and agenda for research from the National Heart and Lung and Blood Institute Aortic Stenosis Working Group. Executive summary: Calcific aortic valve disease-2011 update. *Circulation*, **124**(16), 1783–1791 (2011).
81. J. D. Hutcheson, E. Aikawa, and W. D. Merryman, Potential drug targets for calcific aortic valve disease. *Nat. Rev. Cardiol.*, **11**(4), 218–231 (2014).
82. W. D. Merryman, What modulates the aortic valve interstitial cell phenotype? *Future Cardiol.*, **4**(3), 247–252 (2008).
83. H. Wang, M. W. Tibbitt, S. J. Langer, L. A. Leinwand, and K. S. Anseth, Hydrogels preserve native phenotypes of valvular fibroblasts through an elasticity-regulated PI3K/AKT pathway. *Proceedings of the National Academy of Sciences of the United States of America*, **110**(48), 19336–19341 (2013).
84. S. T. Gould and K. S. Anseth, Role of cell-matrix interactions on VIC phenotype and tissue deposition in 3D PEG hydrogels. *J. Tissue Eng. Regen. Med.*, in press. Available online, Oct. 16, 2013.

Chapter 13

Hydrogels in Vascular Tissue Engineering

Y. J. Blinder,[*,‡] **D. J. Mooney**[†,‡] **and S. Levenberg**[*]

[*] *Technion — Israel Institute of Technology, Israel*

[†] *School of Engineering and Applied Sciences, Harvard University*

[‡] *Wyss Institute for Biologically Inspired Engineering at Harvard University*

Engineering vascular tissue is one of the primary challenges of regenerative medicine, for which hydrogels represent a fundamental and extremely useful set of tools. Understanding the specific requirements necessary for a hydrogel to support vascular development is crucial for the development of new approaches in tissue engineering. This chapter reviews the various hydrogels used to address this challenge, as well as basic principles of new blood vessel network formation and recent developments in hydrogel applications for vascular tissue engineering.

1. Introduction

A major goal of regenerative medicine research is to engineer functional biological substitutes for diseased or injured tissues and organs. The voluminous nature of organs, as well as their complex internal structure, requires an adequate network of blood vessels to serve as an infrastructure for nutrients and oxygen supply from the blood to the organs. This same network is also crucial for the removal of metabolic waste products from the organs. Engineered tissue constructs must therefore also have a well-developed, mature vascular network as an integral component of their functional structure. If formed before transplantation, pre-engineered blood vessel networks must be able to integrate with the native vascular network of the host, thereby enabling the flow of blood through the construct. Long-term viability of these engineered tissue-constructs after implantation will rely on the ability of the engineered vascular network to sustain the transport of oxygen and nutrients throughout the construct.

Hydrogels are used in vascular tissue engineering for an array of different roles; they can serve as a provisional extracellular matrix (ECM) during new blood vessel network formation, provide cells with mechanical support while allowing cell adhesion and migration, act as an injectable cell carrier, or be modified to exhibit controlled release of growth factors (GFs) for induction of new blood vessel network formation. Naturally, not all hydrogels are equally suited for these roles. This chapter contains a review of the most commonly used hydrogels in vascular tissue engineering today. Next, the chapter covers several basic aspects of vascular biology as a necessary foundation for approaching vascular tissue engineering. Lastly, the chapter reviews recent advances in the application of hydrogels for vascular tissue engineering.

2. Hydrogels for Vascular Tissue Engineering

Of the great variety of hydrogels that can be used for tissue engineering and regeneration, a subset of these gels has been shown to be more effective for vascular tissue engineering. For some gels, this effectiveness can be attributed to the putative roles of their constituents in normal neovascular formation processes, while others owe their effectiveness to their ability to be chemically modified or deliver certain blood vessel promoting domains. The following is a description of those hydrogels most commonly used in blood vessel engineering today.

2.1. Fibrin

Fibrin is a natural biopolymer with important roles in natural neovascular processes such as wound healing. During early wound healing, fibrin gel is spontaneously formed and acts as a primary hemostat, clotting the wound and recruiting neovasculature. Fibrin gel contains cell binding motifs that allow cellular migration and promote new blood vessel formation throughout the gel, making it a prime candidate for use in tissue engineering of vascularized tissues.[1–3] Certain commercially produced fibrin gels have received FDA approval for use as surgical sealants.[4]

Fibrin gel is formed when the protein fibrinogen interacts with the protease thrombin and polymerizes into a clear, highly porous, and permeable gel with relatively low mechanical stiffness. Both fibrinogen and thrombin can be isolated from bovine and human plasma and are commercially available at varying degrees of purity.[4] When forming fibrin gel *in vitro*, the initial concentrations of fibrinogen and thrombin affect the gel formation in different ways; as thrombin acts as the catalyzing enzyme for polymerization, increasing its initial concentration will affect the rate of gel formation, while increasing the initial concentration of fibrinogen will change the gel morphology *vis-à-vis* its porosity and density, as well as increase its stiffness. Both factors have been shown to influence neovascular morphology.[5] Cells embedded within fibrin gel naturally degrade the gel by enzymatic proteolysis, causing bulk erosion. For this reason, it is not uncommon for fibrin gel preparations to contain a protease inhibitor such as aprotinin.[1–5]

2.2. Matrigel

Matrigel is a commercially available combination of ECM proteins and GFs extracted from the Engelbreth-Holm-Swarm (EHS) mouse sarcoma. Its main components are laminin, collagen IV, and enactin, and it is defined as a "solubilized basement membrane preparation" (BD). Gel formation from Matrigel is thermally triggered — it is liquid at 4°C and gels at 37°C. This ease of use, as well as the intrinsic ability of Matrigel to promote cell adhesion and the formation of new blood vessels has made it very prevalent in tissue engineering applications, particularly in angiogenesis assays and vascular tissue engineering both *in vitro* and *in vivo*. In addition to the aforementioned ECM proteins, Matrigel naturally contains several GFs such as basic fibroblast growth factor (bFGF), epidermal growth factor (EGF), interleukin factor 1 (ILF1), transforming growth factor beta (TGF-β), platelet-derived growth factor (PDGF), and nerve growth factor (NGF). A "Growth-Factor Reduced" version of Matrigel is also available which is more suitable for applications where a higher biochemical definition is required.[6]

While its biochemical composition makes Matrigel a highly relevant candidate for vascular research, concerns regarding its origin combined with large batch-to-batch variability make it unsuitable for clinical use.

2.3. Hyaluronan/Hyaluronic Acid (HA)

HA is a natural glycosaminoglycan, and a major constituent of the ECM in connective tissues. HA is a linear polysaccharide composed of repeating non-sulfonated, anionic disaccharide units of β(1,4)-D glucuronic acid-β(1,3) N-acetyl-D-glucosamine. Due to its natural biocompatibility, non-animal source HA has been approved by the FDA for a number of applications.

The effects of HA on neovascularization vary with molecular weight. In native, uninjured tissue, high-molecular-weight HA or native HA exhibits stabilizing properties, inhibiting immune response, inflammation, and angiogenesis. Degradation of native HA by hyaluronidases into low-molecular-weight fragments of HA (often referred to as oligosaccharides of hyaluronan or o-HA) results in stimulation of the immune response and inflammation, as well as promotion of angiogenesis. In vascular tissue engineering, HA has been used both as a hydrogel scaffold and as a soluble angiogenic promoter.[7–9]

2.4. Alginate

Alginate is a linear block copolymer of D-mannuronic acid (M) and L-guluronic acid (G) saccharide residues, naturally derived from brown seaweed algae. The M and G residues can combine to form polymer chains of varying composition and distribution (i.e., containing blocks of consecutive G residues, blocks of consecutive M residues and blocks of alternating M and G residues) along the polymer chain.

Alginate stiffness and degradation rates can be controlled by varying the residues' molecular weight and crosslinking method. Crosslinking of alginate polymer chains can be achieved in several different ways, including ionic and covalent crosslinking, using thermal- and photo-crosslinking, or modifying the gel to contain cell binding (Arg–Gly–Asp) RGD motifs, essentially creating a cell mediated crosslinking mechanism. Each of these methods results in hydrogels with varying mechanical properties. Ionic crosslinking is the most common method; when reconstituted in aqueous solution, carboxylic acid groups on the G residues can form ionic bonds with divalent cations such as calcium and barium, allowing the crosslinking of alginate polymer chains and resulting in a biocompatible, hydrophilic, non-immunogenic hydrogel with tunable physical and chemical properties. As alginate can be modified relatively easily during gel preparation, it has been widely studied as a vehicle for controlled delivery. Additionally, alginate can form reversible binding with heparin-binding GFs such as vascular endothelial growth factors (VEGF) and bFGF, allowing a more sustained release profile compared to other proteins and making it particularly interesting for vascular regeneration applications.[10] Alginate has been approved for clinical use as a passive material for wound healing applications. Future clinical application will likely make use of the versatile nature of alginate for potentially more active roles.[10]

2.5. Polyethylene glycol (PEG)

PEG is a synthetic polymer used in a wide variety of biomedical applications. On its own, PEG hydrogel does not support proper tissue formation due to its lack of adhesion domains. However, the relative ease with which PEG can be conjugated to certain proteins such as fibrinogen and collagen, functionalized with RGD (integrin-binding) domains and modified to enable cell mediated degradation enables the production of hybrid biosynthetic hydrogels which can promote cell adhesion, migration and tissue morphogenesis while enabling a high level of control over the mechanical properties of the gel.[11]

3. Basic Principles of New Blood Vessel Network Formation

Familiarization with the key players and processes involved in natural neovascular formation is an essential first step for a methodological approach to engineer new blood vessel networks. The following is an introduction to these fundamental concepts.

3.1. Vasculogenesis, angiogenesis, and other mechanisms of neovascular formation

Vasculogenesis refers to neovascular formation *de-novo* (i.e., independent of pre-existing vasculature) by endothelial progenitor cells, which occurs during embryonic development (embryonic vasculogenesis) and during normal wound healing in the adult (adult vasculogenesis).[12] During both embryonic and adult vasculogenesis, endothelial progenitors migrate and differentiate into endothelial cells (ECs), and interact to form a vascular plexus.[13] These newly formed microvessels then recruit mesenchymal cells which act as pericytes, forming a supporting layer around the new blood vessels. Embryonic vasculogenesis is carried out by angioblast-derived ECs, and the pericytes are derived from mesenchymal stem cells (MSCs). In adult vasculogenesis, circulating bone marrow-derived endothelial progenitor cells are recruited to the wound site. The pericytes in this case are most probably derived either from fibroblasts in the ECM or bone marrow-derived mesenchymal cells.[12] Extraembryonic blood vessels develop from mesenchyme angioblasts in the blood islands of the yolk sac, though some debate currently exists as to whether these precursors can technically be defined as hemangioblasts.[14]

Angiogenesis is the main mechanism by which new blood vessels are formed in the adult. Angiogenesis differs from vasculogenesis in that it typically involves neovascular formation by sprouting of ECs from pre-existing mature blood vessels (a process often referred to as "sprouting angiogenesis"). Sprouting angiogenesis has several stages; it begins when ECs in a blood vessel respond to a pro-angiogenic signal (e.g., a VEGF gradient in the surrounding tissue) by breaking the basement membrane they share with the mural cells.

Next, ECs compete with each other for the tip cell position using a lateral inhibition mechanism.[15] The selected tip cell begins the sprouting process, supported from behind by proliferating ECs acting as "stalk cells". ECs at the leading edge shuffle as they compete for the tip cell position, all the while growing the sprout according to the pro- and anti-angiogenic signals in the tissue. When tip cells reach other existing blood vessels they can form anastomoses. If the anastomosis is successful, the newly formed blood vessel matures by recruiting new pericytes; if it is unsuccessful, the sprout will undergo pruning and regress.[13]

Additional mechanisms of new blood vessel formation are mainly attributed to pathological conditions such as those found in cancer tumors, where the tumor cells can recruit blood vessels from their surrounding tissues, co-opt nearby blood vessels or even mimic vascular structures by differentiation of tumor stem-like cells to ECs. However, tumor blood vessels are generally abnormal in shape and function, forming a disorganized, leaky network.[13,16]

3.2. Principal players in molecular mechanisms governing neovascular formation

Vasculogenesis and angiogenesis require an elaborate system of biochemical and biophysical cell–cell and cell–matrix interactions. Several aspects of the governing molecular mechanisms in neovascular formation have been elucidated over the past four decades, and many more are the subject of ongoing research. Cell–cell interactions during neovascular formation are mediated by a large collection of different chemokines. A list of the most significant molecular interactions for neovascularization includes the family of VEGFs and their receptors (VEGFRs), angiopoietins and their Tie receptors, platelet-derived growth factor beta and its receptor (PDGF-BB and PDGFBR, respectively), the fibroblast growth factor (FGF) family, Notch signaling via the delta-like 4 (Dll4) ligand, and the TGF-β superfamily, among others (See Table 1). Cell–matrix interactions are also of great interest in vascular biology, as these interactions play significant roles in vascular morphogenesis. In particular, the role of EC-secreted matrix metalloproteinases (MMPs) during the early stages of angiogenic sprouting has been shown to be the principal mechanisms by which ECs break up their surrounding ECM prior to sprouting, and is therefore of great relevance when designing hydrogels for vascular tissue engineering.[9,17,18] Cell–matrix interactions also strongly influence aspects of vascular morphogenesis such as sprouting, assembly into cords, vacuole and lumen formation through a variety of ECM-integrin-cytoskeletal interactions.[19]

Table 1. **Important chemokines in vascular morphogenesis.**

Chemokine	Function
VEGFs and VEGFRs	The most studied family of chemokine interactions in neovascular formation. VEGFs play a role in several mechanisms by affecting EC motility, proliferation, and survival, as well as participating in several chemokine pathways.
Angiopoietins and Tie receptors	Ang1 and Ang2 are opposing factors that act at different stages of vascular development. Ang2 enables early sprouting angiogenesis by destabilizing EC-pericyte interactions, while Ang1 stabilizes these interactions at later stages of vessel maturation.
PDGF-BB and PDGFR	PDGF-BB activates and recruits pericytes during blood vessel maturation.
SDF-1	Involved in recruitment of circulating EC and hematopoietic stem cell to sites of neovascularization.
FGF family	Associated with VEGF-dependent vasculogenic responses during wound healing.
Notch-Dll4 signaling	Regulate tip cell selection during angiogenic sprouting.
TGFb superfamily	Implicated in a wide variety of developmental and morphogenic processes, including VEGF-dependent and independent neovascularization, pericyte recruitment, and differentiation.

4. Recent Advances in Hydrogel Application for Vascular Tissue Engineerings

Hydrogels serve as fundamental tools in various experimental protocols for vascular research and tissue engineering. While some protocols have been well-established in the literature, new protocols are continually being developed. The following is a short review of the more prominent protocols and advancements in this field in recent years.

One example of an established hydrogel-based assay for vascular biology research is the so-called 'embryoid body (EB) sprouting' assay. This assay involves culturing a cluster of embryonic stem cells or EB embedded within a collagen gel, and adding specific exogenous GFs of interest. This microenvironment has been shown to be conducive to endothelial differentiation in the EB followed by angiogenic sprouting into the surrounding gel, allowing researchers to study mechanisms such as VEGFR signal transduction,[20] specific local guidance cues during sprouting,[21] and the dynamic shuffling of tip cells at the leading edge of angiogenic sprouts.[15]

Another well-established example of a hydrogel-based *in vitro* assay for studying endothelial sprouting is the sprouting bead assay, originally described by Nehls and Drenckhahn[22] and later optimized by Nakatsu *et al.*[23] The method involves coating microcarrier beads with ECs and then embedding the EC-coated beads inside a fibrin gel within a multiwell plate. Under appropriate culturing conditions, the ECs then sprout outward from the bead surfaces and into the gel volume. As fibrin is highly permeable, different factors can be added to the media on top of the hydrogel, and their effect on EC behavior can be studied using microscopy. Another approach is to seed a second cell type on top of the fibrin gel, allowing the formation of a naturally expressed cytokine gradient throughout the hydrogel. The most prominent example is seeding fibroblasts on top of the hydrogel, which enables one to differentiate between cell–cell contact-dependent effects and soluble cytokine-dependent effects, as well as fibrin matrix density-dependent effects.[24] More recently, human umbilical vein endothelial cells (HUVECs) and neonatal pericytes were co-cultured on microcarrier beads to form a combined angiogenic sprouting/pericyte detachment assay. The cell-coated beads were suspended in fibrin gel and subjected to a variety of protocols for temporally-controlled delivery of VEGF, Ang2, PDGF-BB, and Ang1.[25] Both the EB sprouting assay and the sprouting bead assay are exceptionally useful tools for studying basic EC behaviors such as tip cell selection during sprouting angiogenesis; however they are limited in that the vessels are not perfusable.

While fibrin gel has many advantages as a culturing matrix during neovascularization, the gel alone lacks the mechanical strength to support a stable large scale engineered tissue construct. Cells cultured within fibrin gel exert forces that lead to substantial gel shrinkage after several days, resulting in inadequate tissue formation. To overcome this obstacle, Lesman *et al.* used a macroporous poly-lactic-*co*-glycolic acid (PLLA/PLGA) scaffold for mechanical support, seeding the fibrin gel with a co-culture of HUVECs and human normal skin dermal fibroblasts (HNDFs).[26] When cultured *in vitro*, the co-cultured cells successfully formed new blood vessel networks inside the engineered construct. Upon implantation in nude mice, the vascularized constructs were able to integrate with the host vasculature, with improved results compared to constructs without the added synthetic scaffold and constructs with no cells.[26]

Insights regarding the different roles of cytokines and GFs at different stages of vascular morphogenesis play an important part in the development of new and modified scaffolds and hydrogels for vascular tissue engineering. For example, Hanjaya-Putra *et al.* have developed proangiogenic hydrogels using modified acrylated hyaluric acid (AHA) precursors. AHA precursors conjugated with integrin-binding peptides were polymerized using an MMP-sensitive crosslinker. The resulting hydrogels were supplemented with soluble VEGF, bFGF, and Ang1 as proangiogenic factors, as well as SDF-1 and TNF-α to ensure optimal MMP secretion needed for endothelial sprouting and invasion. Dosage of RGD was shown to regulate vacuole and lumen formation, and the MMP-sensitive crosslinker was necessary for endothelial colony-forming endothelial cell (ECFC) migration.[9] More recently, Hanjaya-Putra *et al.* used ultraviolet (UV) light to locally enhance crosslinking in AHA hydrogels, creating spatially defined areas in the hydrogel where angiogenesis was inhibited.[17]

Alginate has been successfully used as a vehicle for spatially- and temporally-controlled delivery of angiogenic factors. Often intended for use as an injectable therapeutic, these alginate gels can be loaded with different GFs, which may exhibit differential release kinetics based on their specific interactions with the hydrogel. Several strategies have been developed for sequential release of VEGF and PDGF-BB from alginate hydrogels. In one approach by Hao *et al.*, PLG microspheres were used to encapsulate PDGF-BB before incorporation into alginate gels containing free VEGF, allowing the delayed release of PDGF. Sequential or delayed release of PDGF after VEGF was also achieved when both GFs were encapsulated in the hydrogel in free form, due to the greater heparin binding of PDGF.[27]

New hydrogels are constantly being developed and investigated for their ability to properly support or induce neovascular formation. One prominent example of a novel hydrogel which has gained increasing interest in recent years is photocrosslinkable methacrylated gelatin (GelMA). GelMA is synthesized by mixing gelatin with methacrylic anhydride (MA). The concentration of the added MA determines the degree of methacrylation, which in turn determines the mechanical properties of the hydrogel. Addition of a photoinitiator allows the hydrogel to be polymerized by exposure to UV at room temperature. In a recent paper, Chen *et al.* showed that ECFCs co-cultured with bone marrow-derived MSCs within photocrosslinkable GelMA successfully formed microvascular networks throughout the gel. These constructs were also shown to successfully anastomose to host vasculature when implanted in nude mice. Lastly, the group showed that modifying the degree of methacrylation can affect the amount of endothelial sprouting as well as the degree of perivascular interaction.[28] A number of groups have also shown similar success by combining hydrogels known to support neovascular formation to produce hybrid materials, such as fibrin-PEG,[29] collagen-PEG,[30] and collagen-fibrin,[31] among others.

4.1. Microfabricated devices for vascular research

In order to recapitulate vascular phenotypes and behavior *in vitro* in a highly repeatable, controllable, and accurate manner, researchers have turned to microfabrication techniques, similar to those originally developed for microchip fabrication. Using methods such as soft lithography, microfluidic channels can be designed and implemented within elastomeric substrates, an approach pioneered over a decade ago by

J. Vacanti.[32] The most commonly used elastomer for microfluidic applications is polydimethylsiloxane (PDMS), which is silicone-based, bio-inert, and transparent, making it an exceedingly easy material to work with and a prime candidate for biological applications.

Microfabricated devices for vascular research generally contain a hydrogel domain within the PDMS frame, which serves as the ECM into which new blood vessels can sprout. Several approaches have been developed for the hydrogel domain in recent years. Using soft lithography, Zheng *et al.* designed a device containing a perfusable network of HUVEC-lined microvessels within a type I collagen gel. The gel itself was seeded with human brain vascular pericytes (HBVPCs). The HUVEC-lined microvessels were able to sprout into the hydrogel volume and interact with the pericytes, which assembled to form a perivascular layer.[33, 34] Chan *et al.* combined alginate beads within a collagen gel to form a domain into which ECs could sprout within a microfluidic chamber. The alginate beads served to encapsulate GF-secreting cells (human dermal fibroblasts and a human fibrosarcoma cell line were studied) as a proangiogenic stimulus. ECs migrated and sprouted from adjacent microfluidic channels into the hydrogel domain, forming an interconnected 3D capillary bed around the alginate beads.[35] Baker *et al.* used gelatin to form parallel channels within a collagen gel. Once the collagen gel had polymerized around the channels, the gelatin could be extracted, forming perfusable microchannels which were later coated with ECs.[36] In a later paper, Nguyen *et al.* designed a microfluidic device in which the channels were formed by placing two parallel acupuncture needles and forming a collagen gel around them. The needles were then extracted, leaving behind hollow perfusable tubes of 400 μm diameter. One of the channels was then coated with ECs, and the second channel was used to create a spatial gradient of GFs of interest.[37] Using a slightly different microfluidic design, Cuchiara *et al.* have recently shown a similar approach using PEG-based hydrogels, which were modified to contained both RGDS (integrin-binding) domains and MMP-cleavable sites.[38]

Due to the recent technological developments in rapid prototyping and 3D printing techniques, a great interest has developed for applying these increasingly accessible technologies to manufacture engineered tissue, also referred to as "bioprinting". Theoretically, 3D-printed technologies will enable the manufacture of complex 3D architectures in engineered tissue such as those of blood vessel networks, which is a particular area where current microfabrication techniques fall short. One promising approach has been displayed by Wu *et al.*, where "omnidirectional" vascular structures are 3D printed within a variety of gels.[39] In a different approach, Miller *et al.* have applied an open-source approach to rapid manufacturing of vascularized engineered tissue, using a modified MakerBot 3D printer to create patterned channels made of a carbohydrate-based glass. Once printed, these patterned lattices were coated with poly(D-lactide-*co*-glycolide) and embedded within a variety of hydrogels including fibrin, alginate, matrigel, PEG and agarose, and the glass was then allowed to dissolve and removed. The resulting networks were perfusable and interconnected, supported blood flow, and enabled the survival of cells seeded within the surrounding gel.[40]

5. Conclusions

Hydrogels represent an essential element in many vascular tissue engineering endeavors, from basic sprouting assays *in vitro* to the formation of implantable tissue replacements for regenerative medicine. Vascular biology research provides important insights regarding molecular cues needed for proper neovascular formation. Such cues can be incorporated into the design of smarter hydrogels, allowing spatio-temporal control over processes of vascular tissue morphogenesis within the gels. Current research efforts continue to develop and optimize these hydrogels, as well as new microfabricated systems that utilize these hydrogels to recapitulate increasingly accurate vascular phenotypes *in vitro*.[41]

As vascular tissue engineering develops toward more clinically relevant applications, the field of hydrogel design is expected to see continued efforts toward developing gels which allow controlled release of chemical cues, while allowing a high level of control over the mechanical microenvironment to which the cells are exposed.

References

1. J. Weisel, *Adv. Protein Chem.*, **70**, 247–299 (2005).
2. P. Janmey, J. Winer, and J. Weisel, *J. R. Soc. Interface*, **6**, 1–10 (2009).
3. F. Shaikh, A. Callanan, E. Kavanagh, P. Burke, P. Grace, and T. McGloughlin, *Cells Tissues Organs*, **188**, 333–346 (2008).
4. W. Spotnitz, *World J. Surg.*, **34**, 632–634 (2010).
5. V. van Hinsbergh, A. Collen, and P. Koolwijk, *Ann. N. Y. Acad. Sci.*, **936**, 426–437 (2001).
6. C. Hughes, L. Postovit, and G. Lajoie, *Proteomics*, **10**, 1886–1890 (2010).
7. Y. Wang, M. Cao, Y. Liu, Y. He, C. Yang, and F. Gao, *Exp. Biol. Med.*, **236**, 84–90 (2011).
8. M. Slevin, *Int. J. Oncol.*, **35**, 761–773 (2009).
9. D. Hanjaya-Putra, V. Bose, Y.-I. Shen, J. Yee, S. Khetan, K. Fox-Talbot, C. Steenbergen, J. Burdick, and S. Gerecht, *Blood*, **118**, 804–815 (2011).
10. K. Lee and D. Mooney, *Prog. Polym. Sci.*, **37,** 106–126 (2012).
11. D. Seliktar, *Science*, **336**, 1124–1128 (2012).
12. V. Wong and J. Crawford, *Bio. Med. Res. Int.*, **2013**, 190486 (2013).
13. M. Potente, H. Gerhardt, and P. Carmeliet, *Cell*, **146**, 873–887 (2011).
14. S. Nishikawa, Wiley interdisciplinary reviews. *Dev. Biol.*, **1**, 603–608 (2012).

15. L. Jakobsson *et al.*, *Nat. Cell Biol.*, **12**, 943–953 (2010).
16. R. Jain, *Nat. Med.*, **9**, 685–693 (2003).
17. J. Rundhaug, *J. Cell. Mol. Med.*, **9**, 267–285 (2005).
18. D. Hanjaya-Putra, K. Wong, K. Hirotsu, S. Khetan, J. Burdick, and S. Gerecht, *Biomaterials*, **33**, 6123–6131 (2012).
19. D. Senger and G. Davis, *Cold Spring Harb. Perspect. Biol.*, **3** (2011).
20. X. Li, L. Claesson-Welsh, and M. Shibuya, *Methods Enzymol.*, **443**, 261–284 (2008).
21. J. Chappell, S. Taylor, N. Ferrara, and V. Bautch, *Dev. Cell*, **17**, 377–386 (2009).
22. V. Nehls and D. Drenckhahn, *Histochem. Cell Biol.*, **104**, 459–466 (1995).
23. M. N. Nakatsu, J. Davis, and C. C. W. Hughes, Optimized Fibrin Gel Bead Assay for the Study of Angiogenesis. **186**(3) (2007).
24. C. Ghajar, X. Chen, J. Harris, V. Suresh, C. Hughes, N. Jeon, A. Putnam, and S. George, *Biophys. J.*, **94**, 1930–1941 (2004).
25. Y. Brudno, A. Ennett-Shepard, R. Chen, M. Aizenberg, and D. Mooney, *Biomaterials*, **34**, 9201–9209 (2012).
26. A. Lesman, J. Koffler, R. Atlas, Y. Blinder, Z. Kam, and S. Levenberg, *Biomaterials*, **32**, 7856–7869 (2011).
27. X. Hao, E. Silva, A. Månsson-Broberg, K.-H. Grinnemo, A. Siddiqui, G. Dellgren, E. Wärdell, L. Brodin, D. Mooney, and C. Sylvén, *Cardiovasc. Res.*, **75**, 178–185 (2007).
28. Y.-C. Chen, R.-Z. Lin, H. Qi, Y. Yang, H. Bae, J. Melero-Martin, and A. Khademhosseini, *Adv. Funct. Mater.*, **22**, 2027–2039 (2012).
29. B. Jiang, T. Waller, J. Larson, A. Appel, and E. Brey, *Tissue Eng. Pt. A*, **19**, 224–234 (2013).
30. R. Singh, D. Seliktar, and A. Putnam, *Biomaterials*, **34**, 9331–9340 (2013).
31. R. Rao, A. Peterson, J. Ceccarelli, A. Putnam, and J. Stegemann, *Angiogenesis*, **15**, 253–264 (2012).
32. S. Kaihara, J. Borenstein, R. Koka, S. Lalan, E. Ochoa, M. Ravens, H. Pien, B. Cunningham, and J. Vacanti, *Tissue Eng.*, **6**, 105–117 (2000).
33. Y. Zheng *et al.*, *Proceedings of the National Academy of Sciences of the United States of America*, **109,** 9342–9347 (2012).
34. J. Morgan *et al.*, *Nat. Protoc.*, **8,** 1820–1836 (2013).
35. J. M. Chan, I. K. Zervantonakis, T. Rimchala, W. J. Polacheck, J. Whisler, *et al.*, Engineering of *in vitro* 3D capillary beds by self-directed angiogenic sprouting. *PLoS ONE*, **7** (2012).
36. B. Baker, B. Trappmann, S. Stapleton, E. Toro, and C. Chen, *Lab Chip*, **13**, 3246–3252 (2013).
37. D.-H. T. Nguyen, S. Stapleton, M. Yang, S. Cha, C. Choi, P. Galie, and C. Chen, *Proceedings of the National Academy of Sciences of the United States of America*, **110**, 6712–6717 (2013).
38. M. Cuchiara, D. Gould, M. McHale, M. Dickinson, and J. West, *Adv. Funct. Mater.*, **22**, 4511–4518 (2012).
39. W. Wu, A. DeConinck, and J. Lewis, *Adv. Mater.*, **23**, 83 (2011).
40. J. Miller *et al.*, *Nat. Mater.*, **11**, 768–774 (2012).
41. Y. J. Blinder, D. J. Mooney, and S. Levenberg, *Curr. Opin.Chem. Eng.*, **3**, 56–61 (2014).

Chapter 14

Use of Hydrogels in the Engineering of Lung Tissue

Joaquin Cortiella, Jean A. Niles, Stephanie P. Vega, Lissenya B. Argueta,* Adriene Eastaway and Joan E. Nichols

B.S.: Weill Cornell Medical College, Department of Immunology, 420 East 70th Street Apt. 2K-2, NY 10021, USA

**lia2015@med.cornell.edu*

It is important to realize that our knowledge of the field of regenerative medicine is constantly evolving and changing. Progress thus far in the field of lung generation is the result of determined efforts by a vast array of multidisciplinary groups, which include researchers in the fields of chemistry, biotechnology, cell biology, physiology, chemical engineering, and medicine. Most engineering strategies rely on the utilization of a scaffold material to support tissue development, and this is also true of lung engineering. The field of lung regeneration is currently working to produce organ specific biomaterials capable of meeting the needs of the lung. This is an important step towards the development of functional lung tissue worthy of clinical application. Early scientists made many discoveries in scaffold design and production of biomaterials which allow attachment of cells and subsequent formation of tissue. Variable degrees of success have been achieved in the development of hydrogels as scaffolds for engineering a variety of organs, including the lung. Hydrogels vary in their composition architecture, degradability, and mechanical capability. A number of hydrogels have been produced which are structurally similar to the extracellular matrix (ECM) of many tissues. Hydrogels can also be designed to be environmentally sensitive and react to small changes in pH, temperature, or even protein concentration, allowing them to function as sustained release delivery systems for drug or growth factor administration. Synthetic and natural hydrogels are capable of meeting some of the basic requirements for engineering lung tissue. This chapter will concentrate on applications of hydrogels in the tissue engineering of lung in terms of use as sustained release delivery systems, vehicles to enhance cell transfer and attachment, or as scaffolds for direct support of tissue development.

1. Introduction

The design and development of materials to support production of bioengineered lung tissue has been a slow and stepwise process. The ultimate goal has always been to produce tissues that possess functional microvasculature, tissue specific morphology, and specified physiological lung functions. One of the major hurdles we still face in the engineering of lung tissue is the development of lung-appropriate scaffold materials. Hydrogels were the first biomaterials designed for use in the human body and have been used in engineering a variety of tissues.[1,2] There has been limited use of hydrogels in the engineering of lung tissue. Perhaps this is because little work has been done to produce hydrogel scaffolds capable of providing lung-appropriate three-dimensional (3D) architecture with the mechanical elements necessary to support the processes of breathing and gas exchange.

Hydrogels are broadly classified into two major categories: (i) reversible/physical gels and (ii) permanent/chemical gels which are covalently crosslinked (replacing hydrogen bonds with stronger, stable covalent bonds) networks.[3] Reversible/physical gel networks are held together by secondary forces including ionic bonds, hydrogen bonding, or hydrophobic interactions, which can be disrupted by changes in physical conditions of the gel (such as changes in pH or temperature) allowing them to be sol or gel depending on the environmental conditions.[4] Permanent gels attain an equilibrium swelling state which depends on the interaction of the polymerwater component combined with the crosslink density.[4] In physically crosslinked gels, dissolution of the gel is prevented by physical interactions, which exist between the different polymer chains.[3]

Hydrogels are 3D networks of polymer chains where water fills the voids between the chains. Crosslinking provides required mechanical strength to the gel as well as providing for physical integrity of the gel. Hydrogels can be formed from synthetic or natural substances, and, due to their high water content, can possess a high degree of flexibility similar to that of normal tissue, making them good candidates as scaffolds for lung engineering.

2. Hydrogels as Protein and Drug Delivery Systems

The fact that hydrogels allow unrestricted movement of cell products and other materials into developing tissues make them attractive candidates for use in the delivery of selected agents. The delivery of bioactive molecules has been the subject of intensive research, and the dispersal of low molecular weight drugs, nucleic acids, peptides, proteins, and antibiotics remains an emerging field of research. Good reviews of this topic, which provide some insight into the development of hydrogels as delivery systems, have been published.[5–7] Some hydrogels can function as sustained release delivery systems for drug or growth factor administration in close proximity to developing tissues, reducing the need for the addition of large amounts of expensive growth factors into circulating media. Unfortunately, not all hydrogel delivery systems provide for adequate release of the agents they carry. Many protocols for crosslinking hydrogels utilize toxic components, making them unsuitable for use in engineering lung tissue. Despite the issues related to toxicity due to their manufacturing processes, many synthetic hydrogels remain good candidates for use in controlled release of materials during *in vitro* culture of lung tissue.

Polyethylene glycol (PEG) hydrogels are widely used in a variety of biomedical applications, including matrices for controlled release of biomolecules for use in tissue engineering. The design, fabrication, and characterization of PEG hydrogels rely on the understanding of fundamental gelation kinetics as well as the purpose of the applications.[6] Many methods exist which allow for precise control over the gel crosslinking density and subsequent material properties of hydrogels. Control of gelation is quite important in the delivery of therapeutics or growth factors, as it allows for precise mathematical predictions related to delivery, which is highly beneficial for determining appropriate drug or growth factor dosing and release kinetics.[6] One method of producing degradable PEG hydrogels, making use of a conjugate addition reaction (also termed Michael-type addition reaction) between multiacrylated PEG and dithiols, can be carried out at physiological temperature and pH and does not require organic solvents.[8] Using this methodology, a series of degradable PEG-hydrogels has been developed which can be used as cell scaffolds,[9] protein delivery systems[8,10] or for drug delivery.[10] Engineered protein polymers carrying specific key features of the natural extracellular matrix (ECM) were crosslinked with functionalized PEG to form hybrid hydrogel networks. The repeating units of target recombinant protein polymers contained cell-binding sites for ligation of cell-surface integrin receptors and substrates for plasmin and matrix metalloproteinase (MMPs), proteases implicated in wound healing and tissue regeneration[9] as well as differentiation of pluripotent cardio-progenitor cells.[11] Other degradable PEG hydrogels produced using this technology have been shown to be suitable for the delivery of proteins such as human growth hormone using this same procedure which provided for delayed release (4 days) where the mechanism for release appeared to be a combination of slow dissolution of protein in the presence of PEG and hindered diffusion of protein through the gel.[8] Model systems mimicking ECM based on these same degradable synthetic PEG-based hydrogels have been used to examine ECM-based cues for cell migration[12] but could also be used to control movement of cells on 3D matrices. Raeber *et al.* demonstrated that 3D migration of dermal fibroblasts occurred in this type of gel, and quantified the migration by time-lapse microscopy and automated single-cell tracking.[12] Broadband MMP inhibitor and tumor necrosis factor-alpha, a potent MMP inducer in fibroblasts, were used to alter MMP responses. Inhibition in PEG hydrogel led to an almost complete suppression of fibroblast migration, whereas MMP upregulation increased the fraction of migrating fibroblasts significantly.[12] Migration in collagen and fibrin proved to be less sensitive to both MMP modulators. The possibility of molecularly mimicking functions of natural ECM makes PEG hydrogels potential choices for systems to study cell migration and, in the future, for targeting cell migration to specific sites when engineering complex tissues such as the lung.

Although protein delivery is a critical component for support of engineered tissues, there are other factors, such as antibiotics, which are necessary to ensure viability of tissue. It would be best if *in vitro* culture could be done without prophylactic use of antibiotics, but most researchers rely on use of antibiotics and antimycotics to prevent contamination of long-term *in vitro* cultures. Of major concern for engineering tissues such as the lung is the length of time tissues must be maintained in culture and the numerous opportunities for contamination every time new media, growth factors, or antibiotics are added to the developing tissue. Contamination of cultures is particularly problematic when using natural acellular (AC) tissue scaffolds or primary cell cultures for the generation of engineered tissues. It is possible that hydrogel materials developed for use in the treatment of invasive fungal infections in immunocompromised patients could be used to limit the development of contamination in long-term engineered lung cultures and the systemic administration of antibiotics and antimicrobials. One material, a dextran-aldehyde polymer conjugated to amphotericin B (AmB), was produced which retained antifungal efficacy against *Candida albicans*. This was developed as an injectable local antifungal treatment for direct administration into existing or potential sites of fungal infection. The gel provided *in vitro* release of antifungal activity for 11 days, and contact with the gel killed *C. albicans* for 3 weeks.[13] Gels produced by entrapment of a suspension of AmB in CMC-dextran alone without conjugation of drug to polymers were not capable of releasing the antifungal.

Other hydrogel-based delivery systems include injectable thermosensitive hydrogels.[14] Hydrogels of this group are free flowing solutions at temperatures below ambient temperature. Following *in vivo* injection, the hydrogels convert into non-flowing gels at body temperature. Thermosensitive hydrogels have been used as drug delivery systems, as well as cell encapsulation systems. Gong provides an excellent review that summarizes the fundamentals, applications, and advances of injectable thermosensitive hydrogel as drug delivery systems, including chitosan and related derivatives, poly(N-isopropylacrylamide)-based (PNIPAAM) copolymers, poly(ethylene oxide)/poly(propylene oxide) (PEO/PPO) copolymers and derivatives, and PEG/ biodegradable polyester copolymers.[14] Other gels that can flow like a liquid when injected but re-gel once placed in the body include those derived from alginate. Alginate is a natural polymer that is frequently used in a number of biomedical applications and is applicable for use in tissue engineering because of its biocompatibility, low toxicity, low cost, and mild gelation behavior with divalent cations. Alginate is a block copolymer with a linear structure composed of homopolymeric blocks of β-D-mannuronate (M) and α-L-guluronate (G) residues.[15] R: arginine; G: glycine; D: aspartic acid (RGD) peptides have been found to promote cell adhesion. In one study chondrocytes were mixed with either RGD-modified alginate solution in the presence of calcium ions or DMEM/F12 media only, and subcutaneously injected into the dorsum of mice.[15] In this study the viscoelastic properties of alginate hydrogels were manipulated by introducing both cell crosslinking and ionic crosslinking to alginate solutions. The number of cells required to form cell-crosslinked gels was significantly reduced by using a small amount of calcium ions.[15] In addition, the mechanical properties and gelation behaviors of the gels were controlled by varying the concentration of cells and calcium ions in the system. Shear-reversible gelation behavior was also maintained when cells were crosslinked with both cell and ionic crosslinking methods, which is critical when these gels are used as an injectable delivery system for minimally invasive therapy. Although used to support chondrocyte growth and production of cartilage in this study, the controlled gelation behavior of this material could be important in trachea or bronchial tissue regeneration providing mechanical strength necessary to properly support regeneration of these tissues. In an *in vivo* system, there was regeneration of new cartilage tissues.[15]

It is important to realize that a key feature of living systems (cells and tissues) is the ability to sense and react to environmental stimuli. Characterization of biological sensors combined with advances in polymer chemistry has enabled the development of stimulus-sensitive biohybrid materials that translate chemical and biological input into changes in material properties. Synthesis strategies of how biological and chemical polymers can functionally be interconnected have paved the way for future application of these systems for tissue engineering or other biomedical applications.[5]

3. Hydrogels as Cell Delivery Vehicles

With respect to the role of hydrogels as delivery vehicles, hydrogels have the ability to both encapsulate cells and biomolecules and control the release through systematic changes in the gel's physical and chemical structure.[6] Hydrogels support cell seeding and tissue development, but do not impede movement of cells or the interactions of cells in 3D conformations, allowing for appropriate cell-to-cell signaling and cell-to-cell spatial orientation.

Hydrogels may also help to protect cells during transport and enhance cell viability. Hydrogels often adhere to the surface of cells or to damaged spots in cell membranes. Poloxamer-188 (P-188) has been shown to be capable of sealing electroporated and radio-permeabilized cells, including red blood cells[16] and rat skeletal muscle, preventing rapid exhaustion of high-energy cellular compounds and subsequent development of cellular necrosis resulting in loss of cells.[16–18] P-188 has even decreased tissue injury after resuscitation shock, reducing cell loss, and tissue damage.[19] The localization of P-188 in the lipid matrix of cells indicated that P-188 intercalates into the lipid membrane. The presence of poloxamers may induce the expansion of the membrane and the closure of defects ("pores") in the structurally compromised membranes due to physical trauma or diseases.[20] Poloxamer F-127 (PF-127) is a di-block co-polymer of polyoxyethylene (POE) and polyoxypropylene (POP). Aqueous solutions show thermo-reversible gelation, being liquid at temperatures <15°C and robust gels at temperatures >15°C. Poloxamer hydrogels such as Pluronic® maintain cells in a 3D structure that enables them to engage in normal cell signaling. They are a family of more than 50 different amphiphilic nonionic block polymers of hydrophobic propylene oxide (PO) and hydrophilic ethylene oxide (EO), covering a range of liquids, pastes, and solids.[21] Poloxamers consist of a central POP molecule, which is flanked on both sides by two hydrophilic chains of POE.[21] A slightly different structure is exhibited by the poloxamines, which are tetrafunctional block copolymers with four POE–POP blocks joined together by a central ethylene diamine bridge. These surfactants were first introduced in the 1950s (by BASF, NJ, USA) and since then have found a wide range of diverse applications in the pharmaceutical and biomedical fields.[21] There have been indications that PF-127 can also function as a cell sealant, and, in one study, incorporation of hydrocortisone, glucose, and glycerol in PF-127 gel helped to enhance cell viability beyond what resulted from use of the hydrogel alone.[22]

PF-127 has been shown to enhance viability of aerosolized porcine epithelial cells and bovine chondrocytes by reducing the shear stress encountered by the cells and by acting as a cell sealant.[23] More recent studies indicate that immersion of mesenchymal stem cells in PF-127 increased the viability of cells and facilitated transplantation of the cells into the lateral ventricle of rats following traumatic brain injury.[24] In other studies, PF-127 has been used to facilitate cell transport and cell scaffold attachment for somatic lung progenitor cells (SLPCs),[25] murine embryonic stem cells[26] or primary lung cells[27] with excellent results. When PF-127 was used in *in vitro* culture, the hydrogel enhanced viability of cells[25] and was shown in unpublished data to support cell aggregation and therefore enhance tissue formation by SLPCs in plate culture (Figs. 1(A) and 1(B)) or in culture using small (50 mL) (Fig. 1(C)) or large chamber (500 mL) (Figs. 1(D) and 1(E)) rotary bioreactors. The ability of hydrogels to stabilize membranes, reduce cell loss, and

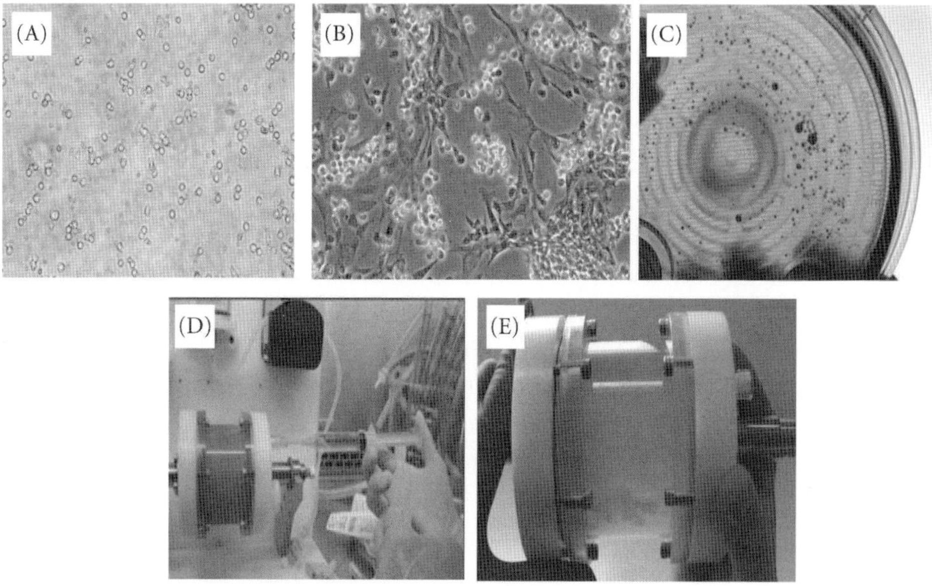

Fig. 1. Ability of PF-127 to aggregate cells and influence engineering of lung tissue. (A) Isolated lung progenitor cells in DMEM F12 with 10% Pluronic F-127 hydrogel. (B) Aggregated cell groupings following *in vitro* culture. (C) SLPC aggregates following culture of cells in DMEM F12 with 10% PF-127 (PF-127) hydrogel in a rotary bioreactor chamber after 14 days in culture. (D) Large bioreactor chamber filled with in DMEM F12 with 15% PF-127 hydrogel showing administration of SLPCs to gel in the chamber. (E) Production of engineered lung tissue constructs after 14 days of culture. Unpublished data J. Cortiella.

enhance cell attachment to scaffolds validates their use as cell vehicles in lung tissue engineering.

4. Hydrogels and Lung Engineering

For lung tissue engineering specifically, it is critical that scaffold materials mimic as closely as possible the natural physiologic environment of lungs and lung ECM. Therefore, mechanical properties such as stiffness and viscoelasticity need to be considered. Mechanical stiffness not only affects gross pulmonary function, but also induces specific lung cell differentiation. The microenvironment, ECM, and soluble signals, also plays a critical role in cell differentiation. The microenvironment in tissue engineered lung should be capable of oxygen diffusion and of therapeutic and growth factor delivery. The mechanical properties of hydrogels and their transport and degradation kinetics make them suitable for use in engineering of lung tissue, although they currently lack the complexity of ECM composition, geometry, and stiffness seen in natural lung. The crosslinked polymer chains possess high water content and allow transport of oxygen, nutrients, and waste within developing tissues.[28] Many hydrogels can be formed under relatively mild, biologically, or cell-compatible conditions and hydrogels can be modified to possess specific gelation requirements, cell adhesion ligands, specified viscoelasticity, or degradation profiles that support cell and tissue survival.[29] Hydrogels used for cell culture can also be formed from a vast array of natural and synthetic materials, offering a broad spectrum of mechanical and chemical properties.

In general, scaffold materials used in engineering any tissue need to be biocompatible and capable of supporting cell attachment and subsequent 3D tissue development. Lung diseases and their subsequent pathology may also influence the choice of scaffold material to be used in the generation of replacement lung tissue. Using scaffold material that is too elastic, incompliant, or one that limits the size of the engineered lung may mimic restrictive lung pathologies resulting in impeded pulmonary function. Restrictive diseases that impede pulmonary function could be exacerbated by scaffolding material that lacks appropriate elasticity or does not degrade sufficiently over time. Obstructive diseases could also be exacerbated by scaffolds that do not degrade properly or that hinder respiratory functions. Collagen I is the main component of lung tissue and provides the basis for stability of the lung structure. Normal lung ECM also contains elastin, fibronectin, laminin, as well as other proteins. Natural gels, which have been used for lung tissue generation, have been formed of ECM components such as collagen, Gelfoam (gelatin) or Matrigel. Hydrogels such as Matrigel, which are derived from natural sources, are often inherently biocompatible and bioactive,[30] and may provide cues for cell function that are lacking in synthetic hydrogels. Synthetic hydrogels, formed of purely non-natural molecules such as PEG,[31] poly (vinyl alcohol),[32] and poly (2-hydroxy ethyl methacrylate)[33] or other materials, have also been produced.

5. Hydrogels in Support of *In Vitro* Engineering of Lung

Hydrogels such as those formed from PF-127, Gelfoam, collagen, or Matrigel have been used in the generation of lung tissue. Natural hydrogel materials that have been used to produce lung tissue include: murine Engelbreth-Holm-Swarm (EHS) tumor membrane, collagen gel, Matrigel, and Gelfoam sponge (these studies are listed in Table 1). EHS tumor membrane is derived from a tumor containing high levels of ECM proteins composed of laminin, collagen IV, and heparin sulfate proteoglycan. Matrigel is a soluble basement membrane material which is isolated from the EHS tumor line. Website reference Matrigel[34] is also commercially available and has been used to culture a wide variety of cell types, including fetal lung cells, lung progenitor cells, and alveolar epithelial cells (AECs) (Table 1). The major components of Matrigel

Table 1. Hydrogels as support material for lung tissue development.

Date published	Scaffold and system	Cell types in construct
Blau, 1988	EHS tumor membrane or Matrigel as scaffolds to support AEC II culture.	AEC II
Sugihara, 1993	Reconstruction of alveolus-like structure using a 3D collagen gel scaffold to seed and culture AEC II.	AEC II
Chinoy MR, 1994	EHS tumor membrane or Matrigel as a scaffold to support AEC II culture and differentiation.	Rabbit fetal lung cells
Chakir, 2001	3D model of engineered human bronchial mucosa (EHBM) using a collagen gel scaffold seeded with primary cells from normal and asthmatic lung tissues cultured with an air-liquid interface.	Human bronchial epithelial cells Fibroblasts and T lymphocytes
Paquette, 2004	Model of bronchial equivalents (BE) using collagen gel as a scaffold for mesenchymal layers seeded with normal and asthmatic human lung primary cells.	Human bronchial epithelial cells Fibroblasts, ciliated and goblet cells
Chen, 2005	Formation of physiologic alveolar-like structures using a type I collagen–glycosaminoglycan (GAG) scaffold seeded with rat fetal lung cells.	AEC II Alpha-smooth muscle actin production
Choe, 2006	3D model of human bronchial mucosa produced by using porous polymeric wells and a collagen gel matrix as a scaffold seeded with human lung cells in an air–liquid interface.	Human bronchial epithelial cells Fibroblasts, ciliated cells, mucus secreting cells, and addition of cytokines
Cortiella, 2006	Production of engineered lung tissue using polyglycolic acid (PGA) and PF-127 as scaffolds to support the culture of SLPCs *in vitro* and *in vivo*.	AEC II Smooth muscle cells, club cells
Mondrinos, 2006	Production of 3D pulmonary tissue constructs using Matrigel hydrogels, poly-lactic-*co*-glycolic acid (PLGA), and poly-L-lactic-acid (PLLA) as scaffolds to support culture of fetal pulmonary cells.	AEC II Pulmonary endothelial cells Fibroblasts
Mondrinos, 2007	Generation of vascularized pulmonary tissue constructs utilizing the Matrigel plug model with polyvinyl sponges to enhance vascularization and support the culture of murine fetal pulmonary cells (FCP).	AECII Pulmonary endothelial cells Fibroblasts

(*Continued*)

Table 1. (*Continued*)

Date published	Scaffold and system	Cell types in construct
Andrade, 2007	Development of physiologic alveolar-like structures using Gelfoam sponge as a scaffold to support culture and differentiation of rat fetal lung cells.	AEC Endothelial-like cells Connective-tissue-like cells
Zani, 2008	Production of (TE) upper respiratory units using implants of the bronchial epithelium and endothelium using Gelfoam collagen scaffolds.	Human bronchial epithelial cells Human aortic endothelial cells Fibroblasts, and addition of cytokines
Cortiella, 2010	Production of TE lung tissue using AC natural lung matrix seeded with mESCs. PF-127 was used as a vehicle for cell transfer and to enhance cell attachment to scaffold.	mESC-derived AEC I and II mESC-derived CD31 mESC-derived club cells
Miller, 2010	Production of human bronchiole mimic using a hydrogel scaffold to support culture of human lung primary cells.	Human AEC Fibroblasts, smooth muscle cells
Kloxin, 2012	Hydrogel comprised of photolabile and enzyme-labile crosslinks and pendant cell adhesion sequences, enabling spatially-specific, *in situ* patterning with light and cell-dictated microenvironment remodeling through enzyme secretion.	Mouse AEC II
Nichols, 2013	Use of PF-127 as a vehicle to enhance attachment of primary lung cells to AC human lung scaffolds for the generation of TE lung.	Primary AEC
Singh, 2013	Blend of hydroxyethyl methacrylate-alginate-gelatin (HAG) cryogel scaffold, which is a mixture of polysaccharide (alginate) and protein (gelatin), and was used as a scaffold for Human lung epithelial cells (L-132).	L-132 which are a human embryonic lung epithelial cell line.

Abbreviations: AEC, alveolar epithelial cells; TE, tissue engineered; mESCs, murine embryonic stem cells.

are laminin, collagen IV, and heparin sulfate proteoglycan. Matrigel matrix forms a gel above 10°C and supports the attachment and differentiation of both normal and transformed anchorage-dependent epithelial and other cell types. Collagen (generally type I) is the main component of connective tissue. Collagen is commercially available from many sources, and has been used for either the development of small AEC-matrix constructs or engineering lung tissue. Gelatin is produced by partial hydrolysis of collagen. Gelatin melts to a liquid when heated and solidifies when cooled. Gelfoam (Pfizer) is a compressed sponge of porcine skin gelatin, and was originally developed as a hemostatic device to arrest bleeding and promote clotting. There are limitations in the *in vitro* use of natural hydrogel scaffolds due to their simplicity of design, mechanical properties, and variable rates of degradation.[35]

Synthetic polymers have the potential to be produced, which possess a wider range of mechanical and chemical properties than that of natural materials. To maintain the elastic nature of the environment of the lung, it is generally accepted that a degradable or an extensively biomodifiable material would be the best overall choice for use as a scaffold for engineering lung tissue. Table 1 presents a list of hydrogel scaffolds and cell types used to engineer lung tissue. One of the earliest attempts to produce alveolar cell-matrix constructs used EHS tumor membrane to support differentiation of rabbit fetal lung cells (29 day gestation)[36] and subsequent development of AEC type II in the culture. In this study, fetal cells attached to thin and thick collagen gels, aggregated on the hydrogel and, following attachment, developed into alveolar-like structures which persisted for 22 days in culture.[36] Similar work was done by Chinoy in 1994 using EHS extract or EHS gel (Matrigel) and fetal rabbit lung cells (22 day gestation) which also showed support of AEC type II development.[37] These studies emphasized the importance of the ECM in the differentiation of fetal cells and utilized cellular attachment to the natural hydrogel to support rudimentary lung tissue development.

Collagen gel has been used to produce both alveolar-like structures and bronchial tissue equivalents. In these early studies, rat AEC II cells were cultured on collagen hydrogel and, following culture, formed large cellular aggregates which were capable of producing surfactant protein (SP).[38] Lung alveolar-like structures and lung tissue were formed when fetal rat lung cells were grown in a GAG tissue-engineered scaffold.[39] In this study, lung cells from fetal Sprague Dawley rats (16–19 days gestation) were harvested, and the dissociated cells were then seeded into type I collagen-chondroitin 6-sulfate matrices, 8 mm in diameter by 2 mm in thickness, crosslinked and sterilized by dehydrothermal treatment.[39] Each sponge-like scaffold piece was seeded with 28 million cells. Histological and immunohistochemical studies were performed after culture of the cell-scaffold constructs for 2 days and 1, 2, and 3 weeks. The mechanical properties of the scaffold induced fetal rat lung cells to form and maintain alveolar-like structures in the collagen-GAG hydrogel scaffold. A novel finding in this study was that all of the cell-seeded collagen-GAG hydrogel constructs underwent cell-mediated contraction, which appeared to be associated with the expression of alpha-smooth muscle actin.[39] Collagen gels have also been used to study development of bronchial mucosa. In order to produce tissue engineered BE, human bronchial epithelial cells and fibroblasts were isolated from asthmatic and non-asthmatic donors and were used to produce human bronchial tissue equivalents.[40,41] The collagen gel supported attachment of cells, differentiation of cell types, and pseudostratified organization of cells in a manner comparable with that of normal bronchi.[40,41] These early respiratory bronchial analogs allowed evaluation of cell-to-cell and cell-matrix interactions, which are key components in both airway remodeling and pathological changes associated with development of asthma. Other groups have also produced engineered components of the respiratory system through manipulation of the scaffold and the culture microenvironment. Type I collagen–GAG scaffold was used to produce a model of human airway that could be induced to undergo matrix remodeling, and demonstrated the importance of cell–matrix interactions and mechano-transduction in airway remodeling.[42] Cells grown on this scaffold produced alveolar-like structures and demonstrated alpha-smooth muscle actin production. Later work by this research group used a porous collagen gel to induce tissue development by human bronchial epithelial cells. Tissue development was evidenced by the presence of tight junctions, cilia formation, and mucus secretion on the collagen gel, which was embedded with human lung fibroblasts.[42] The dynamic strain in this system induced increased fibroblast deposition of types III and IV collagen and secretion of MMPs.[42]

The 3D constructs of murine fetal cells (18 days gestation) have been generated using Matrigel or synthetic polymer scaffolds of PLGA and PLLA fabricated into porous foams and nanofibrous matrices.[43] The Matrigel constructs produced alveolar-like units containing cells which had an AEC type II cellular ultrastructure and ability to produce SP-C. The addition of tissue-specific growth factors to the cell–scaffold constructs induced branching and formation of sacculated epithelial structures consistent with distal lung architecture. Importantly, 3D culture as supported by the hydrogel scaffold was necessary for inducing expression of the morphogenesis-associated distal epithelial gene fibroblast growth factor receptor 2 (FGFr2). PLGA foams and PLLA nanofiber scaffolds facilitated growth of fetal progenitor cells but these matrices did not support the survival of distal lung epithelial cells, despite the presence of tissue-specific growth factors.[43] Although Matrigel supported development of alveolar-like tissue structures and there were indications of production of AEC type II cells, the synthetic scaffolds lacked the necessary cues to induce complex lung tissue development.

Composite scaffolds composed of synthetic and natural materials together have been produced and might be more appropriate for the production of engineered lung tissue in the future. Co-electrospun scaffolds have been made from a blend of a synthetic biodegradable polymer PLGA, 10% solution) and two natural proteins, gelatin (denatured collagen, 8% solution) and alpha-elastin (20% solution).[44] The resulting PLGA-gelatin-elastin (PGE) fibers were shown to be homogeneous in appearance and upon hydration the fibers swelled, however did not disintegrate. These PGE scaffolds were stable in an aqueous environment without crosslinking and were found to be more elastic than those made of pure elastin fibers alone.[44] This type of composite scaffold was successfully used to culture H9c2 rat cardiac cells and rat bone marrow cells, but has not yet been used to engineer lung. However, this is a step in the right direction in terms of producing a highly elastic, yet biologically complex ECM-based scaffold material. Other work by this research group studied the effects of exogenous fibroblast growth factors FGF10, FGF7, and FGF2 on mixed populations of murine FPCS (day 17.5 gestation) cultured in 3D collagen gels. The collagen gels were an excellent scaffold for use in this study due to the ability of the gel to support delivery of growth factors to the developing tissue. The use of the gel scaffold allowed for the morphogenic effects of the FGFs individually and in various combinations without providing extensive cues from the scaffold.[45] FGF10/7 were shown to significantly increase epithelial budding and proliferation of the cells; however, only FGF10 alone induced widespread budding in the cell–matrix constructs. FGF7 alone induced dilation of epithelial structures, however formation of wide spread budding of tissue on the gel was not observed.[45] The combination of FGF10/7/2 was shown to induce robust budding of epithelial structures and the formation of uniform endothelial networks in the collagen gels.

Cortiella *et al.*, 2006, isolated and characterized a population of adult-derived SLPC from adult mammalian lung tissue and generated alveolar tissue using these cells as shown in Fig. 2. After extended *in vitro* culture in the hydrogel PF-127, SLPCs differentiated into cells expressing CC10 (10kDa protein), SP-C, and cytokeratin, but did not form organized lung-specific tissue structures. When lung progenitor cells were combined with the synthetic scaffold and PGA and were maintained in *in vitro* culture for 6 weeks, cells organized into identifiable pulmonary structures (including those similar to alveoli and terminal bronchi) (Fig. 2). Western blot was used to confirm the production of CC10 and SP-C in the engineered lung construct. Later, this research group compared the influence of AC natural rat lung, Gelfoam, Matrigel, or collagen I hydrogel scaffolds on mESC attachment, differentiation, and subsequent formation of complex lung tissue equivalents.[26] PF-127 was used as a cell carrier in these experiments. In this study, AC lung allowed for better retention of cells with increased differentiation efficiency of mESCs into epithelial and endothelial lineages than was supported by the other scaffold materials. In constructs produced on whole AC lung, there were indications of organization of differentiating ESC into 3D structures reminiscent of complex tissues but limited development of these structures was seen in the hydrogel-scaffold constructs.[26] Cells on the AC lung scaffold demonstrated expression of thyroid transcription factor-1, an immature lung epithelial cell marker; pro-SPA C, a type II pneumocyte marker; PECAM-1/CD31, an endothelial cell

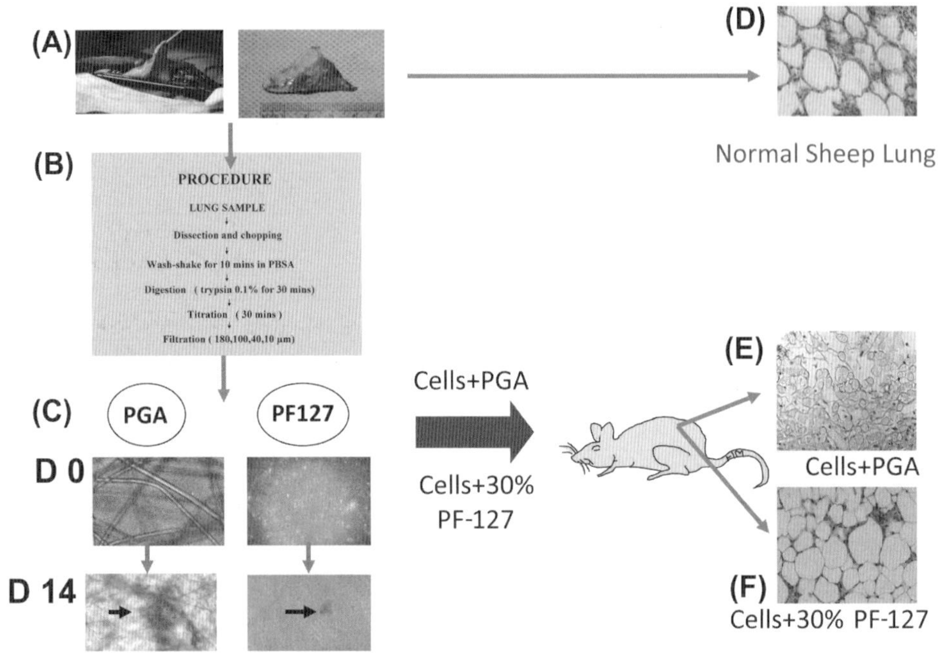

Fig. 2. Schematic diagram of methods for isolation, culture, implantation, and growth of tissue from SLPC (From Cortiella, 2006).

marker; cytokeratin 18; alpha-actin, a smooth muscle marker; CD140a or platelet-derived growth factor receptor-alpha; and club (Clara) cell protein 10. There was also evidence of site-specific differentiation into tracheal tissue with the formation of sheets of cytokeratin-positive cells and club cell protein 10-expressing club cells.[26] These findings support the utility of AC lung as a matrix for engineering lung tissue and highlight the critical role played by matrix or scaffold-associated cues in guiding ESC differentiation toward lung-specific lineages.[26] Later work to engineer lung tissue using AC natural lung matrices utilized PF-127 as a cell carrier to facilitate transplantation of primary AEC onto the AC lung scaffold.[27] These studies indicated that when repopulated with a variety of stem or adult cells including human adult primary alveolar epithelial type II cells, both pig and human AC scaffolds supported cell attachment and cell viability. Unpublished data from our laboratory also indicate that without use of PF-127 as a carrier, attachment was significantly reduced, regardless of the scaffold used.

The work done by Cortiella *et al.*, 2010, supports the concept that cell scaffolds are not just passive support elements in the production of tissue, but play an important role in control and modulation of the cellular microenvironment. This is not a new concept, as Discher and co-workers were one of the first groups to demonstrate that the differentiation of hMSCs was dependent on the mechanical stiffness of the culture substrate that cells were grown on.[46,47] Ingber and his research team have also consistently shown that the degree to which a cell is mechanically distended or "stretched" dictates cell growth and cell loss via apoptosis.[48,49] The critical nature of cell scaffolds and the importance of 3D growth of cells

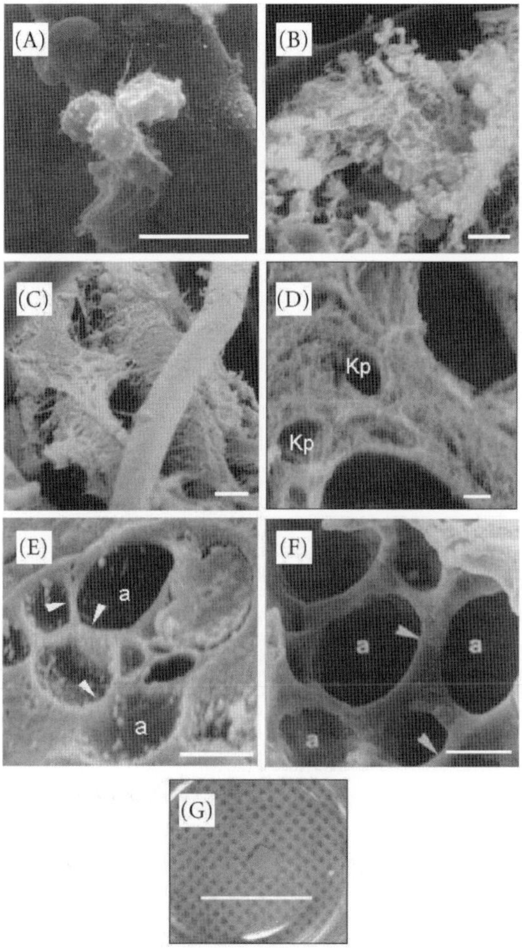

Fig. 3. *In vitro* tissue-engineered lung. (A–F) Scanning electron micrographs of *in vitro* tissue-engineered lung using ovine SLPCs seeded onto PGA scaffolds. Directly after seeding (A) and after 1 week (B), 2 weeks (C), 6 weeks (D), and 8 weeks (E) of culture *in vitro*. By 6 weeks, cultures developed morphologic features of lung tissue. By 8 weeks tissue-engineered lung (E and F) contained structures similar to alveoli (A) with septa (arrows), compared with septa (arrows) in native lung (F) (all bars, 10 m). By 8 weeks, engineered lung tissue grown *in vitro* was macroscopically visible and could be handled easily (G) (bar, 5 mm). For scanning electron microscopic analysis, specimens were pre-fixed in Karnovsky's fixative for 4 h at 4°C, washed with 0.1 M sodium cacodylate buffer (pH 7.4), post- fixed for 1 h in 1% with gold, and observed in a Hitachi scanning electron microscope (From Cortiella, 2006).

was verified by Peterson *et al.* who demonstrated that human breast epithelial cells developed similar to tumor cells when cultured in two-dimensional (2D), but reverted to normal growth behavior patterns when cultured in 3D.[50] Enhanced chondrogenesis of embryonic stem cells has also been observed when cells are cultured as 3D embryoid bodies as compared to monolayer culture.[51] For comprehensive reviews of the influence of hydrogels as ECM mimics, readers are referred to excellent review by Tibbett and Anseth.[52] Hydrogels as a group tend to lack the variation in stiffness or stretch that is found in natural tissues, such as normal lung. The nature of complex organs like the lung may eventually require the development of hybrid scaffolds formed from more than one material or a combination of synthetic and natural materials to support adequate production of lung tissue for pre-clinical and clinical applications.

6. *In Vivo* Use of Hydrogels in Lung Engineering

Currently, there has been limited use of scaffolding material in engineering distal lung-tissue for transplantation *in vivo*. The first reported use of hydrogels for development of engineered lung for transplantation was done in both a large animal model (sheep) as well as a small animal model (nude mouse). Cortiella *et al.*, 2006, produced autologous ovine SLPC/scaffold constructs supported by a scaffold formed from PF-127 alone, PGA alone or a combination of PGA and PF-127. PF-127/SLPC constructs were implanted onto the backs of nude mice. In sheep, PF-127-PGA/SLPC autologous constructs were transplanted directly into the right upper lobe of the lung into a pocket created by a wedge resection. Transplanted PF-127/SLPC constructs were well tolerated and tissue assembly was facilitated *in vivo* by the use of specific growth factors. Transplantation onto the backs of nude mice was done to determine if the growth factors alone could be used to promote differentiation of SLPCs and formation of lung tissue at a site distant from the natural environment of the lung. When PF-127/SLPC constructs were implanted onto the backs of nude mice developing tissue fragments were shown to contain cells which expressed SP-C (Fig. 4), club cell protein 10 (Fig. 5), with evidence of smooth muscle development.[25] In this study there was poor development of tissue in constructs formed from PGA alone as compared to tissue grown in PF-127 hydrogel or normal lung tissue (Fig. 6). Transplantation of fluorescently labeled autologous SLPC/PGA- PF-127 constructs (Fig. 8(A)) into a wedge resection site in the right upper lobe of an adult sheep lung were seen to (Fig. 8(B)) develop into structures resembling alveolar tissue, and histological examination showed that 5-Chloromethylfluorescein Diacetate (CMFDA) labeled cells were forming tissue at the transplant site (Figs. 8(C)–8(E)). Later, immunochemical staining of sections from the same tissue block used in Fig. 8 showed that the tissue contained AEC type II cells based on aquaporin-5 staining (Fig. 8(G)). Transplantation

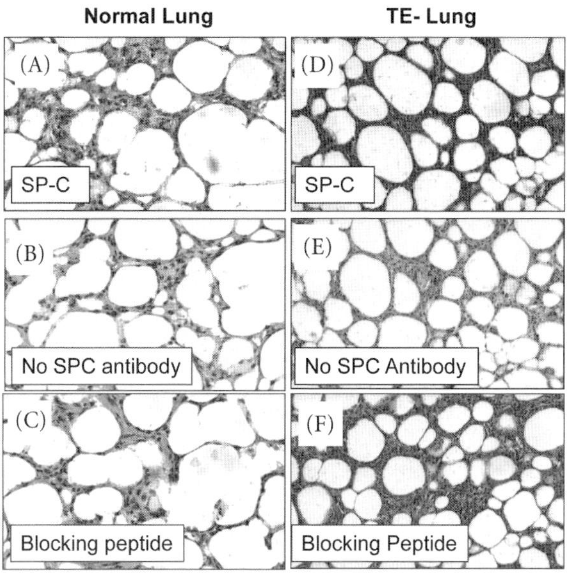

Fig. 4. *In vivo* tissue-engineered lung SP-C. Sections of normal and TE lung produced after implantation of ovine SLPC/PF-127 tissue constructs on the backs of nude mice. Immunohistochemical detection of SP-C in (A–C) sections of normal lung and in (B–F) TE lung. (B and E) Specificity of immunohistochemical analyses was demonstrated by confirming that no evidence of reactivity was obtained in the absence of the primary antibody or (C and F) in the presence of a 30-fold excess of an appropriate blocking peptide (E and F) (From Cortiella, 2006).

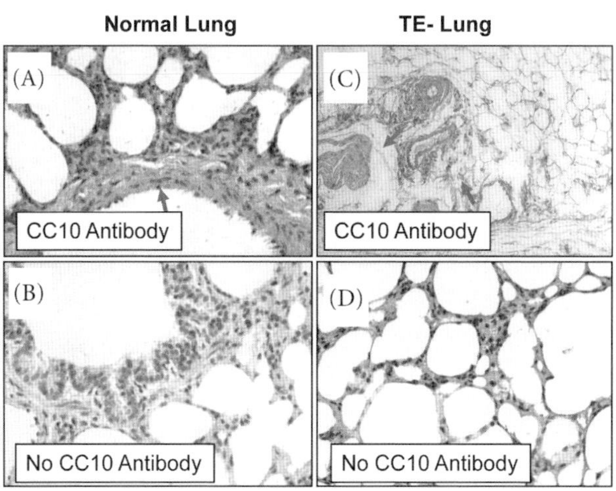

Fig. 5. *In vivo* tissue-engineered lung club cell protein 10. Immunohistochemical detection of club cell 10 in normal lung and TE lung following implantation of ovine SLPC/PF-127 tissue constructs onto the backs of nude mice. (A) Normal Lung with arrows indicating positive staining (magnification, 200×). (B) Normal lung control showing specificity of immunohistochemical analyses confirming that no evidence of reactivity was obtained in the absence of the primary antibody (magnification, 200×). (C) TE lung with arrows indicating positive staining (magnification, 200×). (D) TE lung staining control where no evidence of reactivity was obtained in the absence of the primary antibody (From Cortiella, 2006).

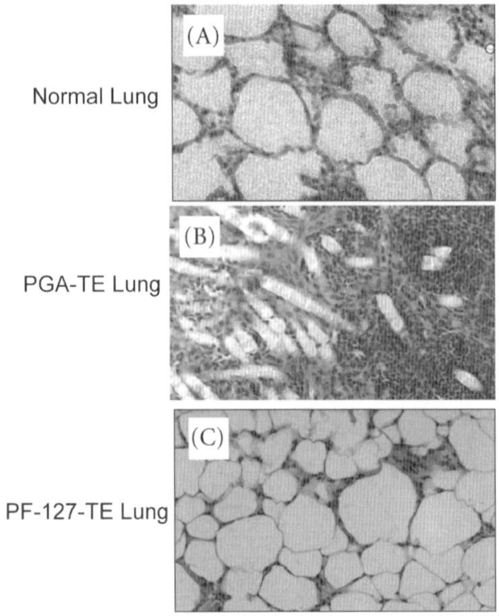

Fig. 6. *In vivo* tissue-engineered lung. Sections of engineered lung produced after implantation of ovine SLPC/PF-127 tissue constructs on the backs of nude mice. Photomicrographs of hematoxylin and eosin stained sections of (A) normal lung and (B and C) TE lung. (B) TE lung grown on PGA and (C) TE lung grown on PF-127 (30%). (magnifications 200×) (From Cortiella, 2006).

of a similar CMFDA-labeled SLPC/scaffold construct into three sheep at the site of a full pneumonectomy produced fleshy tissue pieces, the largest of which was 5×13 cm after 3 months (Fig. 9). When harvested after 3 months, the implants were shown to form soft tissue fragments (Fig. 9). Examination of these pieces of tissue indicated that there was little lung epithelial cell development and no obvious development of lung morphology, but the tissue fragments were highly vascularized.

Gelfoam sponges have been used for production of engineered lung tissue for *in vivo* transplantation as well. Gelfoam sponges were used to form fetal rat lung cell-scaffold constructs which were implanted into rat lungs.[53] CellTracker orange CMTMR [5-(and 6)-{[(4-choromethyl) benzoyl]amino} tetramethylrhodamine (Molecular Probes) labeled fetal lung cells were seeded onto a Gelfoam scaffold and incubated *in vitro* for 7 days prior to injection into the lung parenchyma.[53] The fetal lung grafts were shown to survive for 35 days, and to develop into alveolar-like structures at the border between the sponge and the surrounding lung tissue. Engineered tissues contained cells which produced pro-SP-C, club cell Secreted Protein (CCSP) or von Willebrand Factor (endothelium).[53] Few leukocytes were shown to move into the implanted sponge (using CD45 staining), suggesting that there was no inflammatory response generated as a result of transplantation of the

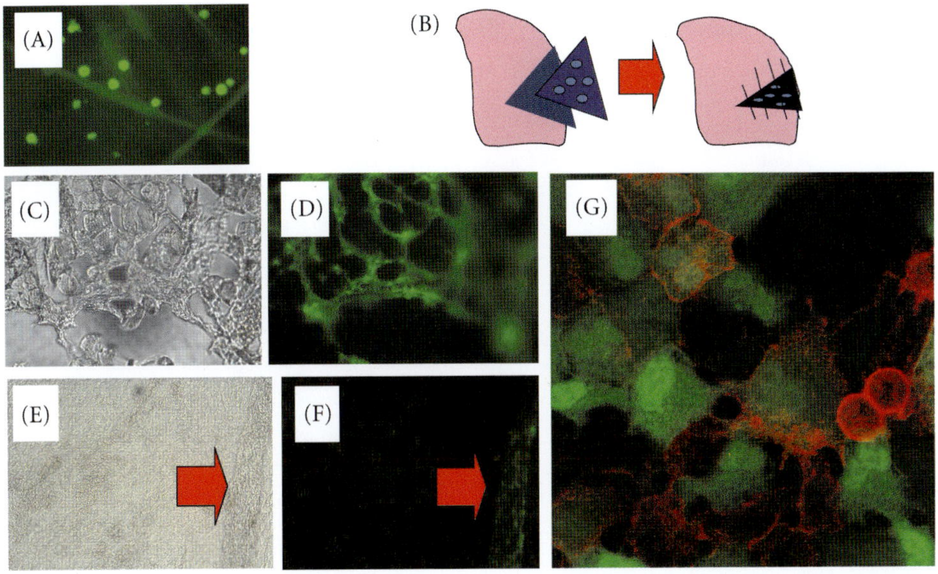

Fig. 7. *In vivo* tissue-engineered ovine lung using PGA plus PF-127, or PF-127 alone to form the cell/polymer constructs. (A) Autologous ovine carboxyfluorescein diacetate, succinimidyl ester (CMFDA)-labeled SLPC prior to implantation showing the CMFDA fluorescent labeling of cells with strands of PGA in the background. (B) Cartoon showing placement of engineered cell-scaffold construct into wedge resection site. (C–F) Sections of engineered lung produced after implantation of ovine SLPC/polymer tissue constructs at the lung wedge resection site. Phase contrast (C and E) and fluorescent micrographs (A, D, F, and G) of tissue sections showing the growth of lung tissue from implanted SLPC/polymer constructs. (C–G) Tissue grown from ovine SLPC/PGA and PF-127 constructs is shown. (E and F) Showing the edge of the wedge resection site after healing and generation of lung tissue at the border between the CMFDA labeled engineered lung and normal lung. Red arrows indicate the edge of the implanted construct (magnifications 200) (From Cortiella, 2006). (G) Unpublished data showing tissue sections produced from the same block as C–F stained for the presence of aquaporin-5, an AEC type I cell marker.

cell/scaffold construct. The Cell Tracker orange label allowed tracking of the donor's cells and indicated that many of the cells within the implanted scaffold were derived from the original engineered graft in this study. India ink perfusion of the pulmonary artery was used to evaluate the level of vascularization of the scaffold, as well as the integrity of the connections between the vascular structures formed in the scaffold material and the pulmonary circulation of the graft recipient.[53] Although there was sufficient labeling with India ink to be able to visualize the blood vessels in the Gelfoam scaffold, there was little evidence to support the development of alveolar-capillary junctions or even areas of close connection between the epithelial cells and the endothelial cells comprising the blood vessels formed in the scaffold.[53]

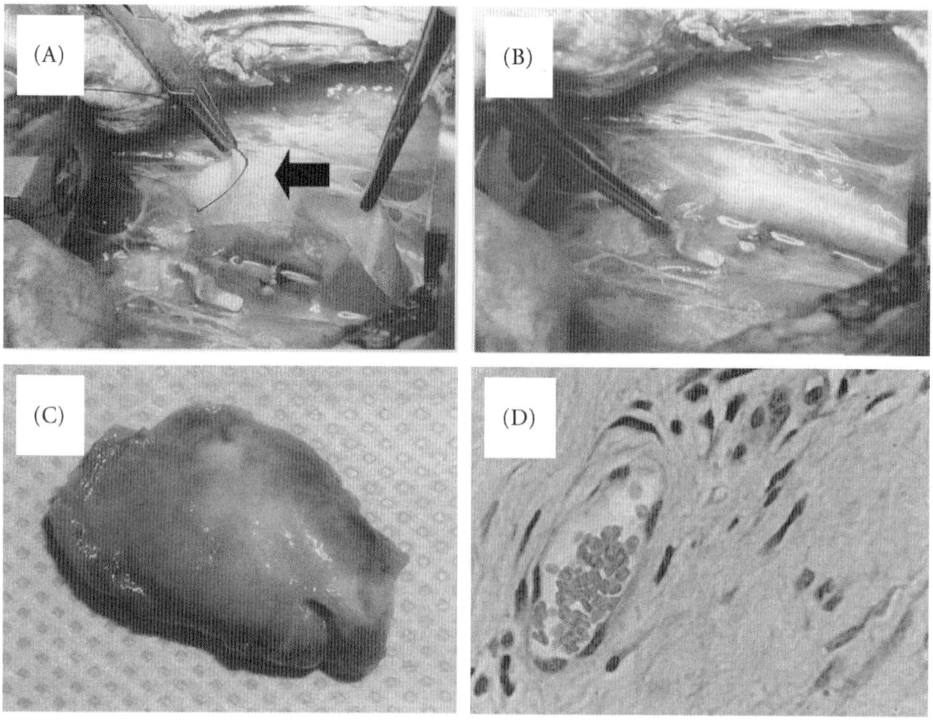

Fig. 8. Implantation of SLPC/PGA construct. *In vivo* tissue-engineered ovine lung using SLPC/PGA construct. (A and B) Surgery showing implantation of construct after a pneumonectomy at the right main stem bronchus site and (C) Subsequent tissue grown after 3 months. (A) Arrow shows actual SLPC/PGA sheet being implanted at pneumonectomy site with attachment to the site of lung removal. (D) Sections of engineered lung, produced 3 months after implantation of ovine SLPC/PGA tissue construct, stained with hematoxylin and eosin. (magnifications 400) (From Cortiella, 2006)

To examine the contribution of donor-derived endothelial cells to tissue construct vascularization, fetal pulmonary cell (FPC, 17.5 day gestation)/Matrigel constructs were injected into the anterior abdominal cavity of adult graft recipients (C57Bl6 mice).[54] In animals receiving transplanted Matrigel alone, there was some infiltration of host cells into the scaffold material. In FPC/Matrigel-FGF2 constructs, the tissues were shown to develop ductal epithelial structures with development of pro-SP-C expressing epithelial cells and patent vasculature. In this case, vascularization of the construct was enhanced by the addition of FGF2, and when FGF2/Matrigel constructs were transplanted alone (no fetal cells), there was a significant increase in capillary density near the construct.[54]

Few hydrogels have been designed to specifically meet the needs of the lung in terms of requirements for mechanical strength, elasticity, controlled degradation kinetics or stiffness. A novel, highly porous scaffold formed from a blend of HAG

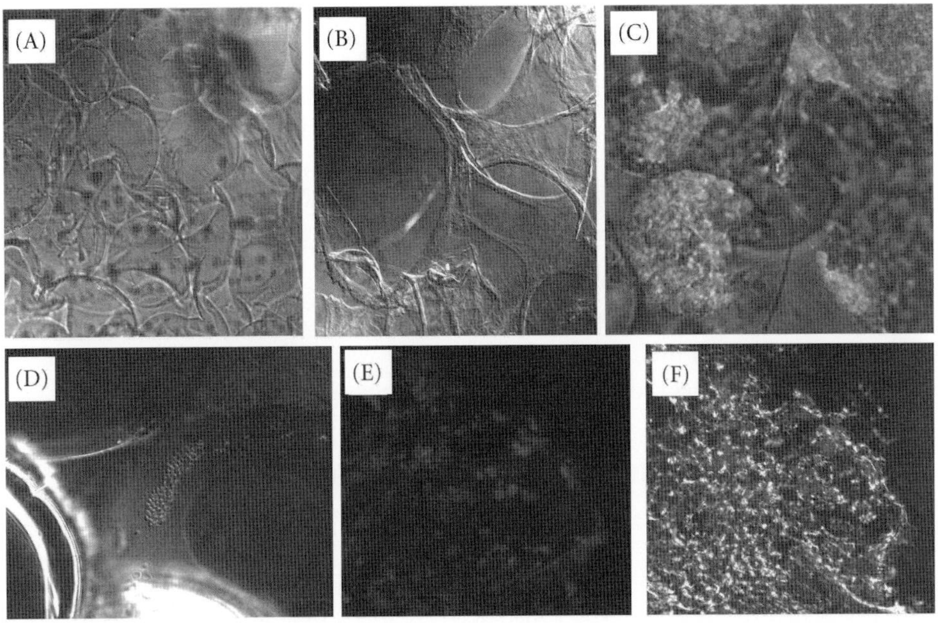

Fig. 9. Inverse colloidal crystal (ICC) scaffolds. (A) Image of 7 um section of modified ICC hydrogel scaffold, magnification 400×. (B) Image of 7 um section of modified ICC hydrogel, magnification 630×. (D) Image of cells attached to wall of the ICC scaffold, magnification 100×. (E) DAPI stained image showing nuclei (blue) of cells in a set of connecting ICC chambers and (F) Same image allowing for visualization of the cell nuclei and the edges of the ICC chamber cells are growing in, magnification 200×. Unpublished data J. E. Nichols.

gel scaffold synthesized using cryogelation technique with the mechanical strength, stiffness, elastic measurements compatible with lung engineering has recently been developed and shown to support L-132 proliferation *in vitro* and *in vivo*.[55] The HAG scaffold-construct was found to be highly elastic with quick hydration, indicating the presence of an interconnected porous network. In general, scaffolds used for the regeneration of tissue and organs should not induce unwanted inflammatory or immune based responses. Following transplantation of HAG- L-132/HAG constructs onto the backs of mice, a small number of infiltrating mast cells were found in the first week, however these cell types subsequently diminished in following weeks. No significant inflammation was seen in animals receiving transplanted constructs, and immunohistochemistry for dendritic cells indicated that few immune cells were present, which confirmed the biocompatible nature of the HAG scaffold *in vivo*.[55] Degradation of the gel occurred slowly, with 80% degradation of the HAG gel by 10 weeks — a reasonable length of time to allow for lung cell attachment and tissue formation and rapid enough to be appropriate for use in the lung. Proliferation of L-132 lung cells on the scaffolds occurred *in vitro* and *in vivo*.

While each of these hydrogel materials was shown to be adequate for the development of lung tissue, the degradation kinetics of a scaffold material is always an important consideration. Most studies do not describe the process or timing of scaffold degradation. Andrade *et al.* demonstrated that the scaffold material Gelfoam, was an excellent supporting material for lung cell attachment, and the timely degradation of the material left the newly formed "alveoli" in place once degradation was complete. This is also true of PF-127 alone or PGA/PF-127 in both *in vitro* and *in vivo* use of these scaffolds, which also degraded as the development of epithelial tissues progressed.[25] HAG-gels were also shown to degrade sufficiently over time so that the developing tissue was well supported, however the scaffold was mostly degraded over a relatively short period of time.[55]

7. Current and Future Use of Hydrogels for Lung Engineering

A significant stumbling block in the creation of functional 3D engineered tissues is the proper vascularization of the constructs. Furthermore, in the context of electrospinning, the development of 3D constructs using this technique has been hindered by the limited infiltration of cells into the scaffold material. In an attempt to address these issues, a hybrid mesh of poly (ε-caprolactone)-collagen blend (PCL/Col) and HA hydrogel, Heprasil™ was created via a dual electrodeposition system.[56] Simultaneous deposition of HA and PCL/Col allowed the dual loading and controlled release of two potent angiogenic growth factors, VEGF and PDGF-BB, over a period of 5 weeks *in vitro*. Furthermore, this manner of loading sustained the bioactivity of the two growth factors. Utilizing an in-house developed 3D co-culture assay model of human umbilical vein endothelial cells and lung fibroblasts, the growth factor-loaded hybrid meshes were shown to not only support cellular attachment, but also their infiltration and the recapitulation of primitive capillary network in the scaffold's architecture.[56] Thus, the creation of a PCL/Col-Heprasil hybrid scaffold is a step forward toward the attainment of any 3D bio-functionalized, vascularized tissue engineering construct.[56]

Most hydrogels have limited mechanical ability or stretchability. Stretchable, tough hydrogels have been produced by mixing two types of crosslinked polymer, ionically crosslinked alginate, and covalently crosslinked polyacrylamide.[57] This alginate–polyacrylamide hybrid gel could be stretched up to >20 times its original length without rupture and remained stable even when a notch was cut into the gel. The combination of relatively high stiffness and toughness, along with an easy method of synthesis, make these materials ideal candidates for further investigation for use in lung engineering.

Recently, a well-based cell culture platform was developed that enables control of the geometry and connectivity of cellular microenvironments spatiotemporally. The base material is a hydrogel comprised of photo-labile and enzyme-labile crosslinks and pendant cell adhesion sequences, enabling spatially-specific, *in situ* patterning with light or by cell-dictated microenvironment remodeling through enzyme secretion.[58] Arrays of culture wells of varying shape and size were patterned into the hydrogel surface using photolithography. The geometry of these devices was subsequently modified through sequential patterning, while changes in cell geometry and connectivity were examined.[58] This type of system will allow for dynamic evaluation of the influence of physical cues on tissue morphogenesis. The effect of well shape and size on lung epithelial cell differentiation (i.e., primary mouse AEC type II cells) was assessed using this model.[58] Shapes inspired by lung alveoli were produced on the hydrogel surfaces. AEC type II cells were then seeded within the well-based arrays and encapsulated by the addition of a top hydrogel layer.[58] Cell differentiation in response to specific geometries was characterized over 7 days of culture with immunocytochemistry evaluation of lung lineages showing production of SP-C and T1α protein by AEC II cells. Individual cell clusters were further connected by eroding channels between wells during culture using controlled two-photon irradiation. Collectively, these studies demonstrate the development and utility of responsive hydrogels that may eventually lead to the production of microenvironment geometries of shape, elasticity, strength, and connectivity with the ability to direct cell function similar to what has been shown for AC natural lung scaffolds.

The classic tissue engineering approach has always been based on the addition of cells to a scaffold to produce a construct which is treated with controlled addition of specific growth factors or growth factor cocktails to support tissue growth *in vitro*. This approach generally utilizes the innate ability of cells to develop into recognizable tissue structures and to respond physiologically in a manner appropriate for the tissue produced. Although this approach has been used by most research involved in the development of tissue engineered, lung tissue development is not efficient and tissues often lack the complex organizational patterns found in normal organs. These tissues also lack appropriate vascularization, and there are very few methodologies which currently support development of vascular systems in engineered tissues. Rudimentary techniques do exist for cell or organ printing following a controlled plan for cell deposition in order to specifically place cells in layer-by-layer constructs to support 3D tissue production. Hydrogels are used as scaffolding materials in organ printing. Organ printing refers to the placement of various cell types into a soft scaffold fabricated according to a computer-aided design template using a single device.[59] The hydrogel matrices can be natural or synthetic polymers and successful cell embedding requires hydrogels that are complemented with biomimetic and ECM components to provide biological cues to elicit specific cellular responses and direct new tissue formation beyond placement of tissue by printing.[60] Computer aided scaffold design has recently gained attention as a viable option for production of complex tissues such as the lung. Simultaneous printing of hydrogels containing cells and biomaterials together will eventually allow for pre-planned precise placement of cells and proteins within 3D structures. This technology will also eventually allow for spatial control of not only the scaffold structure, but also the type and arrangement of cells, the thickness of the tissue and formation of capillaries and vessels in a way that makes this technique a reasonable approach for generation of physiologically functional lung tissues in the future.

Precise control of hydrogel properties, such as porosity, still remains a challenge in scaffold design. Traditional techniques for creating bulk porosity in polymers have demonstrated success in hydrogels for tissue engineering; however, often the conditions are incompatible with direct cell encapsulation.[61] The porosity, pore architecture and pore interconnectivity play a significant role in cell survival, proliferation, and migration in any scaffold.[62,63] Control of hydrogel features such as pore size can influence the movement of cells in a scaffold as well as their ability to attach to a scaffold. Pore interconnectivity supports movement of oxygen and nutrients to developing tissues. Without using an intrinsic capillary network, the maximal thickness of engineered tissue is approximately 150–200 μm because of insufficient oxygen and nutrient transport within the deeper compartments of the biomaterial.[64] Many different techniques exist to produce hydrogel scaffolds, such as solvent casting-particle leaching, freeze drying, gas foaming, and soft lithography to pattern channels. Modified hydrogels, which are biodegradable, highly elastic materials with shape and pore size similar to that found in the alveolar structure itself have been developed. One scaffold with the potential to meet some of the native lung complexity features novel ICC geometry [Zhang, 2005].[65,66] ICC scaffolds that were used to support human hepatocellular carcinoma cell or human bone marrow stromal cell attachment as well as development of a bone marrow analog system were made of 30% acrylamide monomer-5%bis-acrylamide crosslinker-0.1% (N,N,N,N) tetramethylethylene-diamine accelerator [Zhang 2005].[65–67] Primary colloidal crystals are hexagonally packed lattices of spheres with a wide range of diameters from nanometers to micrometers.[66] ICC are similarly organized structures where the spheres are replaced with cavities, while the interstitial spaces are filled.[66] ICC cavities which exceed the diameter of cells can be used as 3D cell scaffolds (Fig. 9) [Zhang, 2005].[65–67] The open geometry of the ICC lattice as well as the high porosity (74% of free space) and large surface area make ICC-based hydrogel scaffolds an attractive structure with the capacity to facilitate development of alveolus formation (Figs. 9(A)–9(C)). The ICC configuration supports cell attachment (Figs. 9(C) and 9(D)) and growth of cells within the individual chambers (Figs. 9(E) and 9(F)). ICC modified hydrogels can also be formulated to have specific layer-by-layer assembly of growth factors whose controlled release can be used to support migration of cells with site specific cell attachment, as well as growth factor-induced cell differentiation and tissue formation [Zhang, 2005].[65–68] ICC topology affords a simple method for biomaterial design, which can allow some degree of control over cellular interactions or cell migration by varying the sphere diameter — an important consideration for development of engineered lung. Normal lung varies in stiffness from the trachea through the bronchi and culminating at the alveoli. ICC scaffolds are not very elastic and possess limited variation in stiffness. In the future it is possible that ICC manufacture could be expanded to allow production of scaffolds with variable regions of stiffness and with greater elasticity to mimic what is seen in the natural lung.

8. Conclusion

It is evident from the information presented in this chapter that there has been limited use of hydrogels in the production of engineered lung tissue. Natural lung contains a complex and exceptionally variable support scaffold in terms of elasticity, strength, and stiffness. Current hydrogel formulations do not provide variation in ECM stiffness, elasticity, or cellular cues inherent in natural lung. At this time, we feel that existing hydrogel formulations do not fully meet the needs of the lung and do not provide an appropriate platform for support of the physiological functions. This does not mean that hydrogels do not have a place in the bioengineering of lung. We have shown that hydrogels are capable of mimicking soft tissue and have been useful in promotion of cell viability, cell attachment and generation of small pieces of lung tissue *in vitro* and *in vivo*. Currently many groups are using natural AC lung scaffolds to engineer lung tissue. While natural lung scaffolds provide the necessary cues to support tissue development, they are formed from damaged human lungs which are in limited supply. In the future we need to work towards the development of lung scaffolds that are biodegradable, non-immunogenic, and able to mimic lung ECM and possess the variable stiffness and elasticity seen in the trachea, bronchi, bronchioles, lung parenchyma and pleura. A scaffold that will not cause or add to existing obstructive or restrictive lung disease is also a requirement. Our experience with lung bioengineering has convinced us that in the future we will need to work towards producing hybrid hydrogel scaffolds with ECM components which provide cells and developing tissues with the stimulus necessary to allow engineered tissues to function as they do in their natural, native *in vivo* environment.

References

1. O. Wicherle and D. Lim, Hydrophilic gels for biological use. *Nature*, **185**, 117 (1960).
2. J. Kopeček and J. Yang, Hydrogels as smart biomaterials. *Polym. Int.*, **56**(9), 1078–1098 (2007).
3. W. E. Hennink and C. F. Nostrum, Novel crosslinking methods to design hydrogels. *Adv. Drug Deliv. Rev.*, **54**, 13–36 (2002).
4. J. M. Rosiak and F. Yoshii, Hydrogels and their medical applications. *Nucl. Instrum. Meth. B*, **151**, 56–64 (1999).
5. N. Hotz, L. Wilcke, and W. Weber, Design, synthesis, and application of stimulus-sensing bio hybrid hydrogels. *Macromol. Rapid Commun.*, **33**(15), 1280–1285 (2013).
6. C. C. Lin and K. S. Anseth, PEG hydrogels for the controlled release of biomolecules in regenerative medicine. *Pharm. Res.*, **26**(3), 631–643 (2009).
7. J. L. Drury and D. J. Mooney, Hydrogels for tissue engineering: Scaffold design variables and applications. *Biomaterials*, **24**(24), 4337–4351 (2003).
8. D. L. Elbert, A. B. Pratt, M. P. Lutolf, S. Halstenberg, and J. A. Hubbell, Protein delivery from materials formed by self-selective conjugate addition reactions. *J. Control Release*, **76**, 11–25 (2001).
9. S. C. Rizzi and J. A. Hubbell, Recombinant protein-co-PEG networks as cell-adhesive and proteolytically degradable hydrogel matrixes. Part 1: Development and physicochemical characteristics. *Biomacromolecules*, **6**, 1226–1238 (2005).
10. P. van de Wetering, A. T. Metters, R. G. Schoenmakers, and J. A. Hubbell, Poly(ethylene glycol) hydrogels formed by conjugate addition with controllable swelling, degradation, and release of pharmaceutically active proteins. *J. Control. Release*, **102**, 619–627 (2005).
11. T. P. Kraehenbuehl, P. Zammaretti, A. J. Van der Vlies, R. G. Schoenmakers, M. P. Lutolf, M. E. Jaconi, and J. A. Hubbell, Three-dimensional extracellular matrix-directed cardio progenitor differentiation: Systematic modulation of a synthetic cell-responsive PEG-hydrogel. *Biomaterials*, **29**, 2757–2766 (2008).
12. G. P. Raeber, M. P. Lutolf, and J. A. Hubbell, Molecularly engineered PEG hydrogels: A novel model system for proteolytically mediated cell migration. *Biophys. J.*, **89**, 1374–1388 (2005).
13. S. P. Hudson, R. Langer, G. R. Fink, and D. S. Kohane, Injectable *in situ* cross-linking hydrogels for local antifungal therapy. *Biomaterials*, **31**(6), 1444–1452 (2010).
14. C. Gong, T. Qi, X. Wei, Y. Qu, Q. Wu, F. Luo, and Z. Qian, Thermosensitive polymeric hydrogels as drug delivery systems. *Curr. Med. Chem.*, **20**(1), 79–94 (2013).
15. H. Park, S. W. Kang, B. S. Kim, D. J. Mooney, and K. Y. Lee, Shear-reversibly crosslinked alginate hydrogels for tissue engineering. *Macromol. Biosci.*, **9**(9), 895–901 (2009).
16. J. Hannig, Polaxamine-1107 sealing of radio-permeabilized erythrocyte membranes. *Int J. Rad. Biol.*, **75**, 379–385 (1999).
17. R. C. Lee, D. Zhang, and J. Hannig, Biophysical injury mechanisms in electrical shock trauma. *Annual Rev. Biomed. Eng.*, **2**(1), 477–509 (2000).

18. J. M. Collins, M. A. Kohanski, D. J. Dwyer, B. Hayete, C. A. Lawrence, and J. J. Collins, A common mechanism of cellular death Induced by bactericidal antibiotics. *Cell*, **130**, 797–810 (2007).
19. R. Zhang, R. L. Hunter, E. A. Gonzalez, and F. A. Moore, Poloxamer 188 prolongs survival of hypotensive resuscitation and decreases vital tissue injury after full resuscitation. *Shock*, **32**(4), 442–450 (2009).
20. G. Wu and K. Y. C. Lee, Effects of poloxamer 188 on phospholipid monolayer morphology: An atomic force microscopy study. *Langmuir*, **25**, 2133–2139 (2009).
21. S. M. Moghimi and A. C. Hunter, Poloxamers and poloxamines in nanoparticle engineering and experimental medicine. *Trends Biotechnol.*, **18**(10), 412–420 (2000).
22. S. F. Khattak, S. R. Bhatia, and S. C. Roberts, Pluronic F127 as a cell encapsulation material: Utilization of membrane-stabilizing agents. *Tissue Eng.*, **11**(5–6), 974–983 (2005).
23. A. Roberts, B. E. Wyslouzil, and L. Bonassar, Aerosol delivery of mammalian cells for tissue engineering. *Biotechnol Bioeng.*, **91**(7), 801–807 (2005).
24. J. E. Nichols, J. A. Niles, D. DeWitt, D. Prough, M. Parsley, S. Vega, A. Cantu, E. Lee, and J. Cortiella, Neurogenic and neuro-protective potential of a novel subpopulation of peripheral blood-derived CD133+ ABCG2+ CXCR4+ mesenchymal stem cells: Development of autologous cell-based therapeutics for traumatic brain injury. *Stem Cell Res. Ther.*, **4**(1), 1–23 (2013).
25. J. Cortiella, J. E. Nichols, K. Kojima, L. J. Bonassar, P. Dargon, A. K. Roy, M. P. Vacant, J. A. Niles, and C. A. Vacanti, Tissue-engineered lung: An *in vivo* and *in vitro* comparison of polyglycolic acid and pluronic F-127 hydrogel/somatic lung progenitor cell constructs to support tissue growth. *Tissue Eng.*, **12**(5), 1213–1225 (2006).
26. J. Cortiella, J. Niles, A. Cantu, A. Brettler, A. Pham, G. Vargas, S. Winston, J. Wang, S. Walls, and J. E. Nichols, Influence of a cellular natural lung matrix on murine embryonic stem cell differentiation and tissue formation. *Tissue Eng. Pt. A*, **16**(8), 2565–2580 (2010).
27. J. E. Nichols, J. A. Niles, and J. Cortiella, Production and utilization of a cellular lung scaffolds in tissue engineering. *J. Cell Biochem.*, **113**(7), 2185–2192 (2012).
28. K. T. Nguyen and J. L. West. Photopolymerizable hydrogels for tissue engineering applications. *Biomaterials*, **23**(22), 4307–4314 (2002).
29. M. P. Lutolf and J. A. Hubbell, Synthetic biomaterials as instructive extracellular microenvironments for morphogenesis in tissue engineering. *Nat. Biotechnol.*, **23**(1), 47–55 (2005).
30. E. Dawson, G. Mapili, K. Erickson, S. Taqvi, and K. Roy, Biomaterials for stem cell differentiation. *Adv. Drug Deliv. Rev.*, **60**(2), 215–228 (2008).
31. A. S. Sawhney, C. P. Pathak, and J. A. Hubbell, Bioerodible hydrogels based on photopolymerized poly (ethylene glycol)-co-poly (.alpha.-hydroxy acid) diacrylate macromers. *Macromolecules*, **26**(4), 581–587 (1993).
32. P. Martens and K. S. Anseth, Characterization of hydrogels formed from acrylate modified poly (vinyl alcohol) macromers. *Polymer*, **41**(21), 7715–7722 (2000).

33. T. V. Chirila, I. J. Constable, G. J. Crawford, S. Vijayasekaran, D. E. Thompson, Y. C. Chen, W. A. Fletcher, and B. J. Griffin, Poly (2-hydroxyethyl methacrylate) sponges as implant materials: *In vivo* and *in vitro* evaluation of cellular invasion. *Biomaterials*, **14**(1), 26–38 (1993).
34. BD Biosciences website. BD Matrigel ™ Basement Membrane Matrix, referenced October 3, 2013 http://www.bdbiosciences.com/ecat/txDetailedTable.jsp?modeCategory=FULL&item=775674&action=SELECT&form=formTree_catBean&size=20.
35. E. Lavik and R. Langer, Tissue engineering: Current state and perspectives. *Appl. Microbiol. Biotechnol.*, **65**(1), 1–8 (2004).
36. H. Blau, D. E. Guzowski, Z. A. Siddiqi, E. M. Scarpelli, and R. S. Bienkowski, Fetal type 2 pneumocytes form alveolar-like structures and maintain long-term differentiation on extracellular matrix. *J. Cell Physiol.*, **136**(2), 203–214 (1988).
37. M. R. Chinoy, M. T. Antonio-Santiago, and E. M. Scarpelli Maturation of undifferentiated lung epithelial cells into type II cells *in vitro*: A temporal process that parallels cell differentiation *in vivo*. *Anat. Rec.*, **240**(4), 545–554 (1994).
38. H. Sugihara, S. Toda, S. Miyabara, C. Fujiyama, and N. Yonemitsu, Reconstruction of alveolus-like structure from alveolar type II epithelial cells in three-dimensional collagen gel matrix culture. *Am. J. Pathol.*, **142**(3), 783–792 (1993).
39. P. Chen, E. Marsilio, R. H. Goldstein, I. V. Yannas, and M. Spector, Formation of lung alveolar- like structures in collagen-glycosaminoglycan scaffolds *in vitro*. *Tissue Eng.*, **11**(9–10), 1436–1448 (2005).
40. J. Chakir, N. Pagé, Q. Hamid, M. Laviolette, L. P. Boulet, and M. Rouabhia, Bronchial mucosa produced by tissue engineering: A new tool to study cellular interactions in asthma. *J. Allergy Clin. Immunol.*, **107**(1), 36–40 (2001).
41. J. S. Paquette, V. Moulin, P. Tremblay, V. Bernier, M. Boutet, M. Laviolette, F. A. Auger, L. P. Boulet, and F. Goulet, Tissue-engineered human asthmatic bronchial equivalents. *Eur. Cell Mater.*, **7**, 1–11 (2004).
42. M. M. Choe, A. A. Tomei, and M. A. Swartz, Physiological 3D tissue model of the airway wall and mucosa. *Nat. Protoc.*, **1**(1), 357–362 (2006).
43. M. J. Mondrinos, S. Koutzaki, E. Jiwanmall, M. Li, J. P. Dechadarevian, P. I. Lelkes, and C. M. Finck, Engineering three-dimensional pulmonary tissue constructs. *Tissue Eng.*, **12**(4), 717–728 (2006).
44. M. Li, M. J. Mondrinos, X. Chen, M. R. Gandhi, F. K. Ko, and P. I. Lelkes, Co-electrospun poly (lactide-co-glycolide), gelatin, and elastin blends for tissue engineering scaffolds. *J. Biomed. Mater. Res. Pt. A*, **79**(4), 963–973 (2006).
45. M. J. Mondrinos, S. Koutzaki, P. I. Lelkes, and C. M. Finck, A tissue-engineered model of fetal distal lung tissue. *Am. J. Physiol. Lung Cell Mol. Physiol.*, **293**(3), L639–L650 (2007).
46. A. J. Engler, F. Rehfeldt, S. Sen, and D. E. Discher, Microtissue elasticity: Measurements by atomic force microscopy and its influence on cell differentiation. *Methods Cell Biol.: Cell Mech.*, **83**, 521–545 (2007).
47. A. J. Engler, S. Sen, H. L. Sweeney, and D. E. Discher, Matrix elasticity directs stem cell lineage specification. *Cell*, **126**(4), 677–689 (2006).

48. C. S. Chen, M. Mrksich, S. Huang, G. M. Whitesides, and D. E. Ingber, Geometric control of cell life and death. *Science*, **276**(5317), 1425–1428 (1997).
49. R. Singhvi, A. Kumar, G. P. Lopez, G. N. Stephanopoulos, D. I. C. Wang, G. M. Whitesides, and D. E. Ingber Engineering cell shape and function. *Science*, **264**(5159), 696–698 (1994).
50. O. W. Petersen, L. Ronnovjessen, A. R. Howlett, and M. J. Bissell, Interaction with basement- membrane serves to rapidly distinguish growth and differentiation pattern of normal and malignant human breast epithelial cells. *Proc. Natl. Acad. Sci. U.S.A.*, **89**(19), 9064–9068 (1992).
51. H. Tanaka, C. L. Murphy, C. Murphy, M. Kimura, S. Kawai, and J. M. Polak, Chondrogenic differentiation of murine embryonic stem cells: Effects of culture conditions and dexamethasone. *J. Cell Biochem.*, **93**(3), 454–462 (2004).
52. M. W. Tibbitt and K. S. Anseth, Hydrogels as extracellular matrix mimics for 3D cell culture. *Biotechnol. Bioeng.*, **103**(4), 655–663 (2009).
53. C. F. Andrade, A. P. Wong, T. K. Waddell, S. Keshavjee, and M. Liu, Cell-based tissue engineering for lung regeneration. *Am. J. Physiol. Lung Cell Mol. Physiol.*, **292**(2), L510–L518 (2007).
54. M. J. Mondrinos, S. H. Koutzaki, H. M. Poblete, M. C. Crisanti, P. I. Lelkes, and C. M. Finck, *In vivo* pulmonary tissue engineering: Contribution of donor-derived endothelial cells to construct vascularization. *Tissue Eng. Pt. A*, **14**(3), 361–368 (2008).
55. D. Singh, S. M. Zo, A. Kumar, and S. S. Han, Engineering three-dimensional macroporous hydroxyethyl methacrylate-alginate-gelatin cryogel for growth and proliferation of lung epithelial cells. *J. Biomater. Sci., Polym. Ed.*, **24**(11), 1343–1359 (2013).
56. A. K. Ekaputra, G. D. Prestwich, S. M. Cool, and D. W. Hutmacher, The three-dimensional vascularization of growth factor-releasing hybrid scaffold of poly (epsilon- caprolactone)/collagen fibers and hyaluronic acid hydrogel. *Biomaterials*, **32**(32), 8108–8117 (2011).
57. J. Y. Sun, X. Zhao, W. R. Illeperuma, O. Chaudhuri, K. H. Oh, D. J. Mooney, J. J. Vlassak, and Z. Suo, Highly stretchable and tough hydrogels. *Nature*, **489**(7414), 133–136 (2012).
58. A. M. Kloxin, K. J. Lewis, C. A. DeForest, G. Seedorf, M. W. Tibbitt, V. Balasubramaniam, and K. S. Anseth, Responsive culture platform to examine the influence of microenvironmental geometry on cell function in 3D. *Integr. Biol.*, **4**(12), 1540–1549 (2012).
59. T. Boland, T. Xu, B. Damon, and X. Cui, Application of inkjet printing to tissue engineering. *Biotech. J.*, **1**(9), 910–917 (2006).
60. N. E. Fedorovich, J. Alblas, J. R. de Wijn, W. E. Hennink, A. J. Verbout, and W. J. Dhert, Hydrogels as extracellular matrices for skeletal tissue engineering: State-of-the-art and novel application in organ printing. *Tissue Eng.*, **13**(8), 1905–1925 (2007).
61. A. Nasim, J. W. Nichol, X. Zhong, C. Ji, S. Koshy, A. Khademhosseini, and F. Dehghani, Controlling the porosity and microarchitecture of hydrogels for tissue engineering. *Tissue Eng. B Rev.*, **16**(4), 371–383 (2010).
62. B. Mandal and S. Kundu, Cell proliferation and migration in silk fibroin 3D scaffolds. *Biomaterials*, **30**, 2956 (2009).

63. S. M. Lien, L. Y. Ko, and T. J. Huang, Effect of pore size on ECM secretion and cell growth in gelatin scaffold for articular cartilage tissue engineering. *Acta Biomater*, **5**(2), 670–679 (2009).
64. C. Fidkowski, M.R. Kaazempur-Mofrad, J. Borenstein, J. Vacanti, R. Langer, and Y. Wang, Endothelialized microvasculature based on a biodegradable elastomer. *Tissue Eng.*, **11**(1–2), 302–309 (2005).
65. N. A. Kotov, Y. Liu, S. Wang, C. Cumming, M. Eghtedari, G. Vagas, M. Motamedi, J. Nichols, and J. Cortiella, Inverted colloidal crystals as three-dimensional cell scaffols. *Langmuir*, **20**(19), 7887–7892 (2004).
66. J. E. Nichols, J. Cortiella, J. Lee, J. A. Niles, M. Cuddihy, S. Wang, J. Bielitzki, A. Cantu, R. Mlcak, E. Valdivia, R. Yancy, M. L. McClure, and N. A. Kotov, *In vitro* analog of human bone marrow from 3D scaffolds with biomimetic inverted colloidal crystal geometry. *Biomaterials*, **30**(6), 1071–1079 (2009).
67. S. H. A. Lee, N. M. Abrams, P. G. Hoertz, G. D. Barber, L. I. Halaoui, and T. E. Mallouk. Coupling of titania inverse opals to nanocrystalline titania layers in dye-sensitized solar cells. *J. Phys. Chem. B*, **112**(46), 14415–14421 (2008).
68. S. R. Pullela, C. Andres, W. Chen, C. Xu, L. Wang, and N. A. Kotov, Permselectivity replication of artificial glomerular basement membranes in nanoporous collagen multilayers. *J. Phys. Chem. Lett.*, **2**(16), 2067–2072 (2011).
69. B. G. Zani, K. Kojima, C. A. Vacanti, and E. R. Edelman, Tissue-engineered endothelial and epithelial implants differentially and synergistically regulate airway repair. *Proc. Natl. Acad. Sci. USA*, **105**(19), 7046–7051 (2008).
70. C. Miller, S. George, L. Niklason, Developing a tissue-engineered model of the human bronchiole. *J. Tissue Eng. Regen. Med.*, **4**(8), 619–627 (2010).

Chapter 15

Hydrogels for Hepatic Tissue Engineering

*Kerim B. Kaylan and Gregory H. Underhill**
Department of Bioengineering,
University of Illinois at Urbana-Champaign, Urbana, IL 61801, USA
**gunderhi@illinois.edu*

1. Summary

Hepatic tissue engineering aims to develop novel therapeutic approaches for liver disease and promote an improved understanding of fundamental characteristics of liver biology. This chapter will review current approaches in hepatic tissue engineering that utilize hydrogels, ranging from *in vitro* model systems to bioartificial livers (BAL) devices and implantable tissue constructs. Reproducing the full complexity of liver architecture and dynamics remains a key barrier to the long-term stabilization and control of hepatic cultures and the development of next-generation therapies for liver failure. Hydrogels provide a valuable means for overcoming these barriers by enabling the systematic analysis and control of hepatocellular environments.

2. Liver Failure and Current Therapies

Liver failure as a result of either chronic or acute disease was the cause of 44,677 deaths in 1998 in the United States.[1] Chronic liver failure alone is currently the 12th most frequent cause of death and accounts for more than half of deaths due to liver failure.[1,2] Categorized into alcohol-induced and non-alcoholic fatty liver diseases, chronic liver failure is associated with a diffuse array of risk factors[3–6] and manifests a range of abnormalities and liver damage, such as steatosis, steatohepatitis with fibrosis, and cirrhosis.[4,6,7] Acute liver failure, caused primarily by viral hepatitis and drug-induced liver injuries, results in rapid loss of hepatic metabolic and immunologic function and often leads to progressive multi-organ failure.[8,9]

Both chronic and acute liver failure are associated with poor survival without orthotopic liver transplantation (OLT), the current standard of care for end-stage liver disease and other forms of liver failure.[10–12] Although OLT has markedly improved mortality for liver failure and related diseases, the persistent (and worsening) shortage of donor livers available for transplant indicates that alternative therapies are needed.[12] Surgical methods to alleviate this shortage have been explored, namely partial OLT from living,[13,14] cadaveric,[13,15] non-heart-beating,[15–18] or xenogeneic[19–22] donors. These methods take advantage of the innate capacity of the liver to regenerate but can result in biliary and vascular complications.[13,23–25] Other concerns regarding OLT include both hepatic and non-hepatic complications,[12] immune rejection,[11,26] costs and difficulties associated with major surgery, and the potential for recurrent primary hepatic disease.[12] Common treatments other than OLT, such as the administration of fluids and serum proteins[27,28] or the use of non-biological extracorporeal systems,[29,30] serve primarily palliative or short-term bridging functions. As no current alternative therapy is able to renew the full range of hepatic functions or match the effectiveness of OLT, more comprehensive approaches to liver support are necessary.

3. Cell-Based Therapies

Cell-based therapies, namely BAL devices, hepatocyte transplantation, and implantable tissue engineered constructs, offer both short- and long-term support of a broader range of hepatic function than the alternatives to OLT identified above. BAL devices, which are extracorporeal bioreactors containing hepatocytes to aid native liver function, have been proposed as a therapeutic bridge to OLT.[29–32] The clinical success of BAL devices depends on the design and control of several different factors, such as long-term maintenance of cell viability and hepatic function, bioreactor scaling, bidirectional mass transfer, bile drainage, and cell sourcing.[29,33–37] Hepatocyte transplantation, which provides more permanent support than BAL devices, has been shown to be effective in some trials and is both less risky and less costly than whole organ transplantation while also retaining the possibility of native liver regeneration.[38–41] Clinical validation and use of hepatocyte transplantation has been restricted by a severe shortage of high quality donor cells, poor cell engraftment, and a lack of suitable preclinical models.[34,42,43] Implantable tissue engineered constructs, the most comprehensive and experimental approach for long-term hepatic support, entails grafting of scaffolds seeded with cells. The difficulties of implantable hepatic constructs are comparable to those of hepatocyte transplantation but additionally include the integration of the vasculature, bile ducts, and other functional components of the graft with those of the host liver.

4. Hepatic Tissue Engineering

There remain several key challenges in hepatic tissue engineering. Paramount among these is the loss of replicative ability and liver-specific functions in hepatocytes cultured *in vitro*. Improved *in vitro* culture models and *ex vivo* platforms capable of dynamic environmental presentation and control of multiple cell types would improve understanding of basic hepatic biology and permit sustained and functional hepatic cultures. The use of these improved platforms in conjunction with stem cell reprogramming and other developing sources of cells, such as hepatic progenitors, would address the acute shortage of hepatocytes for transplantation and other therapeutic uses. Lastly, enhanced control of the parameters governing scale-up, biological transport, biliary and vascular integration, and immunogenicity would enable translation of experimental cell-based therapies to the clinic. Although further investigation and development is needed, hydrogel scaffolds have demonstrated the capacity to address many of these challenges.

5. Hydrogel Platforms

Hydrogels are insoluble three-dimensional (3D) networks of crosslinked hydrophilic polymers capable of absorbing many times their dry weight in water. Applications for hydrogels range from soft contact lenses[44] and surgical adhesives[45] to smart biomaterials[46,47] and scaffolds for tissue engineering.[48–51] Hydrogels are increasingly utilized in tissue engineering and regenerative medicine due to the tunability of their physical, mechanical, chemical, and functional properties to the demands of each application. Considerable effort in tissue engineering has focused on utilizing the myriad capabilities of hydrogels to create biocompatible, modular hydrogel platforms for culturing cells and tissues in arbitrary micro- and macro-scale 3D configurations. These capabilities are crucial for hepatic tissue engineering, as it has long been known that hepatocytes experience loss of function and phenotype in traditional two-dimensional (2D) cultures[52–55] and that different ECM compositions result in disparate effects on hepatic morphology and function.[56,57] Furthermore, an ensemble of other factors dictate the fate and behavior of hepatic cells, such as the extent and nature of cell–cell and cell–extracellular matrix (ECM) interactions, concentration of metabolites or nutrients, and presentation of growth factors. In facilitating control of the factors listed above, hydrogels enhance hepatic function and phenotype *in vitro* and enable 3D cultures that better resemble native liver tissue, simultaneously enabling improved studies of hepatic biology and enhancing prospects for the translation of therapeutic approaches to the clinic.

Many of the advantages of hydrogels that make them appealing for a broad range of tissue engineering applications are relevant to hepatic tissue engineering. Hydrogels are highly biocompatible and structurally similar to many tissues and the ECM.[58] They possess low interfacial tension and minimal mechanical and frictional irritation, further promoting biocompatibility.[59–61] Moreover, their capacity for polymerization in the presence of cells eases seeding and enables full encapsulation of cells for both *in vitro* and *in vivo* studies. Combinations of different chemical moieties, crosslinking methods, copolymer blends, and other properties and parameters enable precision engineering of the cellular microenvironment presented by the hydrogel to further match that of the liver. In addition to the physical and chemical properties of the scaffold, hydrogels offer tailored cell–matrix interactions and functionality, *e.g.*, adhesion, biodegradation, or growth factor delivery, through the addition of various biofunctional groups.[61,62] Hydrogels can also incorporate dynamically controlled physicochemical stimuli and time-dependent factor release to recapitulate the cell-chemical and cell-protein signaling present in liver tissue.[63,64] Microtechnological methods such as 3D photopatterning and microfluidics enable additional spatial and dynamic control of hydrogel architecture and related environmental cues.[51,65–67] As with most other tissue engineering scaffolds, the mass transport of gases, metabolites, nutrients, and waste in hydrogels are important determinants of biocompatibility. Additionally, as the function and morphology of hepatic cells are dependent on the concentration of oxygen, hormones, and nutrients, the transport properties of hydrogel-based hepatic culture systems must be carefully considered for each application.[68–73] These considerations must also be balanced with the mechanical properties of the hydrogel, as both are a function of the crosslinking density and other physical and chemical properties.[51,59]

Choosing the correct scaffold material can be difficult and is especially consequential in hepatic tissue engineering, as hepatocytes are attachment-dependent cells that lose liver-specific function without an appropriate ECM. In this section, we review biologically-derived and synthetic polymer hydrogel scaffolds relevant to hepatic tissue engineering and discuss the advantages, disadvantages, and applications of each.

5.1. Biologically-derived scaffolds

Biologically-derived scaffolds are composed either of protein fibrils and fibers or hydrated networks of glycosaminoglycans, though some materials include both.

Their biocompatibility and macromolecular properties, which are similar to natural ECM, make them an appealing choice for tissue engineering applications. Many types of biologically-derived scaffolds, such as collagen or fibrin, offer inherent biological recognition by cells through the presentation of receptor-binding ligands or moieties and are thus susceptible to adhesion or cell-triggered degradation by collagenase and related enzymes.[50,74] Furthermore, biologically-derived scaffolds often possess natural growth factor and metabolite sequestration, presentation, and release functionalities.[75] Although biologically-derived scaffolds are readily derived from multiple tissue sources, the processes of production and purification can cause high batch-to-batch variability.[51] Depending on scaffold source, there is also the risk of immunogenicity or pathogen transmission, especially acute concerns for clinical applications of hydrogels, such as use in BAL devices. Biologically-derived scaffolds are also subject to the same toxicity concerns as synthetic polymer scaffolds when functionalized, as each variant of a biologically-derived scaffold known to be biocompatible must nonetheless be revalidated.[74]

Collagen, a well-established protein scaffold, exhibits favorable biodegradability, antigenicity, and cell-binding properties and has a history of use in studies of hepatic biology.[48,53,76,77] Notably, so-called "sandwich" (or double-gel) cultures of hepatic cells on collagen brought attention to the effects of matrix composition and architecture on the *in vitro* functionality of hepatocytes.[53,55,78,79] Due to the ubiquity of collagen in the body and the considerable effort directed towards its use as a scaffold, there exist many methods of modifying its physical and biological properties through conjugation with chemical crosslinkers (such as glutaraldehyde and formaldehyde), physical treatments (such as UV irradiation, freeze-drying, or heating), and blending with other polymers.[50,80,81] Furthermore, incorporation of fibronectin, chondroitin sulfate, and other biologically-derived materials into collagen have been employed as a means to encourage cell adhesion.[56,75] Collagen has also been utilized in conjunction with alginate as a means of encapsulating and immunoisolating hepatocytes for autologous transplant.[77] The use of collagen is limited, however, by its low mechanical strength, rapid degradation, and high cost.[48,75]

Alginate, a linear polymer of $\beta(1\rightarrow4)$ linked D-mannuronic acid and $\alpha(1\rightarrow4)$-linked L-guluronic acid from brown seaweed.[59] is also widely used in hepatic biology and tissue engineering.[82–85] It is non-toxic to a variety of hepatic cell types[86] and is easier to functionalize than collagen, though it undergoes a variable degradation process *in vivo*.[87] Although alginate contains repetitive carbohydrate motifs, these do not include integrin-binding domains such as the RGD motifs of ECM proteins.[88] Alginate, either alone or in combination with other polymers, has frequently been used to encapsulate hepatocytes as a means of increasing hepatic function.[89–95] In particular, Dvir-Ginzberg *et al.*, have suggested that the macroporous nature of alginate encourages spheroid formation, cell–cell communication, and prolonged hepatic function.[90] Human hepatocyte cell lines proliferating as cohesive spheroid colonies in alginate exhibit markedly upregulated liver-specific function with regards to synthesis and detoxification.[88] Combinations of alginate with galactosylated chitosan and heparin increased primary mouse hepatocyte viability over alginate alone after 15 days of incubation within the scaffold.[96] Alginate has also commonly been used for immunoisolation of transplanted cells.[97,98]

Chitosan, a linear saccharidic polymer obtained by deacetylation of chitin, is composed of glucosamine subunits with active primary amine subunits. Its similarity to native glycosaminoglycans, which are significant components of the liver ECM, makes chitosan an attractive choice for hepatic tissue engineering.[99] The effects of different types of glycosaminoglycans on cell–cell contacts and cell–matrix interactions modulate cell morphology, differentiation, movement, synthesis, and function.[100–102] Chitosan blended with collagen, albumin, and gelatin has been used to create primitive liver tissue organoids from co-cultures of primary rat hepatocyte and 3T3 fibroblasts.[101] Conjugation of fructose (known to have protective effects on hepatocytes via binding to the asialoglycoprotein receptor[103]) with chitosan induces cellular aggregation and improved albumin secretion and urea synthesis in primary rat hepatocytes.[104] Alginate–chitosan copolymers have also been utilized to observe the effects of pore size and ligand concentration on the liver-specific functions of hepatocyte spheroids.[91–93]

Fibrin is a fibrous, non-globular protein involved in clotting and wound healing. It is a well-established scaffold, both for research and clinical use, in which it is used as a sealant and adhesive in surgery.[45,105] Fibrin can be degraded naturally during migration and wound healing and is thus of some interest for implantable applications. The advantages of fibrin over other scaffolds include high seeding efficiency, more uniform cell distribution, and adhesive capabilities.[48,105,106] Fibrin can also be produced from the blood of a patient and used as an autologous scaffold, making it attractive for translational applications.[107,108]

Other examples of biologically-derived hydrogel scaffolds used in the hepatic tissue engineering include hyaluronic acid (HA),[109–111] heparin,[112–114] gelatin,[115,116] cellulose,[117,118] and liver-derived ECM.[119,120]

5.2. Synthetic polymer scaffolds

Synthetic polymer scaffolds enable rational scaffold design through manipulation of various parameters and functionalities, such as molecular weight, block structure, degradable linkages, and crosslinking modes.[50,58,75] Fine control of these parameters, among others, allow tailoring of the mechanical and chemical properties of scaffold to each application. Although synthetic polymers are bioinert, their biological properties, namely adhesion, proteolytic susceptibility, and factor release, can nonetheless be controlled through the addition of functional moieties or components.[121,122] Functionalization in this manner facilitates applications such as degradable constructs for implantation and drug delivery.[123,124] Synthetic polymers are also more easily and reproducibly synthesized than biologically-derived scaffolds and elicit fewer concerns regarding immunogenic effects, though harsh chemistries may reduce biocompatibility if used.[51,74]

Poly(ethylene glycol) (PEG) is an extensively-utilized synthetic polymer that is non-toxic and non-immunogenic.[49] PEG is easily modified to recreate aspects of the *in vivo* microenvironment that guide cellular morphology, structure, and function.[121,125] Though PEG resists protein adsorption and is not suitable for cell adhesion without motification, PEG hydrogels are widely used for 3D cultures because of the ease and flexibility of modification with functional groups.[62,75,126] When functionalized with diacrylate, PEG can be photocrosslinked *in situ* to create 3D constructs with uniform distributions of cells. Hepatocyte and hepatoblast viability and function have been assessed favorably in functionalized PEG hydrogels.[127,128] Other applications for PEG have included use as a barrier for patterning,[129] a platform to study pore function with regards to viral infection of hepatocytes,[130] or as a micropatterned 3D scaffold.[131,132]

Polylactic acid (PLLA) and poly(lactic-*co*-glycolic acid) (PLGA) together comprise another common synthetic polymer system in hepatic tissue engineering. Both are biocompatible and biodegradable while offering tunable degradation time based on the ratio of PLLA to PLGA. Given these favorable characteristics, it is commonly used in transplantation.[133–137] Composites with poly(vinyl) alcohol have been shown to enhance seeding of primary rat hepatocytes.[133] Similarly, hydrolysis under alkali conditions and coating with various ECM proteins improves scaffold attachment of hepatocytes.[137–139] As the hydrolytic degradation products (namely carboxylic acid monomers and oligomers) of PLGA scaffolds decrease enviromental pH, degrade polypeptides, and cause poor engraftment, there have been efforts to control degradation through the use of pH modifiers (*e.g.*, weakly basic salts), covalently linked modifiers (*e.g.*, PEGylation), or proton sponges.[140,141]

A variety of other synthetic polymers are utilized in hepatic tissue engineering. PVA has demonstrated use for hepatocyte encapsulation and implantation in rat models.[117,133,142] Poly(acrylic acid) or polyacrylamide (PA), possesses easily tuned stiffness by means of crosslinker monomer concentration[143] and has been used to demonstrate mechanosensitivity of hepatocytes to certain material regimes.[144] Synthetic self-assembling peptide scaffolds have also been used with primary rat hepatocytes to demonstrate comparability with sandwich cultures using collagen and Matrigel substrates.[116] Other groups have also used self-assembling peptide hydrogels with biologically active motifs towards similar ends.[145] Compatibility and hepatic functions have been demonstrated using woven and textile scaffolds, namely combinations of poly(ethylene therephtalat) with PLGA and poly[N-p-vinylbenzyl-D-lactoamide].[146,147]

6. Design Considerations

There are several interconnected goals when designing hydrogels for hepatic tissue engineeering. Biocompatibility is key for both *in vitro* and implantable applications, as with any other tissue engineering scaffold. Furthermore, ensuring a limited toxic, immunologic, and foreign body response permits use in translational and clinical applications.[47,51,74] The maximization and maintenance of hepatic function is another key goal, especially for applications involving BAL devices or cells intended for direct transplantation or use in implantable constructs. In designing hydrogel platforms towards these goals, all relevant aspects of cellular behavior should be considered, namely viability, attachment, growth, morphology, differentiation state, and liver-specific function. The physical, chemical, and bioactive characteristics of the scaffold with regards to the microenvironmental factors that control these cellular behaviors must be in alignment with knowledge of the native hepatic environment.[25,148] In this section, we discuss the scaffold design parameters utilized to achieve the goals listed above, namely material properties, biofunctionalization, and architecture at both microscale and macroscale (Fig. 1).

6.1. Material properties

One of the most important properties of a hydrogel is stiffness, often controlled through the percentage of the polymer or crosslinking agent used — higher degrees of crosslinking lead to stiffer gels. Reported values of the elastic modulus of healthy liver tissue range from E~0.4–760 kPa.[149–153] Stiffness has been shown to modulate a number of different aspects of hepatic cells[150] and increases in diseased states such as fibrosis.[153–157] Mechanical signaling via matrix stiffness can also act in concert with other forms of cellular communication. For example, both tension and signaling by transforming growth factor-β are required for portal fibroblast differentiation to myofibroblasts.[152] Studies of stiffness have also been combined with substrate-bound ligand presentation. Primary rat hepatocytes exhibited dedifferentiation, reduced hepatic function, and increased growth on rigid rather than compliant PA hydrogel substrates.[144] Additionally, the extent of cell spreading on the rigid surface was sensitive to different concentrations of PA-conjugated fibronectin, a known hepatic morphogen;[158,159] the same was not true for cells on the compliant surface.[144] Similarly, Chen *et al.* used a tunable polyelectrolyate multilayer system to control the elastic

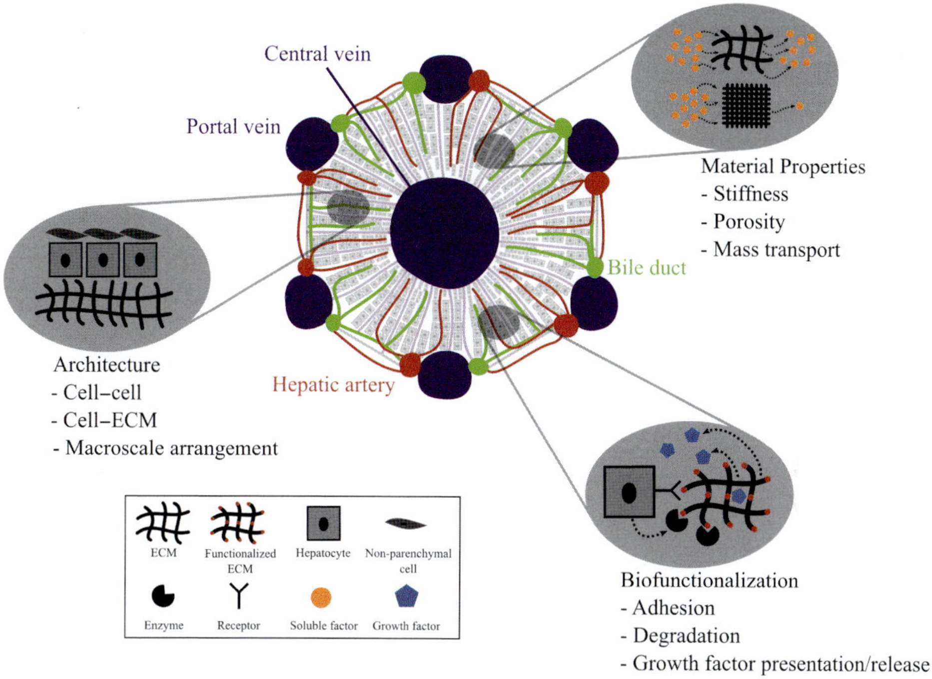

Fig. 1. Design considerations for hepatic hydrogel platforms. The functional unit of the liver is the lobule, shown diagrammatically in this figure. Every polygonal lobule is comprised of a central vein connected to the portal triads (which each include a hepatic artery, portal vein, and bile duct) at the vertices by sinusoids. Plates of hepatocytes and non-parenchymal cells extend between the portal triad and the central vein, occupying regions between sinusoids. The insets illustrate how the properties of hydrogel platforms can be tailored to create liver-like environments for hepatic cultures. Together, the material properties, biofunctionality, and architecture of each particular design have profound effects on biocompatibility, differentiation state, functionality, and potential for therapeutic use.

modulus experienced by primary rat hepatocytes in 2D, substantiating the compliance-mediated effects observed by Semler *et al.* over a much wider range of moduli.[144,149] In discordance with the observations above, HepG2 hepatocellular carcinoma cells encapsulated in a fat-like collagen-PEG hydrogel formed malignant spheroids with limited hepatocyte-like activities while cells encapsulated in a stiffer collagen-PEG hydrogel formed small spheroids with suppressed malignancy and hepatocyte-like phenotype.[160]

Modulation of scaffold porosity is utilized in hepatic tissue engineering as a means of providing mechanical support and regulating biological cues for maintenance of proliferation and morphology. Porosity, like stiffness, is a function of percentage polymer or crosslinking agent used — higher degrees of crosslinking lead to smaller average pore size. Increasing crosslinking density imposes restrictions on mass transport through the hydrogel, a side effect that must be considered during scaffold design. Mass transport properties determine how nutrients, gases, waste products, bioactive agents, growth factors, *etc.*, are exchanged with the scaffold and delivered to cells. Convection does not play a significant role in scaffold transport unless the pore size is very large; diffusion alone is usually the primary driving phenomenon for most applications provided a principally amorphous and homogeneous gel.[74] Scaffolds in which the pore size is larger than individual cells are essentially 2D surfaces from the perspective of the cell, limiting the ability of the scaffold to recapitulate 3D cues.[34] For example, hepatocytes cultured on collagen foams with pore size ranging from 10 μm to 82 μm exhibited sustained differentiation at subcellular and supercellular ranges of pore size and dedifferentiation at intermediate ranges of pore size.[161] Spheroid formation is also controlled by porosity, as demonstrated by the enhanced aggregation of hepatocytes in alginate sponges[89,93,162] and similarly in gelatin microspheres.[163]

6.2. Biofunctionalization

Addressing the bioinert nature of synthetic polymer scaffolds and sometimes limited functionality of biologically-derived scaffolds requires controlled integration of chemical moieties, peptide domains, or proteins to enable adhesion, degradation, and other cell–ECM interactions. Approaches to improving adhesion and facilitating hepatocyte–matrix interactions have also included coating with ECM molecules,[137,138] incorporating sugars,[164,165] and attaching bioactive proteins.[147] Functionalization with adhesive peptides that interact with integrins or laminins, such as RGD,[166,167] LGPA,[168] or YIGSR,[168] is commonly utilized and known to improve long-term stability of hepatocytes.[127,169] Genove *et al.*, modified RAD16-I, a self-assembling peptide, to create synthetic scaffolds containing integrin-, laminin-, or heparin-binding domains.[145] Comparison of primary rat hepatocytes cultured in functionalized RAD16-I scaffolds with standard collagen sandwich cultures by quantitative PCR and proteome analysis of secreted proteins showed that the peptide sandwich configuration maintained similar levels of differentiation and function. Presentation of proteins like fibronectin,[144] fibrinogen,[170] decorin,[149] or hepatocyte growth factor[171] is also known to improve hepatocyte morphology and function *in vitro.*

Biodegradable hydrogel scaffolds present a more biomimetic environment to suspended cells, ideally undergoing the same controlled breakdown and remodeling as most living tissue while also gradually ceding function to native tissue if implanted.[59,168,172,173] Most sythetic hydrogel scaffold materials are not biodegradable but can, however, be modified with degradable crosslinkers or protease-sensitive peptide sequences.[124] Both inherently biodegradable and functionalized hydrogels have been utilized in *in vitro* hepatic models.[101,134,146,165,171,174,175] Alternatively, copolymers with collagen, HA, or other biologically-derived scaffolds can imbue biodegradability.[74] Biodegradation, particularly in synthetic hydrogel systems, also enables the use of hydrogels as drug or growth factor depots to encourage natural processes.[64,74,171]

6.3. Architecture

The functional unit of the liver is the polygonal lobule, which is comprised of a central vein and portal triads at each vertice (Fig. 1). Hepatocytes are arranged in plate-like structures occupying the regions between sinusoids, which stretch from the central vein to the portal triads. The entire lobule is non-uniform with regards to cell type, ECM composition, metabolite and growth factor concentration, mechanical properties, and other parameters. Reconstructing the elements of this complex 3D architecture, *i.e.*, cell viability and function, cell–cell interactions, macroscale properties, *etc.*, is essential to long-term *ex vivo* hepatic cultures and to the development of future therapies for liver failure. Hydrogels offer one means for doing so, as they can be controlled and patterned in either 2D or 3D using a number of different techniques, primarily soft lithography, microcontact printing, micromolding, and photolithography, which are also discussed in detail in the following section.

Classical collagen sandwich configurations, which first highlighted the importance of hepatocellular architecture, are capable of maintaining function and morphology of rat hepatocytes for one week,[55,78,176] though imbalances between Phase I and II detoxification have been observed in sustained cultures.[177] Soft lithography and related methods have since enabled the creation of both 2D and 3D biomaterial scaffolds with feature sizes in the range of micrometers for fundamental tissue engineering studies.[178,179] Micropatterning of PEG was used to create arrays of hepatic spheroids that remained viable after three weeks of culture while sustaining albumin secretion and *CYP1A2* activity.[180] Others have incorporated multiple cell types and protein patterning. For example, Lee *et al.* used a surface micropatterning approach to register protein and hydrogel patterns on a glass substrate for fenced co-culture of hepatocytes and non-parenchymal cells, enabling control of both juxtracrine and paracrine cell signaling.[129] Layered 3D patterning of hydrogels using PEG-diacrylate chemistry has been used to create functional hepatic structures with favorable viability and function in a continuous flow bioreactor compared to unpatterned cultures, a technique that could be expanded to co-cultures.[132,181] Related printing and encapsulation methods for architecture design and fabrication have also been utilized to show the importance of heterotypic cell–cell interactions in the hepatocellular tumor environment,[182] assay the toxicity of Au and CdTe nanoparticles,[183] and layer co-culture spheroids.[184]

7. 3D Hepatocellular Constructs

3D hepatocellular constructs composed of hydrogels have been used primarily for improving understanding of the biology of hepatic cells, especially with regards to phenotypic stability and differentiation state, and (more recently) for furthering understanding of spatiotemporal dynamics and combinatorial effects (Fig. 2). In this section, we review these 3D hydrogel constructs while also integrating discussion of relevant material properties, design considerations, and tissue engineering applications.

7.1. Hepatocyte differentiation and function

Much effort in using hydrogels for hepatic tissue engineering has focused on improving long-term hepatic differentiation and function, especially of primary hepatocytes due to their limited availability.[185,186] As with 2D cultures, the culture medium,[185,187,188] ECM composition,[57] and heterotypic cell–cell interactions with non-parenchymal cells.[127,132,189–192] are central parameters for maintaining hepatocyte differentiation and function.

Both medium and matrix compositions are important parameters to control in order to attain sustained culture and differentiation state of hepatic cells. Supplementing 3D hepatic cultures with various soluble factors, such as

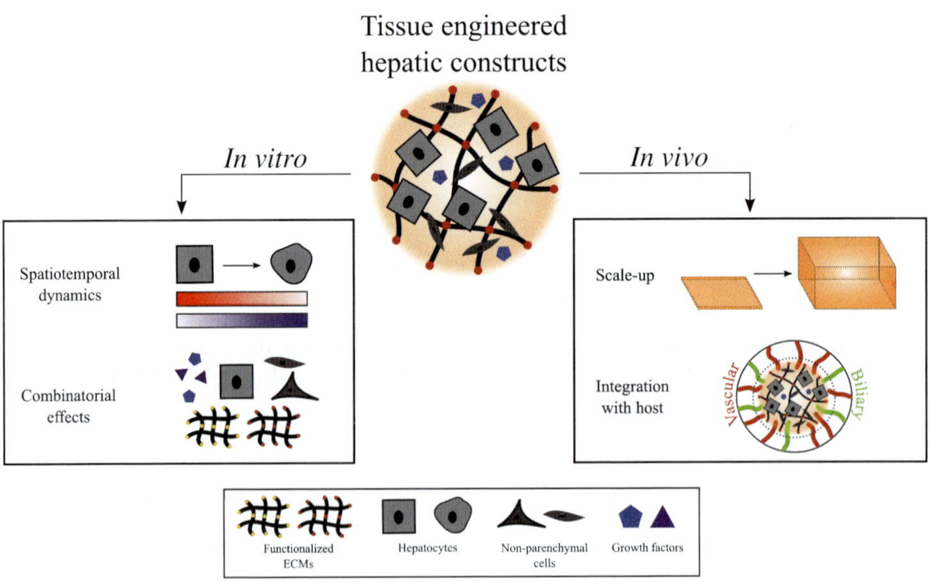

Fig. 2. Challenges in hepatic tissue engineering. Tissue engineered hepatic constructs utilizing hydrogels enable studies of both hepatic complexity *in vitro* and translational challenges *in vivo*.

nicotinamide, dimethyl sulfoxide, and oncostatin M, yields improved liver-specific function.[193] Metabolomic footprinting of both human hepatic stem cells and hepatoblasts using nuclear magnetic resonance showed that growth and differentiation are a function of matrix component, medium composition, and dimensionality (*i.e.*, 2D or 3D).[194] Hydrogel composition was key to maintenance of morphology as hepatoblasts embedded in collagen I rapidly differentiated while those embedded in hyaluronan proliferated more rapidly while maintaining a progenitor phenotype. Both 2D and 3D cultures of hepatocytes on scaffolds containing a combination of extracts prepared from acellular liver ECM and whole liver tissue show marked improvement in cell proliferation and maturation.[113,119,195] Liver function is also dependent on gradients of ECM composition along the sinusoidal tract in the liver. In agreement with knowledge of liver zonation and cellular localization, McClelland *et al.*, found that attachment and expansion of human hepatic progenitors was enhanced on periportal matrix components.[196] In contrast, cultures on a pericentral matrix component, type I collagen, elicited growth arrest and differentiation.

7.2. Cell–Cell interactions

The liver exhibits a great deal of cellular heterogeneity, consisting of both parenchymal and non-parenchymal (*e.g.*, endothelial, stellate, Kupffer, biliary, *etc.*) cells. Both homotypic[127,189,197] and heterotypic[76,127,132,189–191,198–204] cell–cell interactions play a role in the maintenance of hepatic phenotype and function in addition to a number of other processes, such as xenobiotic toxicity, response to oxidative stress, and acute-phase response.[205] Hydrogels provide a means of establishing 3D models of native-like cell–cell communication under various physiologic and pathophysiologic conditions. Hydrogel-based models of cell–cell interactions should be utilized not only for improving hepatocellular function *in vitro* but also for eliciting the exact molecular mechanisms of beneficial interactions, which is vital to the simplification and scaling of tissue engineered liver therapies through removal of unessential supporting cell types.[34,206]

Studies utilizing both spheroid formation and cell encapsulation in hydrogels have clarified the contribution of cell–cell interactions to hepatic function and morphology.[207,208] Primary rat hepatocytes cultured as spheroids exhibit increased homotypic cell–cell contacts, cell–ECM interactions, morphogenesis, and liver-specific function when compared with monolayer cultures.[209,210] Spheroid or aggregate cultures of hepatocytes with non-parenchymal cells such as fibroblasts,[127,191,211,212] endothelial cells,[213] mesenchymal stem cells,[214] and stellate cells[215] also exhibit improvement over monocultures. Pre-aggregation of primary hepatocytes with 3T3 fibroblasts before encapsulation in PEG hydrogels enhances hepatocyte function and viability.[127] Typical issues with 3D spheroid and encapsulation models include fusion of small spheroids into large aggregates, cell death at the center of spheroids due to mass transport limitations, and difficulty of handling,[34] all of which emphasize the importance of technologies to facilitate these cultures.[180,216]

7.3. Microscale fabrication

Microtechnologies capable of both 2D and 3D control over hydrogel architecture and patterning are developing rapidly,[131,217–219] allowing spatiotemporal control of scaffold architecture, cell–cell and cell–ECM interactions at micron-scale.[220] These techniques enable studies of adhesion, shape, and other phenomena in a highly parallel fashion in an environment more conducive to hepatic morphology than classical culture techniques.[221–224] The control of hydrogel properties, distribution of cells, and biofunctional properties afforded by these developing microtechnologies will accelerate understanding of the dynamics of hepatic cultures and inform the design of future tissue engineered constructs.

Photolithographic patterning of hepatic co-cultures have demonstrated that a balance between homotypic and heterotypic interactions provides improved liver-specific function.[205,223] Patterning techniques within hydrogels have enabled a number of different hepatic studies that are otherwise not possible. Photopolymerizable hydrogels can be functionalized with adhesive peptides to support hepatocyte survival and function in arbitrary, multi-layer 3D patterns.[132,181] Similar methods have been utilized by other groups for patterning of fibroblasts and immobolizing adhesion peptides,[131] ECM proteins in hepatic cocultures,[129,225] and heparin-based electroswitchable (releasable) hydrogels on glass.[114] Micropatterned PEG-fibrogen hydrogels have been used to culture hepatocytes in 3D microwells.[170] Inhibition of $\alpha_5\beta_1$ integrin binding to fibronectin in this format resulted in decreased survival, metabolic function, and growth factor sequestration in the 3D hydrogel. Deficiencies in fibrillar fibronectin assembly, aggregate formation, and cell polarization were also observed, emphasizing the importance of cell–ECM contacts and autocrine matrix and growth factors for hepatocyte differentiation and survival *in vitro*. Similarly, dielectrophoresis patterning of suspended hepatocytes and fibroblasts in PEG pre-polymers before encapsulation allows for placement of cells within hydrogel structures at both the microscale and macroscale.[217] Lastly, a cell-laden hydrogel "biopaper" was created through gelation of sol-gel transitional hydrogels on hydrophilic substrates.[226] Hepatic sheets assembled using this method secreted more albumin than standard 2D cultures and exhibited sustained proliferative capacity over five days, showing possible utility as a tool for tissue construction and cell–environment interactions.[226]

Microfluidic hydrogel systems for hepatic cultures have been utilized towards several ends, primarily spatial control through on-chip patterning,[227,228] drug toxicity testing,[229,230] encapsulation,[231,232] cytochrome P450 activity,[233] and scale-up.[234] In particular, Mao *et al.* designed an integrated microchip that used PEG-encapsulated liver microsomes to imitate human drug metabolism,[230] showing toxicity in HepG2 cells by mass spectrometry after exposure to acetomenophen.[230] Recently, Yamada *et al.* co-cultured hepatocytes and fibroblasts in a microfluidic channel to recreate cord-like hepatic micro-organoids, showing viability and albumin secretion at high O_2 perfusion.[235] Microfluidic hydrogels have been used as a means of creating *in vitro* physiological models. Lee *et al.* created spheroids of HepG2 cells encapsulated in alginate using PDMS micromolding, demonstrating viability and liver-specific function in a perfusable BAL chip.[236] Chen *et al.* created microtissue suspensions of hepatocytes and hepatic progenitors in PEG to study knockdown of *BCL-XL*, a gene implicated in chemotoxin-induced apoptosis protection.[231] A combination of matrigel and alginate hydrogels were used to encapsulate cells in microfluidic chambers representing liver, tumor, and bone marrow.[237] This combined model was used to test physiologically-based pharmacokinetic and pharmacodynamic model of anti-cancer drugs. Combining the dynamic capabilities of microfluidics with the biomimetic and patterning capabilities hydrogels enables new studies of liver biology and applications that would otherwise be impossible.

8. Translational Challenges

One of the primary goals of hepatic tissue engineering is to create hepatic tissues *in vitro* for implantation. The principal challenges in translating experimental implantable therapies to the clinic are scale-up and integration of constructs with native tissue (Fig. 2), both of which require suitable animal models for the assessment of safety and efficacy.[238,239] Although hydrogels are readily injected and gelled uniformly *in situ*, implanting pre-fabricated hydrogels is sometimes more difficult due to mechanical weakness, variable seeding efficiency, or poor engraftment. Other related concerns include cell sourcing, incorporation of tissue-inducing substances, sterilization of prefabricated matrices, and toxicity due to unreacted polymer monomers and polymerization initiators. In this section, we discuss current solutions to these translational issues and possible future directions.

8.1. Scale-up

It has been estimated that BAL devices and implantable hepatic constructs will likely require at least 20×10^9 functional hepatocytes (approximately 20% of healthy liver mass) to sustain a patient with no remaining native liver function.[240–243] Although serial transplantation studies have demonstrated the proliferative capacity of mature hepatocytes *in vivo*, long-term differentiation and function remain difficult to achieve *in vitro*.[23,24,185,244] The success of hepatic cultures in hydrogels and other scaffolds is dictated by the material, chemical, and biofunctional properties of the scaffold (as discussed in previous sections) all of which must be considered in order to achieve the requisite cell density and morphology for therapeutic use. Additionally, maintenance of large-scale cultures is hampered by transport considerations limiting nutrient, macromolecule, and metabolite influx and waste efflux, an especially important consideration given the variance of hepatic function along the sinusoidal tracts. Transport of O_2 gas is critical for hepatic cultures, as it is one of the primary determinants of liver zonation and function.[68–70,245] Thus, the transport properties of hydrogels must be designed so as to incorporate knowledge acquired through computational, *in vitro* and *in vivo* models of liver zonation.[68–71,246–248] Pre-vascularization of scaffold constructs prior to implantation addresses some of these limitations and has been demonstrated in gelatin hydrogels[171] and PVA sponges,[142] as has incorporation of hydrogels into BAL devices.[117,249–251] Recently, Erro *et al.*, seeded HepG2 cells encapsulated in ~500 μm alginate beads in a fluidized bed bioreactor and harvested more than 5×10^{10} cells — 25–50% of a healthy adult liver[252] — in a volume of less than 1,700 cm^3 after 11 days of culture.[251] Although the harvested cells expressed a panel of Phase 1 and 2 detoxifying enzymes, metabolized ammonia, and excreted albumin, the use of a tumor cell line is not ideal due to deficiencies in some hepatic functions.[253] Nonetheless, this alginate-encapsulated liver system compares favorably with extracorporeal liver-assist devices and could be adapted for use with other cell sources.[254]

8.2. Integration of implantable systems

A vital concern in the design of implantable systems is the integration of graft with host. To ensure maximal clinical benefit, the scaffold must integrate with the native vascular and biliary systems while gradually ceding function to host tissue as it degrades. The scaffold must also have appropriate material and biological properties, namely biodegradability and growth factor release, for vascular remodeling to occur. Studies in which hepatocytes were encapsulated in hydrogels composed of collagen,[255] alginate,[84,97,256] or a combination of chitosan and gelatin[116] and subsequently implanted in animal models have shown encouraging results with regards to vascular integration of graft with host. Similarly, *in situ* polymerization can improve graft integration, as demonstrated by Bruns *et al.* who used an injection-based approach to orthotopically transplant rat hepatocytes suspended in a fibrin glue, observing remodeling and engraftment with host liver at seven days post-implantation.[108] Although there are a number of *in vitro* models of biliary morphogenesis,[257–261] it is not yet clear whether engineered bile ducts are necessary for integration with the native biliary system.[34] Perez-Alonso *et al.*, implanted a cell-free 3D extrahepatic bile duct construct composed of collagen/agarose hydrogel in a porcine animal model.[262] Four weeks after implantation, constructs displayed morphology similar to that of native biliary system and flow of bile through the duodenum, though no measures of vascular integration were reported.

As with OLT, minimizing inflammatory and immune response is key to successful scaffold integration and clinical translation of implantable liver constructs and other cell-based therapies.[11,26] Due to persistent difficulties in cell sourcing, BAL devices have primarily utilized xenogeneic porcine cells,[263–266] though studies of cryopreservation,[267,268] fetal liver cells,[269] embryonic stem cell-derived cells,[270] pluripotent stem cells,[271–273] and reprogramming of autologous primary cells[274–276] are underway. Use of human cells screened for xenogeneic material would minimize host immune response,[34] as would serum-free conditions during the creation of engineered tissues.[277,278] Immunoisolation of transplanted cells by hydrogel encapsulation — first proposed by Chang[279] — has been adapted for experimental and clinical use.[124,280,281] If the properties of the encapsulating material are appropriately tuned, the exchange of oxygen, waste, nutrients, and therapeutic products is facilitated without inducing an immune response. Wen *et al.* investigated host immune response as a consequence of intraperitoneal implantation of xenogeneic hepatocytes encapsulated in poly(acrylonitrile-sodium methallylsulfonate) (AN69) and showed that encapsulated cells functioned and proliferated for the duration of observation, 28 days, whereas free cells were lysed.[282] Similarly, others have observed sustained viability and function after implantation of hepatocytes encapsulated in a copolymer of AN69,[283,284] PLGA sponges infiltrated with PVA,[133] alginate,[94,285] and thiolated heparin mixed with acrylated PEG.[112] Recently, Chen *et al.* created a reproducible and immunocompetent humanized mouse model of the liver through implantation of PEG-diacrylate encapsulated primary hepatocytes, circumventing the limitations of approaches utilizing transgenic methods and transplantation.[286] The model of Chen *et al.* (and others like it) may prove useful in studies of immune response or as a means to screen donors for transplant.

Controlled release of growth factors and scaffold pre-vascularization have been used to improve graft integration.[37,287–290] Release of factors implicated in hepatic differentiation and proliferation, such as hepatocyte growth factor (HGF), fibroblast growth factor (FGF), and prostaglandin (PG), improves engraftment. Incorporation of HGF in a biodegradable gelatin hydrogel significantly enhanced capillary formation and induction of vascular endothelial growth factor after implantation in a mouse model.[171] Ogawa *et al.* stimulated cells seeded in PVA sponges with FGF and showed that treated cells contained many more blood vessels than control sponges without FGF.[142] Similarly, incorporation of 16,16-dimethyl PG E2 in an implanted nanogel composed of a saccharidic copolymer of L-lactic acid enabled prolonged lifetime and improved immunological tolerance.[291]

9. Conclusion

The use of hydrogels as modular scaffolds has enabled substantial progress in hepatic biology and engineered liver constructs. Consideration of the material properties, biofunctionality, and architecture during the design of hydrogel-based systems allows for the creation of biomimetic environments for hepatocellular cultures. Future experimental and clinical hydrogel systems must combine knowledge of hepatic physiology with developing technologies to advance *in vitro* models and cell-based therapies for liver failure.

Acknowledgments

We would like to thank Viktoriya Ermilova for helpful suggestions regarding figure design.

References

1. W. R. Kim, R. S. Brown Jr., N. A. Terrault, and H. El-Serag, Burden of liver disease in the United States: Summary of a workshop. *Hepatology*, **36**(1), 227–242 (2002).
2. D. L. Hoyert and J. Xu, Deaths: Preliminary data for 2011. *National Vital Statistics Reports*, **61**(6), 1–64 (2012).
3. R. E. Mann, R. G. Smart, and R. Govoni, The epidemiology of alcoholic liver disease. *Alcohol Res. Health*, **27**(3), 209–219 (2003).
4. A. M. Diehl, Liver disease in alcohol abusers: Clinical perspective. *Alcohol*, **27**(1), 7–11 (2002).
5. V. M. Struben, E. E. Hespenheide, and S. H. Caldwell, Nonalcoholic steatohepatitis and cryptogenic cirrhosis within kindreds. *Am. J. Med.*, **108**(1), 9–13 (2000).
6. P. Angulo, Nonalcoholic fatty liver disease. *N. Engl. J. Med.*, **346**(16), 1221–1231 (2002).
7. C. A. Matteoni, Z. M. Younossi, T. Gramlich, N. Boparai, Y. C. Liu, and A. J. McCullough, Nonalcoholic fatty liver disease: A spectrum of clinical and pathological severity. *Gastroenterology*, **116**(6), 1413–1419 (1999).
8. W. M. Lee, Acute liver failure. *N. Engl. J. Med.*, **329**(25), 1862–1872 (1993).
9. W. Bernal, G. Auzinger, A. Dhawan, and J. Wendon, Acute liver failure. *Lancet*, **376**(9736), 190–201 (2010).
10. Liver Disease Subcommittee of the Digestive Diseases Interagency Coordinating Committee. *Action Plan for Liver Disease Research*. Bethesda, MD: National Institutes of Health. NIH Publication No. 04-5491 (2004).
11. T. E. Starzl and F. G. Lakkis, The unfinished legacy of liver transplantation: Emphasis on immunology. *Hepatology*, **43**, 151 (2006).
12. J. L. Dienstag and A. B. Cosimi, Liver transplantation — a vision realized. *N. Engl. J. Med.*, **367**(16), 1483–1485 (2012).
13. K. A. Brown, Liver transplantation. *Curr. Opin. Gastroenterol.*, **21**(3), 331–336 (2005).

14. A. Marcos, J. M. Ham, R. A. Fisher, A. T. Olzinski, and M. P. Posner, Single-center analysis of the first 40 adult-to-adult living donor liver transplants using the right lobe. *Liver Transpl.*, **6**(3), 296–301 (2000).
15. W. Jassem, D. D. H. Koo, P. Muiesan, L. Cerundolo, M. Rela, S. V. Fuggle, *et al.*, Non-heart-beating versus cadaveric and living-donor livers: Differences in inflammatory markers before transplantation. *Transplantation*, **75**(8), 1386–1390 (2003).
16. S. Reddy, M. Zilvetti, J. Brockmann, A. McLaren, and P. Friend, Liver transplantation from non-heart-beating donors: Current status and future prospects. *Liver Transpl.*, **10**(10), 1223–1232 (2004).
17. P. Abt, M. Crawford, N. Desai, J. Markmann, K. Olthoff, and A. Shaked, Liver transplantation from controlled non-heart-beating donors: An increased incidence of biliary complications. *Transplantation*, **75**(10), 1659–1663 (2003).
18. A. Casavilla, C. Ramirez, R. Shapiro, D. Nghiem, K. Miracle, O. Bronsther, *et al.*, Experience with liver and kidney allografts from non-heart-beating donors. *Transplantation*, **59**(2), 197–203 (1995).
19. T. E. Starzl, J. Fung, A. Tzakis, S. Todo, A. J. Demetris, I. R. Marino, *et al.*, Baboon-to-human liver transplantation. *Lancet*, **341**(8837), 65–71 (1993).
20. T. E. Starzl, A. Tzakis, J. J. Fung, S. Todo, I. R. Marino, and A. J. Demetris, Human liver xenotransplantation. *Xenotransplantation*, **1**(1), 4–7 (1993).
21. L. Makowa, D. V. Cramer, A. Hoffman, M. Breda, L. Sher, Eiras-G. Hreha, *et al.*, The use of a pig liver xenograft for temporary support of a patient with fulminant hepatic failure. *Transplantation*, **59**(12), 1654–1659 (1995).
22. B. Ekser, B. Gridelli, M. Veroux, and D. K. C. Cooper. Clinical pig liver xenotransplantation: How far do we have to go? *Xenotransplantation*, **18**(3), 158–167 (2011).
23. G. K. Michalopoulos and M. C. DeFrances, Liver regeneration. *Science*, **276**(5309), 60–66 (1997).
24. N. Fausto, Liver regeneration. *Hepatology*, **32**(Suppl 1), 19–31 (2000).
25. A. Martinez-Hernandez and P. S. Amenta, The extracellular matrix in hepatic regeneration. *FASEB J.*, **9**(14), 1401–1410 (1995).
26. V. Racanelli and B. Rehermann, The liver as an immunological organ. *Hepatology*, **43**(Suppl 2), 54 (2006).
27. M. Plauth, M. Merli, J. Kondrup, A. Weimann, P. Ferenci, M. J. Müller, *et al.*, ESPEN guidelines for nutrition in liver disease and transplantation. *Clin. Nutr.*, **16**(2), 43–55 (1997).
28. D. J. Nompleggi and H. L. Bonkovsky, Nutritional supplementation in chronic liver disease: An analytical review. *Hepatology*, **19**(2), 518–533 (2005).
29. C. Chan, F. Berthiaume, B. D. Nath, A. W. Tilles, M. Toner, and M. L. Yarmush, Hepatic tissue engineering for adjunct and temporary liver support: Critical technologies. *Liver Transpl.*, **10**(11), 1331–1342 (2004).
30. B. Carpentier, A. Gautier, and C. Legallais, Artificial and bioartificial liver devices: Present and future. *Gut.*, **58**(12), 1690–1702 (2009).
31. J.-K. Park and D.-H. Lee, Bioartificial liver systems: Current status and future perspective. *J. Biosci. Bioeng.*, **99**(4), 311–319 (2005).

32. Y.-T. Ding and X.-L. Shi, Bioartificial liver devices: Perspectives on the state of the art. *Front. Med.*, **5**(1), 15–19 (2011).
33. J. W. Allen and S. N. Bhatia, Engineering liver therapies for the future. *Tissue Eng.*, **8**(5), 725–737 (2002).
34. J. Shan, K. R. Stevens, K. Trehan, G. H. Underhill, A. A. Chen, and S. N. Bhatia, Hepatic tissue engineering. In *Molecular Pathology of Liver Diseases*, Ed. S. P. S. Monges, New York: Springer Science+Business Media, LLC, pp. 321–342 (2011).
35. R. A. Chamuleau, Future of bioartificial liver support. *World J. Gastrointest. Surg.*, **1**(1), 21–25 (2009).
36. A. Dhawan, J. Puppi, R. D. Hughes, and R. R. Mitry, Human hepatocyte transplantation: current experience and future challenges. *Nat. Rev. Gastroenterol. Hepatol.*, **7**(5), 288–298 (2010).
37. H. Lee, R. A. Cusick, F. Browne, T. Ho Kim, P. X. Ma, H. Utsunomiya, *et al.*, Local delivery of basic fibroblast growth factor increases both angiogenesis and engraftment of hepatocytes in tissue-engineered polymer devices. *Transplantation*, **73**(10), 1589–1593 (2002).
38. R. A. Fisher and S. C. Strom, Human hepatocyte transplantation: Worldwide results. *Transplantation*, **82**(4), 441–449 (2006).
39. E. Fitzpatrick, R. R. Mitry, and A. Dhawan, Human hepatocyte transplantation: State of the art. *J. Intern. Med.*, **266**(4), 339–357 (2009).
40. K. Fukumitsu, H. Yagi, and A. Soto-Gutierrez, Bioengineering in organ transplantation: Targeting the liver. *Transplant. Proc.*, **43**(6), 2137–2138 (2011).
41. J. A. Wertheim, P. M. Baptista, and A. Soto-Gutierrez, Cellular therapy and bioartificial approaches to liver replacement. *Curr. Opin. Organ. Transplant.*, **17**(3), 235–240 (2012).
42. K. Yamanouchi, H. Zhou, N. Roy-Chowdhury, F. Macaluso, L. Liu, T. Yamamoto, *et al.*, Hepatic irradiation augments engraftment of donor cells following hepatocyte transplantation. *Hepatology*, **49**(1), 258–267 (2009).
43. S. L. Nyberg, Bridging the gap: Advances in artificial liver support. *Liver Transpl.*, **18**, 10 (2012).
44. P. C. Nicolson and J. Vogt, Soft contact lens polymers: An evolution. *Biomaterials*, **22**(24), 3273–3283 (2001).
45. T. B. Reece, T. S. Maxey, and I. L. Kron, A prospectus on tissue adhesives. *Am. J. Surg.*, **182**(Suppl 2), 40–44 (2001).
46. J. Kopeček and J. Yang, Hydrogels as smart biomaterials. *Polym. Int.*, **56**(9), 1078–1098 (2007).
47. D. Seliktar, Designing cell-compatible hydrogels for biomedical applications. *Science*, **336**(6085), 1124–1128 (2012).
48. F. Rosso, G. Marino, A. Giordano, M. Barbarisi, D. Parmeggiani, and A. Barbarisi, Smart materials as scaffolds for tissue engineering. *J. Cell. Physiol.*, **203**(3), 465–470 (2005).
49. N. A. Peppas, J. Z. Hilt, A. Khademhosseini, and R. Langer, Hydrogels in biology and medicine: From molecular principles to bionanotechnology. *Adv. Mater.*, **18**(11), 1345–1360 (2006).

50. K. Y. Lee and D. J. Mooney, Hydrogels for tissue engineering. *Chem. Rev.*, **101**(7), 1869–1880 (2001).
51. R. Gauvin, R. Parenteau-Bareil, M. R. Dokmeci, W. D. Merryman, and A. Khademhosseini, Hydrogels and microtechnologies for engineering the cellular microenvironment. *Wiley Interdiscip. Rev. Nanomed. Nanobiotechnol.*, **4**(3), 235–246 (2012).
52. D. M. Bissell and P. S. Guzelian, Phenotypic stability of adult rat hepatocytes in primary monolayer culture. *Ann. N. Y. Acad. Sci.*, **349**, 85–98 (1980).
53. D. M. Bissell, D. M. Arenson, J. J. Maher and F. J. Roll, Support of cultured hepatocytes by a laminin-rich gel. Evidence for a functionally significant subendothelial matrix in normal rat liver. *J. Clin. Invest.*, **79**(3), 801–812 (1987).
54. J. J. Maher, Primary hepatocyte culture: Is it home away from home? *Hepatology*, **8**(5), 1162–1166 (2005).
55. J. C. Dunn, M. L. Yarmush, H. G. Koebe, and R. G. Tompkins, Hepatocyte function and extracellular matrix geometry: Long-term culture in a sandwich configuration. *FASEB J.*, **3**(2), 174–177 (1989).
56. E. L. LeCluyse, P. L. Bullock, and A. Parkinson, Strategies for restoration and maintenance of normal hepatic structure and function in long-term cultures of rat hepatocytes. *Adv. Drug Deliv. Rev.*, **22**(1–2), 133–186 (1996).
57. C. J. Flaim, S. Chien, and S. N. Bhatia, An extracellular matrix microarray for probing cellular differentiation. *Nat. Methods*, **2**(2), 119–125 (2005).
58. J. L. Drury and D. J. Mooney, Hydrogels for tissue engineering: Scaffold design variables and applications. *Biomaterials*, **24**(24), 4337–4351 (2003).
59. B. Baroli, Hydrogels for tissue engineering and delivery of tissue-inducing substances. *J. Pharm. Sci.*, **96**(9), 2197–2223 (2007).
60. J. Chen, S. Jo, and K. Park, Degradable hydrogels. In *Handbook of Biodegradable Polymers*, Eds. A. J. Domb, J. Kost, and D. M. Wiseman, Amsterdam: Overseas, Publishers Association, pp. 203–230 (1997).
61. A. S. Hoffmann, Hydrogels for biomedical applications. *Adv. Drug. Deliv. Rev.*, **43**(1), 3–12 (2002).
62. J. Zhu, Bioactive modification of poly(ethylene glycol) hydrogels for tissue engineering. *Biomaterials*, **31**(17), 4639–4656 (2010).
63. C.-C. Lin and K. S. Anseth, PEG hydrogels for the controlled release of biomolecules in regenerative medicine. *Pharm. Res.*, **26**(3), 631–643 (2009).
64. K. Y. Lee, M. C. Peters, K. W. Anderson, and D. J. Mooney, Controlled growth factor release from synthetic extracellular matrices. *Nature*, **408**(6815), 998–1000 (2000).
65. A. Khademhosseini and R. Langer, Microengineered hydrogels for tissue engineering. *Biomaterials*, **28**(34), 5087–5092 (2007).
66. A. M. Kloxin, M. W. Tibbitt, A. M. Kasko, J. A. Fairbairn, and K. S. Anseth, Tunable hydrogels for external manipulation of cellular microenvironments through controlled photodegradation. *Adv. Mater.*, **22**(1), 61–66 (2010).
67. C. A. DeForest and K. S. Anseth, Photoreversible Patterning of biomolecules within Click-Based Hydrogels. *Angew. Chemie.*, **124**(8), 1852–1855 (2012).
68. J. W. Allen and S. N. Bhatia, Formation of steady-state oxygen gradients *in vitro*: Application to liver zonation. *Biotechnol. Bioeng.*, **82**(3), 253–262 (2003).

69. J. W. Allen, S. R. Khetani and S. N. Bhatia, *In vitro* zonation and toxicity in a hepatocyte bioreactor. *Toxicol Sci.*, **84**(1), 110–119 (2005).
70. A. W. Tilles, H. Baskaran, P. Roy, M. L. Yarmush, and M. Toner, Effects of oxygenation and flow on the viability and function of rat hepatocytes cocultured in a microchannel flat-plate bioreactor. *Biotechnol. Bioeng.*, **73**(5), 379–389 (2001).
71. K. Domansky, W. Inman, J. Serdy, A. Dash, M. H. M. Lim, and L. G. Griffith, Perfused multiwell plate for 3D liver tissue engineering. *Lab. Chip.*, **10**(1), 51–58 (2010).
72. K. Jungermann and R. G. Thurman, Hepatocyte heterogeneity in the metabolism of carbohydrates. *Enzyme*, **46**(1–3), 33–58 (1992).
73. K. Jungermann and T. Kietzmann, Oxygen: Modulator of metabolic zonation and disease of the liver. *Hepatology*, **31**(2), 255–260 (2000).
74. B. V. Slaughter, S. S. Khurshid, O. Z. Fisher, A. Khademhosseini, and N. A. Peppas, Hydrogels in regenerative medicine. *Adv. Mater.*, **21**(32–33), 3307–3329 (2009).
75. R. DeVolder and H.-J. Kong, Hydrogels for *in vivo*-like three-dimensional cellular studies. *Wiley Interdiscip. Rev. Syst. Biol. Med.*, **4**(4), 351–365 (2012).
76. G. Michalopoulos, F. Russell, and C. Biles, Primary cultures of hepatocytes on human fibroblasts. *In vitro*, **15**(10), 796–806 (1979).
77. V. Dixit, R. Darvasi, M. Arthur, M. Brezina, K. Lewin, and G. Gitnick, Restoration of liver function in Gunn rats without immunosuppression using transplanted microencapsulated hepatocytes. *Hepatology*, **12**(6), 1342–1349 (1990).
78. E. L. LeCluyse, K. L. Audus, and J. H. Hochman, Formation of extensive canalicular networks by rat hepatocytes cultured in collagen-sandwich configuration. *Am. J. Physiol.*, **266**(6 Pt 1), 1764 (1994).
79. Y. H. Kang, F. Berthiaume, B. D. Nath, and M. L. Yarmush, Growth factors and nonparenchymal cell conditioned media induce mitogenic responses in stable long-term adult rat hepatocyte cultures. *Exp. Cell Res.*, **293**(2), 239–247 (2004).
80. I. Rault, V. Frei, D. Herbage, N. Abdul-Malak, and A. Huc, Evaluation of different chemical methods for cross-linking collagen gel, films and sponges. *J. Mater. Sci. Mater. Med.*, **7**(4), 215–221 (1996).
81. H. Schoof, J. Apel, I. Heschel, and G. Rau, Control of pore structure and size in freeze-dried collagen sponges. *J. Biomed. Mater. Res.*, **58**(4), 352–357 (2001).
82. T. Haque, H. Chen, W. Ouyang, C. Martoni, B. Lawuyi, A. M. Urbanska, *et al.*, *In vitro* study of alginate-chitosan microcapsules: An alternative to liver cell transplants for the treatment of liver failure. *Biotechnol. Lett.*, **27**(5), 317–322 (2005).
83. T. Maguire, E. Novik, R. Schloss, and M. Yarmush, Alginate-PLL microencapsulation: Effect on the differentiation of embryonic stem cells into hepatocytes. *Biotechnol. Bioeng.*, **93**(3), 581–591 (2006).
84. S. Hirai, S. Kasai, and M. Mito, Encapsulated hepatocyte transplantation for the treatment of D-galactosamine-induced acute hepatic failure in rats. *Eur. Surg. Res.*, **25**(4), 193–202 (1993).
85. Y. Miura, T. Akimoto, H. Kanazawa, and K. Yagi, Synthesis and secretion of protein by hepatocytes entrapped within calcium alginate. *Artif. Organs*, **10**(6), 460–465 (1986).

86. S. F. Khattak, M. Spatara, L. Roberts, and S. C. Roberts, Application of colorimetric assays to assess viability, growth and metabolism of hydrogel-encapsulated cells. *Biotechnol. Lett.*, **28**(17), 1361–1370 (2006).
87. M. Izydorczyk, S. W. Cui, and Q. Wang, *Polysaccharide Gums: Structures, Functional Properties, and Applications. Food Carbohydrates: Chemistry, Physical Properties, and Applications.* Boca Raton, FL: CRC Press; pp. 263–307 (2005).
88. M. Khalil, A. Shariat-Panahi, R. Tootle, T. Ryder, P. McCloskey, E. Roberts, *et al.*, Human hepatocyte cell lines proliferating as cohesive spheroid colonies in alginate markedly upregulate both synthetic and detoxificatory liver function. *J. Hepatol.*, **34**(1), 68–77 (2001).
89. R. Glicklis, L. Shapiro, R. Agbaria, J. C. Merchuk, S. Cohen, Hepatocyte behavior within three-dimensional porous alginate scaffolds. *Biotechnol. Bioeng.*, **67**(3), 344–353 (2000).
90. M. Dvir-Ginzberg, I. Gamlieli-Bonshtein, R. Agbaria, and S. Cohen, Liver tissue engineering within alginate scaffolds: Effects of cell-seeding density on hepatocyte viability, morphology, and function. *Tissue Eng.*, **9**(4), 757–766 (2003).
91. S.-J. Seo, I.-Y. Kim, Y.-J. Choi, T. Akaike, and C.-S.Cho, Enhanced liver functions of hepatocytes cocultured with NIH 3T3 in the alginate/galactosylated chitosan scaffold. *Biomaterials*, **27**(8), 1487–1495 (2006).
92. J. Yang, T. Woong Chung, M. Nagaoka, M. Goto, C. S. Cho, and T. Akaike, Hepatocyte-specific porous polymer-scaffolds of alginate/galactosylated chitosan sponge for liver-tissue engineering. *Biotechnol. Lett.*, **23**(17), 1385–1389 (2001).
93. T. W. Chung, J. Yang, T. Akaike, K. Y. Cho, J. W. Nah, S. I. Kim, *et al.*, Preparation of alginate/galactosylated chitosan scaffold for hepatocyte attachment. *Biomaterials*, **23**(14), 2827–2834 (2002).
94. J. P. Miranda, A. Rodrigues, R. M. Tostões, S. Leite, H. Zimmerman, M. J. T. Carrondo, *et al.*, Extending hepatocyte functionality for drug-testing applications using high-viscosity alginate-encapsulated three-dimensional cultures in bioreactors. *Tissue Eng. Part C Meth.*, **16**(6), 1223–1232 (2010).
95. C. M. Hwang, S. Sant, M. Masaeli, N. N. Kachouie, B. Zamanian, S.-H. Lee, *et al.*, Fabrication of three-dimensional porous cell-laden hydrogel for tissue engineering. *Biofabrication*, **2**(3), 035003 (2010).
96. S.-J. Seo, Y.-J. Choi, T. Akaike, A. Higuchi, and C.-S. Cho, Alginate/galactosylated chitosan/heparin scaffold as a new synthetic extracellular matrix for hepatocytes. *Tissue Eng.*, **12**(1), 33–44 (2006).
97. H. Wong and T. M. Chang, Bioartificial liver: Implanted artificial cells microencapsulated living hepatocytes increases survival of liver failure rats. *Int. J. Artif. Organs.*, **9**(5), 335–336 (1986).
98. P. de Vos, M. M. Faas, B. Strand, and R. Calafiore, Alginate-based microcapsules for immunoisolation of pancreatic islets. *Biomaterials*, **27**(32), 5603–5617 (2006).
99. J. M. Chupa, A. M. Foster, S. R. Sumner, S. V. Madihally, and H. W. Matthew, Vascular cell responses to polysaccharide materials: *In vitro* and *in vivo* evaluations. *Biomaterials*, **21**(22), 2315–2322 (2000).

100. B. P. Toole, L. Golberg, G. Chi-Rosso, L. B. Underhill, and R. W. Orkin, Hyaluronate–cell interaction. In Ed. R. L. Trelstad, *The Role of Extracellular Matrix in Development.* New York: A.R. Liss, pp. 43–66 (1984).
101. Y. M. Elçin, V. Dixit, and G. Gitnick, Hepatocyte attachment on biodegradable modified chitosan membranes: *In vitro* evaluation for the development of liver organoids. *Artif. Organs.*, **22**(10), 837–846 (1998).
102. D. C. Spray, M. Fujita, J. C. Saez, H. Choi, T. Watanabe, E. Hertzberg, *et al.*, Proteoglycans and glycosaminoglycans induce gap junction synthesis and function in primary liver cultures. *J. Cell Biol.*, **105**(1), 541–551 (1987).
103. B. S. Kim, I. K. Park, T. Hoshiba, H. L. Jiang, Y. J. Choi, T. Akaike, *et al.*, Design of artificial extracellular matrices for tissue engineering. *Prog. Polym. Sci.*, **36**(2), 238–268 (2011).
104. J. Li, J. Pan, L. Zhang, and Y. Yu, Culture of hepatocytes on fructose-modified chitosan scaffolds. *Biomaterials*, **24**(13), 2317–2322 (2003).
105. T. A. E. Ahmed, E. V. Dare, and M. Hincke, Fibrin: A versatile scaffold for tissue engineering applications. *Tissue Eng. Part B Rev.*, **14**(2), 199–215 (2008).
106. D. D. Swartz, J. A. Russell, and S. T. Andreadis, Engineering of fibrin-based functional and implantable small-diameter blood vessels. *Am. J. Physiol. Heart Circ. Physiol.*, **288**(3), 1451 (2005).
107. S. Jockenhoevel, G. Zund, S. P. Hoerstrup, K. Chalabi, J. S. Sachweh, L. Demircan, *et al.*, Fibrin gel — advantages of a new scaffold in cardiovascular tissue engineering. *Eur. J. Cardiothorac. Surg.*, **19**(4), 424–430 (2001).
108. H. Bruns, U. Kneser, S. Holzhüter, B. Roth, J. Kluth, P. M. Kaufmann, *et al.*, Injectable liver: A novel approach using fibrin gel as a matrix for culture and intrahepatic transplantation of hepatocytes. *Tissue Eng.*, **11**(11–12), 1718–1726 (2005).
109. J. Fan, Y. Shang, Y. Yuan, and J. Yang, Preparation and characterization of chitosan/galactosylated hyaluronic acid scaffolds for primary hepatocytes culture. *J. Mater. Sci. Mater. Med.*, **21**(1), 319–327 (2010).
110. E. Schmelzer, F. Triolo, M. E. Turner, R. L. Thompson, K. Zeilinger, L. M. Reid, *et al.*, Three-dimensional perfusion bioreactor culture supports differentiation of human fetal liver cells. *Tissue Eng. Pt. A*, **16**(6), 2007–2016 (2010).
111. O. A. Lozoya, E. Wauthier, R. A. Turner, C. Barbier, G. D. Prestwich, F. Guilak, *et al.*, Regulation of hepatic stem/progenitor phenotype by microenvironment stiffness in hydrogel models of the human liver stem cell niche. *Biomaterials*, **32**(30), 7389–7402 (2011).
112. M. Kim, J. Y. Lee, C. N. Jones, A. Revzin, and G. Tae, Heparin-based hydrogel as a matrix for encapsulation and cultivation of primary hepatocytes. *Biomaterials*, **31**(13), 3596–3603 (2010).
113. A. Skardal, L. Smith, S. Bharadwaj, A. Atala, S. Soker, and Y. Zhang, Tissue specific synthetic ECM hydrogels for 3-D *in vitro* maintenance of hepatocyte function. *Biomaterials*, **33**(18), 4565–4575 (2012).
114. S. S. Shah, M. Kim, E. Foster, T. Vu, D. Patel, L.-J. Chen, *et al.*, Electrochemical release of hepatocyte-on-hydrogel microstructures from ITO substrates. *Anal. Bioanal. Chem.*, **402**(5), 1847–1856 (2012).

115. H. Ijima, Y.-T. Hou, and T. Takei, Development of hepatocyte-embedded hydrogel-filled macroporous scaffold cultures using transglutaminase. *Biochem. J. Eng.*, **52**(2), 276–281 (2010).
116. S. Wang, D. Nagrath, P. C. Chen, F. Berthiaume, and M. L. Yarmush, Three-dimensional primary hepatocyte culture in synthetic self-assembling peptide hydrogel. *Tissue Eng. Part A*, **14**(2), 227–236 (2008).
117. M. B. Yang, J. P. Vacanti, and D. E. Ingber, Hollow fibers for hepatocyte encapsulation and transplantation: Studies of survival and function in rats. *Cell Transplant.*, **3**(5), 373–385 (1994).
118. M. Bhattacharya, M. M. Malinen, P. Lauren, Y.-R. Lou, S. W. Kuisma, L. Kanninen, *et al.*, Nanofibrillar cellulose hydrogel promotes three-dimensional liver cell culture. *J. Control. Rel.*, **164**(3), 291–298 (2012).
119. P. Lin, W. C. W. Chan, S. F. Badylak, and S. N. Bhatia, Assessing porcine liver-derived biomatrix for hepatic tissue engineering. *Tissue Eng.*, **10**(7–8), 1046–1053 (2004).
120. T. L. Sellaro, A. Ranade, D. M. Faulk, G. P. McCabe, K. Dorko, S. F. Badylak, *et al.*, Maintenance of human hepatocyte function *in vitro* by liver-derived extracellular matrix gels. *Tissue Eng. Pt. A*, **16**(3), 1075–1082 (2010).
121. M. P. Lutolf and J. A. Hubbell, Synthetic biomaterials as instructive extracellular microenvironments for morphogenesis in tissue engineering. *Nat. Biotechnol.*, **23**(1), 47–55 (2005).
122. J. L. Ifkovits and J. A. Burdick, Review: Photopolymerizable and degradable biomaterials for tissue engineering applications. *Tissue Eng.*, **13**(10), 2369–2385 (2007).
123. W. M. Saltzman and W. L. Olbricht, Building drug delivery into tissue engineering. *Nat. Rev. Drug. Discov.*, **1**(3), 177–186 (2002).
124. G. D. Nicodemus and S. J. Bryant, Cell encapsulation in biodegradable hydrogels for tissue engineering applications. *Tissue Eng. Pt. B Rev.*, **14**(2), 149–165 (2008).
125. D. L. Hern and J. A. Hubbell, Incorporation of adhesion peptides into nonadhesive hydrogels useful for tissue resurfacing. *J. Biomed. Mater. Res.*, **39**(2), 266–276 (1998).
126. C. R. Nuttelman, M. C. Tripodi, and K. S. Anseth, Synthetic hydrogel niches that promote hMSC viability. *Matrix Biol.*, **24**(3), 208–218 (2005).
127. G. H. Underhill, A. A. Chen, D. R. Albrecht, and S. N. Bhatia, Assessment of hepatocellular function within PEG hydrogels. *Biomaterials*, **28**(2), 256–270 (2007).
128. L. J. Itle, W.-G. Koh, and M. V. Pishko, Hepatocyte viability and protein expression within hydrogel microstructures. *Biotechnol. Prog.*, **21**(3), 926–932 (2005).
129. J. Y. Lee, S. S. Shah, J. Yan, M. C. Howland, A. N. Parikh, T. Pan, *et al.*, Integrating sensing hydrogel microstructures into micropatterned hepatocellular cocultures. *Langmuir*, **25**(6), 3880–3886 (2009).
130. N.-J. Cho, M. Elazar, A. Xiong, W. Lee, E. Chiao, J. Baker, *et al.*, Viral infection of human progenitor and liver-derived cells encapsulated in three-dimensional PEG-based hydrogel. *Biomed. Mater.*, **4**(1), 011001 (2009).
131. M. S. Hahn, L. J. Taite, J. J. Moon, M. C. Rowland, K. A. Ruffino, and J. L. West, Photolithographic patterning of polyethylene glycol hydrogels. *Biomaterials*, **27**(12), 2519–2524 (2006).

132. V. Liu Tsang, A. A. Chen, L. M. Cho, K. D. Jadin, R. L. Sah, S. DeLong, *et al.*, Fabrication of 3-D hepatic tissues by additive photopatterning of cellular hydrogels. *FASEB J.*, **21**(3), 790–801 (2007).
133. D. J. Mooney, S. Park, P. M. Kaufmann, K. Sano, K. McNamara, J. P. Vacanti, *et al.*, Biodegradable sponges for hepatocyte transplantation. *J. Biomed. Mater. Res.*, **29**(8), 959–965 (1995).
134. D. J. Mooney, K. Sano, P. M. Kaufmann, K. Majahod, B. Schloo, J. P. Vacanti, *et al.*, Long-term engraftment of hepatocytes transplanted on biodegradable polymer sponges. *J. Biomed. Mater. Res.*, **37**(3), 413–420 (1997).
135. E. Török, J. M. Pollok, P. X. Ma, C. Vogel, M. Dandri, J. Petersen, *et al.*, Hepatic tissue engineering on 3-Dimensional biodegradable polymers within a pulsatile flow bioreactor. *Dig. Surg.*, **18**(3), 196–203 (2001).
136. J. S. Lee, S. H. Kim, Y. J. Kim, T. Akaike, and S. C. Kim, Hepatocyte adhesion on a poly[N-p-vinylbenzyl-4-O-beta-D-galactopyranosyl-D-glucoamide]-coated poly(L-lactic acid) surface. *Biomacromolecules*, **6**(4), 1906–1911 (2005).
137. H. C. Fiegel, J. Havers, U. Kneser, M. K. Smith, T. Moeller, D. Kluth, *et al.*, Influence of flow conditions and matrix coatings on growth and differentiation of three-dimensionally cultured rat hepatocytes. *Tissue Eng.*, **10**(1–2), 165–174 (2004).
138. V. Hasirci, F. Berthiaume, S. P. Bondre, J. D. Gresser, D. J. Trantolo, M. Toner, *et al.*, Expression of liver-specific functions by rat hepatocytes seeded in treated poly(lactic-co-glycolic) acid biodegradable foams. *Tissue Eng.*, **7**(4), 385–394 (2001).
139. Y. S. Nam, J. J. Yoon, J. G. Lee, and T. G. Park, Adhesion behaviours of hepatocytes cultured onto biodegradable polymer surface modified by alkali hydrolysis process. *J. Biomater. Sci. Polym. Ed.*, **10**(11), 1145–1158 (1999).
140. L. E. Freed, G. Vunjak-Novakovic, R. J. Biron, D. B. Eagles, D. C. Lesnoy, S. K. Barlow, *et al.*, Biodegradable polymer scaffolds for tissue engineering. *Biotechnology*, **12**(7), 689–693 (1994).
141. M. L. Houchin and E. M. Topp, Chemical degradation of peptides and proteins in PLGA: A review of reactions and mechanisms. *J. Pharm. Sci.*, **97**(7), 2395–2404 (2008).
142. K. Ogawa, K. Asonuma, Y. Inomata, I. Kim, Y. Ikada, Y. Tabata, *et al.*, The efficacy of prevascularization by basic FGF for hepatocyte transplantation using polymer devices in rats. *Cell Transplant.*, **10**(8), 723–729 (2001).
143. R.J. Pelham Jr. and Y.-L. Wang, Cell locomotion and focal adhesions are regulated by substrate flexibility. *Proc. Natl. Acad. Sci. USA*, **94**(25), 13661–13665 (1997).
144. E. J. Semler, P. A. Lancin, A. Dasgupta, and P. V. Moghe, Engineering hepatocellular morphogenesis and function via ligand-presenting hydrogels with graded mechanical compliance. *Biotechnol. Bioeng.*, **89**(3), 296–307 (2005).
145. E. Genové, S. Schmitmeier, A. Sala, S. Borrós, A. Bader, L. G. Griffith, *et al.*, Functionalized self-assembling peptide hydrogel enhance maintenance of hepatocyte activity *in vitro*. *J. Cell Mol. Med.*, **13**(9B), 3387–3397 (2009).
146. E. Karamuk, J. Mayer, E. Wintermantel, and T. Akaike, Partially degradable film/fabric composites: Textile scaffolds for liver cell culture. *Artif. Organs.*, **23**(9), 881–884 (1999).

147. J. Mayer, E. Karamuk, T. Akaike, and E. Wintermantel, Matrices for tissue engineering-scaffold structure for a bioartificial liver support system. *J. Control. Rel.*, **64**(1–3), 81–90 (2000).
148. G. K. Michalopoulos, Liver regeneration. *J. Cell Physiol.*, **213**(2), 286–300 (2007).
149. A. A. Chen, S. R. Khetani, S. Lee, S. N. Bhatia, and K. J. Van Vliet, Modulation of hepatocyte phenotype *in vitro* via chemomechanical tuning of polyelectrolyte multilayers. *Biomaterials*, **30**(6), 1113–1120 (2009).
150. R. G. Wells, The role of matrix stiffness in regulating cell behavior. *Hepatology*, **47**(4), 1394–1400 (2008).
151. G. Constantinides, Z. I. Kalcioglu, M. McFarland, J. F. Smith, and K. J. Van Vliet, Probing mechanical properties of fully hydrated gels and biological tissues. *J. Biomech.*, **41**(15), 3285–3289 (2008).
152. Z. Li, J. A. Dranoff, E. P. Chan, M. Uemura, J. Sévigny, and R. G. Wells, Transforming growth factor-beta and substrate stiffness regulate portal fibroblast activation in culture. *Hepatology*, **46**(4), 1246–1256 (2007).
153. F. J. Carter, T. G. Frank, P. J. Davies, D. McLean, and A. Cuschieri, Measurements and modelling of the compliance of human and porcine organs. *Med. Image Anal.*, **5**(4), 231–236 (2001).
154. M. Yin, J. A. Talwalkar, K. J. Glaser, A. Manduca, R. C. Grimm, P. J. Rossman, *et al.*, Assessment of hepatic fibrosis with magnetic resonance elastography. *Clin. Gastroenterol. Hepatol.*, **5**(10), 1207–1213 (2007).
155. P. C. Georges, J.-J. Hui, Z. Gombos, M. E. McCormick, A. Y. Wang, M. Uemura, *et al.*, Increased stiffness of the rat liver precedes matrix deposition: Implications for fibrosis. *Am. J. Physiol. Gastrointest. Liver Physiol.*, **293**(6), 1147 (2007).
156. Z. Liu and L. Bilston, On the viscoelastic character of liver tissue: Experiments and modelling of the linear behaviour. *Biorheology*, **37**(3), 191–201 (2000).
157. W.-C. Yeh, P.-C. Li, Y.-M. Jeng, H.-C. Hsu, P.-L. Kuo, M.-L. Li, *et al.*, Elastic modulus measurements of human liver and correlation with pathology. *Ultrasound Med. Biol.*, **28**(4), 467–474 (2002).
158. C. P. Hodgkinson, M. C. Wright, and A. J. Paine, Fibronectin-mediated hepatocyte shape change reprograms cytochrome P450 2C11 gene expression via an integrin-signaled induction of ribonuclease activity. *Mol. Pharmacol.*, **58**(5), 976–981 (2000).
159. K. Bhadriraju and L. K. Hansen, Hepatocyte adhesion, growth and differentiated function on RGD-containing proteins. *Biomaterials*, **21**(3), 267–272 (2000).
160. Y. Liang, J. Jeong, R. J. DeVolder, C. Cha, F. Wang, Y. W. Tong, *et al.*, A cell-instructive hydrogel to regulate malignancy of 3-D tumor spheroids with matrix rigidity. *Biomaterials*, **32**(35), 9308–9315 (2011).
161. C. S. Ranucci, A. Kumar, S. P. Batra, and P. V. Moghe, Control of hepatocyte function on collagen foams: Sizing matrix pores toward selective induction of 2-D and 3-D cellular morphogenesis. *Biomaterials*, **21**(8), 783–793 (2000).
162. S. Zmora, R. Glicklis, and S. Cohen, Tailoring the pore architecture in 3-D alginate scaffolds by controlling the freezing regime during fabrication. *Biomaterials*, **23**(20), 4087–4094 (2002).

163. X. Tao, L. Shaolin, and Y. Yaoting, Preparation and culture of hepatocyte on gelatin microcarriers. *J. Biomed. Mater. Res. A*, **65**(2), 306–310 (2003).
164. A. T. Gutsche, H. Lo, J. Zurlo, J. Yager, and K. W. Leong, Engineering of a sugar-derivatized porous network for hepatocyte culture. *Biomaterials*, **17**(3), 387–393 (1996).
165. K.-N. Chua, W.-S. Lim, P. Zhang, H. Lu, J. Wen, S. Ramakrishna, *et al.*, Stable immobilization of rat hepatocyte spheroids on galactosylated nanofiber scaffold. *Biomaterials*, **26**(15), 2537–2547 (2005).
166. U. Hersel, C. Dahmen, and H. Kessler, RGD modified polymers: Biomaterials for stimulated cell adhesion and beyond. *Biomaterials*, **24**(24), 4385–4415 (2003).
167. K.-H. Park, K. Na, S. W. Kim, S. Y. Jung, K. H. Park, and H.-M. Chung, Phenotype of hepatocyte spheroids behavior within thermo-sensitive poly(NiPAAm-co-PEG-g-GRGDS) hydrogel as a cell delivery vehicle. *Biotechnol. Lett.*, **27**(15), 1081–1086 (2005).
168. P. N. Patel, A. S. Gobin, J. L. West, and C. W. Patrick Jr., Poly(ethylene glycol) hydrogel system supports preadipocyte viability, adhesion, and proliferation. *Tissue Eng.*, **11**(9–10), 1498–1505 (2005).
169. E. S. Carlisle, M. R. Mariappan, K. D. Nelson, B. E. Thomes, R. B. Timmons, A. Constantinescu, *et al.*, Enhancing hepatocyte adhesion by pulsed plasma deposition and polyethylene glycol coupling. *Tissue Eng.*, **6**(1), 45–52 (2000).
170. C. M. Williams, G. Mehta, S. R. Peyton, A. S. Zeiger, K. J. Van Vliet, and L. G. Griffith, Autocrine-controlled formation and function of tissue-like aggregates by primary hepatocytes in micropatterned hydrogel arrays. *Tissue Eng. Pt. A*, **17**(7–8), 1055–1068 (2011).
171. M. Ozeki, T. Ishii, Y. Hirano, and Y. Tabata, Controlled release of hepatocyte growth factor from gelatin hydrogels based on hydrogel degradation. *J. Drug Target*, **9**(6), 461–471 (2001).
172. M. P. Lutolf, J. L. Lauer-Fields, H. G. Schmoekel, A. T. Metters, F. E. Weber, and G. B. Fields, *et al.*, Synthetic matrix metalloproteinase-sensitive hydrogels for the conduction of tissue regeneration: Engineering cell-invasion characteristics. *Proc. Natl. Acad. Sci. USA*, **100**(9), 5413–5418 (2003).
173. D. Seliktar, A. H. Zisch, M. P. Lutolf, J. L. Wrana, and J. A. Hubbell, MMP-2 sensitive, VEGF-bearing bioactive hydrogels for promotion of vascular healing. *J. Biomed. Mater Res. A*, **68**(4), 704–716 (2004).
174. S. S. Kim, H. Utsunomiya, J. A. Koski, B. M. Wu, M. J. Cima, and J. Sohn, *et al.*, Survival and function of hepatocytes on a novel three-dimensional synthetic biodegradable polymer scaffold with an intrinsic network of channels. *Ann. Surg.*, **228**(1), 8–13 (1998).
175. S. Uyama, P. M. Kaufmann, U. Kneser, H. C. Fiegel, J. M. Pollok, D. Kluth, *et al.*, Hepatocyte transplantation using biodegradable matrices in ascorbic acid-deficient rats: Comparison with heterotopically transplanted liver grafts. *Transplantation*, **71**(9), 1226–1231 (2001).

176. F. Berthiaume, P. V. Moghe, M. Toner, and M. L. Yarmush, Effect of extracellular matrix topology on cell structure, function, and physiological responsiveness: Hepatocytes cultured in a sandwich configuration. *FASEB J.*, **10**(13), 1471–1484 (1996).
177. L. Richert, D. Binda, G. Hamilton, C. Viollon-Abadie, E. Alexandre, D. Bigot-Lasserre, *et al.*, Evaluation of the effect of culture configuration on morphology, survival time, antioxidant status and metabolic capacities of cultured rat hepatocytes. *Toxicol In vitro*, **16**(1), 89–99 (2002).
178. G. Vozzi, C. Flaim, A. Ahluwalia, and S. Bhatia, Fabrication of PLGA scaffolds using soft lithography and microsyringe deposition. *Biomaterials*, **24**(14), 2533–2540 (2003).
179. J. T. Borenstein, E. J. Weinberg, B. K. Orrick, C. Sundback, M. R. Kaazempur-Mofrad, and J. P. Vacanti, Microfabrication of three-dimensional engineered scaffolds. *Tissue Eng.*, **13**(8), 1837–1844 (2007).
180. R. Kojima, K. Yoshimoto, E. Takahashi, M. Ichino, H. Miyoshi, and Y. Nagasaki, Spheroid array of fetal mouse liver cells constructed on a PEG-gel micropatterned surface: Upregulation of hepatic functions by co-culture with nonparenchymal liver cells. *Lab Chip*, **9**(14), 1991–1993 (2009).
181. V. A. Liu and S. N. Bhatia, Three-dimensional photopatterning of hydrogels containing living cells. *Biomed. Microdevices*, **4**(4), 257–266 (2002).
182. W. Lee and J. Park, The design of a heterocellular 3-D architecture and its application to monitoring the behavior of cancer cells in response to the spatial distribution of endothelial cells. *Adv. Mater.*, **24**(39), 5339–5344 (2012).
183. J. Lee, G. D. Lilly, R. C. Doty, P. Podsiadlo, and N. A. Kotov, *In vitro* toxicity testing of nanoparticles in 3-D cell culture. *Small*, **5**(10), 1213–1221 (2009).
184. S. Sakai, H. Inagaki, K. Inamoto, and M. Taya, Wrapping tissues with a pre-established cage-like layer composed of living cells. *Biomaterials*, **33**(28), 6721–6727 (2012).
185. A. Guillouzo, Liver cell models in *in vitro* toxicology. *Environ. Health Perspect.*, **106** (Suppl 2), 511–532 (1998).
186. R. Gebhardt, J. G. Hengstler, D. Müller, R. Glöckner, P. Buenning, B. Laube, *et al.*, New hepatocyte *in vitro* systems for drug metabolism: Metabolic capacity and recommendations for application in basic research and drug development, standard operation procedures. *Drug Metab. Rev.*, **35**(2–3), 145–213 (2003).
187. B. A. Laishes and G. M. Williams, Conditions affecting primary cell cultures of functional adult rat hepatocytes. 1. The effect of insulin. *In vitro.* **12**(7), 521–532 (1976).
188. S. Kidambi, R. S. Yarmush, E. Novik, P. Chao, M. L. Yarmush, and Y. Nahmias, Oxygen-mediated enhancement of primary hepatocyte metabolism, functional polarization, gene expression, and drug clearance. *Proc. Natl. Acad. Sci. USA*, **106**(37), 15714–15719 (2009).
189. S. N. Bhatia, M. L. Yarmush, and M. Toner, Controlling cell interactions by micropatterning in co-cultures: Hepatocytes and 3T3 fibroblasts. *J. Biomed. Mater. Res.*, **34**(2), 189–199 (1997).

190. P. M. Kaufmann, H. C. Fiegel, U. Kneser, J. M. Pollok, D. Kluth, and X. Rogiers, Influence of pancreatic islets on growth and differentiation of hepatocytes in co-culture. *Tissue Eng.*, **5**(6), 583–596 (1999).
191. R. N. Bhandari, L. A. Riccalton, A. L. Lewis, J. R. Fry, A. H. Hammond, S. J. Tendler, *et al.*, Liver tissue engineering: A role for co-culture systems in modifying hepatocyte function and viability. *Tissue Eng.*, **7**(3), 345–357 (2001).
192. C. H. Cho, F. Berthiaume, A. W. Tilles, and M. L. Yarmush, A new technique for primary hepatocyte expansion *in vitro. Biotechnol. Bioeng.*, **101**(2), 345–356 (2008).
193. J. Jiang, N. Kojima, L. Guo, K. Naruse, M. Makuuchi, A. Miyajima, *et al.*, Efficacy of engineered liver tissue based on poly-L-lactic acid scaffolds and fetal mouse liver cells cultured with M. oncostatin, nicotinamide, and dimethyl sulfoxide. *Tissue Eng.*, **10**(9–10), 1577–1586 (2004).
194. W. S. Turner, C. Seagle, J. A. Galanko, O. Favorov, G. D. Prestwich, J. M. Macdonald, *et al.*, Nuclear magnetic resonance metabolomic footprinting of human hepatic stem cells and hepatoblasts cultured in hyaluronan-matrix hydrogels. *Stem Cells*, **26**(6), 1547–1555 (2008).
195. R. Lang, M. M. Stern, L. Smith, Y. Liu, S. Bharadwaj, G. Liu, *et al.*, Three-dimensional culture of hepatocytes on porcine liver tissue-derived extracellular matrix. *Biomaterials*, **32**(29), 7042–7052 (2011).
196. R. McClelland, E. Wauthier, J. Uronis, and L. Reid, Gradients in the liver's extracellular matrix chemistry from periportal to pericentral zones: Influence on human hepatic progenitors. *Tissue Eng. Pt. A*, **14**(1), 59–70 (2008).
197. S. N. Bhatia, U. J. Balis, M. L. Yarmush, and M. Toner, Microfabrication of Hepatocyte/Fibroblast Co-cultures: Role of Homotypic Cell Interactions. *Biotechnol. Prog.*, **14**(3), 378–387 (1998).
198. R. Langenbach, L. Malick, A. Tompa, C. Kuszynski, H. Freed, and E. Huberman, Maintenance of adult rat hepatocytes on C3H/10T1/2 cells. *Cancer Res.*, **39**(9), 3509–3514 (1979).
199. R. J. Thomas, R. Bhandari, D. A. Barrett, A. J. Bennett, J. R. Fry, D. Powe, *et al.*, The effect of three-dimensional co-culture of hepatocytes and hepatic stellate cells on key hepatocyte functions *in vitro. Cells Tissues Organs.*, **181**(2), 67–79 (2005).
200. M. T. Donato, M. J. Gómez-Lechón, and J. V. Castell, Drug metabolizing enzymes in rat hepatocytes co-cultured with cell lines. *In vitro Cell Dev. Biol.*, **26**(11), 1057–1062 (1990).
201. C. Guguen-Guillouzo, B. Clément, G. Baffet, C. Beaumont, E. Morel-Chany, D. Glaise, *et al.*, Maintenance and reversibility of active albumin secretion by adult rat hepatocytes co-cultured with another liver epithelial cell type. *Exp. Cell Res.*, **143**(1), 47–54 (1983).
202. S. Shimaoka, T. Nakamura, and A. Ichihara, Stimulation of growth of primary cultured adult rat hepatocytes without growth factors by coculture with nonparenchymal liver cells. *Exp. Cell Res.*, **172**(1), 228–242 (1987).
203. L. Agius, Metabolic interactions of parenchymal hepatocytes and dividing epithelial cells in co-culture. *J. Biochem.*, **252**(1), 23–28 (1988).

204. E. E. Hui and S. N. Bhatia, Micromechanical control of cell–cell interactions. *Proc. Natl. Acad. Sci. USA*, **104**(14), 5722–5726 (2007).
205. S. N. Bhatia, U. J. Balis, M. L. Yarmush, and M. Toner, Effect of cell–cell interactions in preservation of cellular phenotype: Cocultivation of hepatocytes and nonparenchymal cells. *FASEB J.*, **13**(14), 1883–1900 (1999).
206. S. R. Khetani, A. A. Chen, B. Ranscht, and S. N. Bhatia, T-cadherin modulates hepatocyte functions *in vitro*. *FASEB J.*, **22**(11), 3768–3775 (2008).
207. F. J. Wu, J. R. Friend, R. P. Remmel, F. B. Cerra, and W. S. Hu, Enhanced cytochrome P450 IA1 activity of self-assembled rat hepatocyte spheroids. *Cell Transplant*, **8**(3), 233–246 (1999).
208. L. K. Hansen, C.-C. Hsiao, J. R. Friend, F. J. Wu, G. A. Bridge, R. P. Remmel, *et al.*, Enhanced morphology and function in hepatocyte spheroids: A model of tissue self-assembly. *Tissue Eng.*, **4**(1), 65–74 (1998).
209. C. Yuasa, Y. Tomita, M. Shono, K. Ishimura, and A. Ichihara, Importance of cell aggregation for expression of liver functions and regeneration demonstrated with primary cultured hepatocytes. *J. Cell. Physiol.*, **156**(3), 522–530 (1993).
210. J. Landry, D. Bernier, C. Ouellet, R. Goyette, and N. Marceau, Spheroidal aggregate culture of rat liver cells: Histotypic reorganization, biomatrix deposition, and maintenance of functional activities. *J. Cell. Biol.*, **101**(3), 914–923 (1985).
211. J. Fukuda, A. Khademhosseini, Y. Yeo, X. Yang, J. Yeh, G. Eng, *et al.*, Micromolding of photocrosslinkable chitosan hydrogel for spheroid microarray and co-cultures. *Biomaterials*, **27**(30), 5259–5267 (2006).
212. J. Fukuda, Y. Sakai, and K. Nakazawa, Novel hepatocyte culture system developed using microfabrication and collagen/polyethylene glycol microcontact printing. *Biomaterials*, **27**(7), 1061–1070 (2006).
213. M. Inamori, H. Mizumoto, and T. Kajiwara, An approach for formation of vascularized liver tissue by endothelial cell-covered hepatocyte spheroid integration. *Tissue Eng. Pt. A*, **15**(8), 2029–2037 (2009).
214. J. Gu, X. Shi, Y. Zhang, X. Chu, H. Hang, and Y. Ding, Establishment of a three-dimensional co-culture system by porcine hepatocytes and bone marrow mesenchymal stem cells *in vitro*. *Hepatol. Res.*, **39**(4), 398–407 (2009).
215. K. Harada, T. Mitaka, S. Miyamoto, S. Sugimoto, S. Ikeda, H. Takeda, *et al.*, Rapid formation of hepatic organoid in collagen sponge by rat small hepatocytes and hepatic nonparenchymal cells. *J. Hepatol.*, **39**(5), 716–723 (2003).
216. S. M. Chia, K. W. Leong, J. Li, X. Xu, K. Zeng, P. N. Er, *et al.*, Hepatocyte encapsulation for enhanced cellular functions. *Tissue Eng.*, **6**(5), 481–495 (2000).
217. D. R. Albrecht, G. H. Underhill, T. B. Wassermann, R. L. Sah, and S. N. Bhatia, Probing the role of multicellular organization in three-dimensional microenvironments. *Nat. Methods*, **3**(5), 369–375 (2006).
218. A. M. Kloxin, A. M. Kasko, C. N. Salinas, and K. S. Anseth, Photodegradable hydrogels for dynamic tuning of physical and chemical properties. *Science*, **324**(5923), 59–63 (2009).

219. C. A. DeForest and K. S. Anseth, Cytocompatible click-based hydrogels with dynamically tunable properties through orthogonal photoconjugation and photocleavage reactions. *Nat. Chem.*, **3**(12), 925–931 (2011).
220. S. R. Khetani and S. N. Bhatia, Engineering tissues for *in vitro* applications. *Curr. Opin. Biotechnol.*, **17**(5), 524–531 (2006).
221. C. S. Chen, M. Mrksich, S. Huang, G. M. Whitesides, and D. E. Ingber, Geometric control of cell life and death. *Science*, **276**(5317), 1425–1428 (1997).
222. R. Singhvi, A. Kumar, G. P. Lopez, G. N. Stephanopoulos, D. I. Wang, G. M. Whitesides, *et al.*, Engineering cell shape and function. *Science*, **264**(5159), 696–698 (1994).
223. S. R. Khetani and S. N. Bhatia, Microscale culture of human liver cells for drug development. *Nat. Biotechnol.*, **26**(1), 120–126 (2008).
224. U. A. Gurkan, S. Tasoglu, D. Kavaz, M. C. Demirel, and U. Demirci, Emerging technologies for assembly of microscale hydrogels. *Adv. Healthc. Mater.*, **1**(2), 149–158 (2012).
225. A. Revzin, P. Rajagopalan, A. W. Tilles, F. Berthiaume, M. L. Yarmush, and M. Toner, Designing a hepatocellular microenvironment with protein microarraying and poly(ethylene glycol) photolithography. *Langmuir*, **20**(8), 2999–3005 (2004).
226. W. Lee, C. Y. Bae, S. Kwon, J. Son, J. Kim, Y. Jeong, *et al.*, Cellular hydrogel biopaper for patterned 3D cell culture and modular tissue reconstruction. *Adv. Health. Mater.*, **1**(5), 635–639 (2012).
227. J. C. Zguris, L. J. Itle, W.-G. Koh, and M. V. Pishko, A novel single-step fabrication technique to create heterogeneous poly(ethylene glycol) hydrogel microstructures containing multiple phenotypes of mammalian cells. *Langmuir*, **21**(9), 4168–4174 (2005).
228. W. Tan and T. A. Desai, Microfluidic patterning of cells in extracellular matrix biopolymers: Effects of channel size, cell type, and matrix composition on pattern integrity. *Tissue Eng.*, **9**(2), 255–267 (2003).
229. P. M. van Midwoud, M. T. Merema, N. Verweij, G. M. M. Groothuis, and E. Verpoorte, Hydrogel embedding of precision-cut liver slices in a microfluidic device improves drug metabolic activity. *Biotechnol. Bioeng.*, **108**(6), 1404–1412 (2011).
230. S. Mao, D. Gao, W. Liu, H. Wei, and J.-M. Lin, Imitation of drug metabolism in human liver and cytotoxicity assay using a microfluidic device coupled to mass spectrometric detection. *Lab Chip*, **12**(1), 219–226 (2012).
231. A. A. Chen, G. H. Underhill, and S. N. Bhatia, Multiplexed, high-throughput analysis of 3D microtissue suspensions. *Integr. Biol.*, **2**(10), 517–527 (2010).
232. Y. Ling, J. Rubin, Y. Deng, C. Huang, U. Demirci, J. M. Karp, *et al.*, A cell-laden microfluidic hydrogel. *Lab Chip*, **7**(6), 756–762 (2007).
233. J. H. Sung, J.-R. Choi, D. Kim, and M. L. Shuler, Fluorescence optical detection *in situ* for real-time monitoring of cytochrome P450 enzymatic activity of liver cells in multiple microfluidic devices. *Biotechnol. Bioeng.*, **104**(3), 516–525 (2009).
234. E. Leclerc, Y. Sakai, and T. Fujii, Microfluidic PDMS (polydimethylsiloxane) bioreactor for large-scale culture of hepatocytes. *Biotechnol. Prog.*, **20**(3), 750–755.
235. M. Yamada, R. Utoh, K. Ohashi, K. Tatsumi, M. Yamato, T. Okano, *et al.*, Controlled formation of heterotypic hepatic micro-organoids in anisotropic hydrogel microfibers

for long-term preservation of liver-specific functions. *Biomaterials*, **33**(33), 8304–8315 (2012).

236. K. H. Lee, D. Y. No, S.-H. Kim, J. H. Ryoo, S. F. Wong, and S.-H..Lee, Diffusion-mediated in situ alginate encapsulation of cell spheroids using microscale concave well and nanoporous membrane. *Lab Chip*, **11**(6), 1168–1173 (2011).
237. J. H. Sung and M. L. Shuler, A micro cell culture analog (microCCA) with 3-D hydrogel culture of multiple cell lines to assess metabolism-dependent cytotoxicity of anti-cancer drugs. *Lab Chip*, **9**(10), 1385–1394 (2009).
238. P. N. Newsome, J. N. Plevris, L. J. Nelson, and P. C. Hayes, Animal models of fulminant hepatic failure: A critical evaluation. *Liver Transpl.*, **6**(1), 21–31 (2000).
239. M.-P. van de Kerkhove, R. Hoekstra, T. M. van Gulik, and R. A. F. M. Chamuleau, Large animal models of fulminant hepatic failure in artificial and bioartificial liver support research. *Biomaterials*, **25**(9), 1613–1625 (2004).
240. M. P. van de Kerkhove, R. Hoekstra, R. A. F. M. Chamuleau, and T. M. van Gulik, Clinical application of bioartificial liver support systems. *Ann. Surg.*, **240**(2), 216–230 (2004).
241. M.-P. van de Kerkhove, K. P. de Jong, A. M. Rijken, A.-C. J. M. de Pont, T. M. van Gulik, MARS treatment in posthepatectomy liver failure. *Liver Int.*, **23**(Suppl 3), 44–51 (2003).
242. M. D. McCarter and Y. Fong, Metastatic liver tumors. *Semin. Surg. Oncol.*, **19**(2), 177–188 (2000).
243. R. P. DeMatteo, Y. Fong, and L. H. Blumgart, Surgical treatment of malignant liver tumours. *Baillieres Best Pract. Res. Clin. Gastroenterol.*, **13**(4), 557–574 (1999).
244. K. Overturf, M. Al-Dhalimy, R. Tanguay, M. Brantly, C. N. Ou, M. Finegold, *et al.*, Hepatocytes corrected by gene therapy are selected *in vivo* in a murine model of hereditary tyrosinaemia type I. *Nat. Genet.*, **12**(3), 266–273 (1996).
245. K. Jungermann and T. Keitzmann, Zonation of parenchymal and nonparenchymal metabolism in liver. *Annu. Rev. Nutr.*, **16**(1), 179–203 (1996).
246. C. H. Cho, J. Park, D. Nagrath, A. W. Tilles, F. Berthiaume, M. Toner, *et al.*, Oxygen uptake rates and liver-specific functions of hepatocyte and 3T3 fibroblast co-cultures. *Biotechnol Bioeng.*, **97**(1), 188–199 (2007).
247. A. Rotem, M. Toner, R. G. Tompkins, and M. L. Yarmush, Oxygen uptake rates in cultured rat hepatocytes. *Biotechnol. Bioeng.*, **40**(10), 1286–1291 (1992).
248. R. E. McClelland, J. M. MacDonald and R. N. Coger, Modeling O2 transport within engineered hepatic devices. *Biotechnol. Bioeng.*, **82**(1), 12–27 (2003).
249. K. Yanagi, K. Ookawa, S. Mizuno, and N. Ohshima, Performance of a new hybrid artificial liver support system using hepatocytes entrapped within a hydrogel. *ASAIO Trans.*, **35**(3), 570–572 (1989).
250. J. Park, Y. Li, F. Berthiaume, M. Toner, M. L. Yarmush, and A. W. Tilles, Radial flow hepatocyte bioreactor using stacked microfabricated grooved substrates. *Biotechnol. Bioeng.*, **99**(2), 455–467 (2008).
251. E. Erro, J. Bundy, I. Massie, S.-A. Chalmers, A. Gautier, S. Gerontas, *et al.*, Bioengineering the liver: Scale-up and cool chain delivery of the liver cell biomass for

clinical targeting in a bioartificial liver support system. *Biores. Open Access*, **2**(1), 1–11 (2013).

252. H. Imamura, S. Kawasaki, J. Shiga, Y. Bandai, K. Sanjo, and Y. Idezuki, Quantitative evaluation of parenchymal liver cell volume and total hepatocyte number in cirrhotic patients. *Hepatology*, **14**(3), 448–453 (1991).
253. D. Mavri-Damelin, S. Eaton, L. H. Damelin, M. Rees, H. J. F. Hodgson, and C. Selden, Ornithine transcarbamylase and arginase I deficiency are responsible for diminished urea cycle function in the human hepatoblastoma cell line HepG2. *Int. J. Biochem. Cell Biol.*, **39**(3), 555–564 (2007).
254. S. M. Coward, C. Legallais, B. David, M. Thomas, Y. Foo, D. Mavri-Damelin, *et al.*, Alginate-encapsulated HepG2 cells in a fluidized bed bioreactor maintain function in human liver failure plasma. *Artif. Organs.*, **33**(12), 1117–1126 (2009).
255. Y. Zhao, Y. Xu, B. Zhang, X. Wu, F. Xu, W. Liang, *et al.*, *In vivo* generation of thick, vascularized hepatic tissue from collagen hydrogel-based hepatic units. *Tissue Eng. Pt. C Methods*, **16**(4), 653–659 (2010).
256. T. M. Rahman, C. Selden, M. Khalil, I. Diakanov, and H. J. F. Hodgson, Alginate-encapsulated human hepatoblastoma cells in an extracorporeal perfusion system improve some systemic parameters of liver failure in a xenogeneic model. *Artif. Organs.*, **28**(5), 476–482 (2004).
257. R. Sudo, T. Mitaka, M. Ikeda, and K. Tanishita, Reconstruction of 3D stacked-up structures by rat small hepatocytes on microporous membranes. *FASEB J.*, **19**(12), 1695–1697 (2005).
258. Y. Ishida, S. Smith, L. Wallace, T. Sadamoto, M. Okamoto, M. Auth, *et al.*, Ductular morphogenesis and functional polarization of normal human biliary epithelial cells in three-dimensional culture. *J. Hepatol.*, **35**(1), 2–9 (2001).
259. M. K. Auth, R. E. Joplin, M. Okamoto, Y. Ishida, P. McMaster, J. M. Neuberger, *et al.*, Morphogenesis of primary human biliary epithelial cells: Induction in high-density culture or by coculture with autologous human hepatocytes. *Hepatology*, **33**(3), 519–529 (2001).
260. J. E. Barralet, L. L. Wallace, and A. J. Strain, Tissue engineering of human biliary epithelial cells on polyglycolic acid/polycaprolactone scaffolds maintains long-term phenotypic stability. *Tissue Eng.*, **9**(5), 1037–1045 (2003).
261. M. Miyazawa, T. Torii, Y. Toshimitsu, K. Okada, I. Koyama, and Y. Ikada, A tissue-engineered artificial bile duct grown to resemble the native bile duct. *Am. J. Transplant.*, **5**(6), 1541–1547 (2005).
262. A. J. Pérez Alonso, C. Del Olmo Rivas, I. M. Romero, F. J. Cañizares Garcia, and P. T. Poyatos, Tissue-engineering repair of extrahepatic bile ducts. *J. Surg. Res.*, **179**(1), 18–21 (2013).
263. A. A. Demetriou, R. S. Brown Jr, R. W. Busuttil, J. Fair, B. M. McGuire, P. Rosenthal, *et al.*, Prospective, randomized, multicenter, controlled trial of a bioartificial liver in treating acute liver failure. *Ann. Surg.*, **239**(5), 660–670 (2004).
264. I. M. Sauer, D. Kardassis, K. Zeillinger, A. Pascher, A. Gruenwald, G. Pless, *et al.*, Clinical extracorporeal hybrid liver support — phase I study with primary porcine liver cells. *Xenotransplantation*, **10**(5), 460–469 (2003).

265. M. P. van de Kerkhove, P. P. C. Poyck, T. Deurholt, R. Hoekstra, R. A. F. M. Chamuleau, and T. M. van Gulik, Liver support therapy: An overview of the AMC-bioartificial liver research. *Dig. Surg.*, **22**(4), 254–264 (2005).
266. J. M. Millis, D. C. Cronin, R. Johnson, H. Conjeevaram, C. Conlin, S. Trevino, *et al.*, Initial experience with the modified extracorporeal liver-assist device for patients with fulminant hepatic failure: System modifications and clinical impact. *Transplantation*, **74**(12), 1735–1746 (2002).
267. U. Baccarani, A. Donini, A. Sanna, A. Risaliti, A. Cariani, B. Nardo, *et al.*, First report of cryopreserved human hepatocytes based bioartificial liver successfully used as a bridge to liver transplantation. *Am. J. Transplant.*, **4**(2), 286–289 (2004).
268. U. Baccarani, A. Sanna, A. Cariani, M. Sainz, G. L. Adani, D. Lorenzin, *et al.*, Cryopreserved human hepatocytes from cell bank: *In vitro* function and clinical application. *Transplant Proc.*, **37**(1), 256–259 (2005).
269. N. Shirakigawa and H. Ijima, Nucleus number in clusters of transplanted fetal liver cells increases by partial hepatectomy of recipient rats. *J. Biosci. Bioeng.*, **115**(5), 568–570 (2012).
270. J. S. Odorico, D. S. Kaufman, and J. A. Thomson, Multilineage differentiation from human embryonic stem cell lines. *Stem Cells*, **19**(3), 193–204 (2001).
271. H. Basma, A. Soto-Gutiérrez, G. R. Yannam, L. Liu, R. Ito, T. Yamamoto, *et al.*, Differentiation and transplantation of human embryonic stem cell-derived hepatocytes. *Gastroenterology*, **136**(3), 990–999 (2009).
272. G. J. Sullivan, D. C. Hay, I.-H. Park, J. Fletcher, Z. Hannoun, C. M. Payne, *et al.*, Generation of functional human hepatic endoderm from human induced pluripotent stem cells. *Hepatology*, **51**(1), 329–335 (2010).
273. K. Si-Tayeb, F. K. Noto, M. Nagaoka, J. Li, M. A. Battle, C. Duris, *et al.*, Highly efficient generation of human hepatocyte-like cells from induced pluripotent stem cells. *Hepatology*, **51**(1), 297–305 (2010).
274. A. Soto-Gutiérrez, N. Kobayashi, J. D. Rivas-Carrillo, N. Navarro-Alvarez, D. Zhao, D. Zhao, *et al.*, Reversal of mouse hepatic failure using an implanted liver-assist device containing ES cell-derived hepatocytes. *Nat. Biotechnol.*, **24**(11), 1412–1419 (2006).
275. P. P. C. Poyck, R. Hoekstra, A. C. W. A. van Wijk, C. Attanasio, F. Calise, R. A. F. M. Chamuleau, *et al.*, Functional and morphological comparison of three primary liver cell types cultured in the AMC bioartificial liver. *Liver Transpl.*, **13**(4), 589–598 (2007).
276. J. C. Gerlach, K. Mutig, I. M. Sauer, P. Schrade, E. Efimova, T. Mieder, *et al.*, Use of primary human liver cells originating from discarded grafts in a bioreactor for liver support therapy and the prospects of culturing adult liver stem cells in bioreactors: A morphologic study. *Transplantation*, **76**(5), 781–786 (2003).
277. I. Peiffer, R. Barbet, A. Hatzfeld, M.-L. Li, and J. A. Hatzfeld, Optimization of physiological xenofree molecularly defined media and matrices to maintain human embryonic stem cell pluripotency. *Methods Mol. Biol.*, **584**, 97–108 (2010).
278. M. C. Vemuri, T. Schimmel, P. Colls, S. Munne, and J. Cohen, Derivation of human embryonic stem cells in xeno-free conditions. *Methods Mol. Biol.*, **407**, 1–10 (2007).
279. T. M. Chang, Semipermeable microcapsules. *Science*, **146**(3643), 524–525 (1964).

280. G. Orive, R. M. Hernández, A. R. Gascón, R. Calafiore, T. M. S. Chang, P. De Vos, *et al.*, Cell encapsulation: Promise and progress. *Nat. Med.*, **9**(1), 104–107 (2003).
281. J. T. Wilson and E. L. Chaikof, Challenges and emerging technologies in the immunoisolation of cells and tissues. *Adv. Drug. Deliv. Rev.*, **60**(2), 124–145 (2008).
282. L. Wen, Y. Calmus, J. Honiger, F. Conti, J. Capeau, B. Weill, *et al.*, Encapsulated xenogeneic hepatocytes remain functional after peritoneal implantation despite immunization of the host. *J. Hepatol.*, **29**(6), 960–968 (1998).
283. P. Balladur, E. Crema, J. Honiger, Y. Calmus, M. Baudrimont, R. Delelo, *et al.*, Transplantation of allogeneic hepatocytes without immunosuppression: Long-term survival. *Surgery*, **117**(2), 189–194 (1995).
284. J. Honiger, P. Balladur, P. Mariani, Y. Calmus, M. Vaubourdolle, R. Delelo, *et al.*, Permeability and biocompatibility of a new hydrogel used for encapsulation of hepatocytes. *Biomaterials*, **16**(10), 753–759 (1995).
285. D. Bazou, Biochemical properties of encapsulated high-density 3-D HepG2 aggregates formed in an ultrasound trap for application in hepatotoxicity studies: Biochemical responses of encapsulated 3-D HepG2 aggregates. *Cell Biol. Toxicol.*, **26**(2), 127–141 (2010).
286. A. A. Chen, D. K. Thomas, L. L. Ong, R. E. Schwartz, T. R. Golub, and S. N. Bhatia, Humanized mice with ectopic artificial liver tissues. *Proc. Natl. Acad. Sci. USA*, **108**(29), 11842–11847 (2011).
287. T. P. Richardson, M. C. Peters, A. B. Ennett, and D. J. Mooney, Polymeric system for dual growth factor delivery. *Nat. Biotechnol.*, **19**(11), 1029–1034 (2001).
288. M. K. Smith, M. C. Peters, T. P. Richardson, J. C. Garbern, and D. J. Mooney, Locally enhanced angiogenesis promotes transplanted cell survival. *Tissue Eng.*, **10**(1–2), 63–71 (2004).
289. M. K. Smith, K. W. Riddle, and D. J. Mooney, Delivery of hepatotrophic factors fails to enhance longer-term survival of subcutaneously transplanted hepatocytes. *Tissue Eng.*, **12**(2), 235–244 (2006).
290. A. Kedem, A. Perets, I. Gamlieli-Bonshtein, M. Dvir-Ginzberg, S. Mizrahi, and S. Cohen, Vascular endothelial growth factor-releasing scaffolds enhance vascularization and engraftment of hepatocytes transplanted on liver lobes. *Tissue Eng.*, **11**(5–6), 715–722 (2005).
291. T. Okamoto, T. Saito, Y. Tabata, and S. Uemoto, Immunological tolerance in a mouse model of immune-mediated liver injury induced by 16,16 dimethyl PGE2 and PGE2-containing nanoscale hydrogels. *Biomaterials*, **32**(21), 4925–4935 (2011).

Chapter 16

Agarose Hydrogel Beads for Treating Diabetes

Nguyen Minh Luan, Yuji Teramura and Hiroo Iwata*
Department of Reparative Materials, Institute for Frontier Medical Sciences, Kyoto University, 53 Kawahara-Cho, Shogoin, Sakyo-Ku, Kyoto 606-8507, Japan
**iwata@frontier.kyoto-u.ac.jp*

1. Introduction

Insulin-dependent diabetes mellitus (IDDM) is an autoimmune disease caused by the specific destruction of insulin-producing pancreatic β-cells in islets of Langerhans (islets).[1] As a consequence, insufficient insulin is available, resulting in hyperglycemia. The transplantation of islets isolated from a donated pancreas has been evaluated as a means of long-term treatment for IDDM.[2–4] Recently, a high rate of insulin independence was realized in IDDM patients by islet transplantation followed by appropriate immunosuppressive therapy.[5–9] Immunosuppressive drugs, however, are associated with various side effects and also cause deterioration of islet function. Also, the patient must take the immunosuppressive drugs for life to prevent graft rejection.[10–12]

Islets encapsulated within biocompatible hydrogels, usually termed a bioartificial pancreas, are an attractive option for achieving successful islet transplantation without need of immunosuppressive therapy.[13–15] Our group has examined agarose, a polysaccharide isolated from seaweed, as a material for microencapsulating islets. In this chapter, we will review our work on microencapsulated islets in agarose hydrogel as a bioartificial pancreas.

2. Enclosure of Islets in Agarose Hydrogel

Agarose is mainly composed of β-D-galactopyranosyl and 3,6-anhydro-α-L-galactopyranosyl coupled through 1–3 (Fig. 1). Agarose dissolves in aqueous medium, gives a viscous solution at around 80°C, and forms a hydrogel at around 40°C. Due to this large thermal hysteresis, agarose hydrogel is stable under physiological conditions.[16] Agarose with gelling temperatures below 37°C (low gelling-temperature agarose) is usually employed to encapsulate islets. During the encapsulation procedure, islets isolated from a donated pancreas are suspended in the agarose solution, which is maintained at 40°C. The solution is then molded into the desired shape and exposed to lower temperature to allow the agarose solution to gel. Several types of bioartificial pancreas, e.g., microbead, string, and tablet shapes made from agarose hydrogel were examined for their performance in both *in vitro* and *in vivo* models.[17] The microcapsule type in which a single islet is encapsulated in a spherical microbead is the most suitable shape for bioartificial pancreas because of their large surface area that maximize the influx and outflux of nutrients for islet survival and function, and that enables insulin secretion into the blood.

For the microencapsulation of islets, 0.15 g low gelling-temperature agarose powder is added to 3 mL minimum essential medium (MEM) in a glass centrifuge tube. Sterilized 5% agarose solution is prepared by autoclaving the mixture at 120°C for 20 min. The agarose solution is cooled to 40°C by immersing the tube into a water bath at 40°C for 10 min. Islets are isolated from donated pancreases using the collagenase digestion method, and are suspended in a minimum volume of Hanks' buffered salt solution (HBSS). The islets are added to the agarose solution and mixed well. Then 15 mL of liquid paraffin (Merck, NJ, USA) pre-warmed to 40°C is added to the glass tube. The tube is briefly agitated and then immersed in an ice bath with agitation for 3 min to induce gelation of microdroplets of the agarose solution. Cold HBSS (15 mL) is added to the glass tube and the suspension is centrifuged at 2000 rpm for 5 min at 4°C. Paraffin oil and supernatant are removed, and the microbeads are washed three times with HBSS.

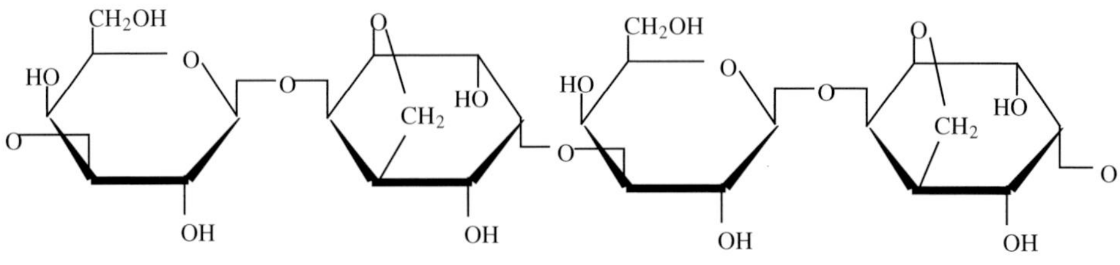

Fig. 1. Molecular structure of agarose.

3. Microencapsulated Islets as a Bioartificial Pancreas

3.1. Allotransplantation

Non-obese diabetic (NOD) mice are a good model for evaluation the function of a bioartifical pancreas given that the diabetes in these animal models are caused by autoimmune reactions, similar to IDDM patient pathophysiology.[18] Naked and microencapsulated C3H mouse islets were intraperitoneally transplanted into diabetic NOD mice. When 1,500 naked allogeneic islets were transplanted, they were rejected within 10 days post-transplantation. In contrast, 1,500 C3H islets enclosed in agarose microbeads reversed the animals' hyperglycemia to normal levels for more than 80 days in four out of five NOD mice.[19] Two mice maintained stable normoglycemia until they were sacrificed at 102 and 192 days (Fig. 2). These data suggest that agarose is a promising material that could be used to prepare immunoisolative membranes to effectively protect allogeneic islet grafts from both autoimmune and alloimmune responses without the use of immunosuppressive drugs.

3.2. Xenotransplantation

A shortage of donors has been the major obstacle in clinical organ and cell transplantation. Xenogeneic islet transplantation, for example, transplantation of porcine islets to a human patient, is a promising method for solving this problem.[20–22] The

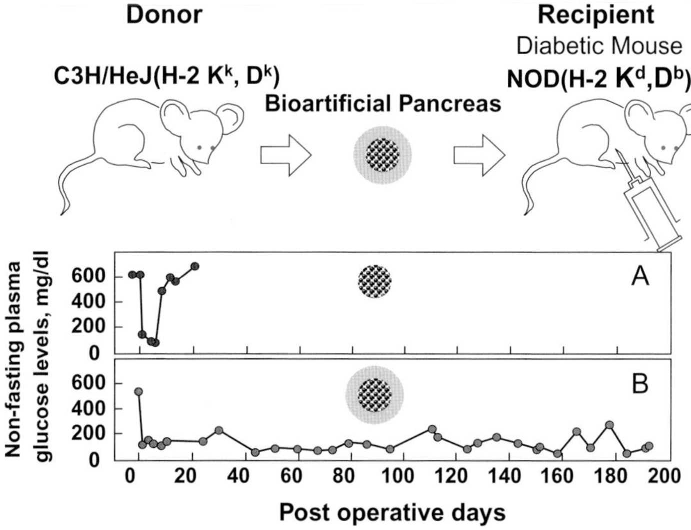

Fig. 2. Non-fasting plasma glucose levels of recipient NOD mice after intraperitoneal transplantation of 1500 C3H mouse islets: (A) naked islets, (B) microencapsulated islets.

feasibility of agarose microencapsulation to prevent rejection of xenogeneic islets was examined using the xenotransplantation model from hamster to diabetic mouse. Unfortunately, while proving effective in an allotransplantation mode between mice, agarose microencapsulation did not significantly prolong xenogeneic graft survival.[23–25] Hamster islets enclosed in 5% agarose microcapsules gradually lost their function within 20 days after intraperitoneal transplantation in streptozotocin-induced diabetic BALB/c mice.[25] This reflects the difference in the mechanism of graft rejection between allotransplantation and xenotransplantation. It is generally accepted that cellular immunity is the predominant cause of allograft rejection. The agarose semipermeable membrane, which physically prevents direct contact between islet cells and host immune cells, can significantly prolong allograft survival. However, in xenotransplantation, humoral immunity, in which antibody and complement play the major roles, is suspected of destroying islet grafts.[26,27] The cytotoxic effects of antibody and complement should be moderated to prolong xenogeneic islet graft survival.

Before our strategies can be introduced, we will briefly discuss the actions of the complement system in rejecting xenogeneic islets in agarose microcapsules. Binding of antibodies to antigens on the islet cell surface alone cannot lethally damage the islet cells. Complement is expected to play important roles in the destruction of xenogeneic islets.[27] The complement system is part of innate immunity, and is activated in a cascade manner when exotic materials or non-self tissues are introduced.[28,29] The complement system is activated through the classical pathway by an antigen–antibody complex or through the alternative pathway by a variety of activators including polysaccharides, bacteria, and damaged tissues. Antibodies and complement proteins, which can diffuse through the agarose membrane, can activate the complement system on the surface of islet cells through the classical pathway. Once activation occurs, C3b bound on the islet surface triggers the amplification loop to generate a vast number of C3b molecules. A number of the membrane attack complexes are formed, resulting in the lysis of islet cells. This activation also leads to the generation of anaphylatoxins (C3a and C5a), which in turn recruit and activate inflammatory cells such as phagocytes and lymphocytes, which surround the graft. Massive numbers of immune cells accumulate around the microcapsules and produce large quantities of various cytotoxic molecules (TNF-α, interferon-γ, interleukin-1β peroxides, and radicals, etc.). These molecules induce apoptosis of islet cells.[30–33] Inhibition of activation of the complement system is therefore necessary to prolong survival of xenogeneic islet cells inside microcapsules.

3.2.1. *Microbeads containing poly(styrene sulfonic acid)*

Polymers carrying sulfonic acid or sulfate groups are known to strongly interact with complement proteins, resulting in a decrease in cytolytic complement activity.[34,35]

Since poly(styrene sulfonic acid) (PSSa) is miscible with agarose, microbeads containing PSSa were prepared and examined for their efficacy in inhibiting complement activation.[24,36]

Preparation of agarose-PSSa microbeads is briefly described here. Agarose (0.15 g) and PSSa (0.3 g) are suspended in 3 mL of serum-free MEM solution in a 50-mL glass centrifuge tube and autoclaved. The mixture is then cooled to 40°C. Islets are added to the agarose/PSSa solution and mixed well. 20 mL of pre-warmed paraffin oil is then added to the centrifuge tube. After the tube is agitated to form agarose/PSSa droplets of the appropriate size in the paraffin oil, the tube is immersed into an ice bath with gentle agitation to induce gelation of the agarose/PSSa droplets. 30 mL of a 1% polycation polybrene solution is added to the tube and gently agitated to form a polyion complex layer on the microbeads. The tube is then centrifuged at 2000 rpm for 5 min at 4°C. Paraffin oil and supernatant are removed. 30 mL of 1% polybrene solution is added to the centrifuge tube again, and the centrifuge tube is immersed in an ice bath with gentle agitation. The polyion complex layer inhibits leakage of PSSa from the microbeads. The microbeads are washed three times with ice-cold HBSS then immersed in a 1% carboxymethyl cellulose solution in serum-free MEM, followed by twice washing the beads with HBSS. The outermost surface of the microbeads is covered by polyanion carboxymethyl cellulose.

The protective effects of an agarose-PSSa membrane were tested by carrying out the xenotransplantation of hamster islets in the microbeads into the intraperitoneal cavity of NOD mice without use of immunosuppressive therapy.[36] Five NOD mouse recipients of 1,000 agarose-PSSa microbeads demonstrated stable normoglycemia for more than 60 days, whereas mice receiving islets in plain agarose reverted to a diabetic state within 2 weeks post-transplantation (Fig. 3). The polyion complex layer inhibits leakage of PSSa from the microbeads. Thus, agarose-PSSa microbeads could protect xenogenieic islets from rejection more effectively than plain agarose microbeads.

3.2.2. *Microbeads carrying sCR1*

Complement receptor 1 (CR1) has the ability to bind complement C3b/C4b, control the dissociation of C3 and C5 convertase, and can also serve as a cofactor for the proteolytic cleavage of C3b/C4b by factor I. Administration of CR1 soluble form (sCR1) has been reported to prolong the survival of allografted and xenografted organs in a dose-dependent manner in several animal models,[37–39] and its use in preclinical trials has been examined.[40,41] Local inhibition of complement is sufficient for islet transplantation. Our group examined the protective effect of agarose microbeads carrying sCR1 on the ability to protect xeno-islets from rejection.[42–46]

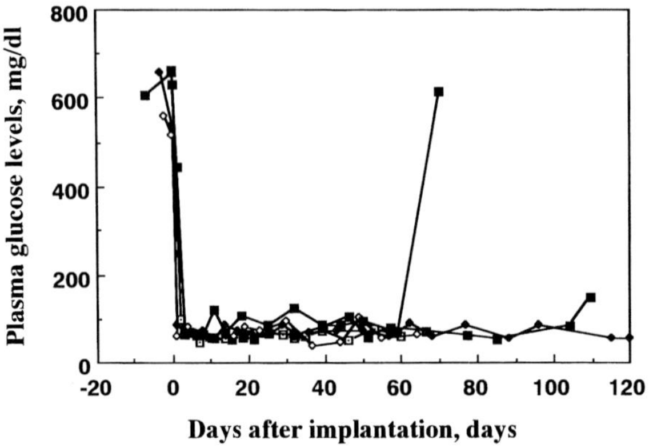

Fig. 3. Changes in non-fasting plasma glucose levels after intraperitoneal implantation of 1000 enclosed islets in the 5% agarose/5% poly(styrene sulfonic acid) microbeads that have the outermost surface covered by carboxymethyl cellulose. One mouse was sacrificed at 60 days. One mouse died at 64 days from bleeding after the glucose tolerance test. Plasma glucose levels of the other mouse increased more than 250 mg/dL at 70 days.

A method to covalently immobilize sCR1 onto agarose microbeads is briefly described.[42] The hydroxyl groups of the agarose are modified to maleimide groups using N-(pmaleimidophenyl) isocyanate (PMPI). Maleimide-functionalized agarose (MAL-agarose) is then used to enclose islets using the conventional method described above. Thiol groups are introduced onto sCR1 by treating the proteins with Traut's reagent. Subsequent incubation of MAL-agarose microbeads in a thiolated sCR1 solution allows the conjugation of sCR1 onto the agarose hydrogel. Islets enclosed in sCR1-agarose microbeads maintain intact morphology and secret insulin normally in response to glucose stimulation (Fig. 4).

The protective effect of microbeads carrying sCR1 with regards to the antibody-complement-dependent destruction of islets was examined *in vitro*. Rat-rabbit is a discordant combination[47]; that is, rabbit carries preformed antibodies against rat tissue. When naked rat islets were exposed to fresh rabbit serum, the islets quickly swelled and were completely destroyed within a few days. On the other hand, no damage was observed when islets in sCR1-agarose microbeads were incubated in rabbit serum over a course of 4 days.[42]

The protective effect of the sCR1-agarose microbeads was also examined *in vivo* using xenotransplantation from rat to mouse. One thousand Wistar rat islets, after encapsulation in either 5% plain agarose microbeads or sCR1-agarose microbeads, were transplanted into the intraperitoneal cavity of streptozotocin-induced diabetic C57BL/6 mice.[45] In the plain agarose microbead group, normoglycemia was maintained for less than 2 weeks, whereas normoglycemia was

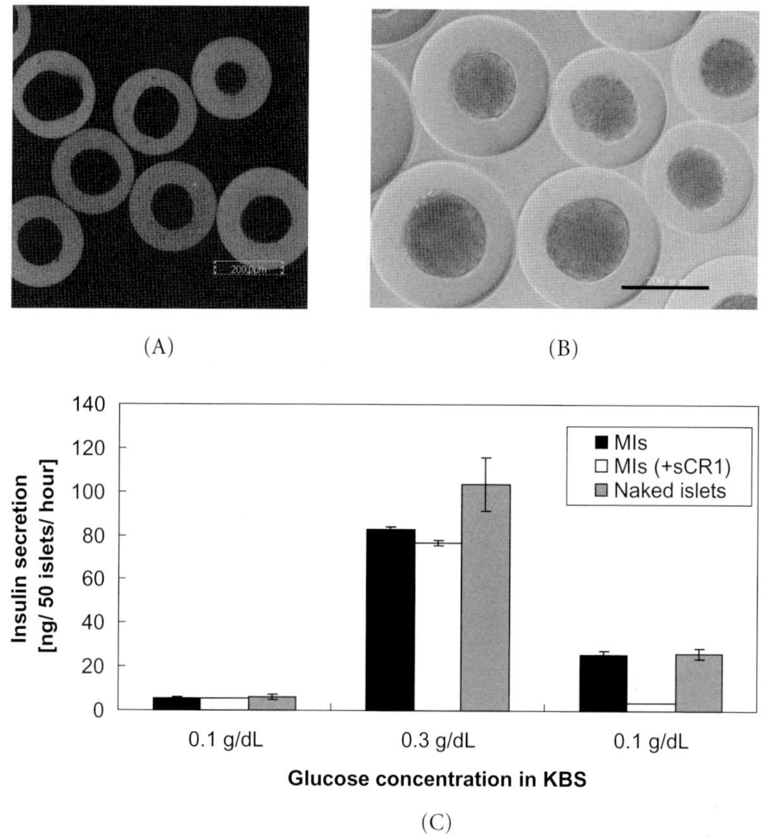

Fig. 4. Islets enclosed in agarose microbeads immobilized with sCR1. (A) Fluorescent image of agarose microbeads immunostained with anti-sCR1 antibodies (J3D3) and counterstained with green fluorescent-labeled Alexa488 antibodies. (B) Phase contrast microscopic image of sCR1-agarose microencapsulated islets after 2 weeks of culture in RMPI 1640 supplemented with 10% FBS. Scale bar = 200 μm. (C) Glucose stimulation tests of naked islets and microencapsulated islets (MIs) with or without sCR1.

maintained for more than 34 days in mice receiving 1000 islets enclosed in sCR1-agarose microbeads (Fig. 5). Islet grafts in two mice maintained partial function for more than 60 days. Viable islets could be found inside the sCR1-agarose microbeads 40 days post-transplantation.

4. Cryopreservation

Banking bioartificial pancreas is definitely required for medical applications because it must be available at any time. Recently, we successfully cryopreserved islets in agarose microbeads using a vitrification solution.[48,49] Vitrification is the solidification of

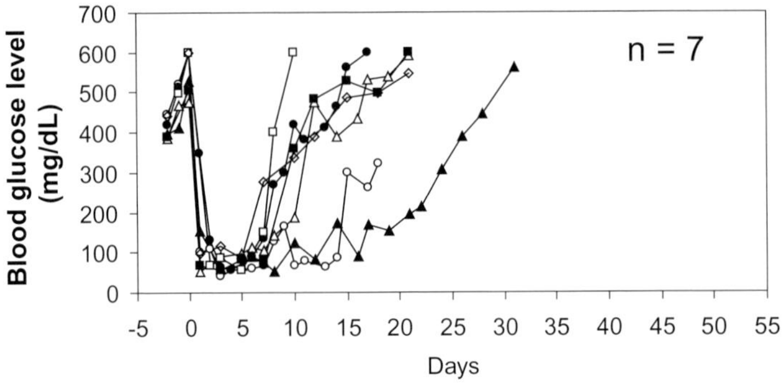

(A)

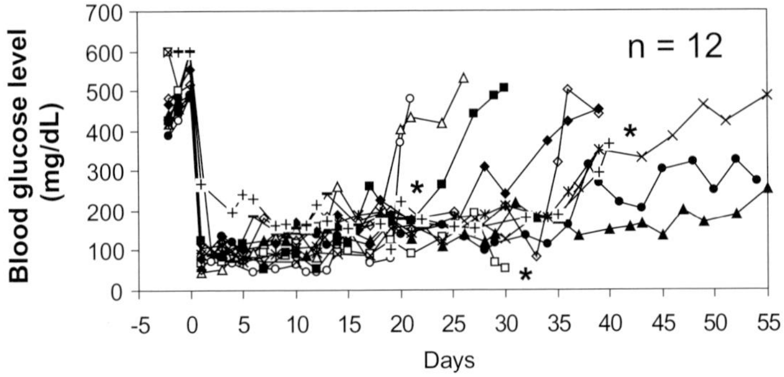

(B)

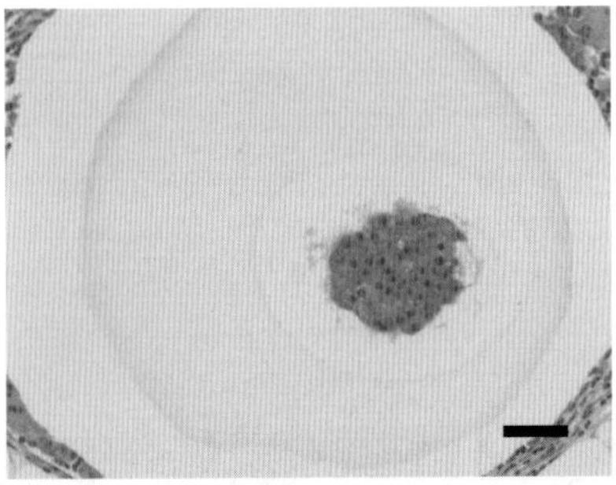

(C)

Fig. 5. Assessment of graft survival after xenotransplantation of sCR1-agarose or non-treated agarose microencapsulated islets. Non-fasting plasma glucose levels after transplantation of (A) 1,000 non-treated agarose microcapsules or (B) 1,000 sCR1-agarose microcapsules into the intraperitoneal cavity of STZ-diabetic mice. *mice killed for immunohistological examination. (C) Hematoxylin and eosin staining of sCR1-agarose microcapsule retrieved 40 days post-transplantation. Scale bars = 50 μm.

a liquid that contains a high concentration of cryoprotectant into a glass state without crystallization. Our vitrification solution consists of 5.38 M ethylene glycol, 2 M DMSO, 0.1 M PEG 1000, and 0.00175 M PVP K10 (MW 10000) in Euro-Collins solution (34.95 g/L dextrose, 7.3 g/L K_2HPO_4, 2.04 g/L KH_2PO_4, 1.12 g/L KCl, and 0.84 g/L $NaHCO_3$).

We cryopreserved islets in agarose microbeads and performed some of the fundamental *in vitro* and *in vivo* assays to examine the functional ability of islets and agarose microbeads after cryopreservation.[48] Figure 6 shows the islets in microbeads before and after cryopreservation. No significant difference was found in the morphologies of the islets or the microbeads. However, insulin secretion from the cryopreserved islets in microbeads was reduced to 78% and 64% that of non-cryopreserved ones at low- and high-glucose concentration, respectively. However, these islets still maintained control of their insulin secretion in response to glucose changes.

In vivo functions of islets in microbeads with/without cryopreservation were compared.[49] Hamster islets in microbeads were transplanted into diabetic mice. Figure 7 shows the graft survival plots of 1,000 islets in microbeads without cryopreservation, and 1,000 and 2,000 islets in microbeads after cryopreservation.

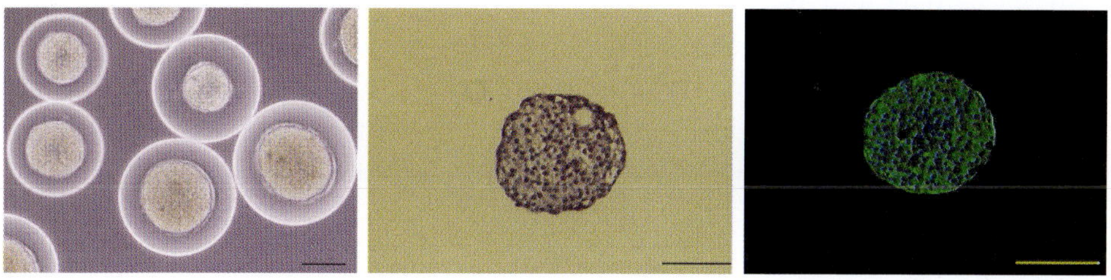

(A)

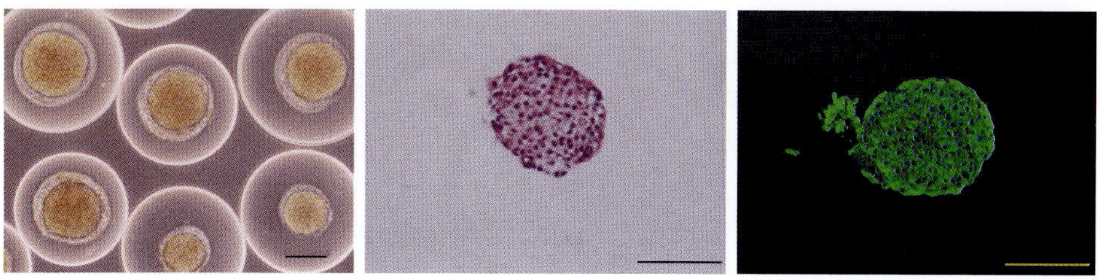

(B)

Fig. 6. Morphology and histology of agarose islets (A) before and (B) after cryopreservation, thawing, and culture for 7 days. The morphology of agarose-islets was examined by phase-contrast microscopy (left). The samples were stained with hematoxylin & eosin staining (center), and both Alexa488-labeled anti-insulin antibody and Hoechst 33342 for nuclear staining (right). The images were merged pictures of insulin and Hoechst 33342 staining.

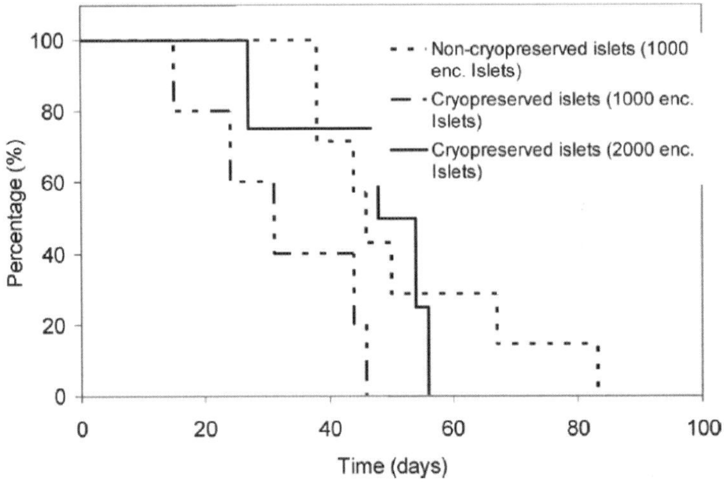

Fig. 7. Percent of surviving grafts in recipients of islets in agarose microbeads with/without cryopreservation. Graft failure was defined as two consecutive plasma glucose determinations that are more than 200 mg/dL.

Recipients of 1,000 and 2,000 cryopreserved islets in microbeads demonstrated 32.0 ± 13.2 days and 46.3 ± 13.3 days of normoglycemia, respectively. Although this was slightly shorter than the 53.2 ± 16.7 days for recipients of 1,000 islets in microbeads without cryopreservation, the results indicate that the cryopreserved islets could control blood glucose levels in the recipients and the agarose hydrogel could protect xenogeneic islets from attack by the host's immune system.

5. Future Prospects

Several issues remain to be resolved before islets in microbeads become an acceptable medical device as a bioartificial pancreas in the clinical setting, although many successful results of islet transplantation have been reported.[4–7] In these, a shortage of donors has been the major obstacle in clinical islet transplantation. Many technologies have been developed to solve this problem. Human embryonic stem cells (ES cells) and induced pluripotent stem cells (iPS cells) are promising cell sources of insulin-releasing cells.[50–52] Difficulties facing the use of insulin-releasing cells differentiated from iPS cells derived from IDDM patients are persistent for treatment of IDDM patients. Certain cell lines have a higher capacity to differentiate into derivatives of some germ layers. This has led to the view that particular cell lines should be selected to achieve efficient differentiation. It is doubtful that iPS cells derived from IDDM patients can meet this criterion. In addition, although a large number of studies have been done to differentiate ES/iPS cells into insulin-releasing cells, challenges remain in derivation of insulin-releasing cells that can control insulin release in response to glucose concentration changes. Further work is needed to establish the methodology to effectively differentiate ES/iPS cells into insulin-releasing cells that can be used to treat IDDM patients.

Another point we should remember is that IDDM is an autoimmune disease caused by the specific destruction of insulin-producing pancreatic β-cells in islets. NOD mice have been used as an IDDM model. Kobayashi *et al.*, transplanted islets isolated from young non-diabetic NOD mice to diabetic NOD mice syngeneically.[53] Naked islet grafts were rapidly destroyed within 1 to 2 weeks after transplantation with strong mononuclear infiltration, suggesting that islet grafts were destroyed by the autoimmune reactions. This result suggests that autoimmune reactions should be carefully controlled when insulin-releasing cells differentiated from iPS cells derived from IDDM patients are used to treat IDDM. They also reported in the same article that 13 out of 16 mice receiving 1,500 to 2,000 islets in agarose microbeads maintained normoglycemia throughout the 150-day observation period. Histologic findings of the retrieved grafts after this follow-up time showed many viable islets with a perfectly intact shape and clear expression of insulin inside the microcapsules. No mononuclear cell infiltration was observed. These data demonstrated that agarose microbeads can protect islets from autoimmune destruction in NOD mice and will be useful in iPS cell therapy of IDDM.

Since stem cell technologies are still in the development stage, xenogeneic islet transplantation is still an attractive method. Although microbeads containing PSSa or protein sCR1 prolonged xenogeneic islet survival, PSSa may lose the potential to inhibit the cytolytic activity of complement when PSSa in the microcapsules is saturated with cationic proteins, or protein sCR1 might be denatured and degraded by proteases that are produced through the inflammatory reactions surrounding the grafts. Further work is clearly needed. For instance, we are making an effort to develop cells that continuously deliver soluble sCR1. Microencapsulation of co-aggregates of the sCR1-releasing cells with islet cells may be a promising method to prolong xenograft survival.

References

1. D. Mathis, L. Vence, and C. Benoist Beta-cell death during progression to diabetes. *Nature*, **414**, 792–798 (2001).
2. J. A. Bluestone, K. Herold, and G. Eisenbarth, Genetics, pathogenesis and clinical interventions in type 1 diabetes. *Nature*, **464**, 1293–1300 (2010).
3. T. L. van Belle, K. T. Coppieters, and M. G. von Herrath, Type 1 diabetes: Etiology, immunology, and therapeutic strategies. *Physiol. Rev.*, **91**(1), 79–118 (2011).

4. R. P. Robertson, Islet transplantation as a treatment for diabetes — a work in progress. *N. Engl. J. Med.*, **350**, 694–705 (2004).
5. A. M. Shapiro, J. R. Lakey, E. A. Ryan, G. S. Korbutt, E. Toth, G. L. Warnock, *et al.*, Islet transplantation in seven patients with type 1 diabetes mellitus using a glucocorticoid-free immunosuppressive regimen. *N. Engl. J. Med.*, **343**, 230–238 (2000).
6. E. A. Ryan, J. R. Lakey, R. V. Rajotte, G. S. Korbutt, T. Kin, S. Imes, *et al.*, Clinical outcomes and insulin secretion after islet transplantation with the Edmonton protocol. *Diabetes*, **50**(4), 710–719 (2001).
7. E. A. Ryan, B. W. Paty, P. A. Senior, D. Bigam, E. Alfadhli, N. M. Kneteman, *et al.*, Five year follow-up after clinical islet transplantation. *Diabetes*, **54**, 2060–2069 (2005).
8. A. M. Shapiro, Strategies toward single-donor islets of Langerhans transplantation. *Curr. Opin. Organ. Transplant.*, **16**, 627–631 (2011).
9. F. B. Barton, M. R. Rickels, R. Alejandro, B. J. Hering, S. Wease, B. Naziruddin, *et al.*, Improvement in outcomes of clinical islet transplantation: 1999–2010. *Diabetes Care*, **35**, 1436–1445 (2012).
10. J. D. Johnson, Z. Ao, P. Ao, H. Li, L. J. Dai, Z. He, *et al.*, Different effects of FK506, rapamycin, and mycophenolate mofetil on glucose-stimulated insulin release and apoptosis in human islets. *Cell Transplant*, **18**(8), 833–845 (2009).
11. A. Hyder, C. Laue, and J. Schrezenmeir, Effect of the immunosuppressive regime of Edmonton protocol on the long-term *in vitro* insulin secretion from islets of two different species and age categories. *Toxicol. In vitro*, **19**(4), 541–546 (2005).
12. R. Gao, J. Ustinov, O. Korsgren, and T. Otonkoski, Effects of immunosuppressive drugs on *in vitro* neogenesis of human islets: Mycophenolate mofetil inhibits the proliferation of ductal cells. *Am. J. Transplant.*, **7**(4), 1021–1026 (2007).
13. G. Basta and R. Calafiore, Immunoisolation of pancreatic islet grafts with no recipient's immunosuppression: Actual and future perspectives. *Curr. Diab. Rep.*, **11**(5), 384–391 (2011).
14. S. Kizilel, M. Garfinkel, and E. Opara, The bioartificial pancreas: Progress and challenges. *Diabetes Technol. Ther.*, **7**(6), 968–985 (2005).
15. N. Kobayashi, Bioartificial pancreas for the treatment of diabetes. *Cell Transplant.*, **17**(1–2), 11–17 (2008).
16. M. W. Agar, In *Handbook of Water-Soluble Gums and Resins,* Ed., R. L. Davidson, McGraw-Hill, Inc, pp. 1–19.
17. H. Yang, H. Iwata, H. Shimizu, T. Takagi, T. Tsuji, and F. Ito, Comparative studies of *in vitro* and *in vivo* function of three different shaped bioartificial pancreases made of agarose hydrogel. *Biomaterials*, **15**(2), 113–120 (1994).
18. S. Darquy, D. Chicheportiche, F. Capron, C. Boitard, and G. Reach, Comparative study of microencapsulated rat islets implanted in different diabetic models in mice. *Horm. Metab. Res. Suppl.* **25**, 209–213 (1990).
19. H. Iwata, T. Takagi, H. Amemiya, H. Shimizu, K. Yamashita, K. Kobayashi, and T. Akutsu, Agarose for a bioartificial pancreas. *J. Biomed. Mater. Res.*, **26**(7), 967–977 (1992).

20. R. N. Pierson 3rd, A. Dorling, D. Ayares, M. A. Rees, J. D. Seebach, and J. A. Fishman, *et al.*, Current status of xenotransplantation and prospects for clinical application. *Xenotransplantation*, **16**(5), 263–280 (2009).
21. Y. G. Yang, and M. Sykes, Xenotransplantation: Current status and a perspective on the future. *Nat. Rev. Immunol.*, **7**(7), 519–531 (2007).
22. D. Dufrane and P. Gianello, Pig islet xenotransplantation into non-human primate model. *Transplantation*, **86**(6), 753–760 (2008).
23. H. Iwata, H. Amemiya, T. Matsuda, H. Takano, R. Hayashi, and T. Akutsu, Evaluation of microencapsulated islets in agarose gel as bioartificial pancreas by studies of hormone secretion in culture and by xenotransplantation. *Diabetes*, **38**(Suppl 1), 224–225 (1989).
24. T. Takagi, H. Iwata, H. Tashiro, T. Tsuji, and F. Ito, Development of a novel microbead applicable to xenogeneic islet transplantation. *J. Control Release*, **31**(3), 283–291 (1994).
25. H. Iwata, K. Kobayashi, T. Takagi, T. Oka, H. Yang, H. Amemiya, *et al.*, Feasibility of agarose microbeads with xenogeneic islets as a bioartificial pancreas. *J. Biomed. Mater. Res.*, **28**(9), 1003–1011 (1994).
26. X. He, I. Teppler, L. Copey, C. Smith, K. Reemtsma, and E. Rose, Incidence of pre-formed antibodies against potential xenodonors in human sera. *Transplantation*, **49**, 1022–1024 (1990).
27. D. K. C. Cooper, E. Kemp, K. Reemtsma, and D. J. G. Whithe, Experimental Xenotransplantation in rodents – I: Plasma exchange and splenectomy. In *Xenotransplantation*, Eds. D. K. C. Cooper, E. Kemp, K. Reemtsma, and D. J. G. Whithe, Berlin, Heidelberg, Germany: Springer-Verlag, pp. 312–322 (1991).
28. M. J. Walport, Complement. First of two parts. *N. Engl. J. Med.*, **344**(14), 1058–1166 (2001).
29. M. J. Walport, Complement. Second of two parts. *N. Engl. J. Med.*, **344**(15), 1140–1144 (2001).
30. T. Mandrup-Poulsen, K. Bendtzen, J. H. Nielsen, G. Bendixen, and J. Nerup, Cytokines cause functional and structural damage to isolated islets of Langerhans. *Allergy*, **40**, 424–429 (1985).
31. A. Rabinovitch, W. L. Suarez-Pinzon, K. Strynadka, J. R. Lakey, and R. V. Rajotte, Human pancreatic islet beta-cell destruction by cytokines involves oxygen free radicals and aldehyde production. *J. Clin. Endocrinol. Metab.*, **81**, 3197–3202 (1996).
32. M. Koulmanda, M. Bhasin, Z. Awdeh, A. Qipo, Z. Fan, and D. Hanidziar, The role of TNF-α in mice with type 1- and 2- diabetes. *PLoS One*, **7**(5), e33254 (2012).
33. R. B. Stevens, J. D. Ansite, C. D. Mills, A. Lokeh, T. J. Rossini, M. Saxena, *et al.*, Nitric oxide mediates early dysfunction of rat and mouse islets after transplantation. *Transplantation*, **61**(12), 1740–1749 (1996).
34. D. Bitter-Suermann, R. Burger, and U. Hadding, Activation of the alternative pathway of complement: Efficient fluid-phase amplification by blockade of the regulatory complement protein beta1H through sulfated polyanions. *Eur. J. Immunol.*, **11**(4), 291–295 (1981).

35. R. Burger, U. Hadding, H. U. Schorlemmer, V. Brade, D. Bitter-Suermann, Dextran sulphate: A synthetic activator of C3 via the alternative pathway. *Immunology*, **29**(3), 549–554 (1975).
36. H. Iwata, T. Takagi, K. Kobayashi, T. Oka, T. Tsuji, and F. Ito, Strategy for developing microbeads applicable to islet xenotransplantation into a spontaneous diabetic NOD mouse. *J. Biomed. Mater. Res.*, **28**(10), 1201–1207 (1994).
37. J. R. Pratt, M. J. Hibbs, A. J. Laver, R. A. Smith, and S. H. Sacks, Effects of complement inhibition with soluble complement receptor-1 on vascular injury and inflammation during renal allograft rejection in the rat. *Am. J. Pathol.*, **149**(6), 2055–2066 (1996).
38. S. K. Pruitt and R. R. Bollinger, The effect of soluble complement receptor type 1 on hyperacute allograft rejection. *J. Surg. Res.*, **50**(4), 350–355 (1991).
39. S. K. Pruitt, A. D. Kirk, R. R. Bollinger, H. C. Marsh, B. H. Collins, J. L. Levin, *et al.*, The effect of soluble complement receptor 1 on hyperacute rejection of porcine xenografts. *Transplantation*, 57, 363–370 (1994).
40. J. Mocco, W. J. Mack, A. F. Ducruet, R. G. King, M. E. Sughrue, A. L. Coon, *et al.*, Preclinical evaluation of the neuroprotective effect of soluble complement receptor type 1 in a nonhuman primate model of reperfused stroke. *J. Neurosurg.*, **105**(4), 595–601 (2006).
41. A. F. Ducruet, J. Mocco, W. J. Mack, A. L. Coon, H. C. Marsh, D. J. Pinsky, *et al.*, Preclinical evaluation of an sLe x-glycosylated complement inhibitory protein in a non-human primate model of reperfused stroke. *J. Med. Primatol.*, **36**(6), 375–380 (2007).
42. N. M. Luan, Y. Teramura, and H. Iwata, Immobilization of the soluble domain of human complement receptor 1 on agarose-encapsulated islets for the prevention of complement activation. *Biomaterials*, **31**(34), 8847–8853 (2010).
43. N. M. Luan, Y. Teramura, and H. Iwata, Immobilization of soluble complement receptor 1 on islets. *Biomaterials*, **32**(20), 4539–4545 (2011).
44. N. M. Luan, Y. Teramura, and H. Iwata, Layer-by-layer co-immobilization of soluble complement receptor 1 and heparin on islets. *Biomaterials*, **32**(27), 6487–6492 (2011).
45. N. M. Luan and H. Iwata, Xenotransplantation of islets enclosed in agarose microcapsule carrying soluble complement receptor 1. *Biomaterials*, **33**(32), 8075–8081 (2012).
46. N. M. Luan and H. Iwata, Inhibition of instant blood-mediated inflammatory responses by co-immobilization of sCR1 and heparin on islets. *Biomaterials*, **34**(21), 5019–5024 (2013).
47. Z. Zhang, E. Bédard, Y. Luo, H. Wang, S. Deng, D. Kelvin, *et al.*, Animal models in xenotransplantation. *Expert. Opin. Investig. Drugs.*, **9**, 2051–2068 (2000).
48. C. A. Agudelo and H. Iwata, The development of alternative vitrification solutions for microencapsulated islets. *Biomaterials*, **29**(9), 1167–1176 (2008).
49. C. A. Agudelo, Y. Teramura, and H. Iwata, Cryopreserved agarose-encapsulated islets as bioartificial pancreas: A feasibility study. *Transplantation*, **87**(1), 29–34 (2009).
50. M. Hansson and O. D. Madsen, Pluripotent stem cells, a potential source of beta-cells for diabetes therapy. *Curr. Opin. Investig. Drugs.*, **11**, 417–425 (2010).

51. R. T. Wagner, J. Lewis, A. Cooney, and L. Chan, Stem cell approaches for the treatment of type 1 diabetes mellitus. *Transl. Res.*, **156**(3), 169–179 (2010).
52. C. Aguayo-Mazzucato and S. Bonner-Weir, Stem cell therapy for type 1 diabetes mellitus. *Nat. Rev. Endocrinol.*, **6**(3), 139–148 (2010).
53. T. Kobayashi, Y. Aomatsu, H. Iwata, T. Kin, H. Kanehiro, M. Hisanaga, *et al.*, Indefinite islet protection from autoimmune destruction in nonobese diabetic mice by agarose microencapsulation without immunosuppression. *Transplantation*, **75**(5), 619–625 (2003).

Chapter 17

Hydrogels in Urogenital Applications

James Turner[†] **and Jiro Nagatomi**[†]
Department of Bioengineering, Clemson University
301 Rhodes Research Center, Clemson, South Carolina, USA
[*]jpturne@clemson.edu
[†]jnagato@clemson.edu

This chapter reviews the use of hydrogels in various urogenital applications ranging from tissue engineering to coating to injected agents. The versatility of hydrogels provide solutions to the variety of complications that arise within the urogenital system, and with further refinement, may be the solution to many of the unanswered problems that still exist.

1. Introduction

Hydrogels are used in a number of applications for treating medical conditions involving urogenital organs at various stages of pre-clinical and clinical tests. Both synthetic and naturally derived hydrogels have been explored as scaffold materials for tissue reconstruction, drug releasing devices, coating for implantable devices, hemostats, and tissue sealants. This chapter will first present a brief review of the anatomy and physiology of the urinary system consisting of the kidneys, the ureters, the bladder, and the urethra. The use of hydrogels in the vagina will also be covered briefly. Specific examples of hydrogel application areas and the medical conditions that require use of hydrogels in these organs are then reviewed to illustrate the current status and future directions.

1.1. Kidneys

In the human body, the two kidneys are located posterior to the peritoneum, partially protected by the 11th and 12th ribs. The kidneys are roughly 12 cm long by 6 cm wide, and approximately 3 cm thick with no significant differences between male and female kidneys. The shape of the kidney is often described as a red bean, or to put it more technically, an elongated cardioid. The three tissue layers surrounding

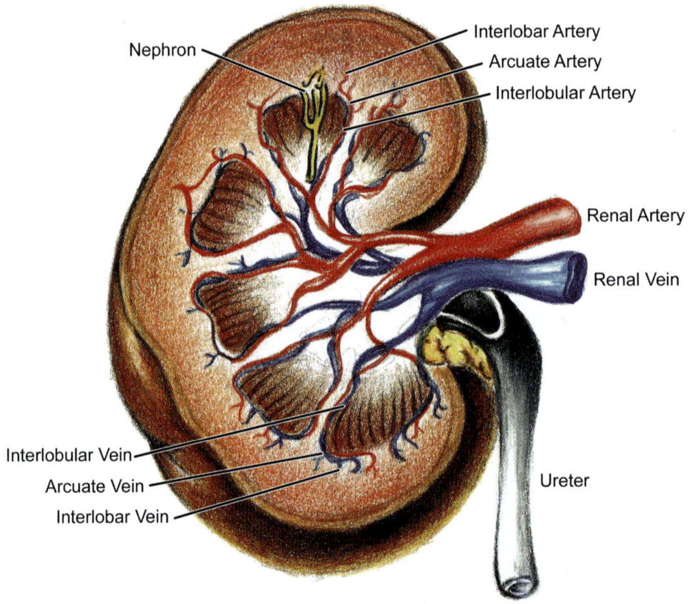

Fig. 1. A cross-section of the kidney. As blood passes from the arteries into the arterioles and capillaries, it undergoes a filtration process in the nephron — the functional unit of the kidney. The filtered blood is ultimately passed out of the kidneys through the renal vein, while the waste products are collected from the nephrons and emptied through the ureter.

the kidney are the renal capsule, the adipose capsule, and the renal fascia. The first two layers maintain the shape of the kidney and provide a thick protective layer, while the fascia holds the kidneys in place. Internally, a dense network of arteries, capillaries, and veins pass through the functional units of kidneys known as nephrons.

The kidneys bear the largest workload in the urinary system, which include regulation of ion composition, pH, volume and pressure of the blood, production of various hormone, as well as filtration and excretion of wastes. Specifically, the filtration, reabsorption, and secretion processes of the kidney occur in the nephrons. Blood is supplied to the kidneys through the renal arteries, which segment into a sequence of different arteries (interlobar, arcuate, interlobular). The interlobular arteries enter the renal cortex and branch off into afferent arterioles that connect into the glomerulus. Efferent arterioles lead away from the glomerulus and form peritubular capillaries. From here, the blood exits through veins in reverse order of the arteries (interlobular, arcuate, interlobar) and passes back into the body through the renal vein.

In conjunction with the blood vessels, renal tubules that originate from the glomerular capsule surrounding the glomerulus travel alongside the arteries and veins. In the capsule, a process called glomerular filtration occurs in which water and solutes are taken from the blood and passed into the renal tubules. Water and solutes are reabsorbed by the tubule cells and returned to the blood in the peritubular capillaries. Meanwhile, waste substances and drugs are secreted back into the renal tubules from the capillaries. The waste produced in this process are collected through the tubules and emptied into collecting ducts. These collecting ducts empty into the minor and major calyces prior to the renal pelvis that connects to the ureters.

1.2. Ureters

The ureters are on average 25 cm long in adults and extend from the renal pelvis of the kidneys and empty into the urinary bladder. The ureters join the urinary bladder through the posterior wall by curving towards the center and entering at a slight angle. Underneath the inner lining of uroepithelium and lamina propria layers, smooth muscle fibers fill the intermediate muscularis layer and contribute to the peristaltic function. The outer adventitia layer contains blood and lymphatic vessels, nerves, and the connective tissue that keeps the ureters in place.

The urine is transported through the ureters through a combination of several effects including peristaltic contractions of the muscle layer, hydrostatic pressure, and natural flow by gravity. As the urinary bladder fills, the expansion of the bladder wall closes the oblique openings of the ureters and prevents backflow of urine. The speed of the peristaltic waves beginning at the renal pelvis corresponds to the rate of urine formation within the kidneys. An increase in wave frequency corresponds to an increase in renal function and urine production.

1.3. Urinary bladder

The bladder is a hollow muscular organ that lies within the pelvic cavity, and directly contacts the uterus and vagina in females and the rectum in males. The urinary bladder

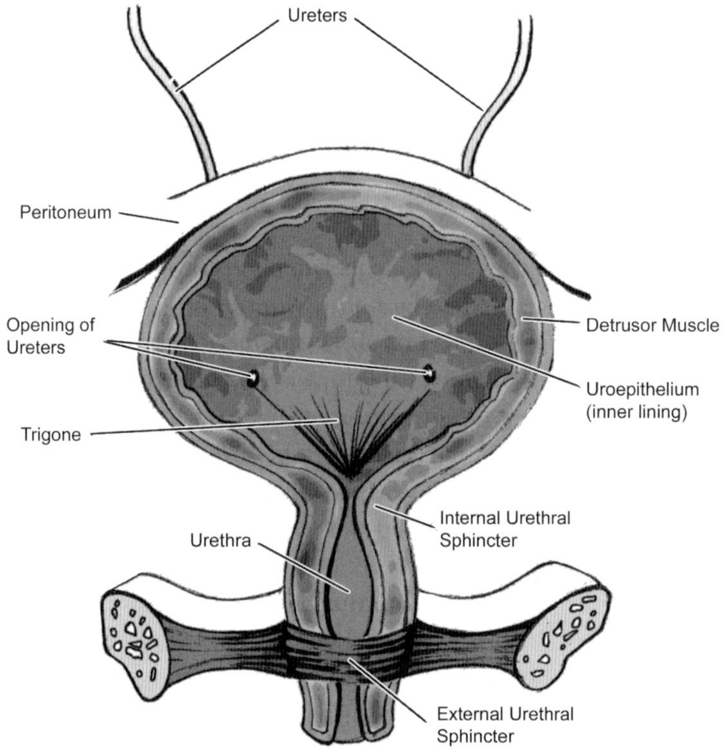

Fig. 2. A cross-section of the urinary bladder. Urine is deposited into the bladder from the ureters in small openings in the rear. As the bladder fills, the ureters are sealed due to the expansion of the bladder wall preventing any backflow. The external urethral sphincter is a voluntary muscle that allows control of urine voiding.

serves as a distensible reservoir for urine collection from a pair of kidneys connected by ureters. Similar to the ureter, the bladder wall tissue consists of four layers. Starting from the interior, the layers consist of the uroepithelium, the lamina propria, the detrusor, and the serosa.

The two main functions of the urinary bladder are to store urine that is generated in the kidney and to void when necessary. During the storage phase, electrolytes are exchanged in the urine due to passive and active ion transport occurring at the inner uroepithelial layer.[2] As the bladder fills, the organ expands and begins to push against other organs in the pelvic cavity. Although the maximum bladder capacity is approximately 600 milliliters, bladder fullness sensation can occur around 150 milliliters with discomfort at 300 milliliters. The external urethral sphincter releases, and contractions in the bladder wall allow for voiding.

1.3.1. *Uroepithelium*

The uroepithelium (also termed urothelium) is a highly specialized transitional epithelial tissue with three distinct layers of cells identified as the basal, intermediate, and umbrella cells. The umbrella cell layer provides a barrier at the interface between the urine and underlying tissue layers and active ion transport functionality of the uroepithelial layer, while the basal and intermediate layer provides progenitor cells for the umbrella layer. The presence of tight junctions and specialized apical membrane lipids contribute to the umbrella cells' ability to prohibit free movement of ions and solutes from the urine to the underlying tissue layers.[2] The apical membrane of the umbrella cells consist of plaque regions in between hinges. The plaque regions appear to be asymmetric, due to the outer leaflet being thicker than the inner leaflet. This asymmetry arises because of a highly crystalline structure formed by asymmetric unit membrane (AUM) particles composed mainly of uroplakins (UP) including UPIa, UPIb, UPII, UPIIIa, and UPIIIb. A glycosaminoglycan (GAGs)-based barrier composed primarily of chondroitin and heparan sulfate also contribute to the protection of tissue from the toxicity of the urine.[3]

What causes the sensation of fullness in the bladder? The mechanisms controlling and signaling the afferent nerves in the bladder are still not fully resolved. Recent studies have shown evidence that the uroepithelium releases diffusible molecules such as adenosine triphosphate (ATP), acetylcholine (ACh), nitric oxide (NO), and prostaglandins which may interact with receptors in the afferent axons. For example, ATP molecules released from urothelium in response to bladder fullness act on P2X3 receptors located in the suburothelial afferent nerve plexus.[32] It is suspected that epithelial sodium channels (ENaC) are responsible for detection of the mechanical stimuli in the uroepithelium as blocking these channels with amiloride suppresses ATP release.[33,34] The afferent signals are carried through myelinated Aδ-fibers found in the lamina propria to the neuronal somata located in the dorsal root ganglia at S2–S4 and T11–L2 spinal segmental levels.

1.3.2. *Lamina propria*

The lamina propria is separated from the uroepithelium by the basal lamina, and acts as barrier between the epithelium and smooth muscle layer of the bladder. It is mainly composed of connective tissue made up of collagen, laminin, and elastin. Within the lamina propria, one can find sensory and motor neurons along with lymphatic and blood vessels. The presence of unmyelinated nerves in the tissue indicate that damage to the lamina propria could potentially lead to disruption in normal electrical response due to mechanical strain in the bladder.[4] The classical view is that the nerves present in the lamina propria respond to stretch in the bladder,[37,38] while recent studies have demonstrated that the signaling molecules produced in the uroepithelial layer could also interact with these nerves. When the bladder fills, tension is transferred between the lamina propria and the detrusor layers at different ranges of capacity during bladder filling.[5] Though the structural mechanisms on how this occurs are not well-defined, it appears that this transfer is what gives the bladder its unique compliant properties.

1.3.3. *Detrusor*

Smooth muscle bundles form an interlaced weave spreading in all directions of the detrusor layer. Collagen (Types I and III), laminin, osteopontin, and hyaluronic acid is present and integrated with the smooth muscle fibers.[3] Fibronectin and collagen types I, III, and IV are produced by the bladder smooth muscle cells.[35] The detrusor muscle also extends towards the urethra forming an internal urethral sphincter allowing for retention of urine until voiding. Lack of collagen type I in this region has been shown to increase stress urinary incontinence in females.[36]

Urination, or micturition, involves afferent signaling pathways that derive from the same region of the spinal cord as bladder sensation. The excitation of the efferent pathways for bladder wall contractions begin from the intermediolateral column of the S_2–S_4 spinal cord via ATP/ACh release. Nerves that terminate at the base of the bladder and urethra contain enzymes called nitric oxide synthase. Nitric oxide synthase enzymes create NO in order to relax the urethral muscle. Motoneurons from Onuf's nucleus of the anterior horn in the S_2–S_4 spinal cord terminate at the striated sphincter muscle and release ACh to act on nicotinic receptors and relax the muscle.[32] This gives voluntary control to hold back urine in cases when urination is unwanted.

1.4. Urethra

The primary function of the urethra, the terminal drainage outlet for the urinary system, is to allow urine to be passed out of the urinary bladder. The urethra differs

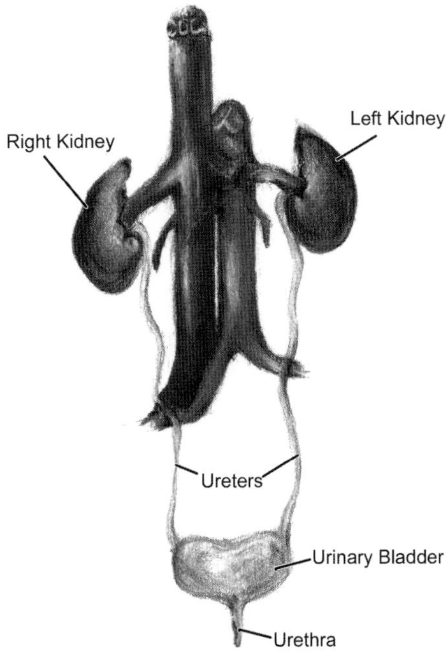

Fig. 3. An overview of the urinary system. Urine is produced in the kidneys and passed through the renal pelvis into the ureters. The urine is stored in the urinary bladder for a period of time before passing through the urethra. The length of the urethra varies based on gender.

considerably between genders. In females, the urethra extends only about 4 cm and exits between the clitoris and vaginal opening. The urethra contains a mucosal layer consisting of the lamina propria and an epithelial layer that differs along the urethral lining. The uroepithelium continues from the urinary bladder and covers the proximal portion of the urethra. Further along the urethra towards the outlet, the epithelium varies from stratified columnar or pseudostratified columnar to stratified squamous epithelium. The muscularis continues from the bladder and consists of circularly arranged smooth muscle fibers.[3]

In males, not only urine, but semen is also discharged through the urethra. The male urethra is approximately 20 cm long and follows a different path from the female urethra. The male urethra is divided into three regions. The prostatic urethra is the portion of the urethra that passes through the prostate. A milky secretion from the prostate (consisting of citric acid, proteolytic enzymes, acid phosphatase, and seminalplasmin as an antibiotic) is carried into the urethra through ducts in this region. Additionally, the vas deferens and seminal glands open into the prostatic urethra. The next and shortest portion, known as the intermediate urethra passes through the deep muscle of the perineum. Around this portion, skeletal muscle fibers are aligned in a circular fashion and form the external urethral sphincter muscle. The longest portion of the male urethra passes through the penis and is known as the spongy urethra. Similar to the changes in the female urethral mucosa, the uroepithelial layer continues from the urinary bladder and changes to stratified columnar epithelium in the intermediate urethra and most of the spongy urethra.

2. Hydrogel Applications in Urologic Tissue

2.1. Kidney

2.1.1. *Kidney Tissue Regeneration*

The high demand for donor kidneys severely limits transplantation as an option for patients suffering from acute kidney injury (AKI), previously known as acute renal failure.[26] For this reason, stem cell therapy has been explored in recent years to provide an alternative solution to AKI,[39,40] although the debate remains on which cells are actually capable of restoring kidney function. In a study using rat models of AKI, hydrogels were shown to assist in stem cell seeding and viability. More specifically, a thermosensitive chitosan chloride hydrogel was used to deliver adipose-derived stem cells (ADSCs) to an ischemia-injured kidney.[27] When compared to cells injected with PBS, significantly less apoptosis and increased cell proliferation of the ones seeded with the hydrogel was observed. Moreover, renal function, measured using the serum levels of creatinine and blood urea nitrogen, improved over a 28-day period with concentrations returning near the normal baseline in the kidneys that received the stem cell seeded with hydrogels. Cell retention within the hydrogel matrix is likely the key mechanism that contributes to the success of the implanted material. Without a substrate for cell adhesion, the cells are susceptible to impaired adhesion due to reactive oxygen species (ROS), which can lead to up to 90% cell death within the first week. In addition to providing a substrate for growth, the chitosan chloride hydrogel has anti-oxidative properties, which may curb the interaction of ROS. With the increased retention of cells, the start of angiogenesis and renal cell proliferation may occur earlier with the use of hydrogel scaffolds. With the apparent difficulty of applying free-floating cells to a targeted site *in vivo,* hydrogels will perhaps find success as a cell scaffold for renal tissue engineering, though there may be many hurdles in finding a cell candidate that will accurately behave as functional renal cells.

An approach to consider for renal tissue regeneration is the response of the host tissue to the natural or synthetic hydrogel materials that might be used for cell scaffolds. In an attempt to restore renal function degenerated by chronic kidney disease, several formulations of hydrogels were used to deliver cells in a rodent model and the host response to the hydrogel material was evaluated.[46] Gelatin beads, crosslinked hyaluronic acid (HA), combination HA/gelatin, polycaprolactone (PCL), and poly(lactic-*co*-glycolic acid) (PLGA) were combined with selected regenerative renal cells to enhance regeneration and promote formation of functional renal tissue. While implantation of acellular gelatin and HA in the renal parenchyma resulted in a minimal immune response after 4 weeks, both PCL and PLGA initiated a severe immune response and promoted little native cell spreading. Only the gelatin-containing hydrogels were able to support cell attachment and spreading *in vitro*. Hyaluronic acid hydrogels did not provide any binding sites, and cells remained in a spheroid shape. Although the authors claim neo-kidney formation *in vivo* in rodents after one week post-implantation, the results of long-term studies are yet to be reported at this time.[46]

2.1.2. *Hemostatic agents for renal surgery*

Laparoscopic renal surgery is arguably the best treatment for renal cell carcinoma since its introduction in 1991.[41] Since then the development of surgical technique and equipment for laparoscopic partial nephrectomy (LPN) has provided similar benefits preserving renal function. However, this technique is not without its challenges especially because control of bleeding requires a great deal of skill from the surgeon. While conventional approaches include suturing and use of hemostatic gauze, hydrogel-based tissue sealants have also been explored as a replacement.[42] When compared to a fibrin sealant (Tisseel; Baxter) and intracorporeal suturing in a porcine model, the PEG-based CoSeal hydrogel (Baxter) was found to be a poor substitute for the fibrin glue noting that the hydrogel could not adhere well to the renal parenchyma.[43] More specifically, one animal in the CoSeal group suffered significant blood loss during surgery and was sacrificed after 8 days due to sepsis. Furthermore, all three kidneys with the CoSeal hydrogel suffered from urinary leakage in acute follow-up studies (after 3 days) and 1 of 4 in chronic follow-up studies (after 6 weeks). CoSeal also encountered bowel adhesion at the site of application, which could lead to further complications in the gastrointestinal tract. Another study, however, demonstrated success using CoSeal in a porcine model though a care must be taken in interpretation as the organ section removed, hydrogel application method, and control of blood loss were different in these two studies.[45] Another hydrogel that was considered for its feasibility was a biodegradable PEG-lactide hydrogel photopolymerized by high intensity xenon light.[61] Although hemostasis was achieved, the necessity for additional tools in the operating room prevent it from being convenient especially for laparoscopic surgery. The contrasting results in these studies could be seen as evidence that factors beyond the hydrogel contribute to the success of the surgery.

A review of 1,347 LPN cases in the United States and Europe conducted in 2005 revealed that most (n = 1042) cases used a variety of different hemostatic agents and/or tissue sealants including gelatin matrix thrombin tissue sealant (FloSeal; Baxter), fibrin glue (Tisseel; Baxter), bovine serum albumin-based adhesive (BioGlue; CryoLife), and cyanoacrylate glue (Glubran; General Enterprise Marketing).[44] Ultimately, the study found that the number of cases requiring transfusion due to blood loss and urine leakage were very low (2.6% and 1.9% respectively). However, it should be noted that many other factors play a role such as improved surgical technique and the use of other methods outside of tissue sealants such as suture bolstering that would improve these rates. It appears that hydrogels have not yet reached a point where they would have significant benefits over established tissue sealants for renal surgery.

2.2. Ureters

Ureteral stents are used to preserve urinary drainage from the kidneys under situations where patency may be compromised. Common problems associated with stent usage include stent migration, infection, encrustation, and urothelial hyperplasia. The use of hydrogels in ureteral stent applications has been shown as an antibacterial coating to prevent infection, to enhance lubricity, and prevent encrustation.[24,25,62] When compared to uncoated control, proprietary hydrogel-coated Boston Scientific Percuflex® ureteral stents dipped in solutions of ciprofloxacin, gentamycin, or cefazolin exhibited an increase in bacterial inhibition activity following *in vitro* assays with activity lasting for more than two weeks in some specimens.[24] Specifically, ciprofloxacin provided the best results up to 15 days after application while gentamycin was only able to maintain bacterial inhibition for 3 days. Although the actual results may vary for different drugs, this study demonstrated that stents coated with a hydrophilic hydrogel could be simply dipped in an antibiotic solution prior to implant to provide protection against infection. Similarly, poly(vinyl pyrollidone) (PVP) hydrogel coating of ureteral stents in a similar manner exhibited decreased bacterial growth of *E. faecalis*, but not *E. coli*, compared to uncoated samples.[25] However, as a lubricant, PVP can decrease friction between implant and tissue and prevent complications due to trauma and irritation during removal. PVP-coated stents also exhibited a significant reduction of hydroxyapatite encrustation on the surface compared to the silicone-coating, which was previously the choice method for prevention of encrustation. Despite the promising results and potential benefits, hydrogel applications in the ureters are currently limited to coatings for stents to prevent encrustation or bacterial growth. Additional biocompatibility studies and *in vivo* studies are necessary for further conclusions to be made and hydrogel-coated stents are moved into clinical trials. Fortunately, developments in the field of cardiovascular stents can often be applied in urological applications, and many ideas can be shared and adapted between the two fields.

2.3. Bladder

2.3.1. *Bladder Tissue Regeneration*

As with the kidney, one potential area of application for hydrogels is bladder tissue engineering. The difficulty of successfully engineering replacement bladder tissue can be attributed to the complexity of mimicking the bladder's unique mechanical properties. Currently, there is little published work with hydrogel compound usage for bladder tissue engineering, and the field remains largely unexplored. While there are some publications on hydrogels using bladder smooth muscle cells (SMCs), there is currently very little on hydrogels associated with uroepithelial tissue engineering, though some other approaches without hydrogels have been considered.

Collectively, bladder diseases affect approximately 400 million people worldwide with approximately 35 million affected in the United States with costs of $16 billion annually.[15] Several congenital abnormalities such as epispadias or bladder extrophy, or developed diseases such as bladder cancer or interstitial cystitis call for surgical intervention and tissue substitution. Typically, gastrointestinal segments are used for tissue transplantation or augmentation.[16,17] However, due to inherent physiological distinctions between intestinal tissue (absorptive) and bladder tissue (excretive, barrier function), long-term complications can include urolithiasis, increased mucous production, and bladder malignancy.[18] Moreover, a shift from a contractile to a synthetic phenotype in bladder SMCs has been associated with scar tissue formation, which, in turn, create a stiffer matrix unable to expand to accommodate the change in volume. Due to the myriad of complications associated with grafting of gastrointestinal segments, tissue engineering with hydrogel matrices has emerged as a potential option for restoring and regenerating the bladder.

A particular challenge in developing tissue *in vitro* is determining how to retain the normal, contractile phenotype of native bladder SMCs. A 3D hydrogel culture system based on poly(ethylene glycol) (PEG) crosslinked with bioactive matrix metalloproteinase (MMP) peptides has been examined for its effect on preserving the contractile phenotype of bladder SMCs and differentiation of mesenchymal stem cells (MSCs) into SMCs *in vitro*.[19] MSCs cultured in the MMP-modified PEG hydrogel exhibited a spread cell morphology and extracellular matrix (ECM) production while cell proliferation was much slower as compared to those in conventional 2D cultures. Moreover, a 1.7-fold increase in αSMA and 4.4-fold increase in myosin expression were observed in SMCs encapsulated in the

PEG hydrogel compared to 2D cultures. In contrast, THY-1, a positive marker for the synthetic phenotype, was significantly down-regulated in 3D culture suggesting a contractile phenotype. MSCs from human bone marrow cultured in the PEG 3D system exhibited significant up-regulation of smooth muscle markers with slight down-regulation of stem cell markers. These results suggest successful differentiation of stem cells towards a contractile smooth muscle phenotype, and that the use of an easily tunable hydrogel matrix is a valuable tool to guide differentiation or retain phenotypes characteristic of the native bladder smooth muscle cells.

Another method for creating controllable biomaterials comes in the emerging field of "bioprinting". Cells encapsulated in various hydrogel compounds are printed onto a cell culture dish or hydrogel substrate through an inkjet printer. Although the technology offers an attractive option for tissue engineering, it has met several undesired limitations. For example, inkjet systems[78] cannot accommodate a viscous hydrogel scaffold where clogging of the print head becomes an issue. Cell viability and functionality may also be a concern considering the high velocity and spray of the print head.[79] As an alternative, a valve-based droplet generator was used to print bladder SMC encapsulated within type-I collagen gel, and cell viability and proliferation were observed over a five day period.[14] Moreover, after a 51-day culture, a thin (100–400 μm) SMC tissue patch was created with high uniformity throughout although the cells are smaller in diameter compared to native bladder and show little alignment along any particular axis possibly due to the absence of cyclic loading that is typically exerted on native bladder tissue.[63] Furthermore, collagen as the sole component of the bioprinted hydrogel makes for a mechanically weak hydrogel.

Much work has been done on the use of natural and synthetic polymeric biomaterials to provide a mechanically sound scaffold with properties similar to bladder tissue. Some of this work goes beyond the use of hydrogels with common synthetic polymers used in tissue engineering such as poly(glycolic acid) (PGA), poly(lactic acid), and poly(ε-caprolactone).[64] Synthetic polymers are beneficial in some cases since they can be fabricated on a larger scale and have controllable mechanical properties. However, they lack inherent biological activity and generally suffer from biological incompatibility.[60] Hydrogels would likely benefit by pairing with a more mechanically stronger material. One application in particular explores the use of two sheets of compressed collagen gel with poly(lactic acid-co-ε-caprolactone) (PLAC) embedded in between.[65] The collagen-PLAC material was easier to handle and was able to be sutured without significant shrinkage unlike its collagen-only counterpart. As such, the combination of (1) materials to promote cell growth and spreading, such as the bioprinting method, with (2) polymers capable of mimicking the bladder compliance may prove to be a viable route. One downside of this method, however, is the potential for release of toxic components from the synthetic polymer as it degrades.[66,67] In the collagen-PLAC material, it was noted that a delayed immune response occurred that corresponded with the degradation time of PLAC.[65]

One challenge to overcome for bladder tissue engineering is developing a material that properly mimics, or at least incorporates well with, the native bladder tissue. Unfortunately at this point, hydrogels alone are too mechanically weak, and many studies turn to more compliant materials such as synthetic polymers or decellularized grafts.[60]

2.3.2. *Uroepithelial tissue regeneration*

One noteworthy endeavor of uroepithelial regeneration did not rely on direct hydrogel application, but used cultured cell sheets on polymer-coated thermosensitive dishes based on the poly(*N*-isopropylacrylamide) (PIPAAm) hydrogel. The polymer is bonded to a tissue culture dish using electron beam irradiation creating a slightly hydrophobic surface at 37°C. On the surface, urothelial cells were cultured for three weeks. After the culture period, the culture dishes are brought below the lower critical solution temperature (LCST) of 32°C, the surface becomes hydrophilic, and the cells spontaneously detach in one contiguous sheet after approximately 30 minutes. Histological sectioning of the cell sheets revealed both monolayered and multilayered sections. Electron microscopy supported the claim that the uroepithelial cells retained umbrella, intermediate, and basal cell morphology.[74] These uroepithelial cell sheets were used in a subsequent canine study in conjunction with demucosalized gastric flaps. Briefly, eight dogs had uroepithelial cells removed by partial cystectomy, which were cultured for a three-week period on the PIPAAm-coated culture dish. After this period, the dogs underwent surgery to have a section of stomach tissue removed and demucosalized. Cell sheets were then seeded and attached to the submucosal surface of the gastric flap, which was placed in a protective latex pouch and returned to the abdominal cavity. After another three weeks, the dogs were sacrificed, and the gastric flaps were examined. Of the eight dogs, five of them had viable regenerated uroepithelial layers covering a majority of the gastric flap surface. There was a change in the morphology of the superficial layer of cells. The remaining three dogs had severe hematomas on the gastric flap, which could possibly be attributed to the cell's exposure to xenogenic material during the culture phase.[75]

Approaching cell culturing from this standpoint offers the advantage of producing a cohesive tissue. Other cell harvesting methods using enzymatic treatment cleaves cell-to-cell junctions associated with the extracellular matrix, reducing cell function. However, this method of culture is also very lengthy and requires near maximum confluence to form a unified cell sheet. Additionally, the detachment upon temperature decrease occurs slowly over 30 min, which can lead to a decrease in cell viability. Several methods have been explored to improve viability and handling of cell sheet.[76]

2.3.3. *Drug delivery for bladder cancer*

In most cases of bladder cell carcinoma, early stages of the tumor formation appear in the uroepithelial layer, and treatment normally consists of transurethral resection of the tumor with high rates of success. Unfortunately, there are also high rates of tumor recurrence, and anti-cancer drugs delivery are commonly administered postoperatively to suppress this. However, limitations arise when the drug is delivered orally. Additionally, large doses would have to be administered systemically in hopes of having a small portion reach the bladder, and thus increasing risks for side effects. Intravesical delivery of the drug by catheter provides a method for local release, yet this technique is not without its disadvantages. Low permeability of the uroepithelium prevents drug uptake, filling of the bladder during catheterization dilutes the drug concentration at the target site, and voiding of urine will carry the drug out of the bladder.[47] Hydrogels may provide a potential solution to prevent repeat catheterization and drug delivery. Ideally, hydrogels loaded with drugs would be able to withstand bladder filling without being washed away too quickly.

Recent studies have explored various types of thermosensitive hydrogel compounds that are liquid at room temperature but undergo gelation at body temperature to improve retention of locally delivered drugs. For example, a FITC dye was used to model drug elution from a tri-block copolymer hydrogel consisting of PEG–PLGA–PEG. The hydrogel was shown to retain FITC dye significantly better than free FITC injected into the bladder in rat models.[48] Urinary cystitis was also induced using an intraperitoneal injection of cyclophosphamide in some rats. This chemically-induced cystitis is marked by increased micturition frequency, development of ulcers, hemorrhages, and erosion. Rats were divided into control groups containing saline, hydrogel, and misoprostol, a therapeutic drug for gastric and duodenal ulcers. The treatment group was injected with a hydrogel and misoprostol compound. The frequency of voiding significantly decreased in rats treated with the drug-loaded hydrogel compared to either the drug or the hydrogel delivered alone. Another similar study explored poly(*N*-isopropylacrylamide) (PNIPAM) grafted with hyaluronic acid (HA) and gelatin.[49] Cisplatin, a drug for the treatment of bladder cancer,[50,51] was added to the compound and drug diffusion and effects on the uroepithelium were observed in rat models. The study found slower release of the hydrogel-encapsulated drug compared to drug dissolved in aqueous solution and minimal disruption of the uroepithelial layer where the hydrogel was applied.

2.3.4. *Therapeutics for bladder diseases*

Another bladder pathological condition that hydrogels might also offer benefits is interstitial cystitis (IC), which is a condition characterized by a disruption or increase in the permeability of the uroepithelial layer causing irritation of the smooth muscle tissue.[52,53] The lack of knowledge in the etiology of IC prevents significant progress on the best method of treatment, and most current methods only attempt to treat the symptoms.[54] Some avenues of treatment have turned to glycosaminoglycans (GAG) such as hyaluronic acid, heparin, pentosan sulfate, and chondroitin sulfate to alleviate the pain of patients suffering from IC.[55–59] Since the mechanisms of IC are not well understood, it was speculated that GAGs administered intravesically are providing relief by replacing the damaged bladder lining. However, research on the mechanisms of how GAGs could adhere to the glycocalyx and interact with the bladder is also lacking. Nonetheless, one recent study examined intravesical instillation of hyaluronic acid and chondroitin sulfate administered weekly for the first 2 months and then every 2 weeks for 6 months.[59] Based on subjective surveys, all patients reported improvement in urinary urgency and bladder pain. A reduction in the times voided and an increase in the volume of urine per voiding were seen post-treatment.

The applications of hydrogels in the bladder have been improving over recent years, but still have much more to go. Though the mechanical strength of hydrogels may not be on par with native tissue yet, they do fare well as a supplement material for application that require longer-term diffusion or degradation like drug delivery or cell encapsulation. Additionally, discoveries in the etiology of interstitial cystitis may lead to more uses of natural hydrogels combined with supplemented medicine to not only relieve symptomatic pain, but to cure the condition altogether.

2.4. Urethra

2.4.1. *Hydrogels as bulking agents in urethral sphincter*

Stress urinary incontinence (SUI) affects approximately 35% of females over the age of 18% and 45% over the age of 60 hampering their quality of life.[20] Economic costs associated with SUI, which is estimated to be \$27 billion dollars annually, includes the purchase of disposable products, laundry costs, injuries resulting from rushing to the bathroom, complications from treatment, requirement of caretakers, and lost time from work.[21]

Currently, the standards of treatment for SUI include pelvic floor exercise to strengthen the muscles surrounding the vagina and urethra and assist in retaining voluntary control. While surgery to implant suburethral slings made of polypropylene mesh is an option, potential complications and pain associated with this procedure deter some patients. Hydrogel bulking agents are gaining popularity as an

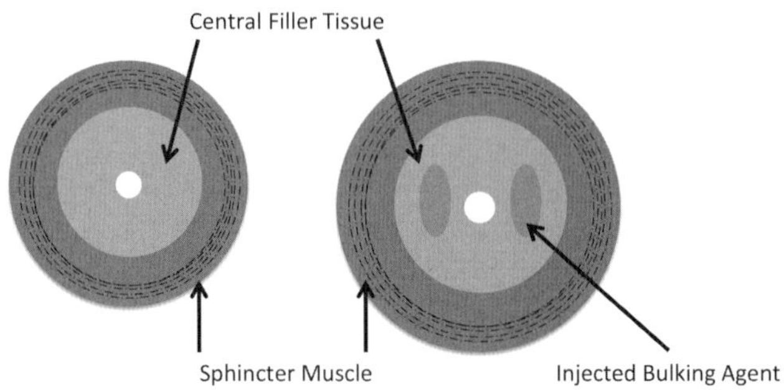

Fig. 4. A schematic of the urethral sphincter. The sphincter muscle consists of concentrically aligned muscle fibers wrapped around central filler tissue of randomly or longitudinally aligned muscle fibers. When the bulking agents are injected into the central volume, the sarcomeres in the sphincter muscle lengthen provide added strength.

alternative for treatment of SUI and its effectiveness and safety were examined in a study on 514 elderly women who were each treated with one of four different bulking agents between 1999 and 2012.[22] The commercially available bulking agents compared in this study were Contigen®, Zuidex®, Tegress®, and Bulkamid® made from crosslinked bovine collagen, hyaluronic acid/dextranomer copolymer, ethylene vinyl alcohol, and polyacrylamide hydrogel, respectively. The treatment for intrinsic sphincter deficiency is performed by periurethral injection of the hydrogel material into the submucosa to thicken the urethral wall (Fig. 4). It is believed that bulking agents work by increasing the central filler volume, which lengthens the sarcomeres in the sphincter muscle and strengthens it.[73] Although the study found significant improvement in maximum urethral closure pressure and reduction in urine leakage, survey taken during the study revealed that patients still feared urine leakage and continued to take the same precautions prior to the treatment. Since the currently available bulking agents only provide a temporary fix to retain continence, and repeat injections are necessary over a period of time, recent work has explored the use of a composite hydrogel polyethylene glycol and carboxymethyl cellulose as a bulking agent for the urethra sphincter. Allowing for little swelling or shrinkage, this particular hydrogel formulation can potentially become a more permanent option. However, additional work is necessary in order to reduce the inflammatory response and urethral degradation before use in humans can be considered.[23] Though bulking agent treatments are considered relatively safe, risks exist for urinary tract infection, urethral erosion and prolapse, and diverticula formation, which can require additional treatments with added costs. With the variability in responses to the bulking agents from patients,[22] it remains to be seen whether a highly successful hydrogel compound will be developed that does not lead to severe complications and eliminates the need for repeat injections.

2.4.2. *Major urethral reconstruction*

In addition to the use as a bulking agent, hydrogels may also contribute to the regeneration of dysfunctional urethra. While current surgical methods rely on implantation of tubes created from buccal mucosa or foreskin tissue, hydrogels may be used instead for treatment of conditions such as hypospadias and urethral stricture. Hypospadias is a congenital defect found in approximately 1 in 300 male births[6] that is characterized by development of the urinary meatus (opening of the urethra where the urine exits) along the shaft of the penis rather than at the glans. One particular study explored the feasibility of high-density collagen gel tubes seeded with rabbit bladder smooth muscle cells for urethral regeneration in rabbit models.[9] At one and three months after implantation, urethrograms revealed that lumen calibers of acellular grafts were 62.69% ± 18.1% and 42.4% ± 34.3% compared to native control urethra, respectively. In contrast, the lumen calibers of seeded grafts were 45.7% ± 43.62% at one month and 96.6% ± 12.5% at three months. The observations that host cell ingrowth occurred in both acellular and seeded scaffolds suggest that use of the acellular scaffold will eliminate the time necessary to harvest and expand smooth muscle cells for seeding and the collagen gel approach may potentially improve short- and long-term complications over current surgical methods. In surgical cases where radical urethral repair or replacement is not necessary, glues or sealants are used to repair damaged tissues. Urethrocutaneous fistula is a post-operative complication of hypospadias repair, and its incidence can be as high as 20%.[10] The use of fibrin glue as a sealant applied along the suture lines during hypospadias repair has been shown to diminish occurrence rate of complications from 53% to 20%.[11] Fibrin glue can also be used to promote wound healing in many cases of urethroplasty, such as those previously described, and to assist with bulbar urethroplasty using buccal mucosa grafts.[12,13]

2.4.3. *Hydrogels as coating for urethral catheters*

Various types of hydrogels are used not only in urethral reconstruction, but also as a coating material for devices that come in contact with the tissue. In both ureteral stents and urethral catheters, problems with infection, lubricity, and encrustation can occur. Several stents and catheters are on the market using proprietary hydrogel coatings such as LubriLAST™, HydroPlus™, BIOCATH®, and LUBRI-SIL® Foley catheter. However, these hydrogels are all advertised to improve only patient comfort by providing adequate lubrication. Further improvements in hydrogels used for stents and catheters will likely be found through developments that prevent bacterial growth and encrustation. For urethral catheters made from segmented polyurethane, surface modifications are used to reduce the complications such as bacterial colonization, urethral trauma, catheter obstruction, encrustation, and infection.[25,68–70] The adhesion and growth of bacteria around the catheter site arises from the protein adsorption cascade.[71] Furthermore, irritation and inflammation due to the friction between the catheter and the urethral lining can exacerbate the complications mentioned. Particularly, one study compared the protein reduction and lubrication properties of a chitosan/poly(vinyl alcohol) (PVA) hydrogel with LubriLAST™, a general solution currently used to create a lubricated surface for several devices including urinary catheters. Long-chain chitosan provides a flexible network for entanglement of PVA that is designed to elute from the chitosan network upon swelling. These particular components were chosen for their antimicrobial capabilities (chitosan) and reduction of protein adsorption and cell adhesion (PVA). Results of the study showed significantly less protein adsorption on the chitosan/PVA-coated catheters compared to uncoated and LubriLAST™-coated samples. Additionally, antibacterial tests measuring areas of bacterial inhibition were performed and large antiseptic regions were demonstrated around the chitosan/PVA-coated samples when incubated with *S. aureus*, *P. aeruginosa*, and *E. coli*.[72]

3. Hydrogel Applications in Reproductive Organs

3.1. *Vagina*

3.1.1. *Hydrogels in prevention of sexually transmitted disease*

The recent rise in sexually transmitted diseases in women has led to an increase in both research and development of female-specific preventative measures such as condoms and microbicides.[28] Specifically, studies on temperature- and pH-sensitive hydrogels have opened up a niche for application in conditionally activated scenarios such as semen-triggered release. When paired with antiviral agents, the hydrogel such as poly(NiPAAm-co-BM-co-AA) (a blend of N-isopropyl acrylamide, butyl methacrylate, and acrylic acid) can be used as a drug delivery system able to easily coat the vaginal wall prior to the transition from liquid to gel.[29] After the hydrogel undergoes thermal gelation at 37°C on the vaginal wall, contact with semen will raise the environmental pH causing it to return to liquid form and rapidly release the drug. When compared to a non-phase shifting homopolymer, hydroxyethylcellulose (HEG), drug release kinetics showed a clear burst release using the terpolymer when exposed to the semen simulant *in vitro*.[29] Another downside of the homopolymers are that they cannot retain the drug for more than a few hours making it a coital-dependent application and, thus, issues of patient compliance arise. Similar studies have been done using other temperature and pH-sensitive hydrogels such as hydrophobically-modified methylcellulose and phenylboronate-salicylhydroxamate

(PBA-SHA) crosslinked hydrogels.[30,31] The methylcellulose compound was not pH sensitive, and thus drug release was not triggered by semen contact but rather slowly released over time.

The prospects of this technology are quite promising, however condoms dominate the field and is arguably the most advocated form of STD prevention other than abstinence. The convenience and effectiveness of traditional methods will create a barrier for acceptance of this hydrogel alternative. Additionally, nonoxynol-9, a drug commonly thought to protect against sexually transmitted infections has recently come under question to its true effectiveness.[77] As such, this drug delivery method cannot be used until a suitable drug is found.

4. Conclusions

The use of hydrogels in the urogenital field has met with varying levels of successes in a wide variety of applications from tissue engineering to drug delivery. As far as urogenital tissue engineering is concerned, a hydrogel material with mechanical properties comparable to native tissue might never be developed, nor does it have to be. Other natural and synthetic polymers have provided the compliance and strengths required to mimic native tissue, and hydrogels have found a niche as a supplemented cell scaffold material. A great deal of research has been conducted and yielded temperature-sensitive, pH-controlled, and tunable timed-release hydrogels for use in urogenital drug delivery applications. Although the challenges of efficient drug uptake and proof of efficacy remain, a number of promising materials provide the necessary benefits to overcome these issues. With additional research and development, hydrogels will continue to provide an invaluable role in many urogenital organ applications.

Acknowledgments

The authors would like to thank Ms. Angela Gomez for providing the illustrations for this chapter. JPT and JN are partially supported by an NIH grant #RR021949.

References

1. S. A. Lewis, *Am. J. Physiol. Renal Physiol.*, 278 (2000).
2. P. Khandelwal, *Am. J. Physiol. Renal Physiol.*, 297 (2009).
3. K. Aitken, *Nat. Rev. Urol.*, 6 (2009).
4. O. J. Wiseman, *BJU Int.*, 91 (2003).
5. S. L. Chang, *Scand. J. Urol. Nephrol.* Suppl., 201 (1999).

6. H. Dolk, M. Loane, and E. Garne, *Rare Diseases Epidemiology*, 686 (2010).
7. A. Bayne and E. Jones, In *Complications of Urologic Surgery*, Ed. S. S. Taneja, Philadelphia, Saunders Elsevier, p. 713 (2010).
8. J. E. Nuininga, R. P. E. De Gier, R. Verschuren, and W. F. J. Feitz, *J. Urol.*, 174 (2005).
9. L. A. Micol, L. F. A. da Silva, P. J. Geutjes, E. Oosterwijk, J. A. Hubbell, W. F. J. Feitz, and P. Frey, *Biomaterials*, 33 (2012).
10. W. T. Snodgrass and A. Lorenzo, *BJU Int.*, 89 (2002).
11. S. C. Gopal, A. N. Gangopadhyay, T. V. Mohan, V. D. Upadhyaya, A. Pandey, A. Upadhyaya, and D. K. Gupta, *J. of Ped. Surg.*, 43 (2008).
12. J. Hick and A. F. Morey, *J. Urol.*, 171 (2004).
13. G. Barbagli, S. De Stefani, and M. C. Sighinolfi, *Eur. Urol.*, 50 (2006).
14. F. Xu, S. J. Moon, A. E. Emre, E. S. Turali, Y. S. Song, S. A. Hacking, J. Nagatomi, and U. Demirci, *Biofabrication*, 2 (2010).
15. L. M. D. Shortliffe, in *Overcoming Bladder Disease: A Strategic Plan of Research*, Report of the Bladder Research Progress Review Group, National Institute of Diabetes & Digestive & Kidney Diseases, National Institutes of Health, 2002.
16. T. W. Hensle and S. M. Gilbert., *Curr. Urol. Rep.*, 8 (2007).
17. G. C. Mingin, J. A. Stock, and M. K. Hanna., *J. Urol.*, 162 (1999).
18. W. S. McDougal, *J. Urol.*, 147 (1992).
19. C. Adelöw, T. Segura, J. A. Hubbell, and P. Frey, *Biomaterials*, 29 (2008).
20. S. Hunskaar, G. Lose, D. Sykes, and S. Voss, *BJU Int.*, 93 (2004).
21. L. A. Kerr, *Rev. Urol.*, 7 (Suppl. 1) 2005.
22. S. Mohr, M. Seigenthaler, M. Mueller, and A. Kuhn, *Int. Urogynecol. J.*, 24 (2013).
23. J. P. Sumner, R. J. Hardie1, J. N. Henningson, R. Drees, M. D. Markel, and D. Bjorling, *Vet. Surg.*, 41 (2012).
24. T. John, A. Rajpurkar, G. Smith, M. Fairfax, and J. Triest, *J. Endourology*, 21 (2007).
25. M.M. Tunney and S.P. Gorman, *Biomaterials*, 23 (2002).
26. N. Lameire, W. V. Biesen, and R. Vanholder, *Lancet*, 365 (2005).
27. J. Gao, R. Liu, J. Wu, Z. Liu, J. Li, J. Zhou, T. Hao, Y. Wang, Z. Du, C. Duan, and C. Wang, *Biomaterials*, 33 (2012).
28. Women and HIV/AIDS: Confronting the Crisis. A Joint Report by UNAIDS, UNFPA, UNIFEM.
29. K. M. Gupta, S. R. Barnes, R. A. Tangaro, M. C. Roberts, D. H. Owen, D. F. Katz, and P. F. Kiser, *J. Pharm. Sci.*, 96 (2007).
30. N. Li, M. Yu, L. Deng, J. Yang, and A. Dong, *J. Mater. Sci. Mater. Med.*, 23 (2012).
31. J. I. Jay, S. Shukair, K. Langheinrich, G. C. Cianci, M. R. Clark, T. J. Hope, L. Rohan, and P. F. Kiser, *Adv. Funct. Mater.*, 19 (2009).
32. I. Araki, S. Du, H. Kobayashi, N. Sawada, T. Mochizuki, H. Zakoji, and M. Takeda, *Int. J. Urol.*, 15 (2008).
33. N. Yoshimura, *Progr. Neurobiol.*, 57 (1999).
34. S. Du, I. Araki, and Y. Mikami, *Urology*, 69 (2007).
35. L. Baskin, P. S. Howard, and E. Macarak, *J. Urol.*, 150 (1993).
36. A. Liapis, P. Bakas, A. Pafiti, D. Hassiakos, M. Frangos-Plemenos, and G. Creatsas, *Urol. Res.*, 28 (2000).

37. J. Morrison, J. Wen, and A. Kibble, *Scand. J. Urol. Nephrol.*, 201 (1999).
38. J. W. Downie and J. A. Armour, *Can. J. Physiol. Pharmacol.*, 70 (1992).
39. O. Pleniceanu, H. S. Orit, and D. Benjamin. *Stem Cells*, 28 (2010).
40. C. Sagrinati, E. Ronconi, E. Lazzeri, L. Lasagni, and P. Romagnani, *Cell*, 14 (2006).
41. R. V Clayman, L. R. Kavoussi, N. J. Soper, *et al.*, *J. Urol.*, 146 (1991).
42. A. Weizer and J. S. Wolf Jr., in *Complications of Urologic Surgery*, Ed. S. S. Taneja Philadelphia, Saunders Elsevier, p. 343 (2010).
43. J. E. Bernie, J. Ng, V. Bargman, T. Gardner, L. Cheng, and C. P. Sundaram, *J. Endourology*, 19 (2005).
44. A. Breta, *et al.*, *Eur. Urol.*, 52 (2007).
45. E. L. Park, J. B. Ulreich, K. M. Scott, *et al.*, *J. Urol.*, 172 (2004).
46. J. Basu, *et al.*, *Cell Transplantation*, 20 (2011).
47. S. GuhaShakar and R. Banerjee, *J. Contr. Rel.*, 148 (2010).
48. P. Tyagi, Z. Li, M. Chancellor, W. C. De Groat, N. Yoshimura, and L. Huang, *Pharm. Res.*, 21 (2004).
49. J. P. Chen, Y. L. Leu, C. L. Fang, C. H. Chen, and J. Y. Fang, *J. Pharm. Sci.*, 100 (2011).
50. R. H. Schoenfeld, W. D. Belville, W. H. Jacob, A. S. Buck, M. L. Dresner, S. J. Insalaco, and G. S. Ward. *J. Am. Osteopath. Assoc.*, 82 (1983).
51. B. A. Hadaschik, M. G. ter Borg, J. Jackson, R. D. Sowery, A. I. So, H. M. Burt, and M. E. Gleave., *BJU Int.*, 101 (2008).
52. S. L. Johansson and M. Fall, *J. Urol.*, 143 (1990).
53. W. L. Lynes, S. D. Flynn, L. D. Shortliffe, and T. A. Stamey, *Am. J. Surg. Pathol.*, 14 (1990).
54. C. H. Hsieh, W. C. Chang, M. C. Huang, T. H. Su, Y. T. Li, and H. S. Chiang, *Taiwanese J. Obst. Gynecol.*, 51 (2012).
55. A. C. Thompson and T. J. Christmas, *Br. J. Urol.*, 78 (1996).
56. J. J. Bade, M. Lasuer, A. Nieuwenburg, L. T. Van der Weele, and H. J. A. Memsink, *Br. J. Urol.*, 79 (1997).
57. D. Porru, M. Cervigni, L. Nasta, F. Natale, R. Lo Voi, C. Tinelli, B. Gardella, A. Anghileri, A. Spinillo, and B. Rovereto, *Rev. Recent. Clin. Trials*, 3 (2008).
58. M. Cervigni, F. Natale, L. Nasta, A. Padoa, R. Lo Voi, and D. Porru, *Int. Urogynecol. J. Pelvic. Floor Dysfunct.*, 19 (2008).
59. D. Porru, F. Leva, A. Parmigiani, D. Barletta, D. Choussos, B. Gardella, M. D. Dacco, R. E. Nappi, M. Allegri, C. Tinelli, C. M. Bianchi, A. Spinillo, and B. Rovereto, *Int. Urogynecol. J.*, 23 (2012).
60. A. M. Turner, R. Subramaniam, D. F. M. Thomas, and J. Southgate, in *Tissue Engineering using Ceramics and Polymers*, Eds. A. R. Boccaccini and J. E. Gough, Boca Raton: CRC Press, p. 445 (2007).
61. S. Ramakumar, W. W. Roberts, O. E. Fugita, P. Colegrove, T. M. Nicol, T. W. Jarrett, L. R. Kavoussi, and M. J. Slepian, *J. Endourol.*, 16 (2002).
62. S. P Gorman, M. M. Tunney, P. F. Keane, K. Van Bladel, and B. Bley, *J. Biomed. Mat. Res.*, 39 (1998).
63. B. S. Kim, J. Nikolovski, J. Bonadio, and D. J. Mooney, *Nat. Biotechnol.*, 17 (1999).
64. P. X. Ma, *Materials Today*, May (2004).

65. E. M. Engelhardt, L. A. Micol, S. Houis, F. M. Wurm, J. Hilborn, J. A. Hubbell, and P. Frey, *Biomat.,* 32 (2011).
66. A. Remes and D. F. Williams, *Biomat.,* 11 (1992).
67. K. Ceonzo, A. Gaynor, L. Shaffer, K. Kojima, C. A. Vacanti and G. L. Stahl, *Tissue Eng.,* 12 (2006).
68. K. D. Park, Y. S. Kim, D. K. Han, Y. H. Kim, E. H. Lee, H. Suh, and K. S. Choi, *Biomater.,* 19 (1998).
69. J. M. Schierholz, N. Yucel, A. F. Rump, J. Beuth, and G. Pulverer, *Int. J. Antimicrob. Agents,* 19 (2002).
70. P. F. Keane, M. C. Bonner, S. R. Johnston, A. Zafar, and S. P. Gorman, *Br. J. Urol.,* 73 (1994).
71. P. Francois, P. Vaudaux, N. Nurdin, H. J. Mathieu, P. Descouts, and D. P. Lew, *Biomaterials,* 17 (1996).
72. S. H. Yang, Y. S. J. Lee, F. H. Lin, J. M. Yang, and K. S. Chen, *J. Biomed. Mat. Res. Pt. B,* 83B (2007).
73. N. Klarskov and G. Lose, *Neurourology and Urodynamics,* 27 (2008).
74. Y. Shiroyanagi, M. Yamato, Y. Yamazaki, H. Toma, and T. Okano, *Tissue Eng.* 9 (2003).
75. Y. Shiroyanagi, M. Yamato, Y. Yamazaki, H. Toma, and T. Okano, *BJU Int.,* 93 (2004).
76. Z. Tang, Y. Akiyama, and T. Okano, *Polymers,* 4 (2012).
77. D. Wilkinson, M. Tholandi, G. Ramjee, *et al., Lancet Infect. Dis.,* 2 (2002).
78. T. Boland, T. Xu, B. Damon, and X. Cui, *Biotechnol. J.,* 1 (2006).
79. T. Xu, C. A. Gregory, P. Molnar, X. Cui, S. Jalota, S. B. Bhaduri, and T. Boland, *Biomaterials,* 27 (2006).

Index

Agarose, 230, 254, 463
Alginate, 180, 230
Alkaline phosphatase, 176
Alkaline phosphatase-mediated mineralization, 176
Alveolar epithelial cells, 404
Amorphous calcium phosphates, 174
Angiogenesis, 188, 389
Annulus fibrosis, 200
Apatitic calcium phosphates, 173
Arginine–Glycine–Aspartate (RGD), 171, 180
Artificial disc replacement, 201
Autologous Chondrocyte Implantation (ACI), 219

Bioactive, 170, 173
Bioactive factor delivery, 301
Bioartificial livers (BAL) devices, 427
Bioartificial pancreas, 463
Biocompatibility, 5
Biodegradability, 222
Bioglasses, 173
Biomacromolecules-driven mineralization, 179
Biomaterials, 222
Biomimetic approach, 173
Bioreactor, 428
Bladder Tissue Regeneration, 489
Bolus, 187
Bone, 165
Bone Biology, 296
Bone marrow, 169
Bone mineralization, 168
Bone morphogenetic proteins, 168
Bone regeneration, 169
Bone sialoproteins, 168
Bone stiffness, 168
Bottom-up approaches, 183
β-tricalcium phosphate, 174

Calcium, 175
Calcium-binding domains, 179
Calcium glycerophosphate, 177
Calcium phosphates, 168
Cancellous/trabecular bone, 167
Cancer, 99
Carbon nanotubes, 174
Carboxyl, 178
Carboxyl groups, 175
Cardiac, 323–327, 329, 331–339, 341–346, 348–350, 352–353
Cardiomyocyte, 98
Cell-based strategies, 247

Cell carriers, 29
Cell-laden microfabrication, 183
Chemical crosslinking, 170
Chitosan, 232
Collagen, 168, 227
Comparative studies, 241
Cornea, 138
Cortical bone, 167
Crosslinking, 224
Cutaneous, 112

3D, 183
Decellularization, 139
Degenerative disc disease, 200
Delivery vehicles, 28
Demineralized bone matrix, 166
Dextran, 111
Diabetes mellitus, 463
Differentiation, 53–54, 56–57, 59–61, 63–68, 71, 73, 77–79
Disc tissue engineering, 201
Drug delivery, 91, 399, 492
Drug releasing systems, 243

Endothelial cells, 170
Endplates, 200
Engineered bone tissue, 295
Extracellular matrix, 167

Fibrin, 228, 254
Fibroblast growth factor, 169
Fibrous constructs, 223
Full thickness defects, 218
Functionalization, 178

Gelatin, 180, 227
Gelation mechanism, 224
Gellan gum, 254
Genipin, 225
GF-mediated mineralization, 182
Glycan components, 172
Gradients in hydrogels, 185
Graphene, 174
Groups, 178
Growth, 218
Growth factors, 166, 243

Hemostatic agents, 487
Heparin, 244
Hepatic tissue engineering, 427
Hydrogels for Cartilage Repair, 223
Hyaluronic acid, 233
Hybrid Hydrogels, 240
Hydrogel, 111, 144, 271–273, 280–287, 295–296, 299–312, 314–315, 323–324, 326–327, 329, 331–335, 338–339, 341–346, 348–350, 352–353
Hydrogel scaffolds, 271–284
Hydroxyapatite, 168
Hydroxyl, 178

Infarction, 323, 325
Injectable, 323–324, 326–327, 329–333, 335, 337–339, 341–342, 345–346, 349, 353
Inorganic components, 173
In situ deposition, 176
Insulin-like growth factor, 170
Interleukin-1, 169
Interpenetrating network, 179
Intervertebral discs, 199

Kidney tissue regeneration, 486

Ligament, 271–280, 282–285
Ligamentogenesis, 272, 281
Ligamentogenic, 272–273, 284, 287
Ligamentum, 274
Lung tissue engineering, 403

Macrophage colony-stimulating factor, 170
Materials, 173
Matrix-Induced Autologous Chondrocytes Implantation (MACI), 220
Matrix mechanics, 186

Matrix metalloproteinases (MMPs), 171, 181
Mediators, 218
Mesenchymal stem cells (MSCs), 250
Mesenchymal/stromal stem cells, 169
Metastasis, 99
Microencapsulated islets, 465
Microengineering, 154
Microenvironment, 53–54, 70–71, 76
Microfabricated hydrogels, 188
Microfabrication, 393
Micromolding, 184
Microscale fabrication, 440
Morphology, 217
Myocardial Infarction, 95

Natural-derived matrices, 222
Naturally-derived hydrogels, 226
Neoligament, 279
Nucleation, 173
Nucleus pulposus, 200

Oligo poly fumarate (OPF), 238
Osteoblasts, 169
Osteocalcin, 168
Osteochondral autograft transplantation, 219
Osteoconductivity, 172
Osteocytes, 169
Osteogenic differentiation, 170
Osteoinductivity, 172
Osteon, 168
Osteonectin, 168
Osteopontin, 168

Partial thickness defects, 218
Patterning hydrogels, 184
Peptide-mediated mineralization, 180
Phosphate, 178
Photolithography, 184
Photoresponsive, 5
Physical, 170
Platelet-derived growth factor, 170
Platelet lysates, 182
Pluronic, 224
Poly ethylene glycol (PEG), 170, 237
Poly ethylene oxide (PEO), 237
Poly glycolic acid (PGA), 237
Poly lactic acid (PLA), 237
Polymeric networks, 170
Poly propylene fumarate (PPF), 238
Polysaccharides, 235
Poly vinyl alcohol (PVA), 235
Porous 3D scaffolds, 223
Printing, 183
Pro-osteogenic factors, 170
Proteoglycans, 216

Regeneration, 111
Regenerative medicine, 26, 53, 55, 57, 79
Regenerative methods, 219
Repair, 323, 325, 327, 329, 331–334, 336–339, 341, 342, 344, 348–349, 352
Repair/replacement therapies, 208
Requirements for cartilage TE scaffolds, 221

Scaffold architectures, 223
Scaffolds, 31, 221, 271–273, 277–287, 296, 300–301, 304, 305, 307, 311, 314, 315
Sequential delivery of GFs, 188
Silanol, 179
Silica, 174
Silk fibroin, 229
Simulated body fluid, 175
Smart polymers, 224
Soaking/incubation treatments, 174
Specific receptor ligands, 222
Stem cells, 53–55, 57–66, 70, 71, 73, 74, 76, 77, 79, 101, 152, 336, 339, 345, 350
Stem cells sources, 246
Structure and functions, 216
Synovium-derived mesenchymal stem cells (SMSC), 250

Synthetic hydrogels, 170
Synthetic materials, 222

Tendon, 271–280, 282–286
Tendon/ligament tissue engineering, 271
Tenocytes, 275
Tenogenesis, 272, 281
Tenogenic, 272–273, 287
Thermoreversible gels, 225
Tissue engineered heart valves, 364
Tissue engineering, 55–58, 79, 166, 220, 271
Tissue models, 87
Tissue regeneration, 96, 302, 305, 306, 309, 314
Toughness, 168
Toxicity, 93
Treatment options, 218

Ureteral stents, 488

Valvular endothelial cells, 366
Valvular interstitial cell, 366
Vascular endothelial growth factor, 170, 388
Vascularization, 186
Vascularized bone te, 188
Vascular tissue engineering, 386
Vasculogenesis, 389
Vertebras, 199

Wound healing, 111–112